ONE WEEK LOAN

APPLIED STATICS, STRENGTH OF MATERIALS, AND BUILDING STRUCTURE DESIGN

Joseph B. Wujek

Front Range Community College

Prentice Hall

Upper Saddle River, New Jersey   Columbus, Ohio

Library of Congress Cataloging-in-Publication Data

Wujek, Joseph B.
    Applied statics, strength of materials, and building structure
design / by Joseph B. Wujek.
       p.   cm.
    Includes index.
    ISBN 0-13-674631-4
    1. Structural design.   2. Statics.   3. Strength of materials.
I. Title.
TA658.W85   1999
624.1′71—dc21

98-23677
CIP

This book is dedicated to my mother and father, Sophie and Joseph Wujek, Sr. Their constant encouragement and continual support make me believe that anything is possible through hard work and dedication. They are the primary reason for any successes I have enjoyed thus far.

Cover photo: ©FPG International
Editor: Ed Francis
Production Coordinator: Tally Morgan, WordCrafters Editorial Services, Inc.
Production Editor: Christine M. Harrington
Design Coordinator: Karrie M. Converse
Text Designer: Tally Morgan
Cover Designer: Dan Eckel
Production Manager: Patricia A. Tonneman
Marketing Manager: Danny Hoyt

This book was set in Times Roman by Carlisle Communications, Ltd., and was printed and bound by R. R. Donnelley & Sons Company. The cover was printed by Phoenix Color Corp.

© 1999 by Prentice-Hall, Inc.
Simon & Schuster/A Viacom Company
Upper Saddle River, New Jersey 07458

Notre Dame: PhotoDisc, Inc., p. 2; Crystal Palace: Royal Institute of British Architects, p. 3; Empire State Building: New York Convention & Visitors Bureau, p. 5; John Hancock Center: Ezra Stoller Associates, Inc./ESTO, p. 6; World Trade Center and Sears Tower: Simon & Schuster/PH College, p. 7; Petronas Towers: AP/Wide World Photos, p. 8.

Printed in the United States of America

10 9 8 7 6 5 4 3 2 1

ISBN: 0-13-674631-4

Prentice-Hall International (UK) Limited, *London*
Prentice-Hall of Australia Pty. Limited, *Sydney*
Prentice-Hall of Canada, Inc., *Toronto*
Prentice-Hall Hispanoamericana, S. A., *Mexico*
Prentice-Hall of India Private Limited, *New Delhi*
Prentice-Hall of Japan, Inc., *Tokyo*
Simon & Schuster Asia Pte. Ltd., *Singapore*
Editora Prentice-Hall do Brasil, Ltda., *Rio de Janeiro*

PREFACE

Applied Statics, Strength of Materials, and Building Structure Design provides an elementary introduction to building structure engineering for the architectural, construction, and civil engineering technical student. Elementary engineering concepts and design principles are introduced on an intermediate mathematics level, requiring that the student have a working knowledge of college algebra and right-angle trigonometry. Mathematics principles are integrated within the body of the text to clarify and support engineering computations. Concepts are well supported by over 200 illustrations and photographs. Nearly 180 examples are provided with helpful explanations. Homework exercises are designed to introduce basic principles with, in most instances, real-world connections.

The primary audience for this book will be the college student studying architectural technology, architectural engineering technology, structural technology, structural engineering technology, civil engineering technology, construction engineering technology, and construction management.

This book is not intended to provide an extensive review of statics, strength of materials, and building structure engineering subjects with the depth and rigor required for an engineering course of study. Such training traditionally includes completing several courses, requiring many texts and other technical resources. Also, this book is *not* written to trivialize the complex field of structural engineering. Instead, it is written to help acquaint the technical student with selected elementary design principles used by engineers and architects so that the future technician can develop a working knowledge of rudimentary engineering and design practices.

This type of book is needed because the roles of the architect, engineer, and drafting technician have changed drastically in the past few decades. This transformation is due to the growing complexity of design standards and regulations, rapid changes in and development of new building technologies, demand for more attention from the design professional, and the introduction of the computer as a tool in drafting and design.

Decades ago, a drafting technician was called upon to produce high-quality technical drawings from design sketches. Today, the technician is required to do more. Simply, there is an emerging reliance on the architectural and engineering technician who can perform rudimentary engineering and design tasks under the direction of an architect or engineer, functions formerly reserved for the architect or engineer. This emerging need presents good career opportunities for a qualified architectural and engineering technician. This book focuses on training the architectural and engineering technician.

Many fortunate persons enter and complete traditional educational programs in architecture or engineering successfully. For other technically focused people, the rigorous nature of a traditional architecture or engineering education limits the likelihood of success. Yet for many others who have good technical aptitude, a hectic lifestyle, societal or financial constraints, or commitments to family make a formal education in architecture or engineering out of reach. With the emerging reliance on architectural and engineering technicians, opportunities exist for a person with "just enough" technical training. This book focuses on providing just enough training in building structure engineering to help this type of person to succeed in a rewarding career as an architectural and engineering technician.

Hopefully, with the presentation of the material in this book, the technically focused person may become successful in a technical career, may begin to appreciate the breadth of the engineering profession, and may become stimulated to search for further information and pursue more advanced training on this subject.

Joseph Wujek

ACKNOWLEDGMENTS

I thank my best friend and wife, Shauna. I recognize her for her patience, assistance and guidance throughout this project. But, mostly I thank her for her love and companionship. I thank my mom for her support of this project. I also recognize the sacrifices of my sons, Blaze and Bryce. They gave their dad the "quiet" time needed for completion of a project of this size. These commitments were all necessary in making this undertaking possible.

My thanks are extended to the following reviewers for their guidance and suggestions:

Saleh Altayeb, Georgia Southern University
Elliott Colchamiro, New York City Technical College
Ron Gallagher, University of Toledo
Mark Hapstack, Greenville Technical College
John Jarchow, PIMA Community College
Marguorite Newton, Niagara Community College
Ron Nichols, State University of New York at Alfred

Special acknowledgments go to Ed Francis, Senior Editor, Prentice Hall, Inc. and Tally Morgan, WordCrafters Editorial Services, Inc.: I thank Ed for his insight, patience and support of my ideas. I thank Tally for her hard work, enthusiasm and creative suggestions.

Last, the many professional organizations that supplied much of the technical information in this book must be acknowledged. Their cooperation and support are greatly appreciated.

CONTENTS

Contents   xi

Chapter 1
INTRODUCTION TO BUILDING STRUCTURAL ENGINEERING

This chapter sets the tone for technician-level training in principles of statics, strength of materials, and building structure engineering. It provides a brief chronicle of building engineering and an introduction that serves to define the study of statics, strengths of materials, and building structure analysis. Since this text may serve as the student's first exposure to engineering computation, in this chapter we provide a review of standard units and data reporting procedures used in engineering.

1.1   THE EVOLUTION OF BUILDING ENGINEERING

The builders of the massive Egyptian pyramids, the elaborate structures of Greek and Roman times and the great European cathedrals of the thirteenth and fourteenth centuries did not rely on sophisticated engineering computation. Instead, design was based upon the knowledge of past accomplishments or failures and the keen eye of an experienced master builder who watched for evidence of structural distress as the structure was being built. When distress was observed during construction, the master builder made changes in an attempt to thwart failure. Sometimes, signs of distress were not recognized until a structural member or system failed catastrophically.

Nearly five millennia ago, the Bent Pyramid in Egypt was begun at a steep slope of 54°, but about halfway through construction the slope was changed to $43\frac{1}{2}°$. It is theorized that after significant settlement was observed, the pyramid's builders resorted to a shallower angle, thereby reducing the upper volume and the weight of the overall pyramid. Similarly, when the arch of a cathedral roof structure placed a lateral thrust on a cathedral's massive masonry walls, the walls began to bulge and the masonry joints began to open. The master builder observed the movement and directed the workers to install buttresses to brace the walls.

The cathedral of Notre Dame. Repetitive buttresses brace the upper walls.

Through the beginning of the Industrial Revolution (before about 1750), building structures were designed and built chiefly by trial-and-error experiences. While primitive computation techniques and rules of thumb may have been handed down from master builders to their apprentices, the capability to build larger and higher structures came principally from trial-and-error experience and not from engineering computation. At the start of the Industrial Revolution (about 1750), the field of engineering was restricted to two branches: military and civil engineering. Military engineers designed fortifications and weapons while civil engineers focused on design of buildings, bridges, harbors, aqueducts, and other structures. During the Industrial Revolution, the role of the engineer changed. The branches of engineering became more specialized, which allowed for daring experimentation in building structure design. Successful innovative uses of iron in bridges opened the door for the use of iron and steel in building structures at the end of the nineteenth century.

The Crystal Palace was designed by Sir Joseph Paxton and constructed to house the Great Exhibition (the first world's fair) in London in 1851. It was the first large building in the world not constructed of solid masonry. Its 140-ft (43-m)-tall

The Crystal Palace was designed and constructed to house the Great Exhibition (the first world's fair) in London in 1851. It was the first large building in the world that was not constructed of solid masonry.

cast iron and wrought iron frame supported large sheets of glass that enclosed the enormous greenhouse-like structure; covering over 800 000 ft^2 (74 300 m^2). The modular frame and prefabricated standardized members allowed the structure to be erected and dismantled in only a few months. At the conclusion of the Great Exposition, the Crystal Palace was taken down and reconstructed as a museum and hall until it was destroyed by fire in 1936. The iron, glass, and wood structure of the Crystal Palace greatly influenced architecture and structural engineering.

Even with demonstrated successes in iron building structures, use of load-bearing masonry continued to dominate building structure design through the end of the nineteenth century. At that time, building codes required masonry walls to be progressively thicker at their base, much like the great cathedrals of centuries before. For example, the 16-story Monadnock Building constructed in Chicago in 1893 had load-bearing masonry walls ranging from about 1 ft thick at the top floor to 6 ft thick at the base. The 555-ft (169-m)-tall Washington Monument in Washington, DC, completed in 1884, remains the world's tallest masonry structure.

The Eiffel Tower, constructed in Paris, France, in 1898 and designed by French engineer Alexandre Gustave Eiffel (1832–1923), was a structural engineering feat.

The 984-ft (300-m) iron tower soared nearly twice as high as the Washington Monument. During the same period, William Le Baron Jenney (1832–1907) designed the Home Insurance Building, a 10-story iron-and-steel framed structure built in Chicago in 1885.

The Home Insurance Building and the Eiffel Tower demonstrated conclusively the value of iron and steel as effective structural materials and their capacity for structures taller than those built of masonry and timber. These iron and steel building structures started the evolution of the modern skyscraper. More important to the development of structural engineering, Eiffel and Jenney based their structural designs on engineering principles. Rules of thumb and the trial-and-error experience of the master builder were being replaced by application of engineering theory.

Portland cement was patented in England by English bricklayer Joseph Aspdin in 1824. Reinforced concrete containing iron bars was patented by French inventor Joseph Monier in 1849. However, the first reinforced concrete building structure of more than two stories was not erected until 1903 when the 16-story Ingalls Building (renamed the Transit Building in 1959) was constructed in Cincinnati, Ohio. Once again, design was based on application of structural engineering theory.

In addition to advancements in structural engineering, the modern skyscraper was made possible by practical inventions such as the safety elevator demonstrated by American inventor Elisha G. Otis in 1853, and the development of commercial electricity for lighting and powering building systems. Advancements in other areas of building engineering were made by other engineers such as Willis H. Carrier, who presented a paper in 1911 to the American Society of Mechanical Engineers that served as the basis of modern air-conditioning systems in buildings.

With refinement of building structure design and engineering came the first true skyscrapers, such as the 792-ft (241-m) Woolworth Building completed in 1913 and the 1046-ft (319-m) Chrysler Building that opened in 1930. Built in 1930–1931, the 102-story 1250-ft (381-m) Empire State Building in New York City became the world's tallest building. Its rigid steel skeleton, designed to meet strict building code requirements, is overdesigned according to the skyscraper standards of today. The heavy structure proved its strength when a U.S. Air Force B-25 bomber crashed into the 72nd and 73rd floors, causing no substantial damage to the structure.

In the 1950s, glass curtain wall construction began to replace heavy masonry cladding and presented new design challenges related to bracing against lateral wind loads. The 100-story 1127-ft (344-m) John Hancock Center opened in Chicago in 1970. It relies on diagonal bracing of the exterior structure to provide wind bracing and even distribution of gravity loads with lighter structural steel requirements than previous skyscrapers. The pyramidlike structural shape and the foundation of the 853-ft (260-m) Transamerica Building in San Francisco is designed to resist lateral movement from an earthquake. Still, the Empire State Building remained the tallest building in the world for four decades.

The Empire State Building built in 1930–31 and located in New York. Its rigid steel skeleton, regulated by strict building codes, was overdesigned according to the skyscraper standards of today. It remained the tallest building in the world for four decades.

In 1973, the 110-story 1368-ft (417-m) and 1362-ft (415-m) rectangular twin towers of the World Trade Center opened in New York City, finally eclipsing the Empire State Building as the world's tallest building. The external steel framing formed an exterior structural frame designed as a rigid tube that eliminated the need for internal diagonal bracing. With the need for bracing minimized, the tubular structure required much less steel than the traditional skeleton structure. The tubular design demonstrated its strength against a new and unforeseen design load; a massive terrorist bomb exploded in a lower-floor parking garage, causing about $500 million in damage but not toppling the structure.

The 110-story 1454-ft (443-m) Sears Tower, was opened in 1974 in Chicago. The building is supported by a bundled tube structure; nine interlocked 75 ft × 75 ft steel-framed tubular structures that brace each other. The tube frames terminate at the 50th, 66th, 90th, and 110th stories.

The world's tallest buildings at the start of the twenty-first century, the 1476-ft (450-m) Petronas Towers in Kuala Lumpur, Malaysia, were completed in 1997.

The John Hancock Center. The building structure relies upon diagonal bracing of the exterior structure to provide wind bracing and even distribution of gravity loads.

Constructed of reinforced cast-in-place concrete, each tower's structure incorporates a core and cylindrical-tube framing system supported by 16 columns that encircle the base of each tower. The buildings combine in size equivalent to nine Empire State Buildings.

Today, complex systems and innovative materials continue to make building structures lighter. Computer-controlled dampening systems sense and shift massive weights located on the upper floors to counteract the swaying movement caused by wind or earthquake. Advanced research in building structural engineering is producing lighter and stronger composite materials. New computer modeling techniques are used to engineer lighter, yet stronger building structures. The future should mean great engineering achievements in building structures that are not envisioned today.

The World Trade Center, opened in 1973 in New York. The building structure uses a tubular structural design that requires much less steel than the traditional skeleton structure.

The Sears Tower, opened in 1974 in Chicago. It is supported by a bundled tube structure, nine interlocked 75-ft by 75-ft steel-framed tubular structures that brace each other. The tube frames terminate at the 50th, 66th, 90th and 110th stories.

The Petronas Towers in Kuala Lumpur, Malaysia, completed in 1997. Constructed of cast-in-place concrete, each tower's structure incorporates a core and cylindrical-tube framing system supported by sixteen columns that encircle the base of each tower.

1.2 BUILDING STRUCTURAL DESIGN AND ENGINEERING

Building structural design and engineering pertains to the analysis and selection of the type of material and size of a structural member under a specific loading condition. The structural member must be large enough to safely support the forces imposed on it but should be no larger than necessary to prevent waste of material. In building structure design, the engineer applies the concepts of mechanics and strength of materials in analysis and selection of structural members.

1.3 MECHANICS AND STRENGTH OF MATERIALS

Mechanics is that branch of physics and engineering that pertains to the study of a body at rest or in motion and the effect that forces have when they act on a body. Sir Isaac Newton (1642–1727), the renowned English mathematician and scientist, derived the three laws of motion and the law of universal gravitation. He is considered to be the founder of modern mechanics. Mechanics is divided into two main branches: statics and dynamics. *Statics* is the study of forces and the effects of the forces that act on a rigid body at rest. It pertains to a body in a state of equilibrium. In a state of equilibrium the resultant force of all outside forces acting on a body is zero, thus keeping the body at rest. *Dynamics* is the study of the forces that act on a rigid body in motion. It may be further divided into kinematics and kinetics. *Kinematics* is the study of the motion of a body or a system of bodies without considering the mass of the body or the forces acting on it. *Kinetics* is the study of forces that produce or change motion in a body or a system of bodies.

Strength of materials is a branch of engineering that pertains to the capacity of materials to withstand stress (the capacity of a material to resist an applied force or system of forces) and strain (the deformation or change in dimension of a body caused by stress). It includes a study of the effects of compressive, tensile, shear, torsional, and bending stresses due to the application of a force or force system on, or due to a change in temperature in, the material of a body.

1.4 BUILDING CODES AND STANDARDS

There are many codes and standards that govern building design and engineering. An engineer must become familiar with and maintain a working-level understanding of current codes and standards. A *building code* establishes the *minimum* requirements for design, construction, use, renovation, alteration, and demolition of a building and its systems. The intent of a building code is to ensure the health, safety, and welfare of the building occupants. Modern building codes began as fire regulations written and enacted by several large cities during the nineteenth century. They have evolved into a code that contains standards and specifications for materials, construction methods, structural strength, fire resistance, accessibility, egress (exiting), ventilation, lighting, energy conservation, and other considerations.

A building code does not become a law until it is formally adopted as a public ordinance by a local governmental entity such as a city or county or until it is enacted into public law by the state government. A few states have adopted a uniform statewide building code, but most states assign code adoption to counties and municipalities within the state. A governmental entity may write its own building code but typically they rely upon adoption of model codes. A *model building code* is a collection of standards and specifications written and compiled by group of professionals and made available for adoption by state and local jurisdictions.

The minimum standard of construction in Canada is the *National Building Code of Canada* (NBC). In the United States, there are three recognized organizations that generate model building codes:

1. The *Standard Building Code* (SBC), published by the Southern Building Code Congress International (SBCCI), is typically used throughout the southeastern states.
2. The *National Building Code* (NBC), published by the Building Officials and Code Administrators International (BOCA), is adopted mostly in the northeast and central states.
3. The *Uniform Building Code* (UBC), published by the International Conference of Building Officials (ICBO), is used throughout the west.

In December 1994, the three model code organizations in the United States created the International Code Council (ICC) to oversee the development of a single set of model codes to replace the corresponding code of each organization. Initial efforts have focused on a common code format, with the ultimate goal of a single national code. The ICC is currently developing the International Building Code, which is scheduled for completion by the year 2000. At that time, the ICC will maintain and disseminate a single set of model code documents, eliminating the need for the private sector to monitor and participate in three separate code development processes.

Professional organizations develop technical standards, specifications, and design techniques that govern the design and construction of buildings and building systems. These standards and specifications form the foundation upon which the model codes are written. Some of these organizations related to building structures include the American Concrete Institute (ACI), American Forest and Paper Association (AF&P), American Institute of Steel Construction (AISC), American Institute of Timber Construction (AITC), American Iron and Steel Institute (AISI), American National Standards Institute (ANSI), APA–The Engineered Wood Association, American Society of Civil Engineers (ASCE), American Society for Testing and Materials (ASTM), and Concrete Reinforcing Steel Institute (CRSI).

Model building codes and technical standards are formally revised periodically, usually every three to six years, to remain current with advancements and new practices in industry. Each time a new model code or standard is revised, it needs to be adopted into law by the governmental entity. As a result, a different code edition may be in effect in neighboring communities. Design professionals must work hard to keep abreast of revisions in the code and standards.

1.5 TRAINING IN STRUCTURAL ENGINEERING

Many centuries ago, it was the master builders who relied on past experiences to engineer and construct buildings. Innovations in building structures that were engineered followed the Industrial Revolution and allowed structural engineering to emerge as a distinct branch of engineering. Engineering of building structures has evolved over thousands of years from reliance upon rules of thumb and trial-and-error experience to use of sophisticated computation techniques and computer models. As engineering has evolved, the roles in design of a building have become more specialized; the builder now concentrates on the details of building construction, the architect focuses on building layout and aesthetics, while the structural engineer concentrates on the design of a safe and efficient building structure.

Traditionally, a career in engineering requires completion of undergraduate (four-year) studies in architectural, civil, or structural engineering or engineering technology, plus several years of internship experience under the supervision of an experienced professional engineer. A course of study in structural engineering includes classes in higher mathematics (advanced calculus), science (physics and chemistry), mechanics (statics and dynamics), and strength of materials. Additionally, it includes one course in timber design, two courses in steel design, two courses in concrete design, and a project-oriented capstone course.

Training in engineering is quite demanding, involving development of computation and critical thinking skills in scientific and engineering theory with limited applied work. Training in engineering technology places somewhat less emphasis on mathematics, scientific, and engineering theory but more emphasis on applying technology. Employment in the engineering profession results in a challenging and rewarding career. People who have an engineering or engineering technology degree as part of their portfolio are in high demand.

1.6 THE EMERGING ROLE OF THE ENGINEERING TECHNICIAN

With the advent of the computer in analysis, design, and drafting, the roles of the architect, engineer, and drafting technician have changed dramatically: Over a century ago, the architect or engineer alone performed research, analysis, and design and communicated the design through their own artistic sketches and technical drawings. Several decades ago, the architect and engineer performed design and employed drafting technicians to produce high-quality technical drawings. In the sophisticated architectural and engineering consulting offices of today, a greater responsibility is placed on the technician. Where the technician was once only called on to produce high-quality drawings, the technician of today must also understand basic engineering principles and perform fundamental design functions. This means that there are good career opportunities in the engineering and architectural fields.

Although a full course of study in engineering or architecture is very demanding and often out of reach for many because of financial, time, or family constraints, a technically oriented person can train as a technician. Such training is typically available in an associate (two-year)-degree program in architectural technology or engineering technology. It requires development of a working knowledge of intermediate mathematics (college algebra and right-angle trigonom-etry) and science (physics and chemistry). People who have completed such train-ing are in high demand and have challenging and rewarding careers; they work with, rather than for, architects and engineers.

1.7 ENGINEERING DATA REPORTING

Expressing Numbers

In the United States and many other countries it is customary to separate numbers having five or more digits into groups of three digits separated by a comma, such as 10,000 and 1,234,567. A period is used to express a number in decimal form, such as 0.0002 and 1.234. These customs may create confusion in other countries where a comma is used as a decimal symbol; in these countries, 0.0002 is written as 0,0002. In our global economy use of the comma in expressing numbers presents obvious problems.

To avoid confusion, numbers having five or more digits on either side of the decimal point should be expressed in groups of three beginning at the decimal point. Each group of three digits should be separated by a space, not a comma. The space between groups of three digits is not needed in four-digit numbers on either side of the decimal place. For example, the following numbers are expressed according to these rules:

| 0.123 456 789 | 1.234 567 89 | 12.345 678 9 | 123.456 789 | 1234.567 89 |
| 12 345.6789 | 123 456.789 | 1 234 567.89 | 12 345 678.9 | 123 456 789 |

Units and Quantities

In any analysis, proper use of units is important. By definition, a *unit* is a stan-dardized measurement on which a physical property or characteristic is gauged. For example, the inch (in), foot (ft), and meter (m) are all standard units of length. Area is expressed in a set of units of length squared, such as square inches (in^2), square feet (ft^2), square meters (m^2), and square kilometers (km^2). In terms of mass, pounds (lb) and kilogram (kg) are commonly used units. A unit is a standard on which a measurement is made. An actual numerical measurement attached to a unit is a *quantity* such as the 10 in 10 feet, 10 inches, or 10 meters. The standard mea-sure on which a quantity is based is the unit.

Use of Units

In engineering analysis, and in everyday life, use of proper units is important. For example, if someone offered you a full-time job at a rate of 100, would it interest you? If it were $100/hour, you would probably accept the offer and go out and celebrate. However, if it were $100/year, it may not be a wise decision to accept the job. How about £100/day, 100 pesos/week, or 100 zlotys/month? Use of units quantifies a physical property or characteristic, so simply saying 100 is not enough. Units must be included to describe the quantity; 100 meters is much different than 100 miles. In expressing a measurement or in stating an answer to a problem, the unit must always be included. Use of units to quantify a measure is just as meaningful in engineering analysis as it is in everyday life.

Systems of Units

A *system of units* is a specified set of base units. A *base unit* is a specific unit within a system of units that describes the measure of a fundamental property such as mass, length, time, electric current, amount of a substance, and light (luminous) intensity. Base units are the building blocks of a system of units that is used to describe all physical properties. The meter (m) or foot (ft) are base units that describe length. An area is quantified in square lengths. The unit of area is expressed by the square of the basic unit for length, such as square feet (ft^2) or square meters (m^2); a volume is described by a cubic length, such as cubic feet (ft^3) or cubic meters (m^3); and so on. *Compound units* such as miles per hour (m/h) or meters per second (m/s) are a combination of two base units, length and time. Many systems of units are in use today. In the United States, only two are important in engineering and technology: the U.S. Customary System and the International System of Units.

The *U.S. Customary System* of ounces, pounds, yards, feet, inches, miles, hours, minutes, pints, quarts, and gallons is a derivative of the English system of measure, where a foot was truly a measurement of the king's foot. Although it continues to see widespread use in the United States, it has not been used in England for decades. It is, however, used sparingly in a handful of other countries. As metrification becomes a true reality in the United States, this measurement system will be called by another name, such as the *inch-pound system*.

In 1960, the Conference of Weights and Measures introduced the *International System of Units* (*SI* for *Système International d'Unités*) as the global standard of weights and measures. SI is a modified metric system that uses base units such as meters, kilograms, and seconds instead of the units expressed in centimeters, grams, and minutes that are found in the old metric system. It is, however, commonly referred to as a metric system. The SI system has many advantages over the U.S. Customary System: It uses only seven base units and two supplementary units; it is organized in a base-10 or decimal system (i.e., 1, 10, 100, 1000, etc.) that is easy to use; and it is understood and widely used globally (except, of course, in the United States). A comparison between the units of the U.S. Customary System and SI is provided in Tables A.1 and A.2 in Appendix A. Upon review of these units, note that in most cases the U.S. Customary System uses many terms to quantify a specific unit; for example, volume can be expressed in units of gallons, cubic feet, or cubic inches. Use of many different units to quantify a single property is cumbersome.

Subdivisions and multiples of SI units are widely used in this system. Each is designated by a *prefix* according to the corresponding power of 10; for example, kilo for 1000, mega for 1 000 000, and milli for 1/1000. The most common prefixes used in engineering are shown in Table 1.1. Prefixes are added to a base unit to increase or decrease its quantity by 10. For example, the *kilo*meter is 1000 meters, a *milli*meter is 1/1000 of a meter, a *nano*meter is 1/100 000 000 of a meter, and so on. Prefixes allow one to work with very large or very small quantities without working with a lot of zeros or decimal places. As a rule, a prefix should be used instead of quantities greater than four digits or decimals: 630 kW is preferred to 630 000 W or 0.630 MW.

As shown in Tables A.1 and A.2, base SI units may be blended to describe a specific physical characteristic in the form of a compound unit such as meters per second (m/s) or newton-meters (N-m). Compound units such as miles per hour (mph) and pound-foot (lb-ft) are also found in the U.S. customary system. Some compound SI units are used so regularly and expressed so frequently that they are known by specific names. For example, the newton (N) has units of $(kg\text{-}m)/s^2$ and the pascal (Pa) is N/m^2. These compound units are derived from SI base units.

As a rule, units should be written in their singular form: for example, kg, not kgs, and lb, not lbs. Units should not be followed by a period, except at the end of a sentence. Units should be written in lowercase letters, such as m, not M, and ft, not FT. Exceptions to the lowercase rule include units with an uppercase prefix, such as the megawatt (MW), and units that are derived from a proper name, such as the joule (J), newton (N), ampere (A), volt (V), and watt (W). Finally, a space should separate the quantity from the unit: 1200 m instead of 1200m.

Table 1.1 PREFIXES USED IN THE SI SYSTEM

SI Multiplying Factor		Prefix (symbol)	Pronunciation
1 000 000 000 000	10^{12}	tera (T)	*ter*race
1 000 000 000	10^{9}	giga (G)	gig-ah
1 000 000	10^{6}	mega (M)	meg-ah
1 000	10^{3}	kilo (k)	kill-oh
0.1	10^{-1}	deci (d)[a]	*deci*mal
0.01	10^{-2}	centi (c)[a]	*centi*pede
0.001	10^{-3}	milli (m)	*milli*pede
0.000 001	10^{-6}	micro (μ)	*micro*scope
0.000 000 001	10^{-9}	nano (n)	nah-no
0.000 000 000 001	10^{-12}	pico (p)	peek-oh

[a]Use of these prefixes should be avoided.

Metrification in the United States

Although the United States is the world's largest producer and user of commerce, it is the only country not fully using SI units. Efforts to drive U.S. industries toward metrification have moved slowly since formal actions were taken in the early 1970s. Most automobile and other manufacturers who export products have made the shift. However, metrification in building design and construction has been slow since it is principally a domestic industry. An Executive Order by President George Bush in 1991 mandated that all federally funded construction projects be designed in metric units and built with metric materials. This action has motivated the building construction industry to move more quickly. Construction that is privately funded is not required to be designed and built in metric units.

Most manufacturers and trade organizations in the construction industry support and have developed metric standards, using soft conversion to SI units. *Soft metric conversion* involves describing a member in SI units but not physically changing its size. *Hard metric conversion* involves physically changing and reengineering a member's size to meet specific or convenient SI units. Some examples of metrification include:

- No. 3 steel rebar having a diameter of 0.375 in (9.5 mm) has a metric designation of No. 10 (10 is approximately 9.5 mm). A soft metric conversion is used.
- A W8×31 steel section has the metric designation of W200×46.1. W200 is a soft conversion from the W8: 8 in · 25.4 mm/in = 203.2 mm, or approximately 200 mm. The 31 lb/ft is a soft conversion to 46.1 kg/m.

The governmental agency charged with the task of coordinating the metrification efforts of the federal government is the Construction Metrification Council (CNC). The CNC is a council of the National Institute of Building Sciences, 1201 L Street, NW, Suite 400, Washington, DC 20005.

Conversion of Units

Use of consistent units in engineering computation is very important. However, often a quantity is expressed in a certain unit but must be converted so that it can be used with another unit of the same type. For example, length may be expressed in miles and must be converted to feet; meters may need to be converted to kilometers, pounds to kilograms, gallons per minute to cubic meters per second, and so on.

Conversion to another unit is treated in exactly the same way as any algebraic quantity. Units can be multiplied algebraically by one another. Much like $y \cdot y = y^2$, $ft \cdot ft = ft^2$, $m \cdot m \cdot m = m^3$, and $N \cdot m = N\text{-}m$. Similarly, units can be divided by one another: $ft^2/ft = ft$, $m^3/m^2 = m$, $(N\text{-}m)/m = N$, and $in/in = 1$. A conversion factor is needed to convert a measurement expressed in one unit or set of units to another unit or set of units. Conversion factors from U.S. Customary units to SI units and from SI units to U.S. Customary units are provided in Tables A.1 and A.2. As shown in Example 1.1, the principle of unit conversion can be used to make a conversion from one unit to another and from units in one system to units in the other.

■ EXAMPLE 1.1

Make the following conversions:
(a) Convert 2.1 miles to feet.
(b) Convert 765.1 kilometers to meters.
(c) Convert 250 kg to lb.
(d) Convert 10 gallons per minute to cubic meters per second

Solution Conversion factors used in the analysis below are shown in parentheses. They are provided in Tables A.1 and A.2.

(a) $2.1 \text{ miles} \cdot (5280 \text{ ft} / \text{ mile}) = 11\ 088 \text{ ft}$

(b) $765.1 \text{ km} \cdot (1000 \text{ m} / \text{ km}) = 765\ 100 \text{ m}$

(c) $250 \text{ lb} \cdot (\text{kg} / 2.2046 \text{ lb}) = 113.4 \text{ kg}$

(d) $10 \text{ gal/min} \cdot (\text{min} / 60 \text{ s}) \cdot (\text{ft}^3 / 7.48 \text{ gal}) \cdot (\text{m} / 3.281 \text{ ft}^3) = 0.006\ 79 \text{ m/s}$

Significant Digits

The number of significant digits expressed in a measurement indicates the accuracy of a measurement. A natural number (1,2,3,4,5, etc.) is always a significant digit. Trailing zeros to the left of a decimal point are not significant if there are no natural numbers to the right of the decimal point; that is, the number 469 000 contains three significant digits, meaning that the true value of the measurement is between 468 500 and 469 500. Trailing zeros to the right of the decimal point are significant; that is, the number 1.000 has four significant digits, meaning that the true value of the measurement is between 0.9995 and 1.0005.

The number of significant digits is counted from left to right beginning with the first natural number (1,2,3,4,5, etc.). A zero counts as a significant digit only if it occurs to the right of a natural number. The quantity 2.693 045 563 reduced to three significant digits is 2.69. Following are examples of quantities expressed in three significant digits:

123	12.3	1.23	0.123	0.0123
102	10.2	1.02	0.102	0.0102
300	30.0	3.00	0.300	0.0300

Do not confuse significant digits with decimal places.

Precision

Precision is defined as the exactness with which a number is specified. It is the number of significant digits with which a measurement is expressed or a computation is made. The use of pocket calculators and computers to scientific and engineering analysis has provided the capability to produce very high mathematical accuracy very quickly. For example, the results of dividing 1.123 by 0.4317 is expressed mathematically as 2.693 045 563. In engineering, the result is expressed as 2.69. As a number with 10 digits, 2.693 045 563 appears very accurate. However, although 2.693 045 563 is more mathematically correct than 2.69, it is not scientifically accurate beyond three significant digits. The digits beyond 2.69 are insignificant because analysis results is restricted by the accuracy of quantities being divided; multiplying two numbers with four significant digits yields a result in three significant digits.

It is frequently necessary to round off the quantity in a measurement. A common practice is to round the last required digit according to the following rules:

1. If the digit immediately to the right of the last required significant digit is less than 5, leave the last required significant digit the same and drop the unneeded digits.

Precision in Structural Engineering

Precision is defined as the exactness with which a number is specified; the number of significant digits with which a measurement is expressed. Although it is possible to fabricate a structural member with an accuracy of 0.001 or even 0.0001 in., an accuracy of 0.01 or 0.1 in. is typically more economical and acceptable. Limited precision is permitted in the structural engineering profession and construction industry;

In the structural engineering computation, it is acceptable to maintain and use the following degrees of precision:

1. Beam spans, column lengths, and sectional dimensions expressed to the nearest 0.01 in. (0.25 mm) in steel and $\frac{1}{16}$ in. (15 mm) in wood and concrete

2. Uniformly distributed loads expressed to the nearest 1 lb/ft^2 and 10 lb/ft (1.5 N/m and 0.4 N/m^2)

3. Concentrated loads expressed to the nearest 100 lb or 0.01 kip (500 N or 0.5 kN)

4. Total loads and reactions expressed to the nearest 10 lb or 0.01 kip (50 N or 0.05 kN)

5. Shear expressed to the nearest 10 lb or 0.01 kip (50 N or 0.05 kN)

6. Moments expressed to the nearest 10 lb-ft or 0.01 kip-ft (50 N-m or 0.05 kN-m)

7. Deflection expressed to the nearest 0.01 in. (0.25 mm)

---

2. If the digit immediately to the right of the last required significant digit is 5 or greater, increase the last required significant digit and drop the unneeded digits.

Data Reporting

The degree of precision in analysis is limited to the number of significant digits that are carried through the equation. Accuracy is restricted to the one digit less than the smallest number of significant digits in any one quantity used in the analysis: for example, $10.0432 \cdot 10.0 = 100$ (not 100.4320). An accuracy of three or four significant digits (not decimal places) is generally acceptable for most practical engineering applications. Additional accuracy may be required in special situations when working in scientific or research testing, but it is not required in engineering work.

Usually, an error that is limited to less than 0.2% is acceptable in practical engineering analysis. Such accuracy can be achieved by performing analysis and reporting results according to the following rules:

1. When the first significant digit in a quantity is 1, use four significant digits.
2. When the first significant digit in a quantity is between 2 and 9, use three significant digits.

Organization and Workmanship

High standards of workmanship are demanded of those who work in the engineering profession. A solution approach founded on a solid understanding of the subject matter, good workmanship, and sound engineering judgment are paramount. Since an engineer may be called upon to review analysis a decade or two after work is performed, all computations must be neat, well organized, and presented in a logical manner according to accepted engineering practices. Effective problem solution and good workmanship go hand in hand. They are described below.

1. *Good organization and documentation.* All work should be neatly documented on engineering quadrille paper (not graph or lined paper). Use of scrap paper should be avoided because it is easily misplaced. Back-of-the-envelope computations may appear expedient at first but are generally undecipherable a day or a month later; they become useless. Handwritten text and computations should be neat and well organized so that it is easily understood by a third party. Diagrams should be complete and drawn neatly with a straightedge. Intermediate results and final results should be underlined. The person's name, the project name, and the date of the computation should appear at the top of all sheets. Related pages should be numbered (i.e., 1/3 means page one of three pages). On large projects, a three-ring loose-leaf binder should be used to store computations.

2. *Identification of a methods of solution.* A clear understanding of the problem must be acquired before beginning work; time spent on research and organization up front means time saved later. Information provided in the statement of problem should be reviewed carefully to establish known quantities and unknown quantities and to identify possible methods of solution. Usually, the final method of solution is not known initially. Research is needed to ascertain the proper method of attack. The solution approach should be formulated and the individual steps required to complete the solution should be identified. With a solution approach formulated, the individual steps required for solution should be executed.

3. *Accurate computation.* Accurate computation is an important component of problem solution. All computations should be made carefully in a sequential manner. Problem solution requires many steps; an error in computation in a beginning step transfers that error to subsequent steps. Work each step individually and reevaluate the relevance of each step as it is worked. Check computations immediately after each step is completed. Units (i.e., ft-lb, N/m^2, etc.) should be used, recorded, and checked throughout the analysis. Errors frequently occur because improper units are used or an incorrect conversion is made. Finally, evaluate the result of each step by estimating approximate results and comparing the approximations to the computation results; simply, does the result of your analysis make sense? If reasonable, intermediate and final results should be certified by underlining.

Recognize that solution of an engineering problem may take several hours or days. Much learning occurs through the solution of a problem, even for the experienced engineer. With time, a person's understanding of a technical subject grows and solving a specific problem becomes easier; that is, the first time a problem is encountered, finding the solution is more time consuming and much more difficult than it is the second or third time. Focus, patience, and persistence are the keys to learning; simply, *never give up!*

EXERCISES

Building Structural Engineering

1.1 Identify and research a pre-1900 architect, engineer, or designer who was instrumental in developing a unique structural design technique or system.
    (a) Describe the technique or system.
    (b) Draw a simple sketch of the building illustrating the design technique or system.

1.2 Identify and research a post-1900 architect, engineer, or designer who was instrumental in developing a unique building structural design technique or system.

(a) Describe the technique or system.

(b) Draw a simple sketch of the building, illustrating the design technique or system.

1.3 Visit a unique structure in your community.

(a) Describe the technique or system.

(b) Draw a simple sketch of the building illustrating the design technique or system.

1.4 Find the model building code that has been enacted in a local or neighboring municipality or county. Review a copy of the code and, in your own words, briefly explain its purpose and contents.

1.5 Research and, in your own words, briefly describe the following branches of engineering:

(a) Mechanics

(b) Statics

(c) Strengths of materials

1.6 Find a university or college with a baccalaureate architectural, civil, or structural engineering program.

(a) Review the curriculum outlined in the college catalog.

(b) Describe the course requirements for a four-year course of study in architectural, civil, or structural engineering.

1.7 Contact the regulatory agency in the state in which you reside and identify the requirements necessary to practice:

(a) Architecture

(b) Structural engineering

1.8 Discuss the process of building structure design with an architect or engineer. Outline the process.

Engineering Analysis and Data Reporting

1.9 Make the following conversions.

(a) 14 ft to m	(b) 10 m to in	(c) 1500 N to lb
(d) 10.5 kips to lb	(e) 15 500 lb-ft to lb-in	(f) 24 000 lb/in^2 to MN/m^2
(g) 135° to rad	(h) 1.00 m^3 to ft^3	(i) 150 000 lb to kN
(j) 1.50 kips/ft to lb/in	(k) 43 560 ft^2 to m^2	(l) 6680 in^4 to m^4

1.10 Express the following numbers in two significant digits.

(a) 3.2549701	(b) 55 009	(c) 16.995
(d) 999.99	(e) 0.000 999	(f) 0.5336

1.11 Express the following numbers in three significant digits.

(a) 3.254 9701	(b) 55 009	(c) 16.995
(d) 999.99	(e) 0.000 999	(f) 0.5336

1.12 Express the following numbers in four significant digits.

(a) 3.2549701	(b) 55 009	(c) 16.995
(d) 999.99	(e) 0.000 999	(f) 0.5336

Chapter 2
FORCES AND BASIC FORCE SYSTEMS

Almost everyone has an intuitive understanding of the concept of force because of everyday work and play experiences. When you push or pull on an object, you exert a force on it. When you kick or throw a ball, you exert a force on it. Although a force can cause motion, it does not follow that a force acting on an object will always cause the object to move. An object can stand still or move, rise or fall, spin or vibrate as a result of a force acting on it.

In this chapter, the effects of a force or group of forces are examined. This presentation is limited to coplaner force systems; that is, a force or group of forces occupying a single plane. Noncoplaner force systems, those with force vectors occupying two planes, are called space vectors. Space vectors involve x, y, and z components of a force. Study of space vectors requires use of a higher level of mathematics and thus is beyond the intended breadth of this text.

2.1 NEWTON'S LAWS OF MOTION

By 1671 and before the age of 30, Sir Isaac Newton (1642–1727) had summarized the effects of a force acting on a body in his three well-known laws of motion. Newton's laws of motion are paraphrased below.

1. A body at rest will remain at rest and a body in motion will move in a straight-line path unless it is influenced by a force.
2. A force acting on a body will act to change the velocity (speed) and/or direction of the body.
3. When one body exerts a force on a second body, the second body will exert an equal but opposite force on the first body.

In his laws of motion, Newton uses the term *body*, which is a single mass of material or a composite of materials that is distinct from other masses. As used in the study of physical science, it is a separate physical entity that has mass. A body is not necessarily an organism such as the body of a human being or animal. Instead, it can be any living being or nonliving object, component, or member. A body is simply an entity having mass.

By Newton's definition, a *force* is an action that produces a change in velocity (an acceleration) of a body in the direction of the applied force. A force can influence a body by changing its velocity and/or by deforming it. A hammer striking a nail causes motion by driving the nail into a block of wood. It also deforms the nail head. If a force causes an insignificant amount of deformation in the body, the deformation is usually neglected in force analysis. A solid body that experiences negligible deformation when a force or group of forces is applied to it is called a *rigid body*. The nail can be treated as a rigid body as long as it does not significantly deform (bend) when it is struck by the force of the hammer.

According to Newton's second law, a force appears to cause motion of a body. Instead, it is better said that a force has a *tendency* to cause motion. For example, gravity on the earth's surface causes a 150-lb person to exert a downward force of 150 lb; that is, when stepping on a bathroom scale, that person has a weight of 150 lb. There is no motion related to the 150-lb force that the person exerts on the scale; the earth surface and the scale react to the force of the person with an upward force of 150 lb. This upward force is equal and opposite to the force of the person. If the person loses or gains weight, the reacting force would increase or decrease with weight change. A reacting force keeps a body from moving and is equal but opposite to the applied force. In this case, Newton's third law applies. In buildings, it is the reacting forces that develop in the building structure that keep the building and its contents from moving.

2.2   FORCE AND MASS

Gravity on the earth's surface causes a person having a mass of 150 lb to exert a downward force of 150 lb; that is, when stepping on a bathroom scale the person weighs 150 lb. It appears that mass and force are equal. However, this relationship is not exactly correct. For example, the acceleration of gravity on the moon is only about one-sixth that found on the earth. The same person having a mass of 150 lb would exert a force of only about 25 lb (one-sixth of 150 lb) on the surface of the moon; the 150-lb person would weigh only 25 lb. Pounds of mass and pounds of force have the same effect only on the earth's surface. In this book when working in U.S. Customary System units, pounds of force (lb) or some derivative of pounds of force will be used.

In SI units, the newton (N) is the basic unit of force. It is derived from the SI base units of kilogram, meter, and second and is expressed in the relationship of $(kg\text{-}m)/s^2$. A newton is that force that when applied to a 1 kg mass gives an acceleration of 1 m/s^2. At the earth's surface the acceleration of gravity is 9.807 m/s^2, so a 1-kg mass exerts a force of 9.807 N. On the moon's surface, a 1-kg mass exerts a force of only about

1.6 N, about one-sixth that found on the earth. Unlike in the U.S. Customary System, where pounds of mass and pounds of force are equal in quantity on the earth's surface, the kilogram (mass) and newton (force) are not equal. When working in SI units in this book, newtons (N) or some derivative of newtons will be used.

2.3 PROPERTIES OF A FORCE

A force must be described by certain characteristics in order to determine what effect it will have on a body. By definition, a force acting on a body has a magnitude (a size), a direction (a line of action and sense), and a point of application. These characteristics are shown graphically in Figure 2.1.

A force must have a *magnitude* or size. In the U.S. Customary System, the standard unit of magnitude of a force is the pound (lb) of force. The kilopound (kip for short) or the ton can also be used to describe a force:

$$1 \text{ kilopound (kip)} = 1000 \text{ pounds (lb)}$$
$$1 \text{ ton} = 2000 \text{ pounds}$$

In SI units, the newton (N) describes the magnitude of a force. The kilonewton (kN) and meganewton (MN) are used to describe large forces. The relationships between SI units of force are

$$1 \text{ kilonewton (kN)} = 1000 \text{ newtons (N)}$$
$$1 \text{ meganewton (MN)} = 1\,000\,000 \text{ newtons (N)}$$

The relationships between base units of force in SI and the U.S. Customary System are

$$1 \text{ pound (lb)} = 4.448 \text{ newtons (N)}$$
$$1 \text{ newton (N)} = 0.2248 \text{ pound (lb)}$$

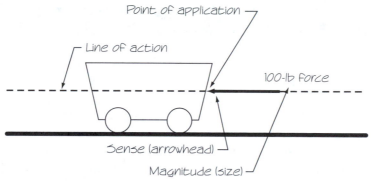

FIGURE 2.1 A force acting on a body has a *magnitude*, a *direction* made up of a line of action and a sense (shown by an arrowhead), and a *point of application*.

A newton is slightly more than one-fifth of a pound of force. As a comparison, an ordinary softball weights about 1 newton, a gallon of water (8.33 lb) weighs about 37 N, a 225-lb person weighs about 1000 N or 1 kN, and a large automobile (4500 lb) weighs about 20 000 N or 20 kN. As shown in Example 2.1, a 112-lb person exerts a force of or weighs about 500 N on the earth's surface.

■ EXAMPLE 2.1

Convert the force exerted by a 112-lb person to SI units.

Solution   From Table A.1 or from above, 1 pound (lb) = 4.448 newtons (N):

$$112 \text{ lb} \cdot (4.448 \text{ N/lb}) = 498 \text{ N}$$

■ EXAMPLE 2.2

Convert a moment of force of 12 kN-m to U.S. Customary units of kip-ft.

Solution   From above and from Table A.2, 1 pound (lb) = 4.448 newtons (N), 1000 lb = 1 kip, 1000 N = 1 kN, and 1 meter (m) = 3.280 840 ft:

$$12 \text{ kN-m}(1000 \text{ N/kN})(1 \text{ lb/4.448 N})(1 \text{ kip/1000 lb})(3.280 840 \text{ ft/m}) = 8.85 \text{ kip-ft}$$

A force may act in any direction: upward, downward, or to the left or the right. The *direction* of a force is made up of a line of action and a sense. The *line of action* is a straight-line path taken by the force. Associated with the line of action is the *sense* of the force; that is, the force may be pushing or pulling. Sense is shown graphically with an arrowhead. Finally, a force must have a *point of application*. The point of application describes the location where the force is applied or presumed to be applied to the body. Sometimes, the designer must rely on intuition to define the characteristics of a force. However, it is only after proper identification of magnitude, direction, sense, and point of application of a force that its effect on a body will be known.

2.4   THE PRINCIPLE OF TRANSMISSIBILITY

The principle known as the *principle of transmissibility* states that the point of application of an external force acting on a body may be transmitted anywhere along its line of action without changing its influence on the body. Simply, a pushing force on the front of a body has the same effect as a pulling force on the rear of a body as long as magnitude, sense, and direction of the forces are the same (Figure 2.2).

Consider the forces acting on a car as shown in Figure 2.3. A pushing force acting horizontally against the rear bumper would cause the car to move forward. A pulling force acting horizontally on the front bumper would have the same effect. Conversely, a pushing force acting alone on the front bumper would have the

FIGURE 2.2   The point of application of an external force acting on a body may be transmitted anywhere along its line of action without changing its effect on the body.

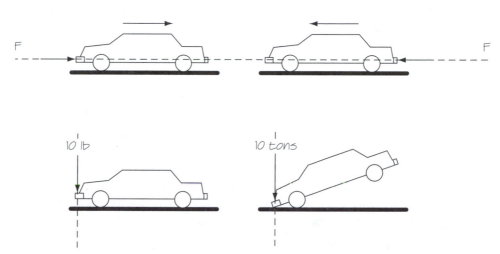

FIGURE 2.3   A change in magnitude, line of action, sense, or point of application of forces acting on a car results in differing effects on the car.

opposite effect. It would cause the car to roll backward. On the other hand, the same force acting downward on the front bumper would lift the rear tires off the ground. A change in magnitude, sense, direction, or point of application greatly influences the effect of the force on the automobile.

Scalar and Vector Quantities

A *scalar quantity* can be expressed in terms of magnitude only, such as miles, dollars, years, square feet, and so on. Scalar quantities can be added, subtracted, multiplied, and divided following the rules of basic arithmetic. A *vector quantity* is described by magnitude and direction (line of action and sense). Vector quantities cannot be added or subtracted arithmetically. Vector quantities must be added graphically or by using vector mathematics. A force is an example of a vector quantity. Two forces of equal magnitude that pull against each other result in a combined force of zero; the opposing forces cancel one another. Vector mathematics is discussed further in Appendix D.

2.5 TYPES OF FORCES

External forces act on a body and result in the development of *internal forces* within the material of the body. Internal forces develop as stress. For example, forces pulling on a rope cause it to straighten and resist the pulling forces. An additional external force applied to the rope causes the material of the rope to resist the stretching force by developing a stress. In this chapter the action of external forces on a body is examined. The concept of internal forces (stresses) is covered in Chapter 6.

External forces can be classified as acting forces and reacting forces. External forces that are applied to a body such as those caused by gravity loads, wind, and water pressure are called *acting forces*. A *reacting force* counters an acting force. A reacting force is frequently called a *reaction*. In this book, the letters *F, P, W,* and *w* are used as symbols that represent an acting force. The letter *R* is used to represent a reacting force.

While an acting force has a tendency to cause motion, a reacting force tends to oppose motion. When a body is at rest, a reacting force develops that opposes the acting force just enough to keep the body still. For example, the seat of a chair develops and exerts a reacting force that opposes the acting force created from the weight of the person seated on the chair. A reacting force has the same point of application, line of action, and magnitude as the acting force it opposes. However, the sense of the reacting force is opposite to the sense of the acting force. In the case of a person seated on a chair, the person applies a force acting downward and the chair reacts by developing and exerting a reacting force that acts upward.

A change in the magnitude of an acting force must result in a change in the magnitude of the corresponding reacting force for the body to remain motionless. For example, for a chair to remain motionless, a 100-lb person seated on the chair results in the chair exerting a reacting force of 100 lb, a 200-lb person results in development of a 200-lb reacting force, and so on as long as the chair remains motionless. The concept of forces acting on bodies at rest (in static equilibrium) is introduced in Chapter 3.

2.6 FORCE SYSTEMS

Two or more constituent forces can be grouped into a *force system*. The configuration of the constituent forces in a force system determines the method of analyzing the combined effect of the constituent forces in the system. Common force systems and basic definitions include:

1. *Concurrent.* All constituent forces have lines of action intersecting at the same point. Nonconcurrent force systems do not have constituent forces that have lines of action intersecting at the same point.

2. *Coplaner.* All constituent forces act in a single plane; forces have lines of action that intersect or lines of action that are parallel. Noncoplaner force systems have three or more constituent forces that *do not* act in a single plane.

3. *Parallel.* All constituent forces have the same direction.

4. *Collinear.* All constituent forces act along the same line of action (i.e., forces pulling on a rope or cable).

Frequently, a designer converts the constituent forces in a force system to a single force to identify the effect the force system has on a body. A *resultant force* is a single theoretical force that will produce the same external effect as the combined effect of the constituent forces in a force system.

2.7   GRAPHIC SOLUTION OF CONCURRENT COPLANAR FORCE SYSTEMS

The *tip-to-tail* graphic solution technique may be used to find the resultant force in concurrent coplaner force systems; a force system where all constituent forces have lines of action intersecting at the same point and act in a single plane. A resultant force in a concurrent coplaner force system will produce the same effect at the point of application as the combined effect of the constituent forces it replaces. A resultant force has its own magnitude and direction (line of action and sense).

The tip-to-tail graphic solution method involves graphically transferring the vectors "tip to tail" to form a polygon. The resultant force is found by drawing a vector from the tail or origin of the first vector to the tip or arrow of the last vector. The tip-to-tail graphic solution technique used to find the resultant of forces is illustrated in Figure 2.4. The line of action and sense of the constituent force vector was drawn accurately at a certain scale. Scale is necessary to relate length to the magnitude of each constituent force vector. In this figure, the line representing the 15-lb constituent force vector is drawn 50% longer than the line representing the 10-lb constituent force vector.

Following is an explanation of the solution for the force system shown in Figure 2.4. Both the 10- and 15-lb constituent force vectors are applied in a concurrent coplaner force system and therefore can be combined as a resultant. Vector **A** pulls up and to the right (at a 30° angle from horizontal) with a force of 15 lb. Vector **B** pulls up and to the left at 45° from horizontal with a force of 10 lb. One should be able to recognize that the combined effect of the constituent forces tends to move the point of application up and to the right. However, determination of the exact combined effect of the two constituent forces requires vector addition; that is, the resultant of the force system must be found.

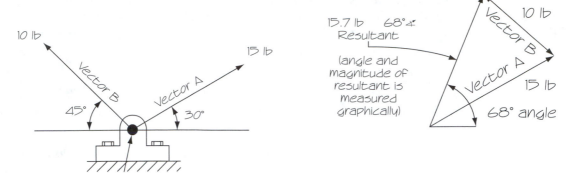

FIGURE 2.4 The tip-to-tail graphic solution technique is used to find the resultant of forces acting on a single point of application.

The graphic solution of the set of constituent forces on top of Figure 2.4 consists of the following approach:

1. Vector **A** is drawn at a 30° angle to scale based on the magnitude of constituent force. A scale of 1 in = 10 lb was selected because it fits neatly on the paper. (A larger scale, say 1 in = 5 lb or 1 in = 1 lb, would yield more accurate results and should be used when working with more vectors.)

2. Vector **B** is drawn at 45° measured from the right horizontal beginning at the "tip" of vector **A.** Once again, a scale of 1 in = 10 lb is used when drawing vector **B.**

3. Determination of the resultant is found by graphically constructing the resultant from the tail of vector **A** to the tip of vector **B.** Measurement of the length of the resultant at a scale of 1 in = 10 lb yields the resulting force of 15.7 lb. The angle of the resultant is 68° from the right horizontal. It is determined by measuring the angle from horizontal with a protractor.

A 15.7-lb force vector acting at 68° "up and to the right" can replace the 10- and 15-lb constituent force vectors without changing the overall effect of the force system. The resultant for this force system is expressed as 15.7 lb 68°∡.

The tip-to-tail graphic solution technique used to find the resultant of the force system is illustrated in Figure 2.5. The force vectors are arranged tip to tail beginning with the 30-lb constituent force acting horizontally and ending with the 15-lb constituent force acting horizontally. The resultant for this force system is expressed as 61.7 lb 59.3°∡.

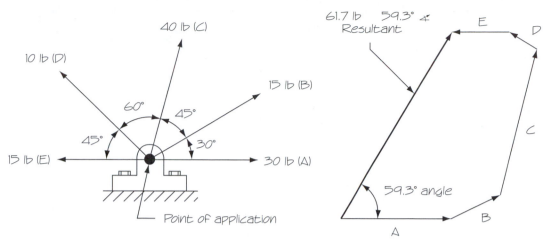

FIGURE 2.5 The tip-to-tail graphic solution technique is used to find the resultant of forces acting on a single point of application.

2.8 COMPONENTS OF FORCE

In force analysis it is sometimes helpful to separate a known force into horizontal (F_x) and vertical (F_y) components of force. Just like the constituent forces in a force system can be added to find the resultant force of the system, a single force may be replaced by two x-and y-components of force that produce the same effect as the force they replace.

The *slope triangle* method of finding the horizontal (F_x) and vertical (F_y) components of force is used when the slope of the single force (F_R) is known:

$$\text{horizontal component of force } F_R, F_x = \frac{a}{c} F_R$$

$$\text{vertical component of force } F_R, F_y = \frac{b}{c} F_R$$

For the force (F_R), F_x is the horizontal component of force F_R, and F_y is the vertical component. The value c is the hypotenuse of a right triangle created by the rise (b) and the run (a) of the slope, as shown in Figure 2.6.

Right-triangle relationships may be found in common a–b–c (leg–leg–hypotenuse) increments, such as 3–4–5, 5–12–13, and 8–15–17, as shown in Figure 2.7. (At least, these right-triangle relationships are conveniently found in "textbook" exercises.) So if the slope is 3 in 4, the hypotenuse must be equal to 5 units because of the 3–4–5 relationship. Similarly, a right triangle having a

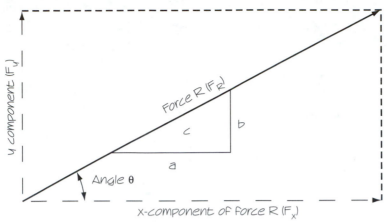

FIGURE 2.6 A single force may be replaced with a horizontal or *x* component of force (F_x) and a vertical or *y* component of force (F_y).

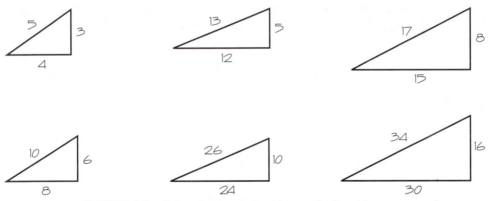

FIGURE 2.7 Right-triangle relationships may be found in common *a–b–c* (leg–leg–hypotenuse) increments, such as 3–4–5, 5–12–13, and 8–15–17. (At least, these right-triangle relationships are conveniently found in textbook exercises.)

12-in-16 slope has a hypotenuse equal to 20, because $3 \cdot 4 = 12$, $4 \cdot 4 = 16$, and thus $5 \cdot 4 = 20$.

If *a*, *b*, or *c* are unknown, as is common in the field, they may be found by algebraic manipulation of the Pythagorean theorem: $a^2 + b^2 = c^2$. The slope of *a/b* is usually known, so it is necessary to find *c* by rearrangement of the Pythagorean theorem:

$$c = \sqrt{a^2 + b^2}$$

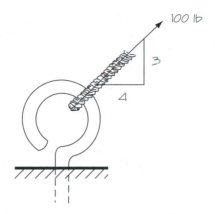

FIGURE 2.8 Force system for Example 2.3.

■ EXAMPLE 2.3

A 100-lb force pulls up and to the right on a body at a slope of 3 in 4 (3 units of rise over 4 units of run), as shown in Figure 2.8. Find the horizontal and vertical components of force.

Solution Since the legs of a 3–4–5 triangle are 3 and 4 as stated, c must equal 5. Or, c may be found by the Pythagorean theorem:

$$c = \sqrt{a^2 + b^2} = \sqrt{4^2 + 3^2} = 5$$

The components of force are found by the slope triangle method ($a = 4$, $b = 3$, and $c = 5$):

$$F_x = \frac{a}{c} F_R = \left(\frac{4}{5}\right)(100 \text{ lb}) = 80 \text{ lb}\rightarrow$$

$$F_y = \frac{b}{c} F_R = \left(\frac{3}{5}\right)(100 \text{ lb}) = 60 \text{ lb}\uparrow$$

In Example 2.3 the x component of force F_R is an 80-lb force acting horizontally to the right. It is expressed as 80 lb →. The 60-lb force acting upward is the y component of force F_R. It is expressed as 60 lb ↑. The effect of these components of force would have the same effect on a body as that of the initial 100-lb force acting at the 3-in-4 slope.

The *right-angle trigonometry* method, sometimes called the right-angle "trig" method, may also be used to determine the x and y components of a force. This method is typically used if the direction of the force is described by an angle:

horizontal component of force, $F_x = F_R(\cos \theta)$

vertical component of force, $F_y = F_R(\sin \theta)$

For a specific force (F_R), F_x is the horizontal component of force F_R, and F_y is the vertical component of force F_R. Theta, θ, is used to describe the angle of F_R from horizontal to the right of the point of application (similar to how angles are measured on a CAD system).

■ EXAMPLE 2.4

A 100-N force pulls a body up and to the *right* at an angle of 30° from horizontal, as shown in Figure 2.9. Find the horizontal and vertical components of force.

Solution

$$F_x = F_R(\cos \theta) = 100 \text{ N } (\cos 30°) = 100 \text{ N } (0.866) = 86.6 \text{ N} \rightarrow$$
$$F_y = F_R(\sin \theta) = 100 \text{ N } (\sin 30°) = 100 \text{ N } (0.500) = 50.0 \text{ N} \uparrow$$

■ EXAMPLE 2.5

A 100-N force pulls a body up and to the *left* at an angle of 30° from horizontal, as shown in Figure 2.10. Find the horizontal and vertical components of force.

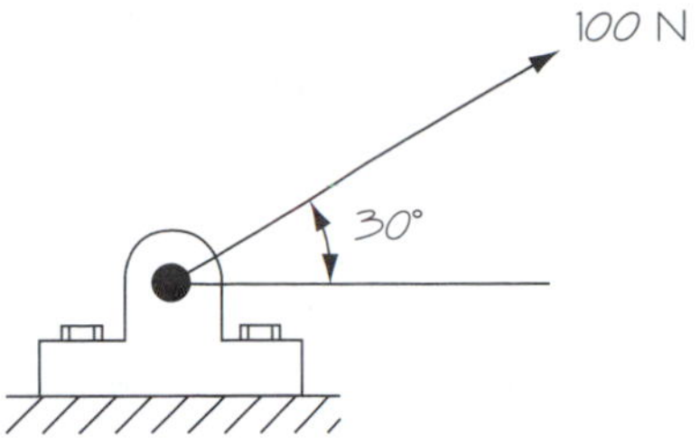

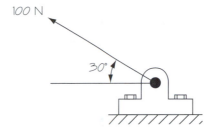

FIGURE 2.9 Force system for Example 2.4.

FIGURE 2.10 Force system for Example 2.5.

Solution

$$\theta = 180° - 30° = 150°$$
$$F_x = F_R(\cos \theta) = 100 \text{ N} (\cos 150°) = 100 \text{ N} (-0.866) = -86.6 \text{ N} = 86.6 \text{ N} \leftarrow$$
$$F_y = F_R (\sin \theta) = 100 \text{ N} (\sin 150°) = 100 \text{ N} (0.500) = 50.0 \text{ N} = 50.0 \text{ N} \uparrow$$

Results of analysis in Examples 2.4 and 2.5 are similar in magnitude, except that in Example 2.5 the x component of force has a negative $(-)$ sign. By sign convention, forces acting horizontally (F_x) are considered positive $(+)$ if they act to the right, and negative $(-)$ if they act to the left. Those forces acting vertically (F_y) are treated as positive if they act upward, and negative if they act downward. Thus, by sign convention, the x and y components of force found in Example 2.5 really have direction (a line of action and a sense):

$$F_x = 86.6 \text{ N} \leftarrow$$
$$F_y = 50.0 \text{ N} \uparrow$$

Sign Convention for Vectors

A sign convention is a custom or practice that assigns a certain sign, either $+$ or $-$, to a specific direction. For example, in engineering an upward force is assigned a positive $(+)$ sign, while a downward force is given the negative $(-)$ sign. Those forces acting vertically (F_y) are treated as positive if they act upward, and negative if they act downward.

2.8 Components of Force

33

As discussed previously, two forces may be substituted with a single force that will produce the same effect as the two forces it replaces. *If the two forces are x and y components* (i.e., acting at right angles to one another), the resultant (F_R) may be found by

$$\text{resultant force, } F_R = \sqrt{F_x^2 + F_y^2}$$

The resultant force, F_R, would produce the same effect if it replaced the combined effect of forces F_x and F_y.

■ EXAMPLE 2.6

With the 86.6-N horizontal force and 50.0-N vertical force described in Example 2.5, calculate the resultant force.

Solution

$$F_R = \sqrt{F_x^2 + F_y^2} = \sqrt{(86.6)^2 + (50.0)^2} = 99.99 \text{ or } 100 \text{ N}$$

In Example 2.6, the resultant force, F_R, is equivalent to the 100-N force in Example 2.5. This demonstrates the relationship between a force and the x and y components of the same 100-N force.

2.9 MATHEMATICAL SOLUTION OF CONCURRENT COPLANAR FORCE SYSTEMS

Constituent forces in a force system that are acting in the same plane may be added mathematically. The *vector addition* solution technique involves mathematically adding the x and y components of each constituent force to find the x and y components of the resultant force. The sign conventions introduced above are used to add and subtract components of the constituent forces as shown in Example 2.7. Once the x component (F_x) and y component (F_y) of the resultant force are determined, the resultant force (F_R) may be found by the following formula:

$$\text{resultant force, } F_R = \sqrt{F_x^2 + F_y^2}$$

The angle of the resultant force (θ) may be found by

$$\theta = \tan^{-1} \frac{F_y}{F_x}$$

By convention, the angle of the resultant force (θ) is measured in a counterclockwise direction about the left endpoint of a line resting horizontally (much as it is measured on a CAD system).

■ EXAMPLE 2.7

Use vector addition to find the resultant of the system of forces illustrated in Figure 2.4.

Solution Vector **A** pulls up and to the right at a 30° angle from horizontal with a force of 15 lb. The x and y components of force **A** are found by

$$F_{Ax} = F_A(\cos \theta) = 15 \text{ lb}(\cos 30°) = 15 \text{ lb}(0.866) = 12.99 \text{ lb}\uparrow$$
$$F_{Ay} = F_A(\sin \theta) = 15 \text{ lb}(\sin 30°) = 15 \text{ lb}(0.50) = 7.50 \text{ lb} \rightarrow$$

Vector **B** pulls up and to the left at 45° from horizontal with a force of 10 lb. By convention, an angle is measured in a counterclockwise direction about the left endpoint of a line resting horizontally (much as it is measured on a CAD system). Therefore, in the case of force **B**, $\theta = 180° - 45° = 135°$. The x and y components of force **B** are found by

$$F_{Bx} = F_B(\cos \theta) = 10 \text{ lb}(\cos 135°) = 10 \text{ lb}(-0.707) = -7.07 \text{ lb} = 7.07 \text{ lb} \leftarrow$$
$$F_{By} = F_B(\sin \theta) = 10 \text{ lb}(\sin 135°) = 10 \text{ lb}(0.707) = 7.07 \text{ lb} = 7.07 \text{ lb} \uparrow$$

The x component (F_x) and y component (F_y) of the resultant force are determined by vector addition:

$$F_x = F_{Ax} + F_{Bx} = 12.99 \text{ lb} + (-7.07 \text{ lb}) = 5.92 \text{ lb} = 5.92 \text{ lb} \rightarrow$$
$$F_y = F_{Ay} + F_{By} = 7.50 \text{ lb} + 7.07 \text{ lb} = 14.57 \text{ lb} = 14.57 \text{ lb} \uparrow$$

The resultant force (F_R) is found by

$$\text{resultant force, } F_R = \sqrt{F_x^2 + F_y^2} = \sqrt{(5.92 \text{ lb})^2 + (14.57 \text{ lb})^2} = 15.73 \text{ lb}$$

The angle of the resultant force (θ) may be found by

$$\theta = \tan^{-1}\frac{F_y}{F_x} = \tan^{-1}\frac{14.57 \text{ lb}}{5.92 \text{ lb}} = \tan^{-1}(2.461) = 67.9°\measuredangle$$

The resultant of 15.73 lb 67.9°$\measuredangle$ found through vector addition in Example 2.7 compares favorably with the resultant found graphically in Figure 2.4.

■ EXAMPLE 2.8

Use vector addition to find the resultant of the system of forces illustrated in Figure 2.5.

Solution Beginning with the first constituent force at the right side of the force system and working counterclockwise: Constituent force **A** is a 30-lb vector that pulls horizontally to the right. The x and y components of this force are found by

$$F_x = 30.00 \text{ lb} \rightarrow = +30.00 \text{ lb}$$
$$F_y = 0$$

Constituent force **B** is a 15-lb force vector that pulls up and to the right at an angle of 30°. The x and y components of this force are found by

$$F_x = F(\cos \theta) = 15 \text{ lb}(\cos 30°) = 15 \text{ lb}(0.866) = 12.99 \text{ lb} \rightarrow = +12.99 \text{ lb}$$
$$F_y = F(\sin \theta) = 15 \text{ lb}(\sin 30°) = 15 \text{ lb}(0.50) = 7.50 \text{ lb} \uparrow = +7.50 \text{ lb}$$

Constituent force **C** is a 40-lb force vector that pulls up and to the right at an angle of 75°. The x and y components of this force are found by

$$F_x = F(\cos \theta) = 40 \text{ lb}(\cos 75°) = 40 \text{ lb}(0.259) = 10.36 \text{ lb} \rightarrow = +10.36 \text{ lb}$$
$$F_y = F(\sin \theta) = 40 \text{ lb}(\sin 75°) = 40 \text{ lb}(0.966) = 38.64 \text{ lb} \uparrow = +38.64 \text{ lb}$$

Constituent force **D** is a 10-lb force vector that pulls up and to the left at an angle of 135°. The x and y components of this force are found by

$$F_x = F(\cos \theta) = 10 \text{ lb}(\cos 135°) = 10 \text{ lb}(-0.707) = 7.07 \text{ lb} \leftarrow = -7.07 \text{ lb}$$
$$F_y = F(\sin \theta) = 10 \text{ lb}(\sin 135°) = 10 \text{ lb}(0.707) = 7.07 \text{ lb} \uparrow = +7.07 \text{ lb}$$

Constituent force **E** is a 15-lb force vector that pulls horizontally to the left. The x and y components of this force are found by

$$F_x = 15.00 \text{ lb} \leftarrow = -15.00 \text{ lb}$$
$$F_y = 0$$

The x component (F_x) and y component (F_y) of the resultant force are determined by vector addition:

$$F_x = 30.00 \text{ lb} + 12.99 \text{ lb} + 10.36 \text{ lb} + (-7.07 \text{ lb}) + (-15.00 \text{ lb}) = 31.28 \text{ lb} \rightarrow$$
$$F_y = 0 + 7.50 \text{ lb} + 38.64 \text{ lb} + 7.07 \text{ lb} + 0 = 53.21 \text{ lb} \uparrow$$

Note: An alternative bookkeeping process that can simplify vector addition is to arrange analysis results in the form of a table by constituent force, as shown below.

Constituent force	F_x	F_y
A (30 lb)	30.00 lb	0
B (15 lb)	12.99 lb	7.50 lb
C (40 lb)	10.36 lb	38.64 lb
D (10 lb)	−7.07 lb	7.07 lb
E (15 lb)	−15.00 lb	0
Resultant components	31.28 lb →	53.21 lb ↑

The resultant force (F_R) is found by

resultant force, $F_R = \sqrt{F_x^2 + F_y^2} = \sqrt{(31.28 \text{ lb})^2 + (53.21 \text{ lb})^2} = 61.7 \text{ lb}$

The angle of the resultant force (θ) may be found by

$$\theta = \tan^{-1} \frac{F_y}{F_x} = \tan^{-1} \frac{53.21 \text{ lb}}{31.28 \text{ lb}} = \tan^{-1}(1.701) = 59.6°$$

The resultant is 61.7 lb 59.6°⦩.

The resultant of 61.7 lb 59.6°⦩ found through vector addition in Example 2.8 compares favorably with the resultant found graphically in Figure 2.5.

A resultant force combines the effect of the constituent forces in a force system into a single theoretical force that will produce the same effect as the force system. A reacting force or reaction reacts to an acting force to keep a body motionless. By definition, a reacting force has the same point of application, line of action, and magnitude as the force it opposes and a sense that is opposite to the sense of the resulting force. Therefore, in a motionless system of forces, a resultant force must be opposed by a single theoretical reacting force.

■ EXAMPLE 2.9

Find the reaction for this force system shown in Figure 2.5 if it is a motionless system.

Solution   In Example 2.8 the resultant was found to be 61.7 lb 59.6°∡. The reacting force has the same point of application, line of action, and magnitude as the force it opposes but a sense that is opposite to the sense of the resulting force; the sense changes by 180°: 59.6°∡ + 180° = 239.6°∡. The reaction for this force system is 61.7 lb 239.6°∡.

2.1   Three cables pull on a flange with the tension forces as shown. Use the tip-to-tail graphic solution technique to find the resultant of constituent forces for the force system shown.

2.2   Three cables pull on a flange with the tension forces as shown. Use the tip-to-tail graphic solution technique to find the resultant of constituent forces for the force system shown.

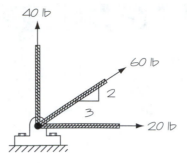

2.3 Four cables pull on a flange with the tension forces as shown. Use the tip-to-tail graphic solution technique to find the resultant of constituent forces for the force system shown.

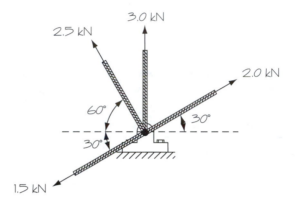

2.4 A cable pulls on a flange with a tension force as shown. Using the slope-triangle method solve for the horizontal (*x* component) and vertical (*y* component) components of the tension force.

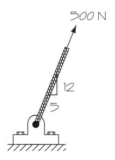

2.5 A cable pulls on a flange with a tension force as shown. Using the slope-triangle method solve for the horizontal (*x* component) and vertical (*y* component) components of the tension force.

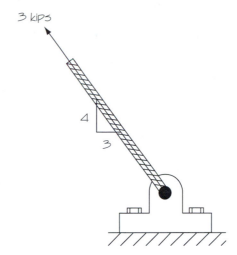

2.6 A cable pulls on a flange with a tension force as shown. Using the slope-triangle method solve for the horizontal (x component) and vertical (y component) components of the tension force.

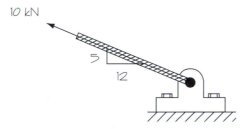

2.7 A cable pulls on a flange with a tension force as shown. Using the right-angle trig method solve for the horizontal (x component) and vertical (y component) components of the tension force.

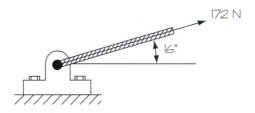

2.8 A cable pulls on a flange with a tension force as shown. Using the right-angle trig method solve for the horizontal (x component) and vertical (y component) components of the tension force.

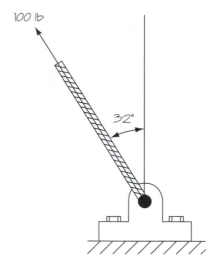

2.9 A cable pulls on a flange with a tension force as shown. Using the right-angle trig method, solve for the horizontal (*x* component) and vertical (*y* component) components of the tension force.

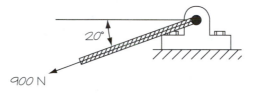

2.10 Cables pull on a flange with the tension forces shown. Using vector addition, find the resultant of the system of forces.

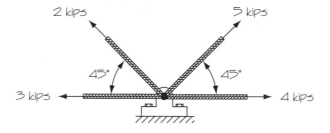

2.11 Cables pull on a flange with the tension forces shown. Using vector addition, find the resultant of the system of forces.

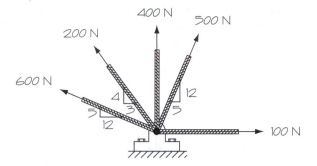

2.12 Cables pull on a flange with the tension forces shown. Using vector addition, find the resultant of the system of forces.

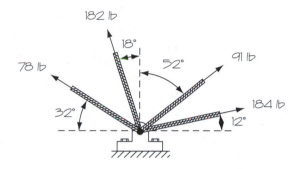

2.13 A joint in a frame is constructed of rigid members attached to a fixed flange as shown. Each member exerts the force shown. Using vector addition, find the resultant of the system of forces.

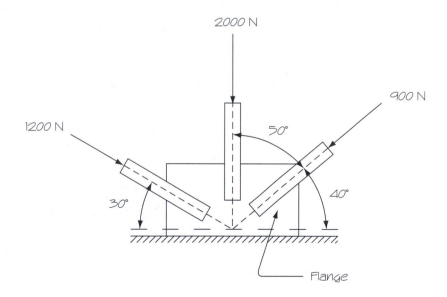

2.14 A joint in a frame is constructed of rigid members attached to a fixed flange as shown. Each member exerts the force shown. Using vector addition, find the resultant of the system of forces.

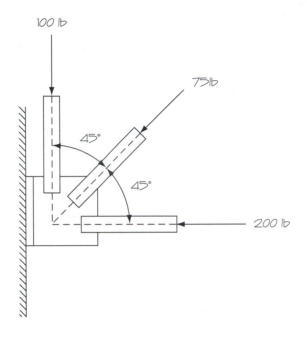

2.15 A joint in a frame is constructed of rigid members attached to a fixed flange as shown. Each member exerts the force shown.
  (a) Using vector addition, find the resultant of the system of forces.
  (b) Find the magnitude and direction of the reaction that will keep the flange motionless.
  (c) Using the slope-triangle method, find the horizontal (x component) and vertical (y component) components of the reaction force.

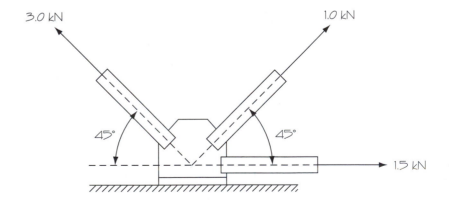

2.16 A joint in the frame of an amusement park ride is constructed of steel angles welded to a steel plate as shown. The plate serves as a flange that is fixed to a surface. Each angle exerts the force shown.

(a) Using vector addition, find the resultant of the system of forces.

(b) Find the magnitude and direction of the reaction that will keep the flange motionless.

(c) Using the slope-triangle method, find the horizontal (x component) and vertical (y component) components of the reaction force.

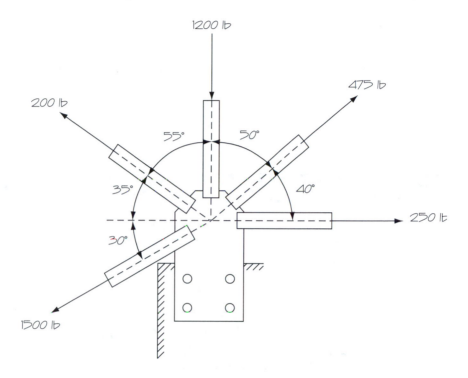

Chapter 3
STATIC EQUILIBRIUM

In this chapter the effects of external forces acting on a body that remains motionless and does not deform when a force or group of forces is applied to it are examined. As in Chapter 2, this presentation is limited to coplaner force systems, force vectors occupying a single plane.

3.1 MOMENT OF A FORCE

A force that causes rotation or has a tendency to cause rotation is a moment. The *moment of a force* (M) about an axis is the product of force (F) and the distance (L) between the point of application to the axis of rotation. It is expressed as

$$\text{moment, } M = FL$$

The distance from the point of application to the axis of rotation (L) is called the *moment arm*. The force under consideration acts perpendicular to the moment arm. The pivot point about which the force rotates or tends to rotate is referred to as the *axis of rotation*. A moment is created whenever a force causes *or has a tendency to* cause rotation about an axis of rotation.

Since a moment is found by multiplying a force by a distance, a moment must be expressed in the units force-distance. In the U.S. Customary System, a moment is expressed in units of pound-foot (lb-ft), pound-inch (lb-in), kip-foot (kip-ft), and kip-inch (kip-in).

$$\text{pound-foot (lb-ft)} = 12 \text{ pound-inches (lb-in)}$$
$$\text{kip-foot (kip-ft)} = 1000 \text{ pound-feet (lb-ft)}$$
$$\text{kip-foot (kip-ft)} = 12 \text{ kip-inches (kip-in)}$$

In the International System of Units, moment is expressed in units of newton-meter (N-m). The relationships between base units of force in the SI and U.S. Customary System are

$$\text{newton-meter (N-m)} = 0.737\ 561 \text{ pound-foot (lb-ft)}$$
$$\text{newton-meter (N-m)} = 8.850\ 75 \text{ pound-inches (lb-in)}$$

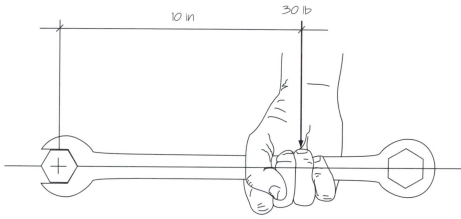

FIGURE 3.1   Force system for Example 3.1.

■ EXAMPLE 3.1

A person tightens a bolt by exerting a force to the handle of a wrench. The person applies the force 10 in. away from the center of the bolt. The 30-lb force is applied perpendicular to the handle of the wrench, as shown in Figure 3.1. Determine the moment that develops.

Solution

$$M = FL = 30 \text{ lb} \cdot 10 \text{ in} = 300 \text{ lb-in}$$

■ EXAMPLE 3.2

A person exerts a 30-lb force to the handle of a wrench. The force is applied 10 in. away from the center of the bolt, but it is applied at an angle of 60° from the handle, as shown in Figure 3.2. Determine the moment that develops.

Solution   Because the force of a moment is applied perpendicular to the moment arm, it is necessary first to solve for the component of the force that is perpendicular to the wrench handle:

$$F_y = F_R(\sin 60°) = 30 \text{ lb} (0.866) = 25.98 \text{ lb}$$
$$M = FL = (25.98 \text{ lb}) (10 \text{ in}) = 259.8 \text{ lb-in}$$

The analysis techniques used in Examples 3.1 and 3.2 are similar, except that in Example 3.2 it was first necessary to find the force acting perpendicular to the moment arm.

It is important to underscore that although it may appear that a moment causes rotation, actual rotation is not a required condition for a moment to develop. A moment must only have a *tendency* to cause rotation about an axis of rotation. Consider the wrench in the previous examples. If the bolt were too tight so

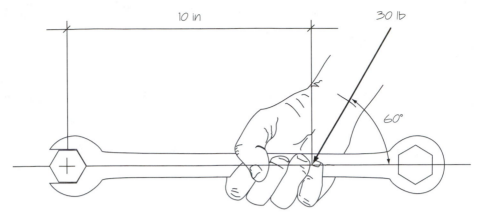

FIGURE 3.2 Force system for Example 3.2.

that the person using the wrench could not loosen it, a moment would be created by the wrench even though rotation would not occur. Rotation is not a required condition for a moment to develop.

■ EXAMPLE 3.3

A 150-lb person sits 8 ft away from the fulcrum (point of rotation) on the left side of a playground seesaw. Determine the moment created by the person.

Solution

$$M = FL = 150 \text{ lb} \cdot 8 \text{ ft} = 1200 \text{ lb-ft}$$

A moment may cause or have a tendency to cause clockwise or counterclockwise rotation. In engineering a common *sign convention* is used to distinguish between the two types of rotation. Clockwise moments are negative ($-$), and counterclockwise moments are positive ($+$). The person in Example 3.3 was observed on the left side of the seesaw, so the moment created was counterclockwise. By convention, a positive ($+$) moment results. However, if the observer viewed from the opposite side of the seesaw, the moment created would be clockwise and thus be negative ($-$) by convention. Clockwise or counterclockwise rotation depends on the viewpoint of the observer. To ensure consistent results, the viewpoint of the observer never changes until analysis is complete.

■ EXAMPLE 3.4

As shown in Figure 3.3, a 150-lb adult sits 4 ft away from the fulcrum (axis of rotation) on the left side of a playground seesaw. A 75-lb child sits on the right side, 8 ft away from the fulcrum. Determine the moments of force created by each person.

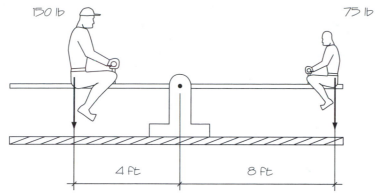

150 lb   75 lb

4 ft   8 ft

FIGURE 3.3 Force system for Example 3.4.

Solution For the 150-lb adult the moment is counterclockwise and therefore positive:

$$M = FL = 150 \text{ lb} \cdot 4 \text{ ft} = +600 \text{ lb-ft}$$

For the 75-lb child the moment is clockwise and therefore negative:

$$M = FL = 75 \text{ lb} \cdot 8 \text{ ft} = -600 \text{ lb-ft}$$

In Example 3.4 the adult and child have similar magnitudes of moment because of the difference in their weight and moment arms. Their tendencies for rotation are, however, opposite: One is clockwise while the other is counterclockwise. The seesaw tends to balance because the moment created by the adult is equal in magnitude but opposite in rotation direction to the moment created by the child.

3.2 COUPLES

A *couple* is a force system that consists of two similar forces that act in opposite directions and that have lines of action that are separate but parallel, as shown in Figure 3.4a. A couple causes or has a tendency to cause rotation, and thus a couple develops a moment. The moment created by a couple (M_{couple}) is equal to the sum of the products of the magnitude of each force (F_a, F_b) multiplied by the perpendicular distance to the point of rotation (L):

$$M_{couple} = F_a\,(L) + F_b\,(L)$$

This simplifies mathematically so that the moment created by a couple (M_{couple}) is equal to the product of the magnitude of a single force (F) multiplied by the perpendicular distance separating the two forces ($L_{separating}$), as shown in Figure 3.4.

$$M_{couple} = F(L_{separating})$$

(a)   (b)

FIGURE 3.4 A *couple* is a force system that consists of two equal forces that act in opposite directions and that have lines of action that are separate but parallel.

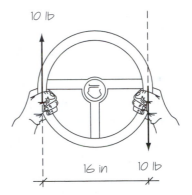

FIGURE 3.5 Force system for Example 3.5.

■ EXAMPLE 3.5

An automobile's steering wheel is 16 in. in diameter. The driver's hands hold the steering wheel at opposite ends. To turn the car's front wheels, each of the driver's hands exerts a force of 10 lb, the left hand pushing up and the right hand pulling down (Figure 3.5). Determine the moment of force that develops.

Solution   There are two methods of solution to this problem. The first solution is theoretically accurate where the perpendicular distance to the point of rotation (L) is measured from the steering wheel to the center point of the wheel:

$$M_{couple} = F_a(L) + F_b(L)$$
$$M_{couple} = 10 \text{ lb}(8 \text{ in}) + 10 \text{ lb}(8 \text{ in}) = 160 \text{ lb-in}$$

The second solution is the mathematically simplified method:

$$M_{couple} = F(L_{separating}) = 10 \text{ lb} \cdot 16 \text{ in} = 160 \text{ lb-in}$$

The Mathematical Symbol Sigma (Σ)

The Greek capital letter sigma, Σ, is a mathematical term that means "the algebraic sum of." For example, the sum of the numbers 1 through 5 may be written as

$$\Sigma(\text{whole numbers 1 to 5}) = 1 + 2 + 3 + 4 + 5 = 15$$

3.3   CONDITIONS OF STATIC EQUILIBRIUM

A body at rest is in *static equilibrium*. Thus any external forces (F) and moments (M) that act on it are completely balanced. For a body to remain in a state of static equilibrium, any external forces and any moments of force acting on that body must equal zero. Mathematically, this is expressed as: $\Sigma F = 0$ and $\Sigma M = 0$.

In the case of a body in static equilibrium, the sum of the external forces acting on that body must equal zero ($\Sigma F = 0$) and the sum of the external moments of force acting on that body must equal zero ($\Sigma M = 0$). Simply, this means that force F_1 (force one) and force F_2 and force F_3, etc., when algebraically added, must negate each other. Similarly, the sum of the moments, M_1 (moment one), moment M_2, moment M_3, and so on, must also cancel each other out. Static equilibrium can exist only when the sum of the external forces acting on that body equals zero ($\Sigma F = 0$) *and* the sum of the external moments of force acting on that body equals zero ($\Sigma M = 0$).

For the two-dimensional problems discussed in this book, the following expressions apply when a rigid body is in or assumed to be in static equilibrium. They are known as the *conditions of static equilibrium*. Each expression is spoken as indicated below.

$^+\!\!\rightarrow\Sigma F_x = 0$ Sum of the horizontal forces equals zero where forces acting to the right are positive.

$^+\!\!\uparrow\Sigma F_y = 0$ Sum of the vertical forces equals zero where forces acting upward are positive.

$\circlearrowright\Sigma M_A = 0$ Sum of the moments of force about point A equals zero where moments with a tendency to cause counterclockwise rotation about the axis of rotation are positive.

By sign convention, forces acting horizontally (F_x) are considered positive ($^+\!\!\rightarrow$) if they act to the right, and forces are negative if they act to the left. Also by sign convention, those forces acting vertically (F_y) are treated as positive ($^{+}\uparrow$) if they act upward, and negative if they act downward. Finally, moments of force (M) that tend to cause counterclockwise rotation about an axis of rotation A are classified positive ($\circlearrowleft$), and moments are negative if the tendency to rotate is clockwise.

3.4   THE FREE-BODY DIAGRAM

A first step in force analysis is constructing a sketch that illustrates the body. This sketch is typically known as a *space diagram.* A space diagram helps the designer visualize the problem. Second, and more important, it is necessary to draw a diagram indicating where and how the forces and moments act on the body. A *free-body diagram* shows the body itself and all known and unknown external forces that are acting on it. The words *external* and *on* cannot be overemphasized. Only those forces that act externally on the body are considered. In essence, the body is isolated from its supports. The supports are represented as force vectors called *reactions* (these are introduced later). The body is assumed to be truly "free"; it is held in place only by the force vectors drawn on it. A free-body diagram for the seesaw discussed in Example 3.6 is shown in Figure 3.7. Note that all forces acting on the seesaw are replaced by force vectors. The body of the seesaw is drawn as a single line.

3.5   CONSTRUCTING THE FREE-BODY DIAGRAM

In drawing free-body diagrams, there are certain accepted conventions. First, the weight of the machine component or structural member under consideration is usually neglected. It is assumed that weight of the body itself is very small in comparison to the forces that act upon it, so the weight of the body is neglected initially. Second, it is assumed that most surfaces in contact with the body are frictionless. This is not true, but it generally introduces a very small error. Consideration of body weight and friction is always a matter of judgment. In initial analysis, body weight and friction are usually neglected.

The free-body diagram serves as the foundation of force analysis; it is an essential first step in applying the equations of static equilibrium. The following steps should be taken in constructing a free-body diagram:

1. Consider the body as an isolated rigid member that is free from the surroundings. Sketch it to a scale of reasonable accuracy.
2. Sketch the force vector of all known external forces acting on the body, such as gravity forces and forces on cables. Indicate the line of action and sense (arrow head) of each force at the point of application on the body. Note the magnitude of each external force.
3. Sketch the force vector of all unknown reactions and external forces acting on the body. Indicate the line of action and sense (arrowhead) of each force at the point of application on the body. Label the magnitude of each external force (i.e., R_1, R_2, F_x, F_y, etc.). If the sense of the force is not known, the sense can initially be assumed. By definition the magnitude of a force is always positive. If analysis shows that the magnitude of the unknown force is negative, the initially assumed sense is improper; the sense (arrowhead) should be reversed.

3.6   TYPES OF REACTIONS

As Newton stated in his second law of motion, external forces applied to a body tend to place the body in motion. However, on a free body in static equilibrium, reacting forces develop that counter this tendency for motion so the body can remain static. When a free body is in static equilibrium, reacting forces oppose the acting forces just enough to keep the body motionless.

As defined in Chapter 2, reactions (R) are external forces that counter the externally applied forces (F) acting on the body. A reacting force has the same point of application, line of action, and magnitude as the acting force it opposes. However, the sense of the reaction is opposite to the sense of the acting force. There are six principal types of reactions that influence a free body. The types of reactions are illustrated in Table 3.1. Types of reactions include:

1. *Roller reaction.* The surface and rollers are assumed to be frictionless. A single reacting force acts perpendicular to the surface on which the roller rests. A moment *cannot* be transferred through this reaction.

2. *Bearing reaction.* The bearing surface is assumed to be frictionless (perfectly smooth). A single reacting force acts perpendicular to the surface on which the body rests. A moment *cannot* be transferred through this reaction.

3. *Tension (cable) reaction.* The cable is assumed to be flexible. With the cable in tension, a single reacting force acts away from the reaction in the direction of the cable. A moment *cannot* be transferred through this reaction.

4. *Linked reaction.* The connections are assumed to be frictionless and free to rotate. A single reacting force acts toward or away from the reaction in the direction of the link. A moment *cannot* be transferred through this reaction.

5. *Hinged or pinned reaction.* The connection is assumed to be frictionless and free to rotate. A single reacting force at an angle acts with x and y components of the force. The force is not necessarily in line with the member imposing the force on the free body. A moment *cannot* be transferred through this reaction.

6. *Fixed reaction.* This is a single reacting force at an angle that acts with the x and y components of the force. The force is not necessarily in line with the member imposing the force on the free body. A moment *can* be transferred through this reaction.

■ EXAMPLE 3.6

The seesaw illustrated in Figure 3.6a is composed of a lever and a fulcrum (axis of rotation). A 200-lb person sits on the lever at 5 ft to the left of the fulcrum. The 100-lb person sits 10 ft away from the fulcrum. Assuming that the persons are not exerting any forces with their feet, determine if the system is in static equilibrium.

Solution   A free-body diagram is drawn as illustrated in Figure 3.6b. Since there are no forces acting horizontally,

$$\Sigma F_x = 0$$

The upward reaction at the fulcrum is unknown, except by intuition. Since the seesaw is isolated in the free-body diagram, the equation $\Sigma F_y = 0$ may be used to solve for the upward reaction (R_f):

$$^{+}\uparrow\Sigma F_y = 0 = R_f + (-200 \text{ lb}) + (-100 \text{ lb})$$
$$0 = R_f + (-200 \text{ lb}) + (-100 \text{ lb})$$
$$R_f = +300 \text{ lb} = 300 \text{ lb} \uparrow$$

Table 3.1 TYPES OF REACTIONS

Type of reaction	Illustration of reaction	Free-body representation	Unknowns
Roller reaction. The surface and rollers are assumed to be frictionless. A single reacting force (R) acts perpendicular to the surface on which the roller rests. A moment *cannot* be transferred through this reaction.			R
Bearing reaction. The surface on which the free body rests is assumed to be frictionless (smooth). A single reacting force (R) acts perpendicular to the surface on which the body rests. A moment *cannot* be transferred through this reaction.			R
Tension (cable) reaction. The cable is assumed to be flexible and fully tensioned. A single reacting force (R) acts away from the reaction in the direction of the cable. A moment *cannot* be transferred through this reaction.			R
Linked reaction. The connections are assumed to be frictionless and free to rotate. A single reacting force (R) acts toward or away from the reaction in the direction of the link. A moment *cannot* be transferred through this reaction.			R
Hinged or pinned reaction. The connection is assumed to be frictionless and free to rotate. A single reacting force (R) at an angle (θ) that acts with x and y components (R_x, R_y) of the reacting force. The reacting force (R) is not necessarily in line with the member imposing the force on the free body. A moment *cannot* be transferred through this reaction.			R_x, R_y or R, θ
Fixed reaction. The free body is fully secured at its end. A single reacting force (R) at an angle (θ) acts with x and y components of reacting force (R_x, R_y). The reacting force is not necessarily in line with or perpendicular to the member imposing the force on the free body. A moment (M) can be transferred through this reaction.			M, R_x, R_y or M, R, θ

3.6 Types of Reactions

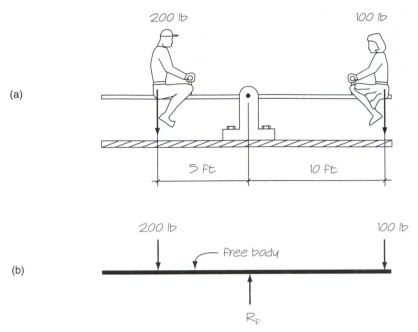

FIGURE 3.6   Space diagram (a) and related free-body diagram (b) of the seesaw discussed in Example 3.6. All forces and reactions acting on the seesaw are replaced by force vectors.

So, by intuition, the fulcrum "reacts" upward to the 300-lb force of the two persons:

$$^+\!\!\uparrow \Sigma F_y = 0 = 300 \text{ lb} + (-300 \text{ lb})$$
$$0 = 300 \text{ lb} + (-300 \text{ lb})$$
$$\Sigma F_y = 0$$

Finally, the moments of force taken about the fulcrum (f) are equal:

$$\circlearrowright \Sigma M_f = 0 = (200 \text{ lb} \cdot 5 \text{ ft}) - (100 \text{ lb} \cdot 10 \text{ ft})$$
$$0 = (200 \text{ lb} \cdot 5 \text{ ft}) - (100 \text{ lb} \cdot 10 \text{ ft})$$
$$= 1000 \text{ lb-ft} - 1000 \text{ lb-ft}$$
$$\Sigma M_f = 0$$

The seesaw is in static equilibrium because it meets the conditions of static equilibrium described above; that is, it is a rigid body where the sum of the forces equal zero and the sum of the moments about the fulcrum equal zero: $\Sigma F_x = 0$, $\Sigma F_y = 0$, $\Sigma M_f = 0$.

The conditions of static equilibrium ($\Sigma F_x = 0$, $\Sigma F_y = 0$, $\Sigma M = 0$) may be used to solve for up to three unknown forces that are acting on a rigid free body as long as the assumption can be made that the body is truly in static equilibrium. The body must be isolated in a free-body diagram before any unknown forces can be determined. Sometimes, a designer assumes that a body is at rest simply to determine the unknown forces that exist at an instantaneous point in time. Following is an approach for solving unknown forces on a free-body diagram using the equations of static equilibrium:

1. Draw a free-body diagram; refer to Section 3.5. Identify all known forces with their magnitude and direction. Identify unknown forces with symbols that identify the point of application of the force (i.e., F_A, F_B, F_C, etc.). Make an assumption on the senses of the unknown forces, if necessary.

2. Where the line of action of a force does not fall parallel with the x and y axes, separate known and unknown forces into horizontal (F_x) and vertical (F_y); refer to Section 2.8. Identify components of unknown forces with symbols (i.e., F_{Ax}, F_{Bx}, F_{Cx}, etc.).

3. Sum all forces in the x direction and set them equal to zero, $^+\!\!\rightarrow \Sigma F_x = 0$. Recall that by sign convention, forces acting horizontally (F_x) are considered positive if they act to the right and are negative if they act to the left. Solve for unknown forces.

4. Sum all forces in the y direction and set them equal to zero, $^+\!\!\uparrow \Sigma F_y = 0$. By sign convention, an upward force is assigned a positive ($+$) sign, while a downward force is given the negative ($-$) sign. Solve for unknown forces.

5. Sum all moments of force and set them equal to zero, $^+\!\!\rightarrow \Sigma F_x = 0$. Recall that by sign convention, moments that have a tendency to cause counterclockwise rotation about the assumed axis of rotation are classified positive, and moments are negative if the tendency to rotate is clockwise. Solve for unknown forces.

Example 3.7 modifies Example 3.6 to demonstrate how unknown forces can be found in a system in static equilibrium.

■ EXAMPLE 3.7

The seesaw illustrated in Figure 3.7a is composed of a lever and a fulcrum (the axis of rotation). A 200-lb person sits on the lever at 5 ft to the left of the

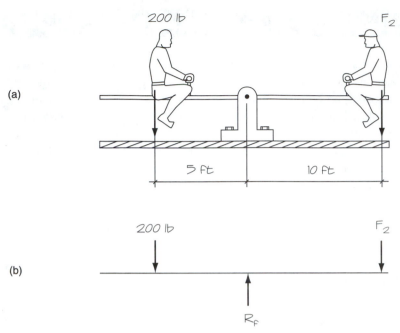

(a)

200 lb

F_2

5 ft

10 ft

(b)

200 lb

F_2

R_f

FIGURE 3.7 Space diagram (a) and related free-body diagram (b) of the seesaw discussed in Example 3.7.

fulcrum. This time a person of unknown weight sits 10 ft away from and to the right of the fulcrum. If the persons are not exerting any forces with their feet, determine the weight (F_2) of the person on the right side of the seesaw and the reaction at the fulcrum (R_f) .

Solution   A free-body diagram is drawn, as illustrated in Figure 3.7b. The moments of force taken about the fulcrum (f) are equal, where F_2 is the weight of the person on the right side of the seesaw:

$$\circlearrowleft \Sigma M_f = 0 = (200 \text{ lb} \cdot 5 \text{ ft}) - (F_2 \cdot 10 \text{ ft})$$
$$0 = (200 \text{ lb} \cdot 5 \text{ ft}) - (F_2 \cdot 10 \text{ ft})$$
$$= 1000 \text{ lb-ft} - F_2 \cdot 10 \text{ ft}$$
$$F_2 \cdot 10 \text{ ft} = 1000 \text{ lb-ft}$$
$$F_2 = 1000 \text{ lb-ft/10 ft}$$
$$= 100 \text{ lb}$$

Since there are no forces acting horizontally,

$$\Sigma F_x = 0$$

Since the seesaw is isolated in the free-body diagram, the equation $\Sigma F_y = 0$ may be used to solve for the upward reaction (R_f):

$$^{+}\!\!\uparrow \Sigma F_y = 0 = R_f + (-200 \text{ lb}) + (F_2)$$

From above, $F_2 = 100$ lb:

$$0 = R_f + (-200 \text{ lb}) + (-100 \text{ lb})$$
$$R_f = +300 \text{ lb} = 300 \text{ lb}\uparrow$$

3.8   CONCENTRATED AND DISTRIBUTED FORCES

Thus far all forces have been assumed to be applied at a specific point, the point of application. However, there are different ways in which a force can be applied. In building construction, a force can be concentrated or distributed.

A *concentrated force* is a force that is applied to an area or length of a free body so small that its point of application may be considered to be a point. Thus it is frequently referred to as a *point force* and designated with the symbol *P.* A force at the end of a chain or rope, the force exerted at the base of a column, and the re-acting forces of a beam are examples of concentrated forces. A concentrated force is expressed in units of force: lb and N or a derivative of these units.

In contrast, a *distributed force* acts on an area or length of free body too large to be considered a point; the force is distributed instead of concentrated. Water pressure exerts a distributed force on the bottom of a storage tank. On a building, the snow load on a roof and the wind load on a wall are distributed forces because the force is distributed over a large area. If a force is *evenly* distributed across an area or length of free body, it is referred to as a *uniformly distributed force.* A snow covering of equal depth or the force at the bottom of a water tank are considered to be uniformly distributed forces. On the other hand, the force exerted at the base of a cone-shaped pile of sand is distributed over the entire area but is not uniformly distributed. When applied over the length of a free body, a distributed force is expressed in units of lb/ft and N/m or some derivative of these units. A distributed force is also expressed in units of N/m^2 or lb/in^2 when applied over an area.

A distributed force must be treated differently than a concentrated force in static equilibrium analysis. Understanding the effect a distributed force has on a free body requires a basic understanding of the concept of center of gravity. Simply, the center of gravity of a body is that point where the total weight of a body may be assumed to act. Therefore, by this definition, *a distributed force can be assumed to act as a concentrated force of equal magnitude at the center of gravity of the distributed force.* This principle of center of gravity will be briefly explained through discussion and examples below. A more detailed presentation of the center of gravity is provided in Chapter 13.

One can explore the effect of the center of gravity of a distributed force by balancing a yardstick (or meter stick) on one finger. The yardstick balances horizontally when a steady finger is positioned at its center point. The yardstick is in static equilibrium when (1) the downward force (weight) of the yardstick is counteracted by the upward-reacting force of the finger, and (2) when moments caused by the weight on one side of the yardstick counteracts the moment on the other side. When in static equilibrium, the yardstick applies a single concentrated force equivalent to its weight to the reacting force (the finger). The line of action of the force passes through the center of gravity of the yardstick and the reaction. Replacing a concentrated force of magnitude equal to the yardstick at the center of gravity of the yardstick will produce the same effect on the reacting force (the finger).

Example 3.8 further demonstrates how, at its center of gravity, a distributed force acts as a concentrated force. By summing the moments about reaction A, two observations can be made when the system is in static equilibrium: (1) the concentrated force F must have a magnitude equal to the magnitude of the uniformly distributed force: both forces have a magnitude of 40 lb; and (2) the forces are the same distance from the reaction: the center of gravity of the uniformly distributed force is 2 in to the right of the reaction while the concentrated force is located 2 in to the left. The same effect is produced by replacing the uniformly distributed force with a concentrated force of equal magnitude at a location 2 in to the right of the reaction, the center of gravity of the uniformly distributed force. Thus a distributed force will act as a concentrated force of equal magnitude at the center of gravity of the distributed force.

■ EXAMPLE 3.8

As shown in Figure 3.8, a weightless horizontally positioned free body is supported by a pinned reaction at the center of a free body. Immediately to the right side of the reaction, a force of 40 lb is uniformly distributed over 4 in of the free body at 10 lb/in. On the left side of the reaction, a force (F) is applied to the body 2 in away from the reaction. Determine the magnitude of force F if the system is in static equilibrium.

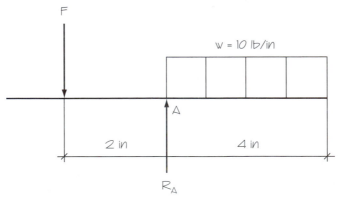

FIGURE 3.8 Free-body diagram for Example 3.8. On the left side of the reaction, a force (*F*) is applied to the free body 2 in away from the reaction. To the right side of the reaction, a force of 40 lb is uniformly distributed over 4 in of the free body at 10 lb/in. A uniformly distributed force is represented by a single or group of rectangles on a free-body diagram.

Solution   Since 40 lb is evenly distributed over 4 in, assume that each force applies a 10-lb force at its center; that is, there are four 10-lb forces acting downward with the point of application of each force at the center of a 1-in interval. A free-body diagram representing this assumption is drawn, as illustrated in Figure 3.8.

The moments of force are taken about the bearing reaction (A):

$$\circlearrowleft \Sigma M_A = 0 = (F \cdot 2 \text{ in}) - (10 \text{ lb} \cdot \tfrac{1}{2} \text{ in}) - (10 \text{ lb} \cdot 1\tfrac{1}{2} \text{ in}) - (10 \text{ lb} \cdot 2\tfrac{1}{2} \text{ in}) - (10 \text{ lb} \cdot 3\tfrac{1}{2} \text{ in})$$

$$0 = (F \cdot 2 \text{ in}) - (10 \text{ lb} \cdot \tfrac{1}{2} \text{ in}) - (10 \text{ lb} \cdot 1\tfrac{1}{2} \text{ in}) - (10 \text{ lb} \cdot 2\tfrac{1}{2} \text{ in}) - (10 \text{ lb} \cdot 3\tfrac{1}{2} \text{ in})$$

$$(F \cdot 2 \text{ in}) = (10 \text{ lb} \cdot \tfrac{1}{2} \text{ in}) + (10 \text{ lb} \cdot 1\tfrac{1}{2} \text{ in}) + (10 \text{ lb} \cdot 2\tfrac{1}{2} \text{ in}) + (10 \text{ lb} \cdot 3\tfrac{1}{2} \text{ in})$$

$$2F = 5 \text{ lb-in} + 15 \text{ lb-in} + 25 \text{ lb-in} + 35 \text{ lb-in}$$

$$= 80 \text{ lb-in}$$

$$\frac{2F}{2} = \frac{80 \text{ lb-in}}{2}$$

$$F = 40 \text{ lb}$$

One might challenge the interpretation of Example 3.8 because the 1-in increments used to divide the uniformly distributed force are large and the center of gravity of each increment was used to calculate moments. However, the uniformly distributed force could easily have been divided into increments much smaller than 1 in. An infinite number of increments could have been used, increments so small that each increment would have become a point force. In this approach the math would be extremely tedious but the result would be the same: In static equilibrium, a distributed force can be assumed to act as a concentrated force of equal magnitude at the center of gravity of the distributed force.

Because a uniformly distributed force is evenly distributed, it is drawn as a single or group of rectangles on a free-body diagram as shown in Figures 3.8 and 3.9. The center of gravity of the total uniformly distributed force is located at half the distance from either end of the force, at the geometric center of the rectangle. The total uniformly distributed force (W) acting on a free body is the uniformly distributed force per unit length (w) multiplied by its length (L):

$$\text{total uniformly distributed force, } W = wL$$

An example of how a total uniformly distributed force (W) is shown on a free-body diagram is shown in Figure 3.9.

■ EXAMPLE 3.9

A free-body diagram of a weightless horizontally positioned member is shown in Figure 3.9. R_B is a pinned connection and R_A is a bearing reaction. Find the unknown reactions R_A and R_B.

Solution R_B is an axis of rotation about which moments can be taken because it is a pinned connection. Moments that tend to cause counterclockwise rotation about R_B are positive ($+$) and clockwise rotation is negative ($-$). In the

3.8   Concentrated and Distributed Forces

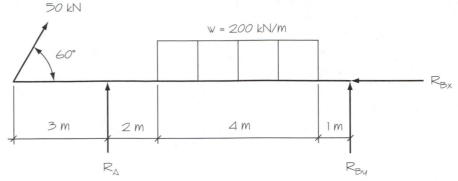

FIGURE 3.9 Free-body diagram for Example 3.9. Because a uniformly distributed force (w) is evenly distributed, it is drawn as a single or group of rectangles on a free-body diagram. In the case of Example 3.9, each block represents 200 kN/m.

case of this beam, the moment created by R_B tends to be positive and moments created by the loads tend to be negative:

$$\circlearrowleft \Sigma M_B = -(R_A \cdot 7 \text{ m}) + (200 \text{ kN/m} \cdot 4 \text{ m} \cdot 3 \text{ m}) - (43.3 \text{ kN} \cdot 10 \text{ m})$$
$$0 = -7R_A + 2400 \text{ kN-m} - 433 \text{ kN-m}$$
$$= -7R_A + 1967 \text{ kN-m}$$
$$7R_{Ay} = 1967 \text{ kN-m}$$
$$R_{Ay} = 281 \text{ kN} \uparrow$$

Note: The 43.3-kN force is the perpendicular (y) component of the 50-kN force. It is found by separating the 50-kN force into x and y components of force:

$$F_y = F(\sin \theta) = 50 \text{ kN}(\sin 60°) = 43.3 \text{ kN} \uparrow$$
$$F_x = F(\cos \theta) = 50 \text{ kN}(\cos 60°) = 25.0 \text{ kN} \rightarrow$$

Summing the vertical forces (F_y) with upward forces assigned a positive (+) value and downward forces given a negative (−) sign:

$$^+\uparrow\Sigma F_y = + 281 \text{ kN} + R_{By} - (200 \text{ kN/m} \cdot 4 \text{ m}) + 43.3 \text{ kN}$$
$$0 = +R_{By} - 475.7 \text{ kN}$$
$$475.7 \text{ kN} = R_{By}$$

Note: The 43.3-kN force is the perpendicular (y) component of the 50-kN force.

Summing the horizontal forces (F_x) with forces acting to the right positive (+) and forces acting to the left assigned a negative (−) sign:

$$^+\rightarrow\Sigma F_x = -25 \text{ kN} - R_{Bx}$$
$$-25 \text{ kN} = R_{Bx}$$

Note: The −25-kN force is the horizontal (x) component of the 50-kN force.

The analysis above finds that reaction R_{Ay} is an upward force of 281 kN. The reaction R_{AX} is zero since there is no horizontal force at R_A. Reaction R_B is broken into two reacting forces: an upward force (R_{By}) of 475.7 kN and a force acting to the left (R_{Bx}) of 25 kN. By sign convention R_{Ay} equals +281 kN and R_{AX} is 0; R_{By} is +475.7 kN and R_{Bx} equals −25 kN.

The resultant force at the reaction R_B (F_{BR}) is found by

$$F_{BR} = \sqrt{F_x^2 + F_y^2} = \sqrt{(-25 \text{ kN})^2 + (475.7 \text{ kN})^2} = 476.3 \text{ kN}$$

The direction of the resultant force (F_{BR}) may be determined by graphic vector analysis (i.e., using the tip-to-tail graphic solution technique to find the resultant of forces R_{Bx} and R_{By}). The angle may also be determined through use of right-angle trigonometry:

$$\theta = \tan^{-1} \frac{y}{x}$$

$$\theta = \tan^{-1} \frac{475.7 \text{ kN}}{-25 \text{ kN}}$$

$$= \tan^{-1}(-19.03)$$

$$= -86.99° \quad \text{or} \quad 180° - 86.99° = 93.01° \measuredangle$$

In Example 3.9, R_B is broken into two reacting forces: an upward force (R_{By}) of +475.7 kN and a horizontal force (R_{Bx}) of −25 kN. By sign convention, a positive sign of a vertical vector indicates an upward sense, and the negative sign of a horizontal force implies that the vector is acting to the left. The resultant of these two forces (F_{BR}) is 476.4 kN. Since R_{By} is a force acting upward and R_{Bx} is a force acting to the left, the resultant of these two forces (F_{BR}) is acting up and to the left. The resultant (F_{BR}) is a 476.4-kN force acting up and slightly to the left. It is expressed as 476.4 kN 93.01° $\measuredangle$.

An evenly increasing distributed force increases (or decreases) evenly in magnitude across the length of the force; it increases from zero to a maximum force per unit length (w_{max}) in a linear and increasing manner. An evenly increasing distributed force is drawn with a triangular shape on a free-body diagram. The hypotenuse (sloped leg) of the triangle represents the increasing (or decreasing) force. The center of gravity of an evenly increasing distributed force is located at the geometric center of a triangle; at *one-third* the distance from the end with the maximum force. When the evenly increasing distributed force increases from zero to the maximum force, the total force (W) exerted is found much like the area of a triangle: its length (L) multiplied by *one-half* of the maximum force (W_{max}) of the evenly increasing distributed force:

$$\text{evenly increasing distributed force, } W = \tfrac{1}{2}(W_{max})L$$

An example of how an evenly increasing distributed force (W) is shown on a free-body diagram is shown in Figure 3.10.

3.8   Concentrated and Distributed Forces   **61**

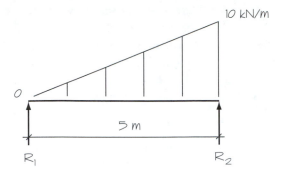

FIGURE 3.10 Free-body diagram for Example 3.10. An evenly increasing distributed force is drawn with a triangular shape or sloping line on a free-body diagram.

■ **EXAMPLE 3.10**

A free-body diagram of a 5-m-long rigid member is shown in Figure 3.10. The member carries an evenly increasing distributed force increasing from zero at R_1 to 10 kN/m at R_2. Reaction R_1 is a pinned reaction and reaction R_2 is a bearing reaction. Find the unknown reactions, R_1 and R_2.

Solution The evenly increasing distributed force increases from zero at R_1 to 10 kN/m at R_2. The total force (W) is found by:

$$W = \tfrac{1}{2}(w_{max})L = \tfrac{1}{2}(10 \text{ kN/m} \cdot 5 \text{ m})$$
$$= 25 \text{ kN}$$

R_1 is the axis of rotation about which moments can be taken because it is a pinned reaction. The total force (W) is assumed to be applied as a concentrated force at its center of gravity. This distance is two-thirds of the length of the member from the axis of rotation:

$$\circlearrowright \Sigma M_{R1} = 0 = +(R_2 \cdot 5 \text{ m}) - (25 \text{ kN})(\tfrac{2}{3} \cdot 5 \text{ m})$$
$$0 = 5R_2 - 83.33 \text{ kN-m}$$
$$5R_2 = 83.33 \text{ kN-m}$$
$$\frac{5R_2}{5} = \frac{83.33 \text{ kN-m}}{5}$$
$$R_2 = 16.67 \text{ kN}$$

Summing the vertical forces (F_y) with an upward force assigned a positive (+) value and a downward force is given the negative (−) sign:

$$^{+}\uparrow\Sigma F_y = 0 = R_1 + R_2 - 25 \text{ kN}$$

From above, $R_2 = 16.67$ kN:

$$0 = R_1 + 16.67 \text{ kN} - 25 \text{ kN}$$
$$= R_1 - 8.33 \text{ kN}$$
$$R_1 = 8.33 \text{ kN}$$

Summing the horizontal forces (F_x) with a force acting to the right assigned a positive ($+$) sign and a force acting to the left assigned a negative ($-$) sign. Since there are no horizontal force vectors on the free-body diagram,

$$^{+}\rightarrow\Sigma F_x = 0$$

The analysis above finds that the reaction R_1 equals 8.33 kN and R_2 equals 16.67 kN.

A combination of a uniformly distributed force and an evenly increasing distributed force covering a specific length of a free-body diagram is represented by a trapezoid. As shown in Figure 3.11, this type of load must be treated separately by dividing the trapezoid into a separate uniformly distributed force and a separate evenly increasing distributed force:

total uniformly distributed force, $W = wL = 30$ N/m $\cdot$ 9 m $= 270$ N

evenly increasing distributed force, $W = \frac{1}{2}(W_{max})L = \frac{1}{2} \cdot 30$ N/m $\cdot$ 9 m $= 135$ N

3.9 STATICALLY DETERMINATE AND INDETERMINATE BODIES

The conditions of static equilibrium ($\Sigma F_x = 0$, $\Sigma F_y = 0$, $\Sigma M = 0$) serve as useful tools that can be used to find unknown external forces. However, not all bodies can be examined under the conditions of static equilibrium alone. A rigid body is *statically determinate* if it can be analyzed under the conditions of static equilibrium; the three unknowns can be determined by the equations of static equilibrium,

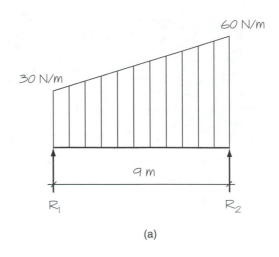

(a)

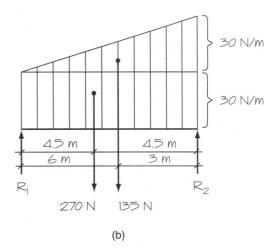

FIGURE 3.11 A trapezoid is used to represent a combination of a uniformly distributed force and evenly increasing distributed force on a free-body diagram (a). This type of load must be treated separately by dividing the trapezoid into a separate uniformly distributed force and a separate evenly increasing distributed force (b).

$\Sigma F_x = 0$, $\Sigma F_y = 0$, $\Sigma M = 0$. Although a body may undergo bending, it can be treated as a rigid body as required by the conditions of static equilibrium; bending is assumed to be negligible.

Not all rigid bodies are statically determinate. A rigid body is *statically indeterminate* if it *cannot* be analyzed under the conditions of static equilibrium. Most bodies having three reactions and all bodies having four or more reactions, beams with two fixed ends, and simple or overhanging beams with one fixed end are statically indeterminate. When a body has one or more very wide reactions (i.e., reactions not concentrated at a small location on the body) it is also statically indeterminate.

Bodies that are statically indeterminate *cannot* be analyzed under the conditions of static equilibrium alone.

Solution of unknown forces acting on statically indeterminate bodies involves examining the conditions of static equilibrium *and* the stiffness of the material (the modulus of elasticity presented in Chapter 7) and section (the moment of inertia presented in Chapter 14) of the body. Such an investigation is beyond the intended breadth of this book.

EXERCISES

3.1 A person exerts a force of 60 lb perpendicular to the end of a 7-ft-long pry bar. The fulcrum (axis of rotation) is 6 ft from the end of the bar upon which the force is applied. Determine the following:
  (a) The moment that develops, in lb-ft
  (b) The moment that develops, in lb-in
  (c) The moment that develops, in kip-ft
  (d) The moment that develops, in kip-in

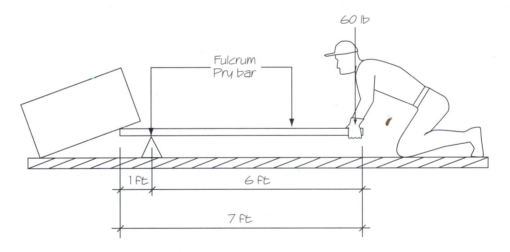

3.2 A 780-N person sits 3 m away from the fulcrum on a seesaw. A second person weighing 825 N sits on the opposite side of the seesaw at a distance of 2.8 m away from the fulcrum.
  (a) Determine the moment created by the 780-N person.
  (b) Determine the moment created by the 825-N person.
  (c) Determine the moment created in this system by both persons on the seesaw, in N-m.
  (d) If any, which person has the tendency to drop toward the ground?
  (e) Is the system in static equilibrium?

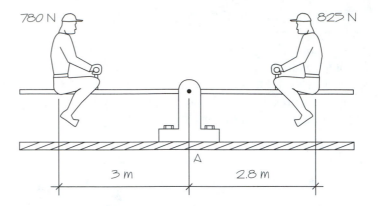

3.3 A rigid body carrying a 25-kN load is supported by a pinned connection and cable as shown. Neglecting the weight of the body, find the following:
(a) Reaction *A*
(b) The tension in the cable at reaction *B*

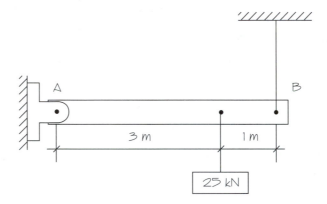

3.4 A rigid body is supported by a cable and pinned connection as shown. It carries two 2000-lb loads. Neglecting the weight of the body, find the following:
(a) Reaction *B*
(b) The tension in the cable at reaction *A*

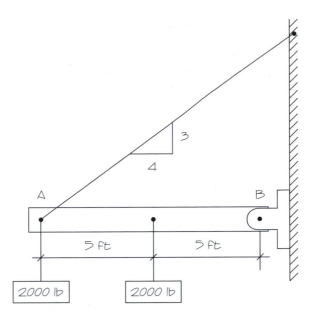

3.5 A rigid body supports a 3.0-kN load as shown. Neglecting the weight of the body, find the following:

(a) The moment that develops at point A as a result of the 3.0 kN force.

(b) The magnitude and direction of reaction A

(c) The magnitude and direction of reaction B

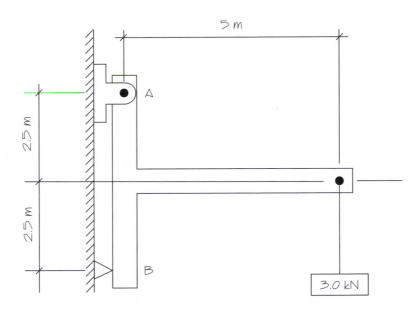

3.6 A 4-m-long rigid body carrying a 10-kN/m uniformly distributed load is supported by a pinned connection and cable as shown.

(a) Neatly draw (with a straightedge) the free-body diagram.

(b) Neglecting the weight of the body, find the force acting at reaction A.

(c) Neglecting the weight of the body, find the tension in cable BC.

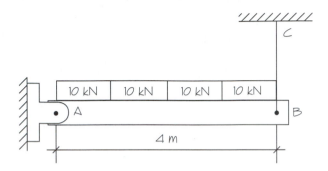

3.7 A 10-ft-long rigid body is supported by a cable and pinned connection as shown. It carries a 10-kip/ft uniformly distributed load across its entire length.

(a) Neatly draw (with a straightedge) the free-body diagram.

(b) Neglecting the weight of the body, find reaction B.

(c) Neglecting the weight of the body, find the tension in the cable at reaction A.

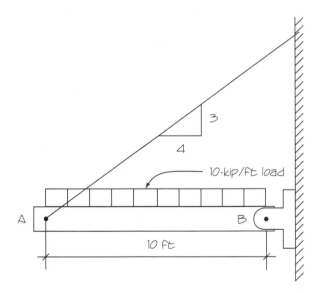

3.8 An 11-ft-long rigid body supports a 15-kip/ft uniformly distributed load as shown.

(a) Neatly draw (with a straightedge) the free-body diagram.

(b) Neglecting the weight of the body, find the reacting moment and reacting force acting at point A.

(c) Assuming that the body has a uniformly distributed weight of 0.5 kip, find the reacting moment and reacting force at point *A*.

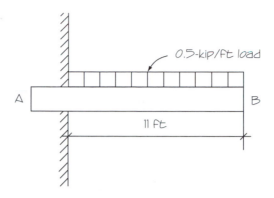

3.9 An 11-ft-long rigid body supports a 15-kip/ft uniformly distributed load across half its length as shown.
  (a) Neatly draw (with a straightedge) the free-body diagram.
  (b) Neglecting the weight of the body, find the reacting moment and reacting force acting at point *A*.
  (c) Assuming that the body has a uniformly distributed weight of 0.5 kip, find the reacting moment and reacting force at point *A*.

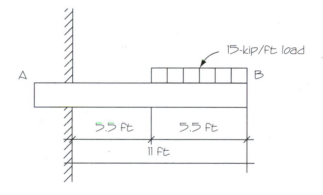

3.10 An 11-ft-long rigid body supports the evenly increasing distributed load as shown.
  (a) Neatly draw (with a straightedge) the free-body diagram.
  (b) Neglecting the weight of the body, find the reacting moment and reacting force acting at point *A*.
  (c) Assuming that the body has a uniformly distributed weight of 0.5 kip, find the reacting moment and reacting force at point *A*.

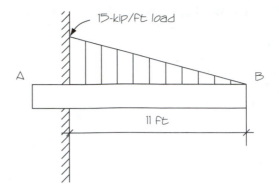

3.11 A truss acts as a rigid body supporting five 5-MN loads, as shown.
  (a) Neatly draw (with a straightedge) the free-body diagram.
  (b) Neglecting the weight of the truss and assuming that the truss is a rigid body, find the reacting forces acting at point *A* and point *B*.

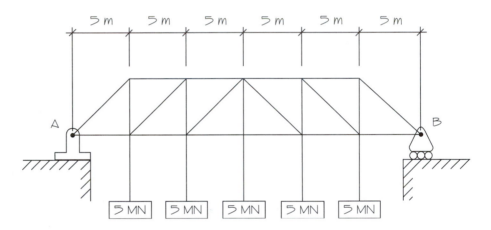

3.12 Boom *AB* is designed to support a maximum load of 1 ton (2000 lb) as shown. Connection *A* is a pinned connection.
  (a) Assuming that the maximum tension force that can develop in cable *BC* is 3000 lb, determine the minimum angle θ required to safely support the load and the reacting force at *A*.
  (b) Assuming that the maximum tension force that can develop in cable *BC* is 4000 lb, determine the minimum angle θ required to safely support the load and the reacting force at *A*.
  (c) Assuming that the maximum tension force that can develop in cable *BC* is 5000 lb, determine the minimum angle θ required to safely support the load and the reacting force at *A*.

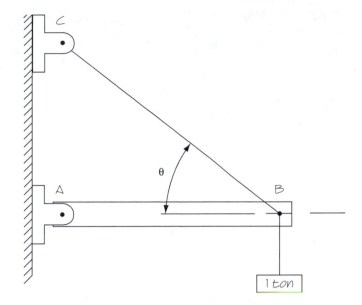

3.13 A crane supports a 1-kip load as shown. Boom *AB* extends from a pinned connection on the crane carriage at an angle of 45° from horizontal. Cable *CD* extends from a sheave near the roof of the cab at an angle of 30° from horizontal. It supports boom *AB* at point *C*.

(a) Neatly draw (with a straightedge) the free-body diagram of the boom.
(b) Find the reaction at point *A*.
(c) Find the tension in cable *CD*.

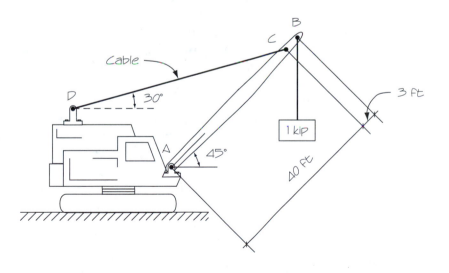

Chapter 4
BEAMS IN STATIC EQUILIBRIUM

A structural engineer or designer typically analyzes a structural member under the assumption that the member is in static equilibrium. Beams used in a building structural system are assumed to be in equilibrium under the applied external and reacting forces. In this chapter the reacting forces acting on beams in static equilibrium are examined by applying the concepts introduced in Chapters 2 and 3. The presentation is limited to statically determinate coplanar (two-dimensional) force systems.

4.1   THE BEAM

A *beam* is a structural member that rests on reactions and is subjected to forces acting normal (perpendicular) to its longitudinal axis, thereby causing it to bend. In the construction industry a beam is a horizontally positioned structural member such as a joist, purlin, girder, header, or lintel. The term *beam* customarily implies a specific type of horizontally positioned, load-bearing structural member used in buildings. In structural design, however, joists, purlins, and girders behave like and are treated as beams.

In building construction, external forces applied to beams and other structural members are called *loads*. Reacting forces at the beam supports counter the applied loads to keep the beam in static equilibrium. These forces are called *reacting forces* or simply *reactions*. Loads and the reactions combine to cause a beam to bend.

There are many different types of beams. Schematic examples of common types of beams are shown in Figure 4.1. A *simple beam* spans between two reactions located at the extreme ends of the beam. A beam that rests on two reactions but extends beyond one or both bearing points is called an *overhanging beam.* A *double overhanging beam* has both beam ends extending beyond the reactions. A *cantilever beam* extends from a single reaction; it is not part of an overhanging beam. Finally, a *continuous beam* is supported by three or more reactions. Commonly, beams carrying gravity loads are coplanar, parallel force systems; their forces act in a single plane and all forces have the same direction.

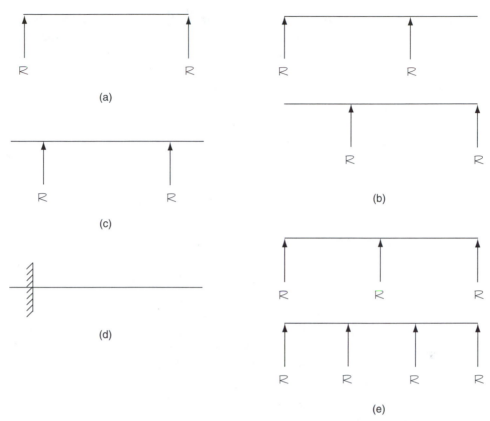

FIGURE 4.1 Schematic examples of common types of beams: (a) a simple beam; (b) an overhanging beam; (c) a double overhanging beam; (d) a cantilever beam; and (e) a continuous beam.

4.2 SOLVING FOR UNKNOWN REACTIONS ON BEAMS

In building structures, a beam generally has known distributed or concentrated loads acting upon it. The reactions, however, are usually unknown. In design, the first step in selection of a beam size is determining unknown reactions. It is also important to know the magnitude of a reacting force because the member supporting the beam at the location of the reaction (i.e., a column or wall) must be capable of developing that reacting force.

Most simple beams, overhanging beams, and cantilever beams are statically determinate; the conditions of static equilibrium can be used to determine unknown reactions, as demonstrated below. Continuous beams having three reactions and all beams having four or more reactions, beams with two fixed ends, and simple or overhanging beams with one fixed end are statically indeterminate. A beam with one or more very wide reactions (i.e., reactions not concentrated at a small location on the body) is also statically indeterminate. Statically indeterminate beams *cannot* be analyzed under the conditions of static equilibrium alone. Analysis involves examining the conditions of static equilibrium *and* the stiffness of the beam material (modulus of elasticity) and section (moment of inertia). Such an investigation is beyond the intended breadth of this book.

Unknown reactions of a beam that is statically determinate can be determined under the conditions of static equilibrium. The equations of static equilibrium, $\Sigma F_x = 0$, $\Sigma F_y = 0$, $\Sigma M = 0$, are used to solve for unknown reactions. The following examples demonstrate the technique. The beams in the examples provided are intentionally kept uncomplicated to present the concept of solving for reactions more effectively.

■ EXAMPLE 4.1

A free-body diagram of a simple beam is shown in Figure 4.2. The 4-m beam carries a concentrated load of 40 kN at its center. Reaction R_1 is a pinned reaction and reaction R_2 is a bearing reaction. Find the unknown reactions, R_1 and R_2.

Solution R_1 is the axis of rotation about which moments can be taken because it is a pinned reaction. R_2 is 4 m from the axis of rotation and is acting with a tendency to cause counterclockwise rotation about R_1 (assumed to be a positive moment). The concentrated load *(P)* is 2 m from the axis of rotation and is acting with a tendency to cause clockwise rotation about R_1 (assumed to be a negative moment):

$$\circlearrowright \Sigma M_{R1} = 0 = + (R_2 \cdot 4 \text{ m}) - (40 \text{ kN} \cdot 2 \text{ m})$$
$$0 = 4R_2 - 80 \text{ kN-m}$$
$$4R_2 = 80 \text{ kN-m}$$
$$\frac{4R_2}{4} = \frac{80 \text{ kN-m}}{4}$$
$$R_2 = 20 \text{ kN}$$

There are no forces in the *x* direction so $\Sigma F_x = 0$. Summing the vertical forces (F_y) with an upward force assigned a positive $(+)$ sign and a down-

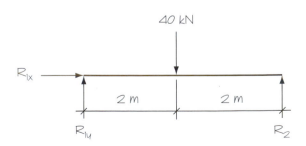

FIGURE 4.2 Free-body diagram for Example 4.1.

ward force given a negative ($-$) sign and setting them equal to zero ($\Sigma F_y = 0$), we have

$$^+\uparrow\Sigma F_y = 0 = R_1 + R_2 - 40 \text{ kN}$$

From above, $R_2 = 20$ kN:

$$0 = R_1 + 20 \text{ kN} - 40 \text{ kN}$$
$$0 = R_1 - 20 \text{ kN}$$
$$R_1 = 20 \text{ kN}$$

■ EXAMPLE 4.2

A free-body diagram of a simple beam is shown in Figure 4.3. The 4-m beam carries a uniformly distributed load of 10 kN/m across its entire length. Reaction R_1 is a pinned reaction and reaction R_2 is a bearing reaction. Find the unknown reactions, R_1 and R_2.

Solution   The uniformly distributed load is 10 kN/m (w). It extends across the entire 4 m length of the beam. The total uniformly distributed load (W) is found by

$$W = wL = 10 \text{ kN/m} \cdot 4 \text{ m}$$
$$= 40 \text{ kN}$$

R_1 is the axis of rotation about which moments can be taken because it is a pinned reaction. R_2 is 4 m from the axis of rotation and is acting with a tendency to cause counterclockwise (positive) rotation about R_1. The uniformly distributed load (W) is assumed to be applied as a concentrated load at its center of gravity. The distance of the center of gravity from the axis of rotation is one-half of the length of the beam (2 m). The load tends to produce a clockwise (negative) moment about the axis of rotation:

$$\circlearrowright \Sigma M_{R1} = 0 = + (R_2 \cdot 4 \text{ m}) - (40 \text{ kN} \cdot 2 \text{ m})$$
$$0 = 4R_2 - 80 \text{ kN-m}$$
$$4R_2 = 80 \text{ kN-m}$$
$$\frac{4R_2}{4} = \frac{80 \text{ kN-m}}{4}$$
$$R_2 = 20 \text{ kN}$$

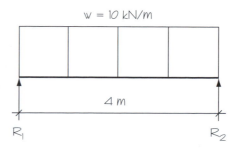

FIGURE 4.3   Free-body diagram for Example 4.2.

There are no forces in the x direction so $\Sigma F_x = 0$. Summing the vertical forces (F_y) with an upward force assigned a positive $(+)$ sign and a downward force given a negative $(-)$ sign and setting them equal to zero $(\Sigma F_y = 0)$, we have

$$^+\!\!\uparrow\Sigma F_y = 0 = R_1 + R_2 - 40 \text{ kN}$$

From above, $R_2 = 20$ kN:

$$0 = R_1 + 20 \text{ kN} - 40 \text{ kN}$$
$$= R_1 - 20 \text{ kN}$$
$$R_1 = 20 \text{ kN}$$

The beams in Examples 4.1 and 4.2 are symmetrically loaded; the magnitude of loads and load and reaction placement on the left side of the beam are a mirror image of those on the right side. Beams that are symmetrically loaded have reactions that are equal $(R_1 = R_2)$. Thus, due to symmetrical loading, the reactions of each beam in these examples are equal in magnitude.

■ EXAMPLE 4.3

A free-body diagram of a simple beam is shown in Figure 4.4. The 4-ft beam carries a 100-kip concentrated load at the center of the beam. Uniformly distributed loads of 10 kip/ft on the left side of the beam and 5 kip/ft on the right side also rest on the beam. Reaction R_1 is a pinned reaction and reaction R_2 is a bearing reaction. Find the unknown reactions, R_1 and R_2.

Solution The uniformly distributed load on the left side of the beam is 10 kip/ft (w_1) for 2 ft. The total uniformly distributed load on the left side of the beam (W_1) is found by

$$W_1 = w_1 L = 10 \text{ kip/ft} \cdot 2 \text{ ft}$$
$$= 20 \text{ kip}$$

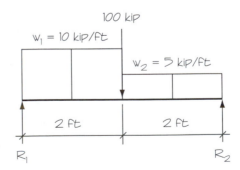

FIGURE 4.4 Free-body diagram for Example 4.3.

The uniformly distributed load on the right side of the beam is 5 kip/ft (w_2) for 2 ft. The total uniformly distributed load on the right side of the beam (W_2) is found by

$$W_2 = w_2L = 5 \text{ kip/ft} \cdot 2 \text{ ft}$$
$$= 10 \text{ kip}$$

R_1 is the axis of rotation about which moments can be taken because it is a pinned reaction. R_2 is 4 ft from the axis of rotation and is acting with a tendency to cause counterclockwise (positive) rotation about R_1. The 100-kip concentrated load acts 2 ft away from R_1 and tends to cause a clockwise (negative) moment about R_1. Each uniformly distributed load (W_1 and W_2) is assumed to be applied as a concentrated load at its center of gravity: at 1 ft from R_1 for W_1 and at 3 ft from R_1 for W_2. Both W_1 and W_2 tend to cause clockwise (negative) rotation about the axis of rotation, R_1:

$$\circlearrowright \Sigma M_{R1} = 0 = + (R_2 \cdot 4 \text{ ft}) - (100 \text{ kip} \cdot 2 \text{ ft}) - (20 \text{ kip} \cdot 1 \text{ ft}) - (10 \text{ kip} \cdot 3 \text{ ft})$$
$$0 = 4R_2 - 200 \text{ kip-ft} - 20 \text{ kip-ft} - 30 \text{ kip-ft}$$
$$0 = 4R_2 - 250 \text{ kip-ft}$$
$$4R_2 = 250 \text{ kip-ft}$$
$$\frac{4R_2}{4} = \frac{250 \text{ kip-ft}}{4}$$
$$R_2 = 62.5 \text{ kip}$$

There are no forces in the x direction, so $\Sigma F_x = 0$. Summing the vertical forces (F_y) with an upward force assigned a positive ($+$) sign and a downward force given a negative ($-$) sign and setting them equal to zero ($\Sigma F_y = 0$), we have

$$^+\uparrow\Sigma F_y = 0 = R_1 + R_2 - 100 \text{ kip} - W_1 - W_2$$

From above, $W_1 = 20$ kip, $W_2 = 10$ kip, and $R_2 = 62.5$ kip:

$$0 = R_1 + 62.5 \text{ kip} - 100 \text{ kip} - 20 \text{ kip} - 10 \text{ kip}$$
$$= R_1 - 67.5 \text{ kip}$$
$$R_1 = 67.5 \text{ kip}$$

In Example 4.3 it was necessary to treat the uniformly distributed loads separately by converting them to unique concentrated forces with unique centers of gravity. Also, because the beam is asymmetrical, R_1 does not equal R_2. By observation R_1 must be slightly greater than R_2 because more of the total load is nearer R_1 than R_2. Analysis results prove that this observation is correct. In making computations it is always helpful for the designer to observe the loading condition, anticipate approximate results, and compare the approximations to analysis results.

■ EXAMPLE 4.4

A free-body diagram of a single overhang beam is shown in Figure 4.5. The 3-m beam carries a concentrated load of 100 kN at the right end of the 1-m overhang. Reaction R_1 is a pinned reaction and reaction R_2 is a bearing reaction. Find the unknown reactions, R_1 and R_2.

Solution R_1 is the axis of rotation about which moments can be taken because it is a pinned reaction. R_2 is 2 m from the axis of rotation and is acting with a tendency to cause counterclockwise rotation about R_1 (assumed to be a positive moment). The concentrated load (P) is 3 m from the axis of rotation and is acting with a tendency to cause counterclockwise rotation about R_1 (assumed to be a negative moment):

$$\circlearrowright \Sigma M_{R1} = 0 = + (R_2 \cdot 2 \text{ m}) - (100 \text{ kN} \cdot 3 \text{ m})$$
$$0 = 2R_2 - 300 \text{ kN-m}$$
$$2R_2 = 300 \text{ kN-m}$$
$$\frac{2R_2}{2} = \frac{300 \text{ kN-m}}{2}$$
$$R_2 = 150 \text{ kN}$$

There are no forces in the x direction so $\Sigma F_x = 0$. Summing the vertical forces (F_y) with an upward force assigned a positive ($+$) sign and a downward force is given the negative ($-$) sign and setting them equal to zero ($\Sigma F_y = 0$):

$$^+\uparrow\Sigma F_y = 0 = R_1 + R_2 - 100 \text{ kN}$$

From above, $R_2 = 150$ kN:

$$0 = R_1 + 150 \text{ kN} - 100 \text{ kN}$$
$$= R_1 + 50 \text{ kN}$$
$$R_1 = -50 \text{ kN}$$

The sense of R_1 must be changed on the initial free-body diagram. Reaction R_1 acts downward.

$$R_1 = 50 \text{ kN}$$

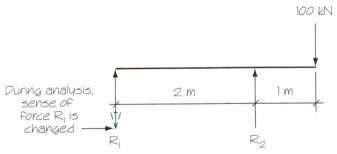

FIGURE 4.5 Free-body diagram for Example 4.4.

In Example 4.4 it was assumed that reaction R_1 acted upward. The analysis results when summing the forces yielded a negative value for R_1. A force with a negative number is an imaginary number; it does not exist. A force cannot be negative, just as length or volume cannot be negative. The sense of R_1 must be changed on the initial free-body diagram so that the negative sign for R_1 can be changed to a positive sign. Reaction R_1 equals 50 kN acting downward, not -50 kN.

■ EXAMPLE 4.5

A free-body diagram of a simple beam is shown in Figure 4.6. The 10-ft-long beam carries a uniformly distributed load of 1000 lb/ft for 3 ft beginning 4 ft to the right of R_1. Reaction R_1 is a pinned reaction and reaction R_2 is a bearing reaction. Find the unknown reactions, R_1 and R_2.

Solution   The uniformly distributed load is 1000 lb/ft (w). It covers 3 ft of the 10-ft beam. The total uniformly distributed load (W) is found by

$$W = wL = 1000 \text{ lb/ft} \cdot 3 \text{ ft}$$
$$= 3000 \text{ lb}$$

R_1 is the axis of rotation about which moments can be taken because it is a pinned reaction. R_2 is 10 ft from the axis of rotation and is acting with a tendency to cause counterclockwise (positive) rotation about R_1. The uniformly distributed load (W) is assumed to be applied as a concentrated load at its center of gravity. The distance of the center of gravity from the axis of rotation is 4 ft plus one-half of the length of the load: 4 ft + 1.5 ft = 5.5 ft. The load tends to produce a clockwise (negative) moment about the axis of rotation:

$$\circlearrowright \Sigma M_{R1} = 0 = + (R_2 \cdot 10 \text{ ft}) - (3000 \text{ lb} \cdot 5.5 \text{ ft})$$
$$0 = 10R_2 - 16\,500 \text{ lb-ft}$$
$$10R_2 = 16\,500 \text{ lb-ft}$$
$$\frac{10R_2}{10} = \frac{16\,500 \text{ lb-ft}}{10}$$
$$R_2 = 1650 \text{ lb}$$

There are no forces in the x direction so $\Sigma F_x = 0$. Summing the vertical forces (F_y) with an upward force assigned a positive (+) sign and a downward

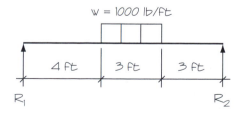

FIGURE 4.6   Free-body diagram for Example 4.5.

4.2   Solving for Unknown Reactions on Beams

79

force given a negative $(-)$ sign and setting them equal to zero ($\Sigma F_y = 0$), we have

$$^+\uparrow\Sigma F_y = 0 = R_1 + R_2 - 3000 \text{ lb}$$

From above, $R_2 = 1650$ lb:

$$0 = R_1 + 1650 \text{ lb} - 3000 \text{ lb}$$
$$= R_1 - 1350 \text{ lb}$$
$$R_1 = 1350 \text{ lb}$$

■ EXAMPLE 4.6

A free-body diagram of a beam with a double (2 ft) overhang is shown in Figure 4.7. The 8-ft-long beam carries a uniformly distributed load of 1 kip/ft over 6 ft of its length. Reaction R_1 is a pinned reaction and reaction R_2 is a bearing reaction. Find the unknown reactions, R_1 and R_2.

Solution The uniformly distributed load is 1 kip/ft (w). It covers 4 ft of the left side of the beam and 2 ft of the right side. Each total uniformly distributed load (W_1 on the left and W_2 on the right side of the beam) is found by

$$W_1 = wL = 1 \text{ kip/ft} \cdot 4 \text{ ft}$$
$$= 4 \text{ kips}$$
$$W_2 = wL = 1 \text{ kip/ft} \cdot 2 \text{ ft}$$
$$= 2 \text{ kips}$$

R_1 is the axis of rotation about which moments can be taken because it is a pinned reaction. R_2 is 4 ft from the axis of rotation and is acting with a tendency to cause counterclockwise (positive) rotation about R_1. The uniformly distributed load on the left side (W_1) is assumed to be applied as a concentrated load at its center of gravity. The center of gravity of the uniformly distributed load on the left side (W_1) is directly over R_1, so the distance from the center of gravity of W_1 to the axis of rotation is zero. The load W_1 cannot cause a moment about R_1. The center of gravity of the uniformly distributed load on the right side of the beam (W_2) is 5 ft from the axis of rotation (R_1). W_2 tends to cause counterclockwise rotation about the axis of rotation, R_1:

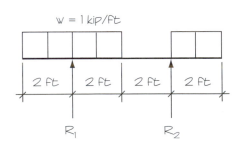

FIGURE 4.7 Free-body diagram for Example 4.6.

$$\circlearrowleft \Sigma M_{R1} = 0 = + (R_2 \cdot 4 \text{ ft}) - (4 \text{ kip} \cdot 0 \text{ ft}) - (2 \text{ kip} \cdot 5 \text{ ft})$$
$$0 = 4R_2 - 10 \text{ kip-ft}$$
$$4R_2 = 10 \text{ kip-ft}$$
$$\frac{4R_2}{4} = \frac{10 \text{ kip-ft}}{4}$$
$$R_2 = 2.5 \text{ kip}$$

There are no forces in the x direction, so $\Sigma F_x = 0$. Summing the vertical forces (F_y) with an upward force assigned a positive (+) sign and a downward force given a negative (−) sign and setting them equal to zero ($\Sigma F_y = 0$), we have

$$^{+}\uparrow \Sigma F_y = 0 = R_1 + R_2 - W_1 - W_2$$

From above, $R_2 = 2.5$ kip:

$$0 = R_1 + 2.5 \text{ kip} - 4 \text{ kip} - 2 \text{ kip}$$
$$= R_1 - 3.5 \text{ kip}$$
$$R_1 = 3.5 \text{ kip}$$

In Examples 4.5 and 4.6, loading is not symmetrical. Therefore, the reactions are not equal ($R_1 \neq R_2$). Additionally, it makes sense that the reaction closest to the largest load is the greatest: Thus, in Example 4.5, R_1 is less than R_2 because the location of the uniformly distributed load is closer to R_2. In Example 4.6, R_2 is greater than R_1 because the location of the load is closer to R_2.

■ **EXAMPLE 4.7**

A free-body diagram of a beam is shown in Figure 4.8. Reaction R_1 is a pinned connection and reaction R_2 is a bearing reaction. Find the unknown reactions, R_1 and R_2.

Solution R_1 is the axis of rotation about which moments can be taken because it is a pinned reaction. By convention, moments that tend to cause counterclockwise rotation about R_1 are classified positive (+). Moments that tend to cause clockwise rotation are negative. In the case of this beam,

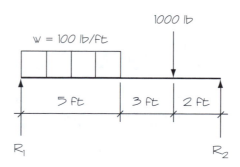

FIGURE 4.8   Free-body diagram for Example 4.7.

moments created by R_2 tend to be positive and moments created by the loads tend to be negative:

$$\circlearrowleft \Sigma M_{R1} = 0 = +(R_2 \cdot 10 \text{ ft}) - (1000 \text{ lb} \cdot 8 \text{ ft}) - (500 \text{ lb} \cdot 2.5 \text{ ft})$$
$$0 = 10R_2 - 8000 \text{ lb-ft} - 1250 \text{ lb-ft}$$
$$0 = 10R_2 - 9250 \text{ lb-ft}$$
$$9250 \text{ lb-ft} = 10R_2$$
$$\frac{9250 \text{ lb-ft}}{10} = \frac{10R_2}{10}$$
$$925 \text{ lb} = R_2$$

Note: The 500-lb load is the total uniformly distributed load of 100 lb/ft for 5 ft. Its moment arm is taken at 2.5 ft because that is the center of gravity of the uniformly distributed load.

Summing the vertical forces (F_y) with an upward force assigned a positive (+) value and a downward force given the negative (−) sign yields

$$^+\uparrow\Sigma F_y = 0 = R_1 + 925 \text{ lb} - 500 \text{ lb} - 1000 \text{ lb}$$
$$0 = +R_1 - 575 \text{ lb}$$
$$575 \text{ lb} = +R_1$$

Summing the horizontal forces (F_x) with a force acting to the right positive (+) and a force acting to the left assigned a negative (−) sign, since there are no horizontal force vectors on the free-body diagram, we have

$$^+\rightarrow\Sigma F_x = 0$$

The analysis above finds that the reaction R_1 equals 575 lb and R_2 equals 925 lb.

The conditions of static equilibrium ($\Sigma F_x = 0, \Sigma F_y = 0, \Sigma M = 0$) require that the body under consideration be isolated with all external forces, whether known or unknown. The analysis technique was introduced in Chapter 3 and was demonstrated in this chapter in Examples 4.1 through 4.7. The equations related to the conditions of static equilibrium are frequently used in engineering to solve for unknown forces that are acting on a body at rest or a body that is assumed to be at rest, that is, a body that is in static equilibrium.

EXERCISES

Eleven sets of illustrations and the corresponding free-body diagrams of building-related beams are provided. These free-body diagrams constitute exercises that will be used here and again in Chapters 11 and 12. *Follow these instructions carefully to save time later:*

(a) On engineering paper, neatly draw (with a straightedge) the free-body diagram for each of the 11 systems at an appropriate scale. Neatly draw one free-body diagram per sheet, using the top third of the individual sheet, leaving the bottom two-thirds of each sheet blank. The blank space is reserved for exercises in Chapters 11 and 12.

(b) Use additional sheets of engineering paper for the analysis. Solve for the unknown reactions, R_1 and R_2. Assume that each system is in static equilibrium, so the conditions of $\Sigma M = 0$ and $\Sigma F = 0$ may be used. Neglect the weight of the member under consideration.

4.1 Member *AB* is the beam under consideration. As shown in the illustration of the loading condition, member *AB* is a simple beam that supports the weight of a floor joist system. Floor joists carry the floor load and transfer a uniformly distributed load of 260 lb/ft to beam *AB* across its entire length.

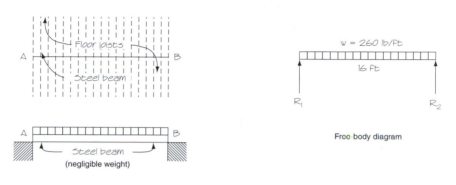

Loading condition

Free-body diagram

4.2 Member *AB* is the beam under consideration. As shown in the illustration of the loading condition, member *AB* is a simple beam that serves as a reaction support for beam *CD*. Beam *CD* supports a floor system load. Beam *CD* transfers a part of this floor load to beam *AB* at point *C*. The load transferred to beam *AB* is equivalent to the reaction of beam *CD* at point *C*. This load exhibits itself as a 2000-lb concentrated load applied to the center of beam *AB*.

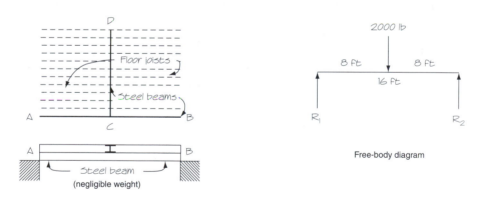

Loading condition

Free-body diagram

4.3 Member *AB* is the beam under consideration. As shown in the illustration of the loading condition, member *AB* is a simple beam. Beam *AB* supports a 260-lb/ft uniformly distributed load from a masonry wall that is bearing across its entire length of the beam. Beam *AB* also serves as a reaction support for beam *CD*. Beam *CD* transfers a part of this floor load to beam *AB* at point *C*. The load transferred from beam *CD* to beam *AB* is equivalent to the reaction of beam *CD* at point *C*. This load exhibits itself as a 2000-lb concentrated load applied to the center of beam *AB*.

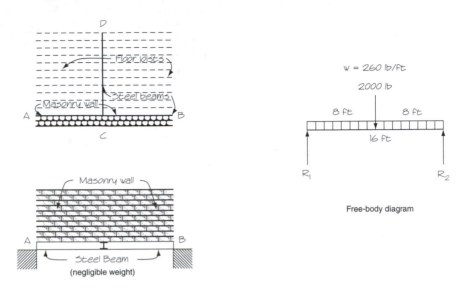

Free-body diagram

Loading condition

4.4 Member *AB* is the beam under consideration. As shown in the illustration of the loading condition, member *AB* is a simple beam that serves as a reaction support for beams *CD* and *EF*. Beams *CD* and *EF* support a floor system. Beams *CD* and *EF* transfer parts of the floor load to beam *AB* at point *C* and point *E*. The loads transferred to beam *AB* are equivalent to the reactions of beam *CD* at point *C* and beam *EF* at point *E*. These end loads are applied to beam *AB* as 2000-lb and 4000-lb concentrated loads.

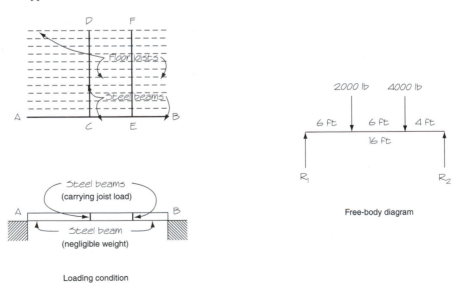

Free-body diagram

Loading condition

4.5 Member *AB* is the beam under consideration. As shown in the illustration of the loading condition, member *AB* supports a 1500-lb/ft uniformly distributed load of a masonry wall that is bearing across the entire length of the beam. Beam *AB* also provides reaction supports for beams *CD* and *EF*. These beams transfer parts of their loads to beam *AB* at point *C* and point *E*. The loads transferred to beam *AB* are equivalent to the reactions of beams *CD* and *EF* at these points. These loads are applied to beam *AB* as 6000-lb and 4000-lb concentrated loads.

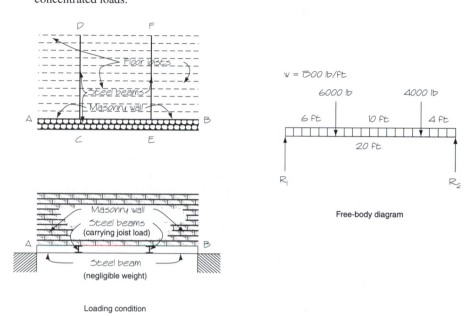

Free-body diagram

Loading condition

4.6 Member *AB* is the beam under consideration. As shown in the illustration of the loading condition, beam *AB* supports a uniformly distributed floor load of 500 lb/ft over part of its length. Beam *AB* also serves as a reaction support for beams *CD* and *EF*. Beam *CD* and *EF* transfer parts of the floor load to beam *AB* at point *C* and point *E*. The loads transferred to beam *AB* are equivalent to the reactions of beam *CD* at point *C* and beam *EF* at point *E*. These reactions are applied to beam *AB* as 4000-lb and 2000-lb concentrated loads.

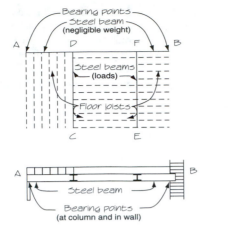

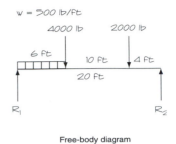

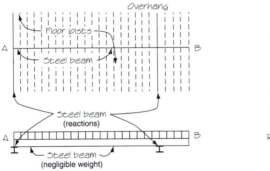

Loading condition

Free-body diagram

4.7 Member *AB* is the beam under consideration. As shown in the illustration of the loading condition, member *AB* is a single overhanging beam. It supports a 500-lb/ft uniformly distributed load from a floor system across the entire length of the beam.

Loading condition

Free-body diagram

4.8 Member *AB* is the beam under consideration. As shown in the illustration of the loading condition, member *AB* is a single overhanging beam. Beam *AB* supports a 500-lb/ft uniformly distributed floor load on the beam overhang only.

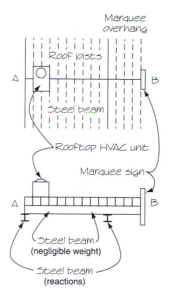

Loading condition

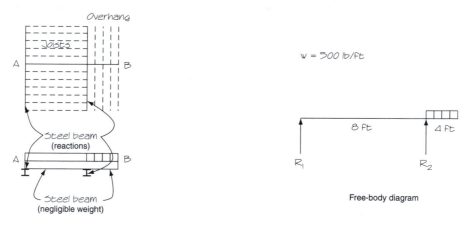

Free-body diagram

4.9 Member *AB* is the beam under consideration. As shown in the illustration of the loading condition, member *AB* is an overhanging beam that supports a uniformly distributed roof load of 500 lb/ft across its entire length. Beam *AB* also supports concentrated loads of 4000 lb and 3000 lb from a rooftop HVAC unit and a marquee sign.

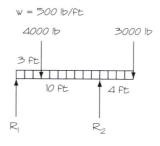

Free-body diagram

Loading conditions

4.10 Member *AB* is the beam under consideration. As shown in the illustration of the loading condition, member *AB* is an overhanging beam that supports a uniformly distributed roof load of 500 lb/ft. It also carries concentrated loads from a rooftop HVAC unit (4000 lb), an interior hanging display support (2000 lb), and a marquee sign (3000 lb).

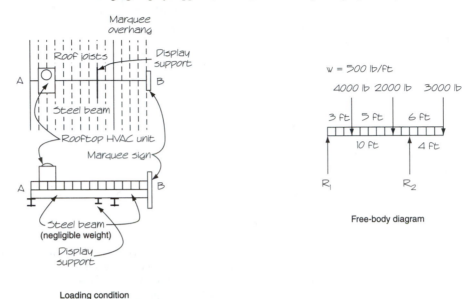

Loading condition

Free-body diagram

4.11 Member *AB* is the beam under consideration. As shown in the illustration of the loading condition, member *AB* is a double overhanging beam that supports a uniformly distributed roof load of 500 lb/ft except where the roof is open to the floor below. It also carries a concentrated load of 3000 lb from a marquee sign.

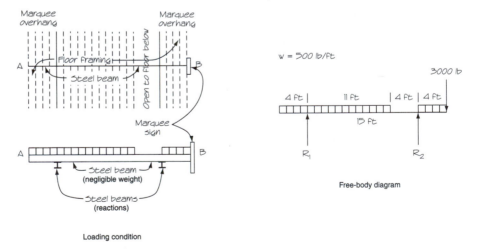

Loading condition

Free-body diagram

Chapter 5
ANALYSIS OF SIMPLE FRAME, TRUSS, AND CABLE SYSTEMS

When used in a building structural system, trusses and cables are frequently assumed to be in equilibrium under the applied external forces. In this chapter, forces acting in frames, trusses, and cables are examined by applying the conditions of static equilibrium that were introduced in Chapters 2 and 3. The presentation in this chapter is limited to coplanar systems, those occupying a single plane.

5.1 THE TRIANGULAR STRUCTURAL UNIT

The triangle is the only true dimensionally stable shape when used as a basic structural unit (Figure 5.1). It has joints that will not rotate when an external force is applied to the unit; even with pinned-end connections that are free to rotate, a triangular structural unit is dimensionally stable. On the other hand, polygonal shapes having four or more members, such as squares, rectangles, pentagons, and octagons, tend to be dimensionally unstable as a basic structural unit. When connected with *pinned connections,* the members (sides) of these shapes will rotate freely if an external force is applied to the unit.

To sustain the stability of a structural unit having four or more members, rigid joints called *moment connections* are needed to keep the joints from rotating when an external force is applied to the unit. Fabrication of a true moment connection is possible but more difficult than a pinned connection. Additionally, with a moment connection, the structural unit itself needs more design attention because shear and bending forces can be transferred through the connection. Thus, because of its dimensional stability, the triangular unit or groups of triangular units are used in many practical structural applications, such as power transmission and radio towers, bridges, aircraft frames, and building structural elements such as open web joists (Figure 5.2).

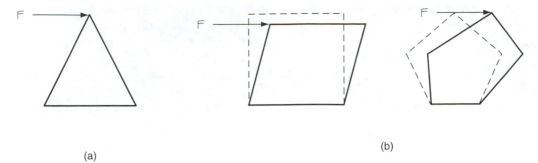

(a)

(b)

FIGURE 5.1 (a) The triangle is the only true dimensionally stable shape. (b) Other polygonal shapes having four or more members (sides) tend to be dimensionally unstable as a basic structural unit. The members of these shapes when connected with pinned end connections will rotate freely if an external force is applied to the unit.

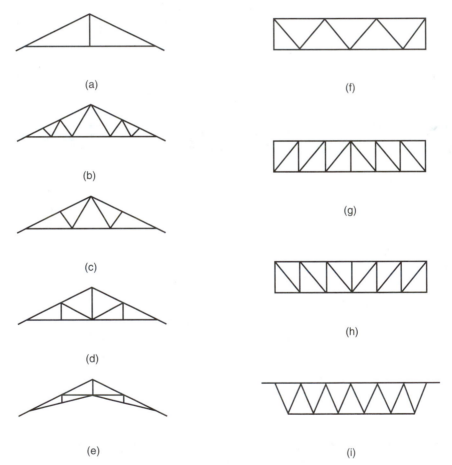

FIGURE 5.2 Examples of truss and frame systems: (a) King post; (b) Belgian or double W; (c) W or Fink; (d) Howe; (e) Scissors; (F) Warren; (g) Howe flat; (h) Pratt flat; (i) Open-web joist.

Trusses are used to form structural systems, such as in a steel joist and joist girder system.

5.2   TRUSSES AND FRAMES

A *frame* is composed of individual members that are joined together in an open structural system that provides support. A *planar truss* is a two-dimensional frame system composed of individual members arranged in groups of structural units that are positioned along one plane. In frame and truss design, a *member* is a long slender element such as a bar, angle or similar structural member that is joined to other members only at the member ends. A *joint* is a connection between two or more members.

Planar trusses are formed of triangular structural units that are very stable when external forces are applied along the plane of the system. If a force applied to the truss is not within the plane of the truss, the truss may fail laterally. Simply, a planar truss used as a structural element is only stable along one plane.

A *space frame* is a trusslike system of structural units arranged in three-dimensional structural groups. The three-dimensional arrangement of structural units makes the space frame more dimensionally stable because it can be designed to support forces acting in any direction. Design of space frames involves application of complex analysis because the members are positioned in three dimensions. The complex concepts related to space frame design are not covered in this book; however, the theory and analysis technique associated with coplaner frames and trusses design relates somewhat to space frame design theory.

Under an applied external force, each member of a planar truss is in tension or compression, and thus each member applies *either* a pushing or a pulling force on the joint. These forces pass through the longitudinal axis of the member and thus are referred to as *axial forces*. A member developing an axial tensile force is called a *tension member.* A *compression member* develops a compressive force. Truss analysis is necessary to determine the type and magnitude of the force acting within each member of a truss. Results of truss analysis allow the designer to design the individual members of the truss and the truss as a system.

Two analysis techniques can be used to analyze the type and magnitude of forces acting in a truss: the method of joints and the method of sections. The method of joints technique is introduced in this book because it is similar to the analysis technique introduced for concurrent coplaner force systems in Chapter 2.

In the *method of joints* truss analysis technique, it is assumed that the individual members that make up the truss are positioned along a single plane. It is also assumed that external forces are applied only to the truss joints. Additionally, connections are assumed to be frictionless pins, so moment and torsion cannot be transmitted through the joint to the attached member(s).

This truss analysis technique assumes that the force system at each joint is in equilibrium. An easy way to envision the condition is to assume that the pin in the pinned connection of the joint is a free body in static equilibrium. The pin is held in equilibrium by the forces exerted on the pin by the members attached to the joint under consideration. A force applied by a member either pulls away or pushes toward the pin. A *tension member pulls from the pin;* the sense (arrowhead) of the force vector points away from the joint. A *compression member pushes toward the pin;* the sense (arrowhead) of the force vector points toward the joint. Forces acting on the pin are always in line with the member exerting the force.

When a force is calculated in a member, that force exists at the joint at the other end of the member: A tension member that applies a pulling force at one joint also applies an equal tensile (pulling) force at the joint at the other end of the member. Conversely, a compression member applies equal pushing forces toward the joints at which it is connected.

In analysis, each joint is isolated; it is treated as a separate force system much like the concurrent coplaner force systems introduced in Chapter 2. Each joint is in equilibrium with the pin assumed to be a free body, so $\Sigma M = 0$, $\Sigma F_x = 0$, and $\Sigma F_y = 0$ apply. However, the condition of $\Sigma M = 0$ can be eliminated from consideration; forces from the members act concurrently and the joint is assumed to be a frictionless pin that is free to rotate. Therefore, only $\Sigma F_x = 0$ and $\Sigma F_y = 0$ apply at each isolated joint. Note that under the conditions of static equilibrium, $\Sigma M = 0$ can only be used to solve for unknown *external* reactions that are influencing the entire truss. The truss itself is a rigid body, so $\Sigma M = 0$ applies.

■ EXAMPLE 5.1

A simple truss is composed of a single isosceles triangle (a triangle having equal legs and interior angles of 60°) as shown in Figure 5.3. Reaction R_A is a pinned reaction and reaction R_B is a bearing reaction. A 50-kN load is hung from the top of the triangle at joint C. Determine the forces in the members.

Solution A diagram of force systems for each joint is shown in Figure 5.4. $R_{Ay} = R_{By}$ because loading is symmetrical:

$$R_{Ay} = R_{By} = \frac{50 \text{ kN}}{2} = 25 \text{ kN}$$

Analysis begins at joint C. Joint C is isolated and the forces acting on the joint are analyzed through $\Sigma F_x = 0$ and $\Sigma F_y = 0$:

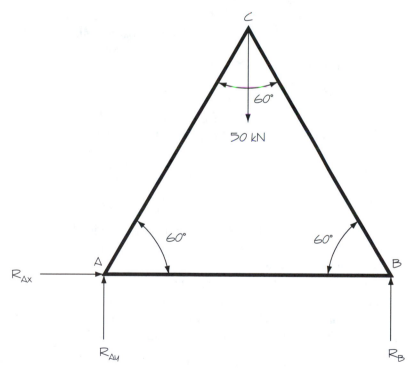

FIGURE 5.3 Simple truss for Example 5.1.

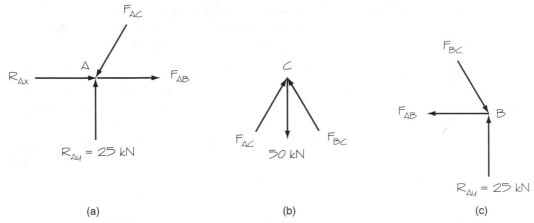

(a)   (b)   (c)

FIGURE 5.4 Force systems of the joints of the simple truss for Example 5.1: (a) joint A; (b) joint C; (c) joint B.

$$^+\rightarrow \Sigma F_x = 0 = F_{AC} \cos 60° - F_{BC} \cos 60°$$
$$0 = F_{AC} \cos 60° - F_{BC} \cos 60°$$
$$F_{AC} \cos 60° = F_{BC} \cos 60°$$
$$F_{AC} = F_{BC}$$
$$^+\uparrow \Sigma F_y = 0 = -50 \text{ kN} + F_{AC} \sin 60° + F_{BC} \sin 60°$$
$$0 = -50 \text{ kN} + F_{AC} \sin 60° + F_{BC} \sin 60°$$
$$F_{AC} \sin 60° + F_{BC} \sin 60° = 50 \text{ kN}$$

From above: $F_{AC} = F_{BC}$:

$$F_{BC} \sin 60° + F_{BC} \sin 60° = 50 \text{ kN}$$
$$2 F_{BC} \sin 60° = 50 \text{ kN}$$
$$F_{BC} = \frac{50 \text{ kN}}{2 \sin 60°}$$
$$F_{BC} = 28.87 \text{ kN}$$
$$F_{AC} = 28.87 \text{ kN}$$

Members *AC* and *BC* are both experiencing compression forces of 28.87 kN. These are compression forces because they are pushing toward the joint (the arrowhead is pointed toward the joint).

Joint *B* is isolated and the forces acting on it are analyzed through $\Sigma F_x = 0$ and $\Sigma F_y = 0$. Member *BC* imposes a compression force (F_{BC}) of 28.87 kN, so the sense of F_{BC} (arrowhead) is toward joint *B*. The reacting force R_{By} at joint *B* is upward; it acts perpendicular to the surface because it is a bearing reaction.

$$^+\rightarrow \Sigma F_x = 0 = 28.87 \text{ kN} \cos 60° - F_{AB}$$
$$0 = 28.87 \text{ kN} \cos 60° - F_{AB}$$
$$F_{AB} = 28.87 \text{ kN} \cos 60°$$
$$= 14.44 \text{ kN}$$

$$^+\uparrow \Sigma F_y = 0 = R_{By} - 28.87 \text{ kN } \sin 60°$$
$$0 = R_{By} - 28.87 \text{ kN } \sin 60°$$
$$R_{By} = 28.87 \text{ kN } \sin 60°$$
$$= 25.0 \text{ kN}$$

Member *AB* is a tension member experiencing a force of 14.44 kN. It is a tension force because it is pulling away from the joint.

■ EXAMPLE 5.2

A 56-ft roof truss is shown in Figure 5.5. Reaction R_A is a pinned reaction and reaction R_B is a bearing reaction. Under the loading condition shown, determine the forces in each member.

Solution   Through the Pythagorean theorem, $a^2 + b^2 = c^2$:

$$5^2 + 12^2 = c^2$$
$$c^2 = 169$$
$$c = 13$$

Joint A. Joint *A* is selected for analysis because there are only two unknown forces acting on the joint, F_{AF} and F_{AD}. The joint is isolated and a free-body diagram with known and unknown forces is drawn (Figure 5.6). Two unknown forces can be determined by summing the forces in two equations $\Sigma F_x = 0$ and $\Sigma F_y = 0$. Analysis begins with ΣF_y because the unknown force F_{AD} is the only unknown force:

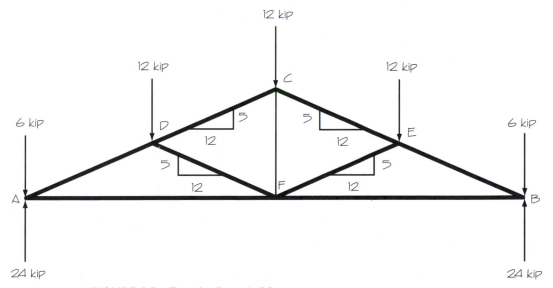

FIGURE 5.5   Truss for Example 5.2.

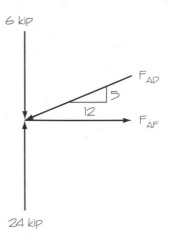

FIGURE 5.6 Force system on joint A in Example 5.2.

$$+\uparrow\Sigma F_y = 0 = +24 \text{ kip} - 6 \text{ kip} - \tfrac{5}{13}F_{AD}$$
$$0 = +24 \text{ kip} - 6 \text{ kip} - \tfrac{5}{13}F_{AD}$$
$$0 = 18 \text{ kip} - \tfrac{5}{13}F_{AD}$$
$$\tfrac{5}{13}F_{AD} = 18 \text{ kip}$$
$$F_{AD} = \tfrac{13}{5}(18 \text{ kip})$$
$$= 46.8 \text{ kip-compression}$$

Because F_{AD} is pushing at the joint (the sense of the force vector was drawn toward joint A), F_{AD} is a compression force and member AD is a compression member.

With F_{AD} known, $\Sigma F_x = 0$ may be used to find F_{AF}:

$$+\rightarrow\Sigma Fx = 0 = -\tfrac{12}{13}F_{AD} + F_{AF}$$
$$0 = -\tfrac{12}{13}F_{AD} + F_{AF}$$
$$F_{AF} = \tfrac{12}{13}F_{AD}$$

From above, $F_{AD} = 46.8$ kip:

$$F_{AF} = \tfrac{12}{13}(46.8 \text{ kip})$$
$$= 43.2 \text{ kip-tension}$$

Because F_{AF} is pulling from the joint (the sense of the force vector was drawn away from joint A), member BF is a tension member undergoing a tensile force (F_{BF}) of 43.2 kip.

Joint D. Joint D is the next joint selected for analysis because it now has only two unknown forces acting on it: F_{CD} and F_{DF}. The force $F_{AD} = 46$ kip was found above. The force system for joint D with known and unknown forces is as shown in Figure 5.7.

Since there are two unknown forces (F_{CD} and F_{DF}) that are acting on joint D in both the x and y directions, two equations are needed to solve for the unknown forces, $\Sigma F_x = 0$ and $\Sigma F_y = 0$. Beginning with $\Sigma F_x = 0$:

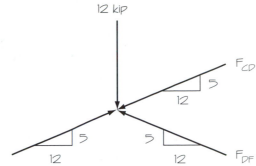

FIGURE 5.7 Force system on joint D in Example 5.2.

$F_{AD} = 46.8 \text{ kip}$

$$^+\!\rightarrow\!\Sigma F_x = 0 = \tfrac{12}{13}(46.8 \text{ kip}) - \tfrac{12}{13}F_{CD} - \tfrac{12}{13}F_{DF}$$
$$0 = \tfrac{12}{13}(46.8 \text{ kip}) - \tfrac{12}{13}F_{CD} - \tfrac{12}{13}F_{DF}$$
$$= 43.2 \text{ kip} - \tfrac{12}{13}F_{CD} - \tfrac{12}{13}F_{DF}$$
$$\tfrac{12}{13}F_{DF} = 43.2 \text{ kip} - \tfrac{12}{13}F_{CD}$$
$$F_{DF} = \tfrac{13}{12}(43.2 \text{ kip} - \tfrac{12}{13}F_{CD})$$
$$= 46.8 \text{ kip} - F_{CD}$$

F_{DF} is now expressed in terms of F_{CD}. With two unknowns, it is necessary to solve for F_{CD} using $\Sigma F_y = 0$:

$$^+\!\uparrow\!\Sigma F_y = 0 = -12 \text{ kip} - \tfrac{5}{13}F_{CD} + \tfrac{5}{13}F_{DF} + \tfrac{5}{13}(46.8 \text{ kip})$$
$$0 = -12 \text{ kip} - \tfrac{5}{13}F_{CD} + \tfrac{5}{13}F_{DF} + \tfrac{5}{13}(46.8 \text{ kip})$$
$$0 = 6 \text{ kip} - \tfrac{5}{13}F_{CD} + \tfrac{5}{13}F_{DF}$$
$$\tfrac{5}{13}F_{CD} = 6 \text{ kip} + \tfrac{5}{13}F_{DF}$$
$$F_{CD} = \tfrac{13}{5}(6 \text{ kip} + \tfrac{5}{13}F_{DF})$$
$$= 15.6 \text{ kip} + F_{DF}$$

As found above, $F_{DF} = 46.8 \text{ kip} - F_{CD}$. By substituting $46.8 \text{ kip} - F_{CD}$ for F_{DF}, the unknown force F_{CD} can be solved:

$$F_{CD} = 15.6 \text{ kip} + (46.8 \text{ kip} - F_{CD})$$
$$= 15.6 \text{ kip} + 46.8 \text{ kip} - F_{CD}$$
$$2F_{CD} = 62.4 \text{ kip}$$
$$\frac{2F_{CD}}{2} = \frac{62.4 \text{ kip}}{2}$$
$$F_{CD} = 31.2 \text{ kip-compression}$$

Because F_{CD} is pushing at joint D, it is a compression force.

Force F_{DF} is still not known. The known force $F_{CD} = 31.2 \text{ kip}$ is substituted into $F_{DF} = 46.8 \text{ kip} - F_{CD}$ from above to solve for F_{DF}:

$$F_{DF} = 46.8 \text{ kip} - F_{CD}$$
$$= 46.8 \text{ kip} - 31.2 \text{ kip}$$
$$= 15.6 \text{ kip-compression}$$

5.3 Planar Truss Analysis

Because F_{DF} is pushing at joint D, it is a compression force.

Joint C. Joint C now has two unknown forces acting on it: F_{CE} and F_{CF}. The force $F_{CD} = 31.2$ kip was found above. The force system for joint C with known and unknown forces is as shown in Figure 5.8. Beginning with $\Sigma F_x = 0$:

$$+\rightarrow \Sigma F_x = 0 = \tfrac{12}{13}(31.2 \text{ kip}) - \tfrac{12}{13}F_{CE}$$
$$0 = \tfrac{12}{13}(31.2 \text{ kip}) - \tfrac{12}{13}F_{CE}$$
$$\tfrac{12}{13}F_{CE} = \tfrac{12}{13}(31.2 \text{ kip})$$
$$F_{CE} = 31.2 \text{ kip-compression}$$
$$+\uparrow \Sigma F_y = 0 = \tfrac{5}{13}(31.2 \text{ kip}) + \tfrac{5}{13}(31.2 \text{ kip}) - 12 \text{ kip} - F_{CF}$$
$$0 = \tfrac{5}{13}(31.2 \text{ kip}) + \tfrac{5}{13}(31.2 \text{ kip}) - 12 \text{ kip} - F_{CF}$$
$$= 12 \text{ kip} - F_{CF}$$
$$F_{CF} = 12 \text{ kip-tension}$$

Joint F. Joint F has two unknown forces acting on it, F_{EF} and F_{BF}. The force system for joint F with known and unknown forces is as shown in Figure 5.9. Beginning with $\Sigma F_y = 0$ because that eliminates force F_{BF} from consideration (it acts horizontally), we have:

$$+\uparrow \Sigma F_y = 0 = 12 \text{ kip} - \tfrac{5}{13}(15.6 \text{ kip}) - \tfrac{5}{13}F_{EF}$$
$$0 = 12 \text{ kip} - \tfrac{5}{13}(15.6 \text{ kip}) - \tfrac{5}{13}F_{EF}$$
$$= 12 \text{ kip} - 6 \text{ kip} - \tfrac{5}{13}F_{EF}$$
$$= 6 \text{ kip} - \tfrac{5}{13}F_{EF}$$
$$\tfrac{5}{13}F_{EF} = 6 \text{ kip}$$
$$F_{EF} = \tfrac{13}{5}(6 \text{ kip})$$
$$= 15.6 \text{ kip-compression}$$

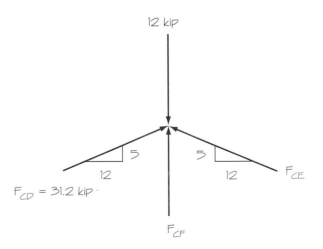

FIGURE 5.8 Force system on joint C in Example 5.2.

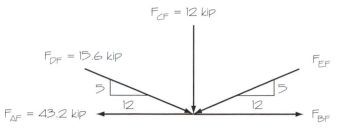

FIGURE 5.9 Force system on joint F in Example 5.2.

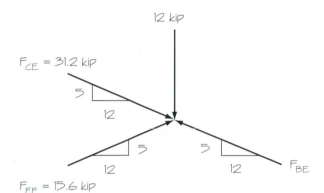

FIGURE 5.10 Force system on joint E in Example 5.2.

Joint E. Joint E has only one unknown force acting on it (F_{BE}). The force system for joint E with known and unknown forces is as shown in Figure 5.10. With $\Sigma F_x = 0$:

$$^+\!\rightarrow\Sigma F_x = 0 = \tfrac{12}{13}F_{CE} + \tfrac{12}{13}F_{EF} - \tfrac{12}{13}F_{BE}$$
$$0 = \tfrac{12}{13}F_{CE} + \tfrac{12}{13}F_{EF} - \tfrac{12}{13}F_{BE}$$
$$= \tfrac{12}{13}(31.2 \text{ kip}) + \tfrac{12}{13}(15.6 \text{ kip}) - \tfrac{12}{13}F_{BE}$$
$$= 28.8 \text{ kip} + 14.4 \text{ kip} - \tfrac{12}{13}F_{BE}$$
$$= 43.2 \text{ kip} - \tfrac{12}{13}F_{BE}$$
$$\tfrac{12}{13}F_{BE} = 43.2 \text{ kip}$$
$$F_{BE} = \tfrac{13}{12}(43.2 \text{ kip})$$
$$= 46.8 \text{ kip-compression}$$

Joint B. The force system for joint B with known and unknown forces is as shown in Figure 5.11. While all the forces have been solved, it is helpful to check work at joint B.

$$^+\!\rightarrow\Sigma F_x = 0 = -F_{BF} + \tfrac{12}{13}F_{BE}$$
$$0 = -F_{BF} + \tfrac{12}{13}F_{BE}$$
$$F_{BF} = \tfrac{12}{13}F_{BE}$$
$$43.2 \text{ kip} = \tfrac{12}{13}(46.8 \text{ kip})$$
$$= 43.2 \text{ kip}$$

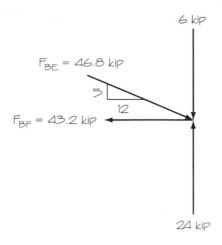

FIGURE 5.11 Force system on joint B in Example 5.2.

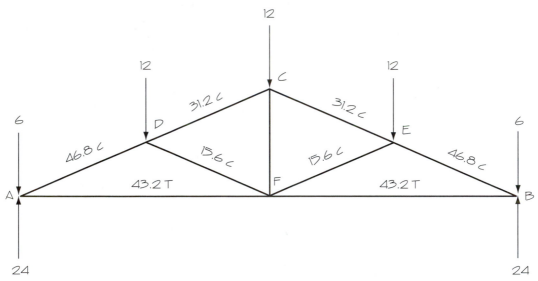

FIGURE 5.12 Tensile and compressive forces in the members in the truss in Example 5.2. All values are expressed in kip (1000 lb): C, compression; T, tension.

$$^+\uparrow\Sigma F_y = 0 = 24 \text{ kip} - 6 \text{ kip} - \tfrac{5}{13} F_{BE}$$
$$0 = 24 \text{ kip} - 6 \text{ kip} - \tfrac{5}{13} F_{BE}$$
$$= 24 \text{ kip} - 6 \text{ kip} - \tfrac{5}{13}(46.8 \text{ kip})$$
$$= 18 \text{ kip} - \tfrac{5}{13}(46.8 \text{ kip})$$
$$18 \text{ kip} = \tfrac{5}{13}(46.8 \text{ kip})$$
$$= 18 \text{ kip}$$

See Figure 5.12.

Ropes and cables are very flexible. If the total weight of the cable is very small in comparison to the load hung from it, the weight of the cable can be neglected. A weightless cable has no sag, so the cable is straight. If a cable meets or approximates these conditions, it can be treated as a true tension member. The force system of a system of *weightless* cables connected at one point and acting is both concurrent (all forces have lines of action intersecting at the same point) and coplaner (forces act in a single plane). Analysis assumes that any cable is in static equilibrium. Therefore, $\Sigma F_x = 0$ and $\Sigma F_y = 0$ apply. The condition of $\Sigma M = 0$ does not apply because the member is fully flexible; any resistance to bending is so small that a moment created by forces in a cable can be neglected.

■ EXAMPLE 5.3

A weightless cable hangs between two supports as shown in Figure 5.13. The cable supports a load of 20 kN at its center. Approximate the tension force in the cable.

Solution Connection B at the center of the span is isolated and the unknown forces F_{AB} and F_{BC} are determined through summation of the forces, $\Sigma F_x = 0$ and $\Sigma F_y = 0$:

$$^+\!\rightarrow\Sigma F_x = 0 = -\tfrac{4}{5}F_{AB} + \tfrac{4}{5}F_{BC}$$
$$0 = -\tfrac{4}{5}F_{AB} + \tfrac{4}{5}F_{BC}$$

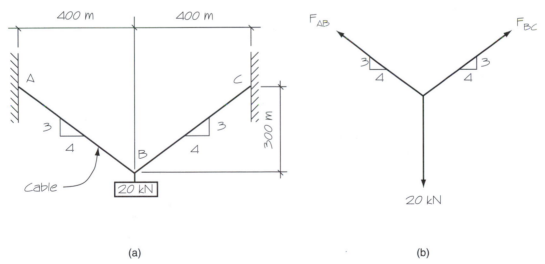

(a)   (b)

FIGURE 5.13   (a) Weightless cable supporting a weight at the center of its spans (b) associated force system at point B (Example 5.3).

$$\tfrac{4}{5}F_{AB} = \tfrac{4}{5}F_{BC}$$
$$F_{AB} = F_{BC}$$
$$+\uparrow \Sigma F_y = 0 = -20 \text{ kN} + \tfrac{3}{5}F_{AB} + \tfrac{3}{5}F_{BC}$$
$$0 = -20 \text{ kN} + \tfrac{3}{5}F_{AB} + \tfrac{3}{5}F_{BC}$$

From above; $F_{AB} = F_{BC}$:

$$0 = -20 \text{ kN} + \tfrac{3}{5}F_{BC} + \tfrac{3}{5}F_{BC}$$
$$\tfrac{6}{5}F_{BC} = 20 \text{ kN}$$
$$F_{BC} = \frac{20\text{kN}}{\tfrac{6}{5}}$$
$$= F_{AB} = 16.67 \text{ kN}$$

■ EXAMPLE 5.4

A weightless cable hangs between two supports 1000 ft apart and at the same elevation as shown in Figure 5.14. The cable is assumed to support a load of 10 kip at its center. The cable will support a roof structure, so sag must be minimized. Assume that the load hangs only 6 in (0.5 ft) below the elevation of the two supports. Approximate the tension force in the cable.

Solution   Through the Pythagorean theorem, $a^2 + b^2 = c^2$:

$$0.5^2 + 500^2 = c^2$$
$$c^2 = 250\,000.25$$
$$c = 500.000\,25 \text{ ft}$$

Connection B is isolated and the unknown forces F_{AB} and F_{BC} are determined through summation of the forces, $\Sigma F_x = 0$ and $\Sigma F_y = 0$:

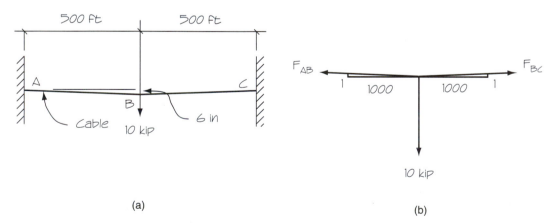

(a)   (b)

FIGURE 5.14   (a) Weightless cable supporting a weight at the center of its spans (b) associated force system at point B (Example 5.4).

$$+\rightarrow \Sigma F_x = 0 = \frac{-500}{500.000\ 25\ \text{ft}\ F_{AB}} + \frac{500}{500.000\ 25\ \text{ft}\ F_{BC}}$$

$$0 = \frac{-500}{500.000\ 25\ \text{ft}\ F_{AB}} + \frac{500}{500.000\ 25\ \text{ft}\ F_{BC}}$$

$$\frac{500}{500.000\ 25\ \text{ft}\ F_{AB}} = \frac{500}{500.000\ 25\ \text{ft}\ F_{BC}}$$

$$F_{AB} = F_{BC}$$

$$+\uparrow \Sigma F_y = 0 = -10\ \text{kip} + \frac{0.5}{500.000\ 25\ \text{ft}\ F_{AB}} + \frac{3}{5} F_{BC}$$

$$0 = -10\ \text{kip} + \frac{0.5}{500.000\ 25\ \text{ft}\ F_{AB}} + \frac{0.5}{500.000\ 25\ \text{ft}\ F_{BC}}$$

From above; $F_{AB} = F_{BC}$:

$$0 = -10\ \text{kip} + \frac{0.5}{500.000\ 25\ \text{ft}\ F_{BC}} + \frac{0.5}{500.000\ 25\ \text{ft}\ F_{BC}}$$

$$\frac{1}{500.000\ 25\ \text{ft}\ F_{BC}} = 10\ \text{kip}$$

$$F_{BC} = \frac{10\ \text{kip}}{1/500.000\ 25\ \text{ft}}$$

$$= F_{AB} = 5000\ \text{kip}$$

Example 5.4 shows that a small force hanging from the center of a weightless cable can develop a significant tension force, a force that presents concerns in design.

Cables are not weightless; they have weight. A hanging rope or cable, such as a power transmission line, develops tension as it carries its own weight. This weight is uniformly distributed over the entire length of the cable. Tensioned roof and suspension bridge structures can also carry a uniformly distributed load. A freely hanging cable carrying a uniformly distributed load and supported at ends at an equal elevation will hang in the shape of a parabola. The maximum tension force (F_{tension}) that develops in such a cable is based on the cable's length (L), maximum sag (y), and uniformly distributed load (w) as illustrated in Figure 5.15 and as expressed in the equation:

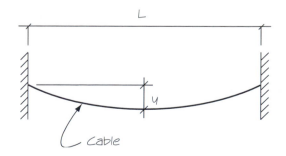

FIGURE 5.15 The maximum tension force ($F_{tension}$) that develops in such a cable is based on the cable's length (L), maximum sag (y), and uniformly distributed load (w).

5.4 Cable Systems

$$\text{maximum tension force, } F_{\text{tension}} = \frac{wL}{2} \sqrt{\frac{L^2}{16y^2} + 1}$$

Derivation of this formula is beyond the intended breadth of this book. It is, however, useful in developing an understanding of the effect of a uniformly distributed force acting on a cable.

■ EXAMPLE 5.5

A power transmission cable hangs freely between two towers of equal elevation that are spaced 100 m apart. The cable has a uniform weight of 30 N/m.

(a) Determine the maximum tension force that develops in a cable with a maximum sag of 10 m.

(b) Determine the maximum tension force that develops in a cable with a maximum sag of 5 m.

(c) Determine the maximum tension force that develops in a cable with a maximum sag of 1 m.

Solution (a)

$$F_{\text{tension}} = \frac{wL}{2} \sqrt{\frac{L^2}{16y^2} + 1}$$

$$= \frac{30 \text{ N/m} \cdot 100 \text{ m}}{2} \sqrt{\frac{(100 \text{ m})^2}{(16 \cdot (10 \text{ m})^2} + 1}$$

$$= 4039 \text{ N} = 4.0 \text{ kN}$$

(b)

$$F_{\text{tension}} = \frac{wL}{2} \sqrt{\frac{L^2}{16y^2} + 1}$$

$$= \frac{30 \text{ N/m} \cdot 100 \text{ m}}{2} \sqrt{\frac{(100 \text{ m})^2}{(16 \cdot (5 \text{ m})^2} + 1}$$

$$= 7649 \text{ N} = 7.6 \text{ kN}$$

(c)

$$F_{\text{tension}} = \frac{wL}{2} \sqrt{\frac{L^2}{16y^2} + 1}$$

$$= \frac{30 \text{ N/m} \cdot 100 \text{ m}}{2} \sqrt{\frac{(100 \text{ m})^2}{(16 \cdot (1 \text{ m})^2} + 1}$$

$$= 37\,530 \text{ N} = 37.5 \text{ kN}$$

As shown in Example 5.5, sag in a cable can significantly influence the tension that develops in a cable; a small sag in a freely hanging cable produces greater tension. In design of overhead power transmission lines, the designer must evaluate overhead clearance and the sagging cable as it relates to height and strength of the tower.

■ **EXAMPLE 5.6**

A cable system will support the roof system of a sports center. It hangs between two supports at the same elevation. The supports are horizontally spaced 1000 ft apart. A single cable in the system will support a uniformly distributed load of 50 lb/ft of cable, including the weight of the cable.
(a) Determine the maximum tension force in the cable if the cable will hang a maximum of 15 ft below the elevation of the two supports.
(b) Determine the maximum tension force in the cable if the cable will hang a maximum of 30 ft below the elevation of the two supports.

Solution (a)

$$F_{tension} = \frac{wL}{2}\sqrt{\frac{L^2}{16y^2} + 1}$$

$$= \frac{50 \text{ lb/ft} \cdot 1000 \text{ ft}}{2}\sqrt{\frac{(1000 \text{ ft})^2}{(16 \cdot (15 \text{ ft})^2} + 1}$$

$$= 417\,416 \text{ lb} = 417 \text{ kip}$$

(b)

$$F_{tension} = \frac{wL}{2}\sqrt{\frac{L^2}{16y^2} + 1}$$

$$= \frac{50 \text{ lb/ft} \cdot 1000 \text{ ft}}{2}\sqrt{\frac{(1000 \text{ ft})^2}{(16 \cdot (30 \text{ ft})^2} + 1}$$

$$= 209\,828 \text{ lb} = 210 \text{ kip}$$

As shown in Example 5.6, sag in a tension roof system can significantly influence the tension that develops in a cable; a small sag in a cable produces greater tension.

Trusses

5.1 Determine the forces in the members of the planar truss shown.

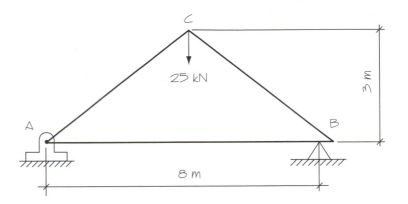

5.2 Determine the forces in the members of the planar truss shown.

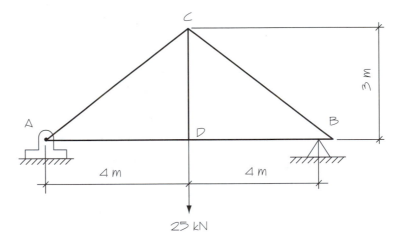

5.3   Determine the forces in the members of the planar truss shown.

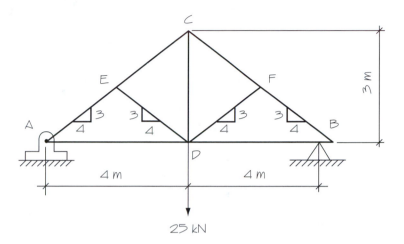

5.4   Determine the forces in the members of the planar truss shown.

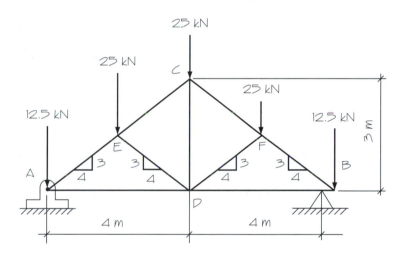

5.5 Determine the forces in the members of the planar truss shown.

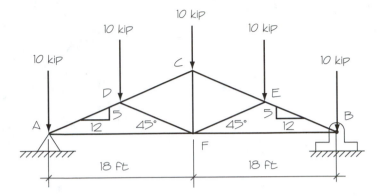

5.6 Determine the forces in the members of the planar truss shown.

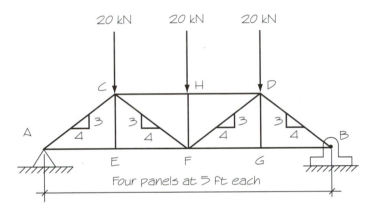

5.7 Determine the forces in the members of the planar truss shown.

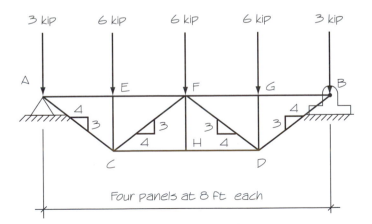

5.8 Determine the forces in the members of the planar truss shown.

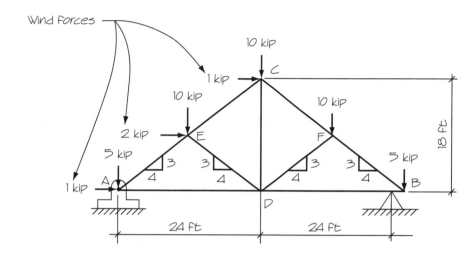

5.9 Determine the forces in the members of the planar truss shown.

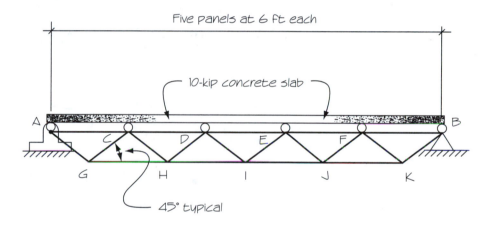

Cables

5.10 A weightless cable hangs between two supports as shown. The cable supports a load of 200 N at its center. Approximate the tension force in the cable.

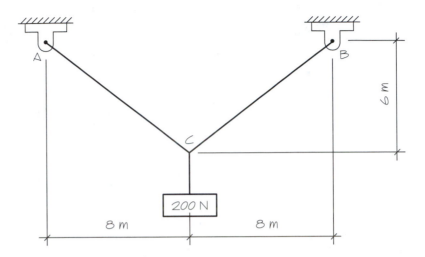

5.11 A weightless rope hangs between two supports as shown. The rope supports a load of 10 lb at its center. Approximate the tension force in the cable.

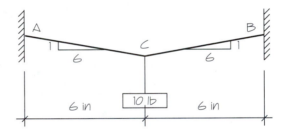

5.12 As shown, a weightless cable hangs between two supports that are 1000 ft apart and at the same elevation. The cable supports a load of 4 kips at its center.

(a) Approximate the tension force in the cable if the load hangs 100 ft below the elevation of the two supports.

(b) Approximate the tension force in the cable if the load hangs 10 ft below the elevation of the two supports.

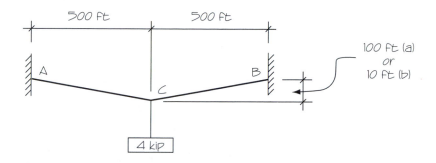

5.13 A power line is supported by an insulator hung from a transmission tower as shown. Approximate the tension force in the insulator.

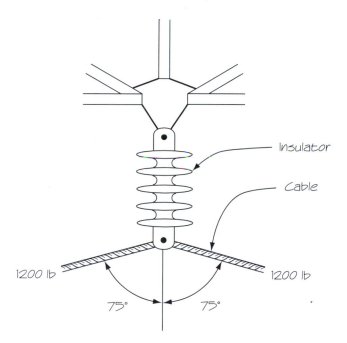

5.14 As shown, a power transmission cable hangs freely between two towers of equal elevation that are spaced 200 m apart. The cable has a uniform weight of 25 N/m.

    (a) Approximate the maximum tension force that develops in a cable with a maximum sag of 15 m.

    (b) Approximate the maximum tension force that develops in a cable with a maximum sag of 30 m.

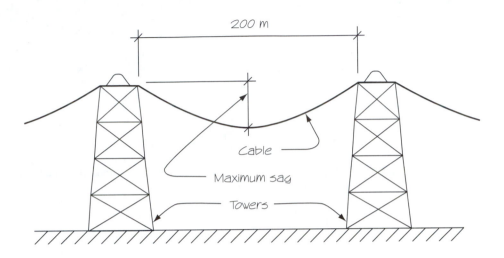

5.15 As shown, six power transmission cables hang freely between two towers to carry power across a canyon. The towers are spaced 465 ft apart and the bases of the towers are at the same elevation. Each cable has a uniform weight of 6 lb/ft and a maximum sag of 30 ft.

(a) Determine the maximum tension force that develops in each cable.

(b) Assuming that the cables pull on the tower arms at an angle of 10° from horizontal, determine the moment that develops at the base of the towers as a result of the tension forces in the cables.

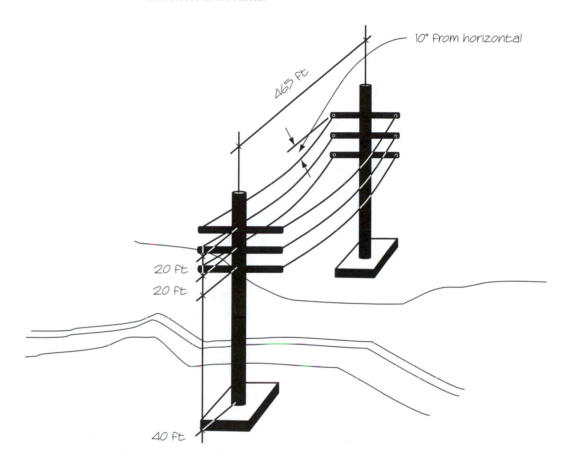

5.16 As shown, a power transmission cable hangs freely between two towers that are spaced 260 ft apart. The bases of the towers are at the same elevation. The cable has a uniformly distributed weight of 16 lb/ft. Tension force in a cable is limited to a maximum of 15000 lb. Determine the minimum height the towers must be to maintain a ground clearance of 40 ft.

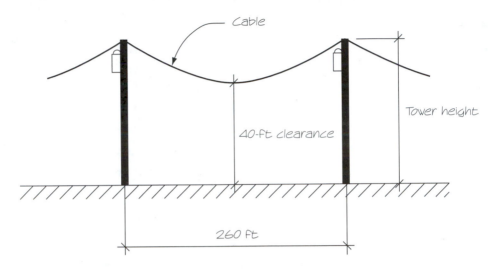

5.17 As shown, an electrical cable hangs freely between power masts of an office building and a warehouse. The masts are at the same elevation and are spaced 60 ft apart. The cable has a uniformly distributed weight of 2 lb/ft. Tension force in a cable is limited to a maximum of 100 lb. Determine the minimum height the masts must be to maintain a ground clearance of 24 ft.

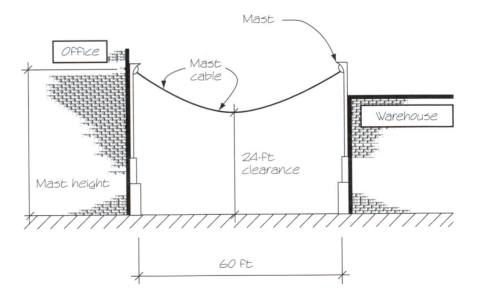

Chapter 6
SIMPLE STRESS

A knowledge of the different behaviors of materials is necessary for effective selection and use. Concepts in strengths of materials are introduced in upcoming chapters. Strength of materials is a branch of engineering that is a study of the effects of compressive, tensile, shear, torsional, and bending stresses due to the application of a force or force system on or due to a change in temperature in the material of a body. In this chapter the concept of simple stress is introduced. Since this book is written to acquaint the technician with basic principles, an exhaustive study of these engineering principles is beyond the intended breadth of the book.

6.1   MECHANICAL PROPERTIES OF MATERIALS

A *property* distinguishes a characteristic trait or specific behavior of a material. Standard tests and methods that are used to specify and validate mechanical properties of many types of materials have been refined over the years. One organization that has established specific methods of testing techniques is the *American Society for Testing and Materials (ASTM)*. Some mechanical properties include:

- *Strength:* the capacity of a material to resist stress without excessive deformation or fracture.
- *Elasticity:* the capability of a material to return to its original size and shape after removal of a force or stress.
- *Stiffness:* the capacity of a material to resist deformation, such as resistance against bending or stretching.
- *Ductility:* the capacity of a material to undergo large deformation without failure.
- *Malleability:* the capability of a material to be shaped by compression without failure, such as by hammering or by pressure from rollers.
- *Hardness:* the capacity of a material to resist indentation or abrasion.
- *Resilience:* the capability of a material to recover its shape without being deformed permanently.

- *Toughness:* the capacity of a material to absorb energy from impact without fracture.
- *Creep:* that property of a material under constant stress to deform slowly but progressively over a long period of time.

In this chapter the focus is on the properties of strength, elasticity, and stiffness. The remaining mechanical properties, although important in design, are not covered, due to their complex nature. The interested reader is referred to the many strength of materials texts that are readily available.

6.2 STRESS

As an external load or force is applied to a body, the material from which the body is composed resists that force. If the external force increases, so does the internal resisting force that develops in the material. If the external force was removed, the internal resisting force would no longer remain. This capacity to develop an internal resisting force is the characteristic of a material that gives it strength. A stronger material has the capacity to develop a greater internal resisting force before failing compared with a weaker material.

In engineering, a need exists to define the effect that an external force has on a material when acting upon it. The term *stress* is used to describe the internal resisting force that develops as a result of applying an external force to a body. *Stress* (σ) is defined as the external force acting on the body divided by the cross-sectional area of the body:

$$\text{stress, } \sigma = \frac{F}{A}$$

Force (F) is expressed in units of pounds, kilopounds or kips (1000 lb), tons (2000 lb), and newtons (N). The resisting area (A) is generally expressed in square inches or square meters. In the U.S. Customary System, stress is usually expressed in units of pounds per square inch (lb/in^2 or psi) or kips per square inch (ksi). In the SI system, stress is expressed in units of pascal (Pa). The pascal is a compound unit that is comprised of the SI base units of newtons and meters in the form of newtons per square meter (N/m^2).

When designing components or members of different cross sections, stress analysis averages the force that is applied through the cross-sectional area of the component being analyzed. It permits use of the same design constraints pertaining to a material's strength. For example, a short 1 in $\times$ 1 in bar carrying a load of 10 000 lb develops a stress of 10 000 lb/in^2: $\sigma = F/A = 10\,000$ lb/1 in^2 = 10 000 lb/in^2. On the other hand, a 5 in $\times$ 5 in bar supporting the same 10 000-lb load develops a stress of only 400 lb/in^2: $\sigma = F/A = 10\ 000\ \text{lb}/25\ \text{in}^2 = 400\ \text{lb/in}^2$. The larger bar has the same 10 000-lb force distributed over a much greater area, so the stress developed by the larger bar is significantly less than the stress that develops in the smaller bar.

Steel can safely develop greater stresses than wood, concrete, and most plastics. Steel can easily withstand a 10 000 lb/in^2 stress. A 1 in $\times$ 1 in bar made of steel can safely carry the 10 000-lb load described above. The designer can safely specify this bar because its material can withstand the development of the anticipated stress for a 1 in $\times$ 1 in cross-sectional size. On the other hand, wood can only safely handle stresses in the range 500 to 1000 lb/in^2. If the member carrying the 10 000-lb load is to be fabricated from wood, a cross section of 5 in $\times$ 5 in is needed.

Two basic classifications of simple stress exist: normal stress and shearing stress. *Normal stress* exists when a force acts perpendicular or normal to the stressed area under consideration. Normal stresses may be either tensile or compressive in nature. A compressive stress shrinks the material on which it acts, and a tensile stress stretches the material on which it acts. *Shearing stress* occurs when a force acts parallel to the stressed area. A shearing force attempts to slice through the area under consideration. The common types of stress are illustrated in Figure 6.1.

Other types of stress exhibit behaviors similar to a single basic stress or act as a combination of these stresses. A twisting shaft or axle develops a *torsion* stress. Similar to a shearing stress, a torsional stress is caused by twisting that develops in the material. Torsional stress is covered in Chapter 8. A beam under load bends, causing it to develop bending stress. *Bending stress* is a combination of tensile and compressive stresses. A beam under load also experiences shearing stresses; vertical

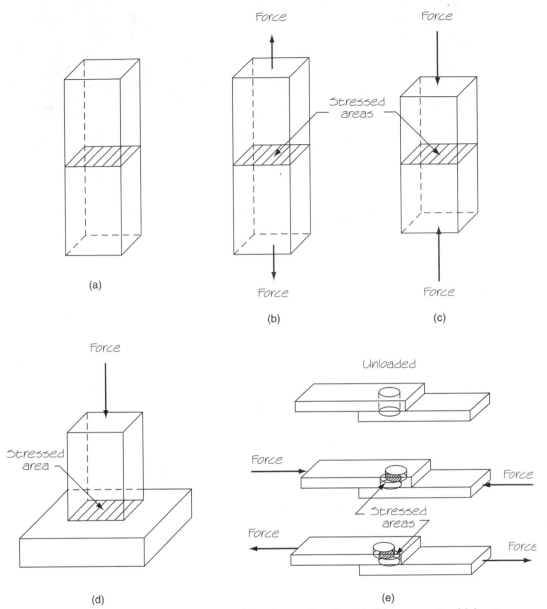

FIGURE 6.1 Unloaded area (a) and examples of tensile (b), compressive (c), bearing (d), and shear (e) stresses. The areas used to calculate each type of stress are hatched.

shearing stresses as the beam tends to drop through its supports and horizontal shearing stresses as horizontal beam fibers tend to slip past one another. The types of stresses in beams are more complex and are discussed in Chapters 11 and 12.

6.3 TENSILE STRESS

Tensile stress occurs when a pair of forces pull on the body under study and tend to stretch or elongate the material from which it is composed. Tensile stress develops as the molecular structure of the material resists being pulled apart. The stressed area is generally considered to be the cross-sectional area or plane that is lying perpendicular to the tensile forces, as shown in Figure 6.1b. A material undergoing tensile stress is said to be in *tension*.

■ EXAMPLE 6.1

A 1 in × 1 in square steel bar is used to hang a load of 10 000 lb, as shown in Figure 6.2. Determine the average tensile stress that develops in the bar.

Solution

$$\sigma = \frac{F}{A} = \frac{10\ 000\ \text{lb}}{1\ \text{in}^2} = 10\ 000\ \text{lb/in}^2$$

Note that the tensile stress developed by the steel bar in Example 6.1 is independent of length; that is, the stress throughout the bar length is assumed equal to 10 000 lb/in². Since the bar is being pulled apart, stress is tensile in nature. The cross-sectional area of the bar was used in the analysis since the force is tensile in nature. The area (A) used in the tensile stress calculation must be perpendicular to the force applied.

■ EXAMPLE 6.2

A 10-mm wire is used to hang a 10-N load, as shown in Figure 6.3. Determine the average tensile stress that develops in the wire.

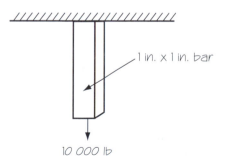

FIGURE 6.2 Illustration for Example 6.1.

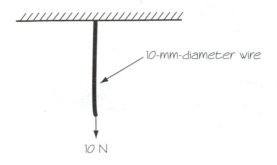

FIGURE 6.3 Illustration for Example 6.2.

10-mm-diameter wire

10 N

Solution

$$\frac{10 \text{ mm}}{1000 \text{ mm/m}} = 0.010 \text{ m}$$

$$A_{\text{rivet}} = \frac{\pi D^2}{4} = \frac{\pi (0.010 \text{ m})^2}{4} = 0.000\ 0785 \text{ m}^2$$

$$\sigma = \frac{F}{A} = \frac{10}{0.000\ 078\ 5 \text{ m}^2} = 127\ 388 \text{ Pa} = 127.4 \text{ kPa}$$

6.4 COMPRESSIVE STRESS

Compressive stress occurs when a force pushes on a body and tends to compress or shorten it. These stresses are produced as a material's molecular structure resists being compressed. A material undergoing compressive stress is said to be in *compression.* The area (A) used in the compressive stress calculation is the cross-sectional area or plane that is perpendicular to the forces causing the material to be compressed, as shown in Figure 6.1c.

Bearing stress is a type of compressive stress that affects the external surface of a body when it presses against another body, as shown in Figure 6.1d. It differs from compressive stress in that it is not an internal stress but a contact stress between two bodies or members. The area under consideration in bearing stress analysis is the contact area where the force is transferred from one body to the next, usually where both bodies are in contact with one another. A building footing resting on soil and a beam end transferring its load to a plate develop a bearing stress at the contact area between the two bodies.

■ EXAMPLE 6.3

A short post with a 40 mm × 80 mm cross section carries a gravity load of 7500 N, as shown in Figure 6.4. Determine the compressive stress that develops in the post.

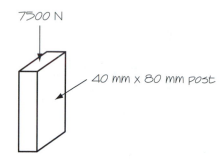

FIGURE 6.4 Illustration for Example 6.3.

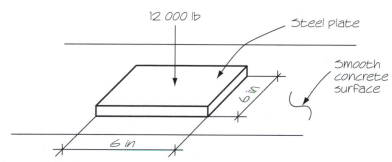

FIGURE 6.5 Illustration for Example 6.4.

Solution

$$\sigma = \frac{F}{A} = \frac{7500 \text{ N}}{0.040 \text{ m} \cdot 0.080 \text{ m}} = 2\ 343\ 750 \text{ N/m}^2 = 2.34 \text{ MPa}$$

Note: A N/m^2 is a pascal and a megapascal (MPa) is 1 000 000 Pa.

■ EXAMPLE 6.4

As shown in Figure 6.5, a column exerts a force on a 6 in × 6 in steel bearing plate . The plate rests on a smooth concrete surface. The plate transmits a 12 000-lb load to the concrete. Determine the bearing stress.

Solution

$$\sigma = \frac{F}{A} = \frac{12\ 000 \text{ lb}}{6 \text{ in} \cdot 6 \text{ in}} = 333 \text{ lb/in}^2$$

Since the post in Example 6.3 supports a gravity load acting downward on the member, the stress is compressive in nature. Again, the cross-sectional area used in the analysis is perpendicular to the force applied. In Example 6.4, bearing stresses developed at the steel and concrete surfaces in contact. The bearing area is that area where both bodies are in contact with one another.

6.5 SHEAR STRESS

Shear stress differs from normal (tensile and compressive) stresses. The forces causing shear stress are offset and act parallel to the direction of the stressed area (called the *shear plane*). When forces tend to slice through a material, a shear stress is created. Tin snips, scissors, and a hole punch apply a shear stress to the material they are cutting. A bolt fastening two metal straps develops shear stresses if the straps are pushed together or pulled apart (Figure 6.1e). The cross-sectional area of the bolt is taken as the stressed area.

■ EXAMPLE 6.5

A $\frac{1}{2}$-in-diameter rivet is used to join two steel bars as illustrated in Figure 6.6. Once fastened, the bars are pulled apart with a force of 500 lb. Determine the stress that develops in the rivet.

Solution

$$A_{\text{rivet}} = \frac{\pi D^2}{4} = \frac{\pi(0.50 \text{ in})^2}{4} = 0.1963 \text{ in}^2$$

$$\sigma = \frac{F}{A} = \frac{500 \text{ lb}}{0.1963 \text{ in}^2} = 2546 \text{ lb/in}^2$$

The shear area in Example 6.5 is parallel with the force applied. It is equivalent to the cross-sectional area of the rivet shank, which is equal to the area of a circle. The formula for the area of a circle (A) is $\pi D^2/4$, where D is the diameter. Also, had the bars been pushed together (without bending), the shear stress developed in the rivet would have been the same.

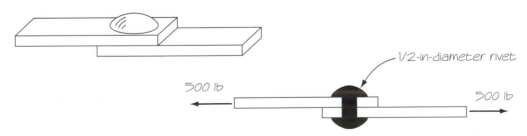

FIGURE 6.6 Illustration for Example 6.5.

In the real world, structural members and components have distinctive shapes that are not simple. Even simple cross sections such as the bars used in many of the previous examples become more complex when a hole is drilled through them. In stress analysis the designer must ascertain where stress is likely the greatest and then analyze the body accordingly. This cross section is usually the section in the member where the cross-sectional area of the stress plane is smallest.

■ EXAMPLE 6.6

A round tapered bar is used to hang a load of 7.5 N. The round bar has a 50-mm diameter at the top end and tapers uniformly to 25 mm at the bottom end (Figure 6.7). Determine the maximum tensile stress that develops in the bar.

Solution The largest average stress occurs at the smallest cross-sectional area, at the smallest diameter of a tapered bar:

$$25 \text{ mm} = 0.025 \text{ m}$$
$$A = \frac{\pi D^2}{4} = \frac{\pi (0.025 \text{ m})^2}{4} = 0.000\ 491 \text{ m}^2$$
$$\sigma = \frac{F}{A} = \frac{7.5 \text{ N}}{0.000\ 491 \text{ m}^2} = 15\ 750 \text{ N/m}^2 \text{ Pa} = 15.75 \text{ kPa}$$

In Example 6.6 the 25-mm end is where the cross-sectional area of the tapered bar is the smallest. Maximum stress occurs where the cross-sectional area is the smallest. It is at this end of the bar that the stress will be the greatest.

■ EXAMPLE 6.7

A 25 mm × 25 mm square steel bar is used to hang a load of 7500 N, as shown in Figure 6.8a.

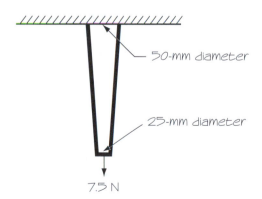

FIGURE 6.7 Illustration for Example 6.6.

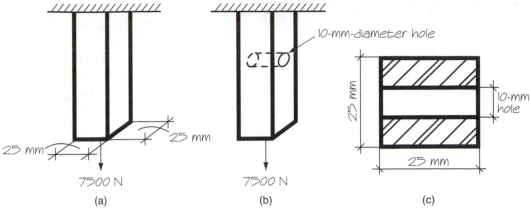

FIGURE 6.8 Illustrations for Example 6.7.

(a) Determine the average tensile stress that develops in the solid portion of the bar.

(b) A 10-mm hole is drilled perpendicularly through the 25 mm × 25 mm square steel bar as shown in Figure 6.8b. Determine the maximum tensile stress that develops in the bar.

Solution

(a) $A = (25 \text{ mm} \cdot 25 \text{ mm}) = 625 \text{ mm}^2 = 0.000\ 625 \text{ m}^2$

$$\sigma = \frac{F}{A} = \frac{7500 \text{ N}}{0.000\ 625 \text{ m}^2} = 12\ 000\ 000 \text{ N/m}^2 = 12\ 000\ 000 \text{ Pa} = 12 \text{ MPa}$$

(b) Maximum tensile stress occurs at the hole with the cross section as shown in Figure 6.8c.

$$A = (25 \text{ mm} \cdot 25 \text{ mm}) - (10 \text{ mm} \cdot 25 \text{ mm}) = 375 \text{ mm}^2 = 0.000\ 375 \text{ m}^2$$

$$\sigma = \frac{F}{A} = \frac{7500 \text{ N}}{0.000\ 375 \text{ m}^2} = 20\ 000\ 000 \text{ N/m}^2 = 20 \text{ MPa}$$

Maximum stress develops where the cross-sectional area is the smallest. In Example 6.7b, the hole decreases the cross-sectional area of the bar and thus the material in which stress develops. The stressed area is smallest at this section, so maximum stress occurs at this point (Figure 6.8c). If the bar is made of a material of consistent strength, this section is the weakest point in the bar. If stresses exceed the strength of the material, the bar will probably fail at this point.

In allowable stress design, the engineer designs the member so that the anticipated maximum stress does not exceed the stress at which the material deforms excessively or fails. With observation and intuition, the engineer carefully evaluates the physical configuration and loading of the member to determine what type of stresses will develop and how the member may be modified to improve its strength against these stresses. If high stresses are anticipated, the engineer may revise the design by increasing the cross-sectional area where stress is the greatest or by changing to a stronger material.

6.7 AVERAGE STRESS VERSUS REAL STRESS

Although it may appear that calculation of stress is fairly simple, the analysis techniques described above actually determine the *average stress*. An abrupt change in cross section radically changes the distribution of the force and thus the stresses at that section. Therefore, the assumption of average stress and its use in analysis may be somewhat misleading. For example, concentrations of stress typically occur above and below rivet or bolt holes positioned perpendicular to the force. A drastic change in cross section also causes a concentration of stress. Furthermore, shearing and bearing stresses frequently occur unevenly because the force is not distributed uniformly across the stress plane. Such concentrations exceed the average stress.

■ EXAMPLE 6.8

A 12-in-long member is used to hang a load of 15 000 lb as shown in Figure 6.9. For the first 6 in of member length, the member is 1.000 in. in diameter. At the center of its length, the member is machined so that its diameter changes abruptly. The other half of the member is 0.500 in. in diameter. Determine the maximum tensile stress that develops in the member.

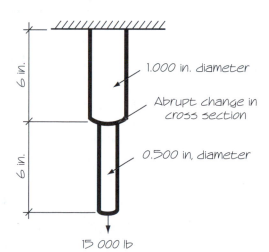

FIGURE 6.9 Illustration for Example 6.8. Concentrations of stress develop at the abrupt change in cross section; in this case, at the center of the length of the member.

Solution

$$A = \frac{\pi D^2}{4} = \frac{\pi (0.500)^2}{4} = 0.1963 \text{ in}^2$$

$$\sigma = \frac{F}{A} = \frac{15\,000 \text{ lb}}{0.1963 \text{ in}^2} = 76\,414 \text{ lb/in}^2$$

In Example 6.8, the maximum average stress of 76 414 lb/in² approaches the maximum strength of a common type of steel. It appears that maximum stress occurs throughout the narrow half of the member. If failure occurs, one would presume that it would occur anywhere in the 0.500-in-diameter portion of the member. In practice, however, concentrations of stress develop at the abrupt change in cross section, in this case at the center of the length of the member. These concentrations will exceed the calculated maximum average stress of 76 414 lb/in². The member will probably fail at the abrupt change in cross section. A gradual transition from the larger diameter to the smaller diameter would significantly reduce these concentrations and make for a stronger member.

While further discussion regarding stress concentrations is beyond the intended breadth of this book, it is an important concern with potentially severe consequences. Due to this concern, components are frequently designed such that their anticipated maximum average stress is significantly lower than the stress at which the material was found to fail in testing; that is, they are designed with a built-in safety factor. Application of factors of safety is covered in Chapter 10.

EXERCISES

6.1 A long, round bar with a cross section 10 mm in diameter is used to hang a pipe. The maximum axial tensile load carried by the bar is anticipated to be 1.25 kN. Determine the maximum stress that develops in the bar, in MPa.

6.2 A long steel bar with a cross section of 20 mm × 20 mm is used to carry an axial load of 9.5 kN. Determine the stress that develops in the bar, in MPa.

6.3 A 6-in-diameter concrete cylinder carries an axial compressive load of 141 300 lb just before failure. Determine the compressive stress in the concrete just before failure, in lb/in².

6.4 A 100 mm × 150 mm steel plate rests on a smooth concrete surface. The plate transfers an axial compressive load of 44 kN to the concrete. Determine the bearing stress, in MPa.

6.5 As shown, four $\frac{1}{8}$-in-diameter rivets per linear foot of seam secure the seam of two layers of sheet metal. The rivets are positioned to resist a shearing effect that develops in the seam.

(a) The maximum tensile force that develops is anticipated to be 350 lb per linear foot of seam. Find the maximum tensile stress that develops in the rivets, in lb/in².

(b) The maximum shearing force that develops is anticipated to be 200 lb per linear foot of seam. Find the maximum shear stress that develops in the rivets, in lb/in².

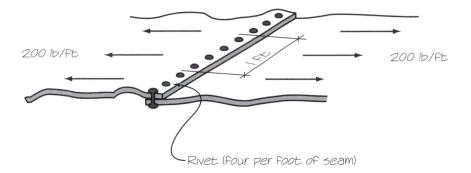

200 lb/ft   1 ft   200 lb/ft

Rivet (four per foot of seam)

6.6 Calculate the stress that develops in brass rods of the following diameters when carrying an axial tensile load of 1750 N:

(a) 5 mm

(b) 7.5 mm

(c) 10 mm

(d) 15 mm

(e) Brass can safely develop a stress of up to 27.6 MPa. Identify the smallest-diameter rod of those diameters indicated above that can safely support the 1750-N load.

6.7 A short wooden post must be used to carry an axial compressive force of 60 kN. The compressive stress of the wood should not exceed 2.0 MPa. Assume that the load is equally distributed across the cross section of the post. Determine the dimensions of the smallest square cross section to the closest 10-mm increment that can carry this load.

6.8 As shown, a steel bearing plate must transfer an axial load of 10 000 lb to a concrete pad. Stress in the steel must not exceed 20 000 lb/in^2. Compressive stress in the concrete must not exceed 400 lb/in^2. Assume that the load is equally distributed across the plate. Find the minimum required area of the plate so that these stresses are not exceeded.

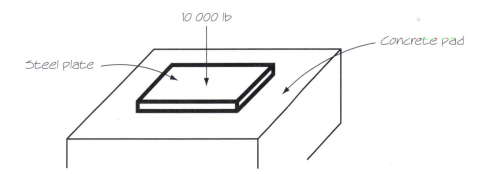

10 000 lb

Steel plate

Concrete pad

6.9 As shown, a pipe serves as a short column. The 4-in-diameter steel pipe has 6 in × 6 in steel plates welded to its ends. The top column plate carries a $5\frac{1}{8}$-in-wide wood beam that exerts an axial load on the column of 11 800 lb, including the weight of the beam. The bottom plate bears on a concrete pad. The weight of the column is 53.5 lb. Compressive stress in the steel must not exceed 20 000 lb/in^2. In wood and concrete, compressive stress must not exceed 400 lb/in^2.

(a) Determine the bearing stress that develops between the beam and the top plate, in lb/in².

(b) Determine the bearing stress that develops between the concrete pad and the bottom plate, in lb/in².

(c) Is the maximum stress of the wood exceeded?

(d) Is the maximum stress of the concrete exceeded?

(e) Determine the maximum axial compressive load the column can support without exceeding the maximum stresses in the wood and concrete.

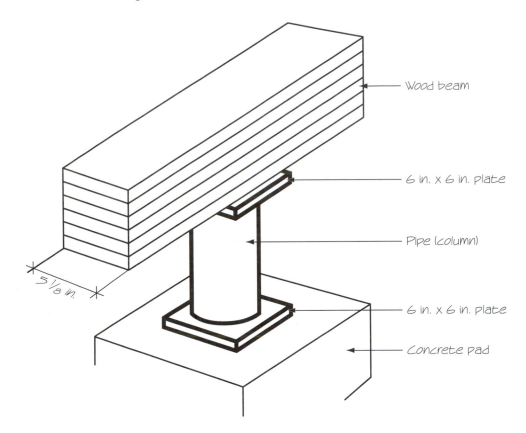

6.10 An anchor bolt connection secures a wood plate to a foundation wall as shown. The $\frac{3}{8}$-in-diameter steel anchor bolt has a minor thread dimension of 0.31 in diameter. A steel washer with a $\frac{7}{8}$-in outside diameter and a $\frac{3}{8}$-in-diameter hole is placed between the bolt nut and the wood plate before the nut on the bolt is tightened. Compressive stress must not exceed 20 000 lb/in² in the steel and 625 lb/in² in the wood.

(a) Determine the maximum compressive force that can be exerted on the wood by the washer.

(b) Determine the tensile stress that develops in the threaded area of anchor bolt if the connection is tightened to the maximum compressive force.

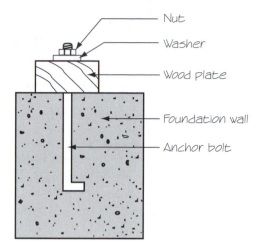

6.11 A steel strap is secured to a steel flange with a single steel pin as shown. The strap is 15 mm thick and is rectangular in section profile throughout. The pull on the strap is 125.0 kN. Shearing stress in the steel must not exceed 100 MPa. Tensile stress in the steel must not exceed 165.5 MPa.

  (a) Determine the minimum diameter of the steel pin to the closest 2.0-mm increment (i.e., 20 mm, 22 mm, 24 mm, etc.) without exceeding the maximum specified shear stress.

  (b) Determine the minimum width of the steel bar at cross section A–A without exceeding the maximum specified tensile stress.

  (c) Using the diameter of the pin found in part (a), determine the minimum width of the steel bar at cross section B–B without exceeding the maximum specified tensile stress.

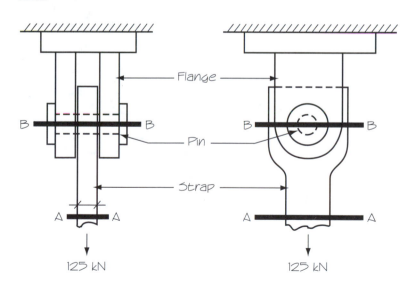

6.12 A steel guy cable hangs vertically to support a roof system at the lower end of the cable. The 240-ft-long wire has a net cross section area of 0.75 in² and weighs 1.50 lb/ft. Tensile stress in the cable must not exceed 20 000 lb/in².

  (a) Determine the maximum tensile stress that develops in the cable due to its own weight.

  (b) Determine the maximum load that can be carried at the lower end of the cable without exceeding the maximum specified tensile stress.

6.13 A 4-m-long rigid beam carrying a 10-kN/m uniformly distributed load is supported by a pinned connection and cable as shown. The beam weighs 2.0 kN. Tensile stress in the steel cable must not exceed 150 MPa. Neglecting the weight of the cable, determine the minimum diameter of the steel cable to the closest 1.0-mm increment (i.e., 10 mm, 11 mm, 12 mm, etc.) without exceeding the maximum specified tensile stress.

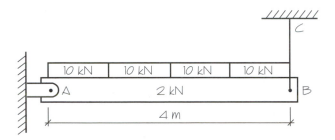

6.14 A 10-ft-long rigid beam is supported by a steel rod and pinned connection as shown. It carries a 10-kip/ft uniformly distributed load across its entire length. The beam weighs 120 lb/ft. Tensile stress in the steel rod must not exceed 20 000 lb/in². Neglecting the weight of the cable, determine the minimum diameter of the steel rod to the closest $\frac{1}{8}$-in increment (i.e., $\frac{1}{8}$-in, $\frac{1}{4}$-in, $\frac{3}{8}$-in, etc.) without exceeding the maximum specified tensile stress.

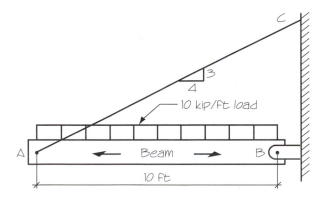

Chapter 7
DEFORMATION
AND STRAIN

Observation and experimentation have shown that a tensile force applied to a body produces an increase in the length of the body. A compressive force applied to a body produces a reduction in length of the body. As a body stretches, its cross section contracts. Conversely, as a body is compressed, its cross section expands. In this chapter deformation and strain that result from tensile and compressive stress are explored.

7.1   DEFORMATION

A body that is subjected to an external force becomes deformed. *Deformation* occurs when the shape of a body is altered due to an applied force. Deformation can be elastic; the distortion of the body is removed and the body returns to its original shape when the stress is relieved. Deformation can also be permanent; when the applied force is removed and the stress is relieved, the deformation remains. A straight length of wire becomes permanently deformed when it is shaped into a paper clip. When a few sheets of paper are slipped into a paper clip and then removed, the paper clip behaves elastically; it returns to its original coiled shape. However, several sheets of paper forced into a paper clip will permanently deform the coil; the coil becomes plastically deformed.

Under tensile stress, the interatomic distance of the molecules of the material increase longitudinally. A tensile stress stretches the material and displays itself physically by an elongation or lengthening of the body. Conversely, a compressive stress tends to compress the material. Under tensile stress, shortening of the longitudinal interatomic distance of molecules of the material results. Compressive stress reduces the longitudinal dimension of the body. Elongation or contraction of a body in the direction of the applied force is called *longitudinal deformation.*

An elongation of a body due to a tensile stress is always accompanied by a reduction of the lateral cross-sectional area of the body. For example, as a rubber band is stretched, a reduction in cross-section is visibly evident. Similarly, a longitudinal compression of a body results in a lateral increase in cross-sectional area.

A *lateral deformation* is an elongation or contraction of a body cross section. Longitudinal deformation of a body due to an applied force typically results in lateral deformation unless the body is constrained laterally.

7.2 LONGITUDINAL STRAIN

Longitudinal strain (ϵ) is the relationship between longitudinal deformation under stress and the original longitudinal dimension when the body is not under stress; it describes linear deformation. Such strain is expressed as

$$\text{longitudinal strain, } \epsilon = \frac{\Delta L}{L_o} \quad \text{or} \quad \epsilon = \frac{\Delta L}{L_o} \cdot 100\%$$

As shown in Figure 7.1, L_o is the original length of the material before it is under load, and ΔL is the longitudinal deformation (change in length) as a result of a specified force. Longitudinal strain may be expressed in the form of a percentage. In this case, simply multiply ϵ by 100, as shown in the second equation above.

As discussed previously, all materials subjected to external forces develop stress. A consequence of these stresses is deformation, and thus, strain. The amount of deformation developed by a material is related to the stiffness of the material and the stresses developed by the material: that is, steel as a stiff material will deform much less than a synthetic rubber.

■ EXAMPLE 7.1

As illustrated in Figure 7.2, an ordinary drafting eraser is 2.00 in long. When it is placed between two fingers and squeezed, it deforms. Under the force of

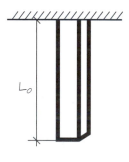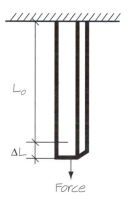

FIGURE 7.1 Longitudinal strain (ϵ) is the relationship between longitudinal deformation from stress and the original dimension when the body is not under stress. L_o is the original length of the material before the force is applied and ΔL is the longitudinal deformation as a result of the applied force.

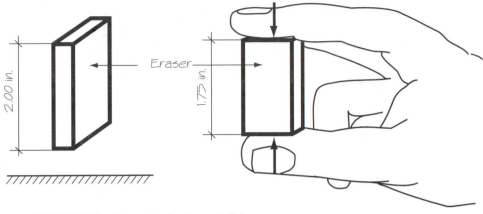

FIGURE 7.2 Illustration for Example 7.1.

The Mathematical Symbol Delta (Δ)

The Greek capital letter delta (Δ) is a mathematical term that means "change in." For example, if your wage changed from $10 to $12 per hour, the change in your wages (Δ$/hr) would be expressed as

$$\Delta\$/hr = \$12/hr - \$10/hr = \$2/hr$$

---

the fingers, it is 1.75 in long. Determine the longitudinal strain under this force.

Solution

$$L_o = 2.00 \text{ in}$$
$$\Delta L = 2.00 \text{ in} - 1.75 \text{ in} = 0.25 \text{ in}$$
$$\epsilon = \frac{\Delta L}{L_o} = \frac{0.25 \text{ in}}{2.00 \text{ in}} = 0.125 \text{ in/in}$$

or

$$\epsilon = \frac{\Delta L}{L_o} \cdot 100 = \frac{0.25 \text{ in}}{2.00 \text{ in}} \cdot 100\% = 12.5\%$$

■ EXAMPLE 7.2

A steel bar has an original length of 3.000 m. When a 2250-N load is hung from the bar, it elongates (stretches) an additional 10.516 mm, as shown in Figure 7.3. Determine the longitudinal strain under this load.

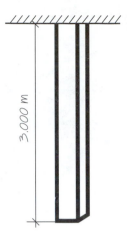
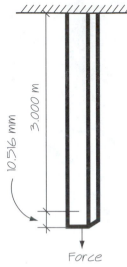

FIGURE 7.3   Illustration for Example 7.2.

Solution

$$L_o = 3.000 \text{ m}$$
$$\Delta L = 10.516 \text{ mm} = 0.010\,516 \text{ m}$$
$$\epsilon = \frac{\Delta L}{L_o} = \frac{0.010\,516 \text{ m}}{3.000 \text{ m}} = 0.003\,505 \text{ m/m}$$

or

$$\epsilon = \frac{\Delta L}{L_o} \cdot 100 = \frac{0.010\,516 \text{ m}}{3.000 \text{ m}} \cdot 100\% = 0.350\%$$

In Example 7.2, when the tensile load was applied, the bar stretched 0.003 505 m per meter of bar length. It stretched 0.350% of its original length. The longitudinal interatomic distance of the molecules in the bar increased by 0.350%, on average, throughout the bar.

7.3   ELASTICITY

A material is *elastic* if it is capable of full recovery of its original shape upon removal of an external force applied to it. As a material, rubber is popularly viewed as being elastic: an elastic rubber band stretches when a force is applied to it and returns to its original shape once the force is removed. The rubber band is elastic because it returns to its original shape when unloaded, not because it stretches easily.

Most materials are somewhat elastic. Technically, both steel and glass are elastic; that is, they return to their original shape once stresses are removed. The property of elasticity may not be as evident with a naked eye as it is in the case of the rubber band. With most materials it is difficult to observe except under careful testing and observation, where measurements of 0.0001 in (0.0025 mm) may be required.

In Example 7.2 it was found that under tensile stress the steel bar became deformed; it stretched when the load was applied. By simply unloading the bar and removing the stresses, the bar would return to its original length. The deformation in the bar exists only when it is carrying load. The bar behaves much like a spring: it elongates when a load is applied and returns to its original shape when the load is removed. This type of longitudinal strain is referred to as elastic deformation. *Elastic deformation* occurs when any deformation in the body is removed by unloading the force and removing the stress.

7.4 HOOKE'S LAW

In 1678, Robert Hooke (1635–1703), an English physicist, inventor, and mathematician, discovered through experimentation that strain is proportional to the total force applied to a specimen. This physical phenomenon is known formally as *Hooke's law*. It implies that stress (σ) is proportional to longitudinal strain (ϵ) based on a *constant of proportionality* (E) that differs with type of material:

$$\sigma \propto E\epsilon$$

With Hooke's discovery that stress is proportional to longitudinal strain, it can be said that twice the stress produces twice the longitudinal strain, three times the stress causes three times the longitudinal strain, and so on. According to Hooke's law, had the load on the bar in Example 7.2 been doubled, deformation (elongation) and longitudinal strain would have been twice that associated with the original load.

When relating the proportionality of stress and strain through the $\sigma \propto E\epsilon$ expression, E is a constant of elastic proportionality called the modulus of elasticity. It will be found that proportionality, and thus E, applies to the $\sigma \propto E\epsilon$ relationship only when a material behaves in an elastic and linear manner. It will also be found that equal increments of stress produce equal increments of longitudinal strain *only to a certain point*. This point is at the value of stress known as the *proportional limit*. At stresses above the proportional limit, proportionality between stress and strain does not or is assumed not to occur. The limitation of elastic behavior is discussed below.

Materials have limited physical stiffness. To preserve the elastic behavior of a material, stress developed by the material must not exceed a certain limit. The stress at which a material no longer behaves truly elastically is called the *elastic limit*. If a material is stressed beyond its elastic limit, it will return only partially to its original size and shape once the force is removed; that is, it becomes permanently deformed.

A material can be assumed to act like an ordinary spring. If a force stretches a spring to a stress below its elastic limit, the spring will snap back to its original shape when unloaded. Doubling the force doubles the elastic strain. However, if the spring was stretched so that its stress is beyond its elastic limit, it would only recover some of its original shape once the load was removed. The spring would not return to its original shape. It is no longer fully elastic and has become permanently deformed.

Permanent deformation occurs when a material is stressed beyond its elastic limit. It is like a spring that is stressed too far. When a material is stressed beyond its elastic limit, it is stressed beyond the *elastic range* and into the *plastic range*. The material springs back only partially to its original shape; it is no longer fully elastic. The result is some permanent deformation; the material gains *permanent set*.

Stress is frequently assumed to be proportional to strain within the elastic range; it is assumed that proportionality between stress and strain exists and thus Hooke's law applies. Laboratory testing roughly confirms the phenomenon of proportionality between stress and strain with some steels. But with other materials, proportionality between stress and strain does not apply directly. Proportionality is a good approximation but not exact. In fact, engineers and designers use the theory behind stress-strain proportionality in design and use of materials.

Generally, structural members and machine components are designed so they are not stressed beyond their elastic limit or what is assumed to be their elastic limit. They are designed to function so they only elastically deform; so that they do not become plastically deformed. In this design procedure, members are designed such that under the maximum anticipated loading condition they do not develop stresses above the elastic limit. Additionally, to ensure safety, these members are designed so that the maximum anticipated stresses are well below the elastic limit.

In some instances, it may be desirable to stress a material beyond its elastic limit so that the material becomes plastically deformed; that is, it remains deformed after the force is removed. For example, during manufacture, the wire used to produce a paper clip or a coat hanger is stressed well into the plastic range to create its final shape. Corrugated steel decking is formed from a flat metal sheet by stressing the material into the plastic range. It is only when the steel sheet is stressed beyond its elastic limit that it begins to retain the shape of the decking. So, sometimes stressing the material too far is just enough!

7.6 MODULUS OF ELASTICITY

The relationship of stress to longitudinal strain below the proportional limit is referred to as the *modulus of elasticity* (*E*):

$$\text{modulus of elasticity,} \quad E = \frac{\sigma}{\epsilon}$$

The modulus of elasticity, *E,* is a constant of proportionality between stress and strain, a measure of elastic proportionality. It can be thought of as a measure of a material's stiffness below the proportional limit. A low modulus of elasticity relates to a low degree of stiffness in the material. Materials with a high modulus of elasticity display a high degree of stiffness. Engineers and scientists formally refer to the modulus of elasticity as *Young's modulus.*

Values of moduli of elasticity are given in Tables A.3 and A.4 in Appendix A. These values are for materials at room temperature. The modulus of elasticity will vary with change in temperature of the material; for example, the modulus of elasticity of steel decreases with change in temperature of the steel. Steel is a relatively stiff but elastic material with a modulus of elasticity of 200 000 MPa (29 000 000 lb/in^2 or 29 000 kip/in^2). Aluminum has a modulus of elasticity of about 69 000 MPa (10 000 000 lb/in^2 or 10 000 kip/in^2). Aluminum is not as stiff as steel. The relationship between the moduli of the two materials is 2.9: 200 000 MPa/69 000 MPa = 2.9. This relationship suggests that an aluminum bar under a specific loading condition will deform (elongate) 2.9 times greater than the same bar made of steel under the same load; this is true because the modulus of elasticity of steel is 2.9 times greater than aluminum and steel is a stiffer material. It is also true that under the same loading conditions an aluminum beam will deflect (bend) about three times more than a steel beam of the same cross section. Of course, a limitation of this relationship is that the material not be stressed beyond its proportional limit and certainly not beyond its ultimate strength.

Many materials become less rigid with a temperature increase. Therefore, the moduli of elasticity of many materials such as metals and thermoplastics decreases with temperature. Values presented in this book are for materials at approximate room temperatures.

7.7 LONGITUDINAL DEFORMATION

The relationship between stress (*s*), longitudinal strain (ϵ), and the modulus of elasticity (*E*) provides for a method of estimating the longitudinal deformation (ΔL) of a body under load:

$$E = \frac{\sigma}{\epsilon}$$

Stress, $\sigma = F/A$ and longitudinal strain, $\epsilon = \Delta L/L_o$; therefore,

$$E = \frac{F/A}{\Delta L /L_o}$$

By algebraically rearranging terms and solving for ΔL, we obtain

longitudinal deformation, $\Delta L = \dfrac{FL_o}{EA}$

Longitudinal deformation (ΔL) may be estimated based on the applied force (F), original length of the body (L_o), modulus of elasticity of the material (E), and cross-sectional area (A) of the body (Figure 7.1). This expression may be used with reasonable accuracy as long as the body has a constant cross-sectional area and is axially loaded (in true tension or compression), the material is homogeneous, and the stress developed by the material does not exceed its proportional limit.

■ EXAMPLE 7.3

A 1-in-diameter steel ($E = 29\,000$ kips/in^2) bar has an original length of 10.00 ft. It is used to hang a load of 25 kips. Estimate the change in its overall length. Neglect the weight of the bar.

Solution

$$A = \frac{\pi D^2}{4} = \frac{\pi(1 \text{ in})^2}{4} = 0.785 \text{ in}^2$$

$$L_o = 10.00 \text{ ft} \cdot 12 \text{ in/ft} = 120.00 \text{ in}$$

$$\Delta L = \frac{FL_o}{EA} = \frac{25 \text{ kips} \cdot 120.00 \text{ in}}{29\,000 \text{ kips/in}^2 \cdot 0.785 \text{ in}^2}$$

$$= 0.1318 \text{ in}$$

Under a 25-kip load, the steel bar in Example 7.3 would stretch 0.1318 in, about $\frac{1}{8}$ in. Longitudinal deformation (elongation) of the bar is due to the tensile stresses caused by carrying the 25-kip load. It may appear that doubling the load to 50 kips would result in doubling the change in its original length (ΔL) to 0.2636 in, because $2 \cdot 0.1318$ in $= 0.2636$ in. However, at a load of 50 kips, the 1-in-diameter steel has developed a stress of 63.7 kips/in^2: 50 kips/0.785 in$^2 = 63.7$ kips/in^2. This stress is well above the elastic limit of most steels, so the elastic deformation equation of FL_o/EA no longer applies. The equation for elastic deformation can be employed only within the proportional range of a material; that is, it can be used only when stress developed by the material does not exceed its proportional limit.

As discussed earlier in this chapter, the interatomic distance of molecules increases longitudinally in a material under tensile stress. With this increase in longitudinal interatomic distance there is a decrease in lateral interatomic distance. The decrease in lateral interatomic distance presents itself as a thinning of the cross section of the body. Conversely, compressive stress reduces the longitudinal dimension of the body and increases the lateral dimension. A compressive stress decreases the longitudinal interatomic distance of the molecules of the material under stress. With this decrease in longitudinal interatomic distance there is an increase in the lateral interatomic distance. The increase in lateral interatomic distance presents itself as a widening of the cross section of the body. Simply, as a material is stretched, its cross section contracts, and as a material is compressed, its cross section expands.

Lateral strain ($\epsilon_{lateral}$) is the relationship between lateral deformation (expansion or contraction) of the cross section of a body to the original lateral dimension. Lateral strain is expressed as

$$\text{lateral strain, } \epsilon_{lateral} = \frac{\Delta L_{lateral}}{L_{o\,lateral}}$$

or

$$\epsilon_{lateral} = \frac{\Delta L_{lateral}}{L_{o\,lateral}} \cdot 100\%$$

As shown in Figure 7.4, $L_{o\,lateral}$ is the original cross-section length of the material before it is under load, $\Delta L_{lateral}$ is the change in cross section length as a result of a specific force. Lateral strain ($\epsilon_{lateral}$) may be expressed in the form of a percentage. In this case, simply multiply $\epsilon_{lateral}$ by 100, as shown in the second equation above.

■ EXAMPLE 7.4

As shown in Figure 7.5, a plastic drafting eraser has a cross section that measures $\frac{1}{4}$ in by $\frac{3}{4}$ in. When squeezed between two fingers (under compressive load) the eraser height decreases and the eraser cross section expands. Under the force of the fingers, one side of the eraser increases from a dimension of $\frac{3}{4}$ in to $\frac{13}{16}$ in. Estimate the lateral strain.

Solution

$$\Delta L_{lateral} = \tfrac{13}{16} \text{ in} - \tfrac{3}{4} \text{ in} = 0.8125 \text{ in} - 0.75 \text{ in} = 0.0625 \text{ in}$$
$$L_{o\,lateral} = 1\tfrac{1}{2} \text{ in} = 1.50 \text{ in}$$
$$\epsilon_{lateral} = \frac{\Delta L_{lateral}}{L_{o\,lateral}} = \frac{0.0625 \text{ in}}{1.50 \text{ in}} = 0.0417 \text{ in/in}$$

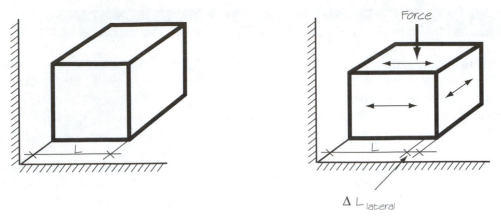

FIGURE 7.4 Lateral strain ($\epsilon_{\text{lateral}}$) results from lateral deformation ($\Delta L_{\text{lateral}}$) of the cross section of a member.

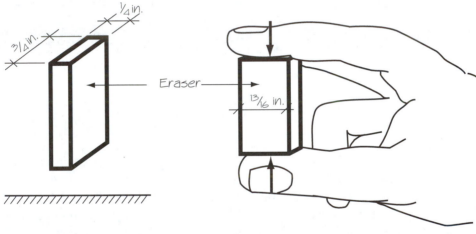

FIGURE 7.5 Illustration for Example 7.4.

7.9 POISSON'S RATIO

A relationship between lateral strain ($\epsilon_{\text{lateral}}$) and longitudinal strain (ϵ) within the elastic range is an assumed constant known as *Poisson's ratio* (μ), defined as

$$\text{Poisson's ratio, } \mu = \frac{\epsilon_{\text{lateral}}}{\epsilon}$$

Poisson's ratio is determined by physical observation through testing, relating elastic deformation (elongation) to change in cross section. Poisson's ratio lies somewhere between 0 and 0.5 for all materials. Thus longitudinal strain (the stretching

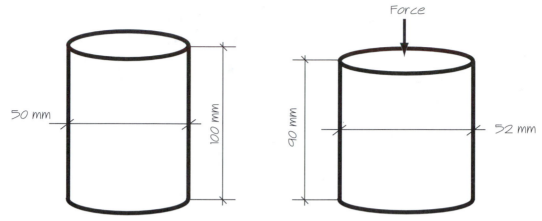

FIGURE 7.6 Illustration for Example 7.5.

or compressing in the direction of the forces) is significantly greater than the resulting lateral strain (thinning or fattening of the cross section). Typically, Poisson's ratio falls between 0.25 and 0.33 for metals, but there are exceptions; for example, μ is 0.42 for gold. Tables A.3 and A.4 in Appendix A list values for selected engineering materials.

■ EXAMPLE 7.5

As shown in Figure 7.6, a 50-mm-diameter elastomeric cylinder is 100 mm tall without a load applied. Under compressive load in a testing apparatus, the cylinder height decreases from 100 mm to 90 mm. As the cylinder length contracts under load, the cylinder diameter increases from 50 mm to 52 mm.
(a) Estimate longitudinal strain.
(b) Estimate lateral strain.
(c) Estimate Poisson's ratio for the elastomeric material.

Solution

(a)
$$L_o = 100 \text{ mm}$$
$$\Delta L = 100 \text{ mm} - 90 \text{ mm} = 10 \text{ mm}$$

$$\epsilon = \frac{\Delta L}{L_o} = \frac{10 \text{ mm}}{100 \text{ mm}} = 0.10 \text{ mm/mm} = 0.10 \text{ m/m}$$

or

$$\epsilon = \frac{\Delta L}{L_o} \cdot 100 = \frac{10 \text{ mm}}{100 \text{ mm}} = \cdot 100 = 10\%$$

(b)
$$L_{o \text{ lateral}} = 50 \text{ mm}$$
$$\Delta L_{\text{lateral}} = 52 \text{ mm} - 50 \text{ mm} = 2 \text{ mm}$$

7.9 Poisson's Ratio

141

$$\epsilon_{lateral} = \frac{\Delta L_{lateral}}{L_{o\,lateral}} = \frac{2\ mm}{50\ mm} = 0.04\ mm/mm = 0.04\ m/m$$

or

$$\epsilon_{lateral} = \frac{\Delta L_{lateral}}{L_{o\,lateral}} \cdot 100\% = \frac{2\ mm}{50\ mm} \cdot 100\% = 4\%$$

(c)
$$\mu = \frac{\epsilon_{lateral}}{\epsilon} = \frac{0.04\ m/m}{0.10\ m/m} = 0.40$$

Poisson's ratio relates lateral strain ($\epsilon_{lateral}$) to longitudinal strain (ϵ):

$$lateral\ strain,\ \epsilon_{lateral} = \mu\epsilon$$

or

$$\epsilon_{lateral} = \mu\epsilon(100\%)$$

7.10 LATERAL DEFORMATION

Poisson's ratio allows the designer to approximate the change in cross section as a result of longitudinally stretching or compressing a material. *Total lateral deformation* ($\Delta L_{lateral}$) may be estimated by

$$lateral\ deformation,\ \Delta L_{lateral} = \mu\epsilon b$$

Lateral deformation ($\Delta L_{lateral}$) may be estimated by applying Poisson's ratio (μ) to longitudinal strain (ϵ) and the original lateral dimension (b). This expression may be used with reasonable accuracy as long as the material is axially loaded (in true tension or compression) and the stress developed by the material does not exceed its elastic limit.

■ EXAMPLE 7.6

A 50-mm-diameter elastomeric cylinder is 100 mm tall. Under compressive load, the cylinder height decreases from 100 mm to 90 mm. Estimate the lateral deformation assuming a Poisson's ratio of $\mu = 0.4$.

Solution

$$L_o = 100\ mm$$
$$\Delta L = 100\ mm - 90\ mm = 10\ mm$$
$$\epsilon = \frac{\Delta L}{L_o} = \frac{10\ mm}{100\ mm} = 0.10\ mm/mm = 0.10\ m/m$$
$$\Delta L_{lateral} = \mu\epsilon b$$
$$= 0.4(0.10\ m/m)\,(50\ mm) = 2\ mm$$

In Example 7.5 it was found that Poisson's ratio for the elastomeric material was approximately 0.40. This value was used in Example 7.6 to approximate the change in lateral dimension.

■ EXAMPLE 7.7

A 6.0000-in-diameter copper cylinder under compressive load develops a longitudinal strain of 0.00 0221 in/in.

(a) Estimate the lateral deformation.

(b) Estimate the lateral dimension under load.

Solution (a) From Table A.3 in Appendix A, for copper, $\mu = 0.333$:

$$\Delta L_{lateral} = \mu \epsilon b$$
$$= 0.333(0.000\ 221\ in/in)\ (6\ in) = 0.000\ 442\ in$$

(b) $L_{under\ load} = L_{o\ lateral} - L_{lateral} = 6.0000\ in - 0.000\ 442\ in = 5.9996\ in$

Because the cylinder in Example 7.7 is under compression, the cylinder cross section increased. Based on the analysis in Example 7.7, the lateral dimension of the 6-in diameter increased by 0.000 442 in under the described load. Had the copper cylinder developed tensile stresses of the same magnitude—that is, if it were being pulled apart instead of being compressed—the lateral dimension of the copper cylinder would have decreased by 0.000 442, to 5.9996 in.

Lateral strain and the resulting lateral deformation is very small. In cases where tolerances are low, lateral strain is considered to be negligible and the associated lateral deformation is neglected. Lateral deformation is an important factor in design when very tight tolerances are required; an engine component that moves freely under little stress may wear excessively under large lateral strain. Excessive wear can cause premature failure if it is neglected in design.

EXERCISES

Longitudinal Strain and Deformation

7.1 Under load, a copper bar measuring 1.00 m long stretches 0.310 mm.
   (a) Determine the strain in the bar, in m/m.
   (b) Determine the strain in the bar, in percent.

7.2 Under load, a 48.00-in-long aluminum bar stretches 0.0650 in.
   (a) Determine the strain in the bar, in in/in.
   (b) Determine the strain in the bar, in percent.

7.3 A 25 mm × 50 mm steel bar is 1.25 m long. It carries a load of 175 kN. Steel has a modulus of elasticity of 200 000 MPa. Estimate how much the bar will stretch. Neglect the weight of the bar.

7.4 A steel surveyor's tape has a cross-sectional area of 0.004 in². To provide adequate tension during measuring, it is stretched with a pull of 16 lb. Steel has a modulus of elasticity of

29 000 000 lb/in². Approximate how much the 100-ft-long tape will stretch when under this tensile force. Neglect the weight of the tape.

7.5 A copper wire measuring 10.00 m in length is designed to support a load. A load of 1750 N will be hung from the end of the wire. The modulus of elasticity of copper is 110 000 MPa. Determine the required diameter of the wire if tensile stress does not exceed 55 MPa and elongation does not exceed 2.5 mm. Neglect the weight of the wire.

7.6 A 24-in rigid horizontal support is hung by two cables as shown. One cable is made of brass (E = 9 000 000 lb/in²) and the other cable is made of steel (E = 29 000 000 lb/in²). Each cable has a diameter of 0.250 in. When hanging without a load, the cables measure 7.50 ft vertically. When a 350-lb load is placed on the center of the horizontal support, find the following:

(a) The elongation in each cable

(b) The change in elevation of points A, B, and C

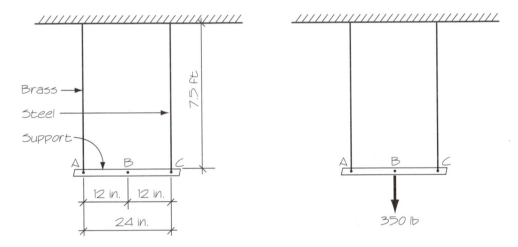

7.7 A 4-m-long rigid beam is supported by a pinned connection and cable as shown. The beam weighs 2.0 kN and hangs horizontally when unloaded. When the beam is not loaded the solid cable is 4.00 m long and has a solid area of 175 mm². Steel used in the cable has a modulus of elasticity of 200 000 MPa. When loaded, the beam carries a 10-kN/m uniformly distributed load. Neglecting the weight of the cable, approximate the following:

(a) The change in length of the cable when the beam is under load

(b) The change in elevation of point B when the beam is under load

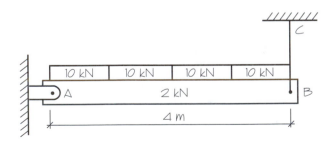

7.8 A 10-ft-long rigid beam is supported by a round steel rod and pinned connection as shown. The beam hangs horizontally when unloaded. The 2.50-in-diameter steel rod is exactly 13'-9" long when connected but before the beam is placed under load. Steel used in the rod has a modulus of elasticity of 29 000 000 lb/in^2. When loaded, the beam carries a 10-kip/ft uniformly distributed load across its entire length. Neglecting the weight of the cable, approximate the change in length of the rod when the beam is placed under load.

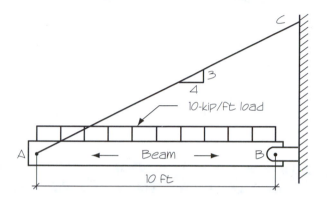

Lateral Strain and Deformation

7.9 A 100.00-mm-diameter solid aluminum ($\mu = 0.340$) component experiences a longitudinal strain of 0.0014 m/m under tensile load.
(a) Estimate the lateral strain experienced by the component.
(b) Estimate the lateral dimension of the component when under tensile load.

7.10 A 1.000-in-diameter solid steel ($\mu = 0.303$) bar is 10.0 ft long. It stretches 0.1300 in under a tensile load.
(a) Estimate the lateral strain experienced by the bar.
(b) Estimate the cross-sectional dimension of the bar when under tensile load.

7.11 A 2.500-in-diameter round steel bar ($E = 29\,000\,000$ lb/in^2 and $\mu = 0.303$) slides tightly through a hole. When not under load, there is a clearance of 0.0001 in on all sides of the bar. A 125-kip compressive load is applied to the end of the bar.
(a) Estimate the cross-sectional dimension of the bar when under load.
(b) Minimum adequate clearance is 0.000 05 in on all sides. Does the bar still have adequate clearance when the load is applied to it?

Chapter 8
TORSIONAL STRESS AND STRAIN

In many cases, building members experience an axial twisting effect known as *torsion*. The concept of torsion and the resulting torsional stress, deformation, and strain are introduced in this chapter.

8.1   TORSION

Torsion occurs when one end of an axlelike body is twisted in one direction along its linear axis while the other end remains fixed or is twisted in the opposite rotational direction. Torsion will also develop when the ends of a body are fixed and a moment is applied perpendicular to and about the longitudinal axis of the member, thereby inducing an axial twisting effect in the body. In buildings, girders supporting secondary beams on one side only and beams supporting eccentric loads (loads positioned off-center to the longitudinal axis of the beam) develop torsion.

The result of torsion is a twisting deformation of the body as the material winds or coils about the linear axis of the member. A torsional stress develops as the material resists torsional deformation. Under light torsional stresses, deformation occurs elastically; that is, torsional deformation is released when the torsion is removed. Under increasing torsional stress, the material develops a plastic deformation; that is, it remains twisted about its longitudinal axis when the torsion is removed. Torsion failure results when the body twists freely or the body shears perpendicular to the longitudinal axis.

8.2   TORQUE

A *torque (T)* is a moment that causes axial twisting when it is applied through the axis of a linear body, as shown in Figure 8.1. Torque and moment produce similar but different effects; a moment produces a rotation or tendency of rotation about an axis and a torque produces a rotation or tendency of rotation about the longitudinal

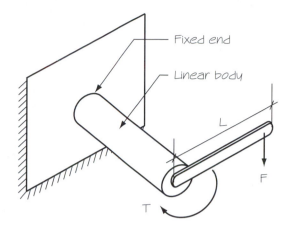

FIGURE 8.1 A torque (T) is a moment that causes axial twisting when it is applied through the axis of a linear body. A torque is the product of force (F) and distance (L).

axis of a shaftlike body. Like a moment, a torque is the product of force (F) and distance (L). It is expressed as

$$\text{torque, } T = FL$$

The symbol for torque is T. In this book the arrangement of units will distinguish between torque and moment: A torque will be expressed in units of distance-force and a moment is expressed in units of force-distance. In the U.S. customary system, a torque is expressed in units of foot-pounds (ft-lb), inch-pounds (in-lb), foot-kips (ft-kips), and inch-kips (in-kips). In the SI, torque is expressed in units of newton-meters (m-N) or some SI derivative. The relationship between base units of force in SI and the U.S. customary system is

$$\text{meter-newton (m-N)} = 0.737\ 561 \text{ foot-pound (ft-lb)}$$
$$\text{newton-meter (m-N)} = 8.850\ 75 \text{ inch-pounds (in-lb)}$$

■ EXAMPLE 8.1

The end of a long steel rod is fixed to an immovable plate. As shown in Figure 8.2, a worker attaches a large wrench to the opposite end of the rod so that the wrench is perpendicular to the longitudinal axis of the rod. She applies a 300-N force to the handle of the wrench a distance of 400 mm away from the center of the rod without deforming the rod. Determine the torque that develops in the rod.

Solution

$$T = FL = 300 \text{ N} \cdot 400 \text{ mm} = 120\ 000 \text{ mm-N} = 120 \text{ m-N}$$

In Example 8.1, a 120-N-m moment couple is applied to the end of the rod by twisting the wrench. The rod is subjected to a torque of 120 m-N. It was created by the worker applying a 300-N force to the handle of the wrench at a distance of

8.2 Torque

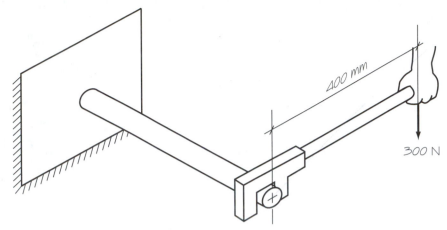

FIGURE 8.2   Illustration for Example 8.1.

400 mm (0.40 m) away from the center of the rod. The rod would be subjected to the same torque if the worker applied a 200-N force to the handle of the wrench at a distance of 600 mm (0.60 m) away from the center of the rod: A 200 N · 0.60 m = 120 N-m moment would subject the rod to a torque of 120 m-N.

8.3   TORSIONAL STRESS

The torque that develops when a moment is applied to the end of the fixed rod results in a twisting action down through the longitudinal axis of the rod. This twisting effect is countered by a *resisting torque*. Resisting torque is equal but opposite in rotation to the torque caused by a twisting moment. The result is development of a *torsional stress* (τ). The symbol for torsional stress is τ, the Greek lowercase letter tau. Similar to a shearing stress, a torsional stress is caused by the twisting effect and resistance to the twisting effect that develops in a linear body.

Assume that the cross section of the rod in Example 8.1 did not deform in cross-sectional and axial shape when the torque existed; it remained circular in cross section and perpendicular to the longitudinal axis of the rod, and the axis of the rod remained straight when the moment was applied. Then, as shown in Figure 8.3, the rod can be assumed to be composed of a series of very thin disks with each disk rotating slightly relative to each other disk, thereby generating a shearlike stress between the disks in the form of a slight angular rotation. The shearlike stress that develops between the theoretical disks increases linearly from zero at the center of the disk to a maximum value at the surface of the disk.

Maximum torsional stress (τ_{max}) that develops in a cylindrical shaft may be found by the following formula:

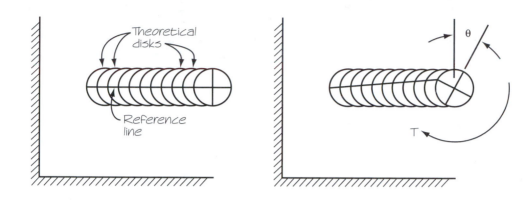

(a)   (b)

FIGURE 8.3 (a) A rod can be assumed to be composed of a series of very thin theoretical disks. Each theoretical disk rotates slightly relative to the adjacent disks when a torque is applied. (b) Torsional strain (θ) is the slight angular rotation of the very thin theoretical disks that make up the shaft.

$$\tau_{max} = \frac{TR}{J}$$

where torque (T), the radius of the shaft (R), and the polar moment of inertia (J), a property of a rotating section, influence the maximum torsional stress. The concept of the polar moment of inertia is introduced in Chapter 14. The polar moment of inertia for a solid circular shaft may be found by $J = \pi D^4/32$, where D is the shaft diameter. The formula for maximum torsional stress of a solid cylindrical shaft may be simplified by

$$\tau_{max\ solid} = \frac{TR}{J} \quad \text{where } J = \frac{\pi D^4}{32} \text{ and } R = \frac{D}{2}$$
$$= \frac{T(D/2)}{\pi D^4/32}$$
$$= \frac{16T}{\pi D^3}$$

Therefore, *maximum torsional stress ($\tau_{max\ solid}$) in a solid cylindrical shaft* is related to torque (T) and the shaft diameter (D). It may be calculated by the formula

$$\tau_{max\ solid} = \frac{16T}{\pi D^3}$$

Maximum torsional stress ($\tau_{max\ solid}$) in a hollow circular shaft like a pipe with an outside diameter (D_o) and an inside diameter (D_i) may be calculated by

$$\tau_{max\ hollow} = \frac{16TD_o}{\pi(D_o^4 - D_i^4)}$$

Both formulas assume that the linear body under consideration remains circular in cross section and perpendicular to the longitudinal axis of the body and that the axis of the body remains straight when a torque is applied. They apply only when the elastic limit of the material is not exceeded.

■ EXAMPLE 8.2

Determine the maximum torsional stress that develops in the rod described in Example 8.1. Assume that the rod has a diameter of 25 mm.

Solution In Example 8.1 it was found that the torque (T) that developed in the rod was 120 m-N. Maximum torsional stress is found by

$$25 \text{ mm diameter} = 0.025 \text{ m diameter}$$

$$\tau_{\text{max solid}} = \frac{16T}{\pi D^3} = \frac{16(120 \text{ m-N})}{\pi (0.025 \text{ m})^3} = 39\,113\,919 \text{ N/m}^2 = 39.11 \text{ MPa}$$

8.4 TORSIONAL STRAIN

When a shaftlike member is subjected to a torque, the shaft develops a rotational spiral or coil-like deformation. The actual twisting that occurs in the entire member is known as *torsional deformation. Torsional strain (θ)* is the slight angular rotation of the very thin theoretical disks that make up the shaft. The symbol for torsional strain is θ, the Greek lowercase letter theta.

Torsional strain may be calculated by the following formula, where torsional strain is as an angle of rotation in radians. The abbreviation for radians is *rad;* by definition there are 2π rad in one revolution, so 2π rad $= 360°$, or 1 rad $= 57.30°$.

$$\theta = \frac{TL}{JG}$$

The polar moment of inertia (J) for a solid shaft of circular cross section may be found by $J = \pi D^4/32$, where D is the shaft diameter. Length (L) is the longitudinal length of the shaft that is subjected to the torque. The shear modulus (G), sometimes called the modulus of rigidity, of a material relates to its modulus of elasticity (E) and Poisson's ratio (μ):

$$G = \frac{E}{2(1 + \mu)}$$

■ EXAMPLE 8.3

Determine the torsional strain that develops in the 25-mm-diameter rod described in Examples 8.1 and 8.2. Assume that the rod is composed of steel and is 1.00 m long.

Solution   The rod is subjected to a torque (T) of 120 m-N. The 25-mm diameter of the rod is 0.025 m. The polar moment of inertia (J) is found by

$$J = \frac{\pi D^4}{32} = \frac{\pi (0.025 \text{ m})^4}{32} = 0.000\,000\,038 \text{ m}^4$$

In Table A.4 in Appendix A, the modulus of elasticity (E) of steel is found as 200 000 MPa and Poisson's ratio (μ) of steel is found as 0.303. The shear modulus (G) of steel is calculated by

$$G = \frac{E}{2(1 + \mu)} = \frac{200\,000 \text{ MPa}}{2(1 + 0.303)} = 76\,750 \text{ MPa} = 76\,750\,000\,000 \text{ N/m}^2$$

Torsional strain is found by

$$\theta = \frac{TL}{JG}$$

$$= \frac{120 \text{ m–N} \cdot 1.00 \text{ m}}{0.000\,000\,038 \text{ m}^4 \cdot 76\,750\,000\,000 \text{ N/m}^2}$$

$$= 0.0411 \text{ rad}$$

Converting radians to degrees yields

$$0.0411 \text{ rad} \cdot 57.30° = 2.36°$$

In Example 8.3, torsional strain in the rod reveals itself as a 2.36° rotation of the wrench located at the end of the rod. As shown in Figure 8.4, rotation is about the longitudinal axis of the rod in a plane that is perpendicular to the axis of the rod.

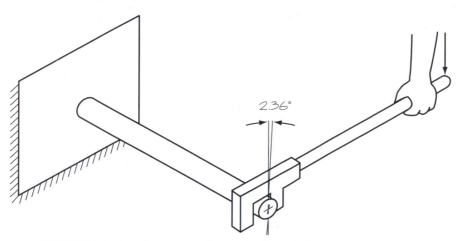

FIGURE 8.4   Resulting torsional strain for Example 8.3.

■ EXAMPLE 8.4

A W33 × 201 steel girder is fixed at rigid end connections and spans 40′-0″. It supports load-carrying secondary beams that extend perpendicular from the girder on one side only. The beams induce a torque equivalent to 15 000 in-lb applied at the center of the girder. The beam has a torsional constant (J) of 20.5 in^4. Structural steel has a shear modulus (G) of 11 200 000 lb/in^2. Determine the torsional strain that develops in the girder.

Solution Assume that each end of the girder resists the torque, so T = 15 000 in-lb/2 = 7500 in-lb. The torque is assumed to be applied at the center of the girder, so L = 40 ft/2 = 20 ft = 240 in. Torsional strain is found by

$$\theta = \frac{TL}{JG}$$

$$= \frac{7500 \text{ in–lb} \cdot 240 \text{ in}}{20.5 \text{ in}^4 \cdot 11\,200\,000 \text{ lb/in}^2}$$

$$= 0.0078 \text{ rad}$$

Converting radians to degrees gives

$$0.0078 \text{ rad} \cdot 57.30° = 0.447°$$

In buildings, girders supporting secondary beams on one side only and beams supporting loads positioned off-center to the longitudinal axis of the beam develop a torsional strain. A designer can limit torsional strain by selecting a member with a higher torsional constant (J), by reinforcing the member against torsion, or by limiting eccentric (off-center) loading that causes torsion in the member.

EXERCISES

8.1 The end of a 0.75-in-diameter rod is fixed to an immovable plate as shown. The length of the rod is 18.0 in. A 50-lb weight is attached to a 16.0-in lever at the opposite end of the rod so that the weight is hanging perpendicular to the longitudinal axis of the rod. For steel, E = 29 000 000 lb/in^2 and μ = 0.303. The polar moment of inertia for a solid circular shaft may be found by $J = \pi D^4/32$, where D is the shaft diameter.
(a) Determine the torque that develops in the rod, in in-lb.
(b) Calculate the maximum torsional stress that develops in the rod.
(c) Determine the torsional strain that develops in the steel rod, in degrees.

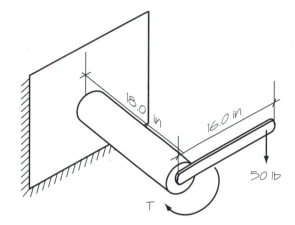

8.2 Same as Exercise 8.1 except that a 25-lb-weight is attached to the lever.

8.3 Same as Exercise 8.1 except that a 1.50-in-diameter rod is fixed to the immovable plate.

8.4 Same as Exercise 8.1 except that the length of the rod is 36 in.

8.5 A 75-mm-diameter brass rod has a 25-mm hole bored concentrically through the axis of the rod. The length of the rod is 450 mm. The end of the rod is fixed to an immovable plate. A 1.5-km-N torque is applied to the opposite end of the rod. Calculate the maximum torsional stress that develops in the rod.

8.6 Much as in Exercise 8.1, the end of a 0.75-in-diameter rod is fixed to an immovable plate. The length of the rod is 18.0 in. A 50-lb weight is attached to a 16.0-in lever at the opposite end of the rod so that the weight is hanging perpendicular to the longitudinal axis of the rod.

    (a) Determine the minimum diameter rod so that maximum torsional stress does not exceed 8000 lb/in^2.

    (b) Determine the minimum diameter rod so that torsional strain does not exceed 1.00°.

8.7 A W24×103 steel girder is fixed at rigid end connections and spans 21'-8". It supports load-carrying secondary beams that extend perpendicular from the girder on one side only. The beams induce a torque equivalent to 6000 in-lb at the center of the girder. The beam has a torsional constant (J) of 7.11 in^4. Structural steel has a shear modulus (G) of 11 200 000 lb/in^2. Determine the torsional strain that develops in the girder.

Chapter 9
THERMAL DEFORMATION AND STRESS

An increase in temperature brings about increased molecular agitation within any material. Simply, atoms of a material vibrate more vigorously at a high temperature and less vigorously at a low temperature. As a result, most materials expand when heated and contract when cooled. Water is one of very few materials that are exceptions to this phenomenon; below 4°C (39°F) water actually increases in volume as it is cooled. In this chapter we review the basic behavior of solid materials experiencing changes in temperature.

9.1 THERMAL DEFORMATION

A solid material *without restraints* will expand freely as it is heated and will contract freely as it is cooled. If a body is allowed to expand or contract freely with temperature change, the material will not develop any stresses; only deformation, a change in length, results from temperature change in a body without restraints.

Thermal deformation ($\Delta L_{\text{thermal}}$) is a change in length of a body caused by a change in temperature of the body (Figure 9.1). Thermal deformation may be found by

$$\text{thermal deformation, } \Delta L_{\text{thermal}} = \alpha L_o \, (\Delta T)$$

The *coefficient of thermal expansion* (α) is a value that relates change in length per unit length per degree change in temperature; for example, inches of change per inch of original length per degree change in temperature (in/in-°F). It quantifies a change in length that results when an unrestrained material is heated or cooled. Typical values of the coefficient of thermal expansion for certain engineering materials are given in Appendix A2. These values are for materials at room temperature. Rate of thermal expansion varies with the temperature of the material; for example, the coefficient of thermal expansion of steel increases with temperature increase.

Length (L_o) is the material's original length, in inches or meters, at the temperature before the temperature change occurs. Change in temperature is the difference between the final and original temperatures in °F or K:

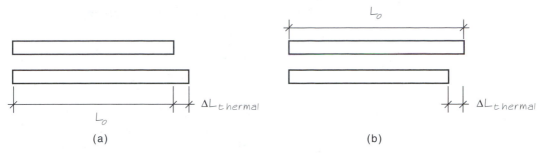

FIGURE 9.1 Thermal deformation ($\Delta L_{thermal}$) is a change in length of a body caused by a change in temperature of the body. As most material are heated they expand (a) and as they are cooled they contract (b).

$$\Delta T = T_{\text{final}} - T_{\text{original}}$$

Change in temperature (ΔT) is the final temperature (T_{final}) minus the original temperature (T_{original}). Change in temperature can be a positive or negative number. For a material being heated, the body under consideration expands and ΔT is positive. When ΔT is negative, the body under consideration is contracting because the material is being cooled.

■ EXAMPLE 9.1

A 100-ft-long concrete slab experiences a temperature increase from -30 to $+90°F$. Determine its change in length over this temperature range.

Solution From Table A.3, for concrete, $\alpha_{\text{concrete}} = 0.000\ 006$ in/in-°F:

$$L_o = 100 \text{ ft} \cdot 12 \text{ in/ft} = 1200 \text{ in}$$
$$\Delta T = T_{\text{final}} - T_{\text{original}} = 90°F - (-30°F) = 120°F$$
$$\Delta L_{\text{thermal}} = \alpha L_o (\Delta T) = (0.000\ 006 \text{ in/in-°F})(1200 \text{ in})(120°F) = 0.864 \text{ in}$$

As shown in Example 9.1, a 100-ft-long concrete slab that is free to expand or contract changes dimensionally by 0.864 in or about $\frac{7}{8}$ in when exposed to an outdoor temperature range of -30 to $+90°F$. This temperature range is common in many areas of the United States and Canada. Design provisions such as expansion joints in masonry and concrete walls and slabs must be made in design to accommodate this dimensional change.

■ EXAMPLE 9.2

The steel crankshaft of an automobile engine is exactly 24.000 in long when at its operating temperature of 250°F. The engine cools to 0°F when parked overnight on a cold winter night. Determine the length of the crankshaft once it has cooled.

Solution From Table A.3, for steel, $\alpha_{steel} = 0.000\ 007$ in/in-°F:

$$\Delta T = T_{final} - T_{original} = 0°F - 250°F = -250°F$$
$$\Delta L_{thermal} = \alpha L_o\ (\Delta T) = (0.000\ 007\ \text{in/in-°F})\ (24.000\ \text{in})\ (-250°F) = -0.042\ \text{in}$$
$$= 24.000\ \text{in} + (-0.042\ \text{in}) = 23.958\ \text{in}$$

In Example 9.2, the steel crankshaft contracts 0.042 in (about $\frac{3}{64}$ in) as the engine cools. Provisions must be made in the design of the engine to allow for this expansion. Note that ΔT is a negative value because the material was cooled; it went from a higher temperature to a lower temperature. With change in temperature (ΔT) negative, ΔL is also negative, indicating a decrease in original length.

■ EXAMPLE 9.3

As shown in Figure 9.2, a boiler in a power plant is 14.50 m tall when installed. The boiler is secured to roof beams at its top end and hangs freely. At a room temperature of 20°C, the boiler is 0.350 m from the floor surface. Due to fire regulations, the minimum clearance during operation is 0.15 m from the floor. Determine if there is adequate floor clearance when the steel boiler shell reaches an average operating temperature of 1000°C.

Solution From Table A.4, for steel, $\alpha_{steel} = 0.000\ 013$ m/m-°C:

$$\Delta T = T_{final} - T_{original} = 1000°C - 20°C = 980°C$$
$$\Delta L_{thermal} = \alpha L_o\ (\Delta T)$$
$$= (0.000\ 013\ \text{m/m-°C})\ (14.50\ \text{m})\ (980°C) = 0.1847\ \text{m}$$
$$\text{clearance during operation} = \text{original clearance} - \Delta L_{thermal}$$
$$= 0.350\ \text{m} - 0.1847\ \text{m} = 0.1653\ \text{m}$$

Because the boiler hangs more than 0.15 m off the floor when at its operating temperature, there is adequate floor clearance.

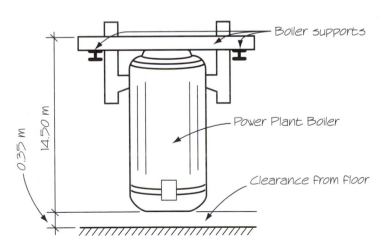

FIGURE 9.2 Illustration for Example 9.3.

Chap. 9   Thermal Deformation and Stress

In Example 9.3 the boiler is estimated to expand 0.1847 m as it is heated from room temperature to 1000°C. Thermal expansion of the boiler reduces the floor clearance from 0.350 m to 0.1653 m. Clearance is acceptable.

A change in the cross-sectional area and overall volume of a body will also result from temperature change. The diameter of a pipe changes with increase and decrease in temperature, as shown in Example 9.4

■ **EXAMPLE 9.4**

A 10-in-diameter steel pipe has an outside diameter of exactly 10.75 in at room temperature (70°F). Determine the diameter of the pipe if it carries high-pressure steam and is heated to 250°F.

Solution   From Table A.3, for steel, $\alpha_{steel} = 0.000\,007$ in/in-°F:

$$\Delta T = T_{final} - T_{original} = 250°F - 70°F = 180°F$$

The outer wall circumference ($L_{o\,circumference}$) is found by πD, where D is the diameter:

$$L_{o\,circumference} = \pi D = \pi(10.75\text{ in}) = 33.77\text{ in}$$
$$\Delta L_{thermal} = \alpha L_o\,(\Delta T) = (0.000\,007\text{ in/in-°F})\,(33.77\text{ in})\,(180°F)$$
$$= 0.043\text{ in}$$
$$L_{new\,circumference} = L_{o\,circumference} + \Delta L_{thermal} = 33.77\text{ in} + 0.043\text{ in} = 33.81\text{ in}$$

By algebraic rearrangement of the terms in the equation $L_{circumference} = \pi D$, we obtain

$$D = \frac{L_{new\,circumference}}{\pi} = \frac{33.81\text{ in}}{\pi} = 10.76\text{ in}$$

Provisions for lateral dimensional change must be made in design in a manner similar to that demonstrated in Example 9.4. A presentation of the effect of lateral dimensional change of complex cross sections such as a cam shaft involves a higher level of mathematics and is thus beyond the scope of this book.

9.2   THERMAL STRESS

A solid body that is restrained so that it cannot change length due to a temperature change will develop internal stresses as it is heated or cooled. Simply, since the partially or fully restrained material cannot expand or contract freely, the material develops a stress. This stress is similar in nature to the stress produced as an external force is applied ($\sigma = F/A$). Stress produced by temperature change is called *thermal stress*. Thermal stress ($\sigma_{thermal}$) caused by temperature change of a material in a *fully restrained* body is found by

$$\text{thermal stress, } \sigma_{\text{thermal}} = E\alpha \, (\Delta T)$$

A material's modulus of elasticity (E), coefficient of thermal expansion (α), and change in temperature (ΔT) influence the thermal stress that develops. A *fully restrained body* is restrained at its ends such that it cannot change length due to a temperature change. If a material is *not* fully restrained, it will expand or contract freely or only partially with temperature change. Thermal stress that develops in a material is greatest when the body is fully restrained.

The $\sigma_{\text{thermal}} = E\alpha(\Delta T)$ equation is derived by first equating longitudinal deformation (ΔL) and thermal deformation ($\Delta L_{\text{thermal}}$):

$$\Delta L = \Delta L_{\text{thermal}}$$

$$\frac{FL_o}{EA} = \alpha L_o(\Delta T)$$

$$\frac{FL_o}{A} = E\alpha L_o(\Delta T)$$

$$\frac{F}{A} = E\alpha(\Delta T)$$

$$\sigma_{\text{thermal}} = E\alpha(\Delta T)$$

Maximum thermal stress (σ_{thermal}) exists only when the material is fully restrained at its ends. Actual stresses that develop if it is partially restrained will be less than the thermal stress that develops if it were fully restrained. If it is not restrained at all and expands or contracts freely with temperature change, a thermal stress does not develop.

■ EXAMPLE 9.5

As shown in Figure 9.3, a solid concrete masonry wall is constructed between two rigid steel columns such that the wall is fully restrained. During installation at a temperature of 40°F, the concrete masonry is placed very tightly against the columns. Assume that $E = 2\,000\,000$ lb/in^2 for concrete masonry. Determine the thermal stress developed in the concrete masonry if its temperature increased to 90°F.

Solution From Table A.3, for concrete, $\alpha = 0.000\,006$ in/in-°F:

$$\Delta T = T_{\text{final}} - T_{\text{original}} = 90°F - 40°F = 50°F$$

$$\sigma_{\text{thermal}} = E\alpha(\Delta T) = (2\,000\,000 \text{ lb/in}^2)\,(0.000\,006 \text{ in/in-°F})\,(50°F)$$
$$= 600 \text{ lb/in}^2$$

The solid concrete masonry wall in Example 9.5 developed a stress of 600 lb/in^2. The masonry attempted to expand with temperature increase, but the steel columns prevented expansion. The material developed compressive stress.

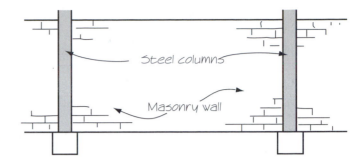

FIGURE 9.3 Illustration for Example 9.5.

When a member is fully restrained, thermal stress is equal to the applied stress, which was found to be calculated by dividing the applied force (F) by a cross-sectional area (A):

$$\sigma = \sigma_{thermal} = E\alpha(\Delta T)$$

$$\sigma_{thermal} = \frac{F}{A}$$

■ EXAMPLE 9.6

Assume that the concrete masonry wall in Example 9.6 is composed of 8-in solid concrete block ($7\frac{5}{8}$ in wide) and that the wall is 8 ft (96 in) high. Determine the force exerted at the intersection of the wall and the steel column.

Solution $\sigma_{thermal} = F/A$. By rearrangement of terms, $F = \sigma_{thermal} \cdot A$

$$F = \sigma_{thermal} \cdot A = 600 \text{ lb/in}^2 \, (7.625 \text{ in} \cdot 96 \text{ in}) = 439\,000 \text{ lb} = 439 \text{ kip}$$

As shown in Example 9.6, the resulting force of 439 kip in the masonry wall is substantial. Initially, the stress of 0.6 kip/in^2 in Example 9.5 appeared insignificant. If ignored by the designer, forces exerted by thermal stress can have catastrophic results.

9.3   EVALUATING THERMAL STRESSES

The designer must develop an intuitive feel for how a material will behave as it is heated or cooled. This behavior is related to temperature change and whether the material is already experiencing tensile or compressive stresses. The following explanations address the considerations.

Effect of cooling a fully restrained body   As a body cools (ΔT is negative), the material attempts to contract. If *initially under tension,* a fully restrained body that is cooled will develop additional tensile stresses. The material may fail if tensile stresses become excessive (i.e., exceed the ultimate tensile strength of the material). If *initially under compression,* stresses are relieved as the fully restrained body cools. Cooling may relieve all stresses in the body. It may slip from its restraints as it contracts. Additionally, if cooled sufficiently, a material initially under compressive stress can contract and begin to develop tensile stresses if it is connected at its restraints.

Effect of heating a fully restrained body   As a body is heated (ΔT is positive), the material attempts to expand. If *initially under compression,* a fully restrained body that is heated will develop additional compressive stresses as the material attempts to expand. The material may fail if compressive stresses become excessive (i.e., exceed the ultimate compressive strength of the material). If the material is *initially under tension* when heated, the fully restrained body attempts to expand and tensile stresses decrease. Heating may relieve all tensile stresses and cause the body to fall or hang freely from its restrains. If heating causes expansion beyond just relieving tensile stresses, the material will begin to develop compressive stresses.

■ EXAMPLE 9.7

A copper wire is stretched tightly between two supports. It is fully restrained. Assume that $E = 16\ 000$ kip/in^2 and that the tensile stress of the wire is 4.5 kip/in^2 at 70°F.
(a) Determine the total stress if the copper wire cools to 50°F.
(b) Determine the total stress if the copper wire is heated to 80°F.
(c) Determine the temperature at which stresses are relieved in the copper bar.
Solution   (a) From Table A.3, for copper, $\alpha = 0.000\ 010$ in/in-°F:

$$\Delta T = 50°F - 70°F = -20°F$$
$$\sigma_{thermal} = E\alpha(\Delta T) = (16\ 000 \text{ kip/in}^2)\ (0.000\ 010 \text{ in/in-°F})\ (-20°F)$$
$$= -3.2 \text{ kip/in}^2$$

The negative sign in the answer indicates that the material is contracting. In this case, the temperature decrease causes contraction of a material under tensile stress and therefore additional stress.

Total stress at 50°F = 4.5 kip/in^2 (at 70°F) + 3.2 kip/in^2 = 7.7 kip/in^2

(b) $\Delta T = 80°F - 70°F = 10°F$
$$\sigma_{thermal} = E\alpha(\Delta T) = (16\ 000 \text{ kip/in}^2)\ (0.000\ 010 \text{ in/in-°F})\ (10°F)$$
$$= 1.6 \text{ kip/in}^2$$

The positive answer indicates that the material is expanding. In this case the temperature increase causes expansion and relieves some of the original tensile stress.

Total stress at 80°F = 4.5 kip/in^2 (at 70°F) − 1.6 kip/in^2 = 2.9 kip/in^2

(c) The existing tensile stress must be reduced by 4.5 kip/in^2 to relieve stresses in the bar:

$$\sigma_{thermal} = E\alpha(\Delta T) \quad \text{where } \Delta T = T_{final} - T_{original}$$
$$= -4.5 \text{ kips/in}^2 = E\alpha(T_{final} - T_{original})$$
$$-4.5 \text{ kips/in}^2 = E\alpha T_{final} - E\alpha T_{original}$$
$$E\alpha T_{final} = E\alpha T_{original} + 4.5\text{kip/in}^2$$
$$T_{final} = \frac{E\alpha T_{original} + 4.5 \text{ kip/in}^2}{E\alpha}$$
$$= T_{original} + \frac{4.5 \text{ kip/in}^2}{E\alpha}$$
$$= 70°F + \frac{4.5 \text{ kip/in}^2}{(16\,000 \text{ kip/in}^2)(0.000\,010 \text{ in/in-°F})}$$
$$= 70°F + 28.1°F$$
$$= 98.1°F$$

A simpler solution may be to recognize that to relieve the tensile stress the bar must be heated; that is, ΔT must be positive:

$$\sigma_{thermal} = E\alpha(\Delta T)$$
$$\Delta T = \frac{\sigma_{thermal}}{E\alpha}$$
$$= \frac{4.5 \text{ kip/in}^2}{(16\,000 \text{ kip/in}^2)(0.000\,010 \text{ in/in-°F})}$$
$$= 28.1°F$$
$$T_{final} = T_{original} + \Delta T$$
$$= 70°F + 28.1°F$$
$$= 98.1°F$$

As the temperature of the wire in Example 9.7 decreased, stress in the wire increased because the wire was originally in tension. Cooling caused the molecules to pull together, thereby increasing tensile stresses. In the second part of Example 9.7, the material expanded as the wire temperature increased, thereby relieving some of the original stresses. Above a temperature of 98.1°F, the stresses are relieved.

EXERCISES

Thermal Deformation

9.1 A 100-ft-long loose-laid, roof membrane ($\alpha = 0.000\,150$ in/in-°F) will be installed on a flat roof at a temperature of 45°F. On a hot summer day, the temperature of the roof membrane can rise to 130°F. In the winter, the temperature of the roof membrane can fall to $-20°F$.

(a) Determine how much the roof membrane will expand if it is not restrained.

(b) Determine how much the roof membrane will contract if it is not restrained.

9.2 At a fabrication temperature of 16°C, steel tubes in a boiler measure 2.500 m. When the boiler is fired up, its operating temperature reaches 250°C. Determine by how much the steel tubes will expand if they are not restrained.

9.3 Pipe in the plumbing system of a large commercial building will carry domestic hot water at a temperature of 125°F. The pipe is installed at a temperature of 50°F. Determine the change in length per 100 ft of pipe for the following pipe materials:
(a) Copper ($\alpha = 0.000\ 010$ in/in-°F)
(b) Plastic ($\alpha = 0.000\ 067$ in/in-°F)

9.4 A 1.0-m rigid horizontal support is hung by two cables as shown. One cable is brass and the other cable is high-density polyethylene plastic. At room temperature (21°C), the cables measure 3.50 m vertically. When the temperature drops to −15°C, find the following:
(a) The change in length in each cable
(b) The change in elevation of points *A, B,* and *C.*

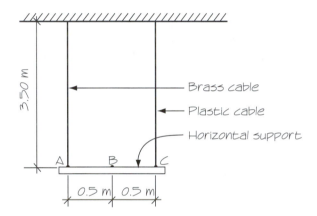

9.5 A concrete masonry ($\alpha = 0.000\ 006$ in/in-°F) wall is 256 ft long. It will be designed to endure a 90°F temperature change. Pliable expansion joints are placed vertically in the masonry wall to limit damage from thermal expansion and contraction. A $\frac{1}{2}$-in vertical expansion joint allows a maximum 0.25 in change in dimension.
(a) Determine the maximum change in length of the masonry wall with this temperature change.
(b) Determine the minimum number of vertical expansion joints required in the wall to limit damage from thermal expansion and contraction.
(c) Determine the maximum spacing of evenly spaced vertical expansion joints required to limit damage from thermal expansion and contraction.

9.6 As shown, a curtain wall constructed of brick masonry ($\alpha = 0.000\ 009$ m/m-°C) bears on a steel angle that extends from a steel spandrel beam at each floor. The brick curtain wall measures 4.0 m vertically between bearing points. The wall will be designed to endure a 45°C temperature change. Dimensional change due to thermal expansion and contraction is limited to 2.0 mm between steel angles.
(a) Determine the maximum change in height of the masonry wall with this temperature change.

(b) Is dimensional change from the 45°C temperature change acceptable?

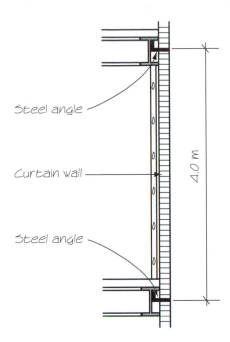

Steel angle

Curtain wall

Steel angle

4.0 m

9.7 A roof membrane ($\alpha = 0.000\,120$ in/in-°F) will be installed on a flat roof. Once installed, the membrane will be allowed to expand and contract freely. On a hot summer day, the temperature of the roof membrane can rise to 140°F. In the winter, the temperature of the roof membrane can fall to 0°F. Due to the configuration of the roof expansion joint, expansion is limited to 4.5 in and contraction is limited to 3.0 in.
   (a) Determine the maximum length the membrane can be if it is installed at a temperature of 35°F.
   (b) Determine the maximum length the membrane can be if it is installed at a temperature of 95°F.

Thermal Stress

9.8 At a fabrication temperature of 60°F, steel tubes in a boiler measure 86.50 in. When the boiler is fired up, its operating temperature reaches 320°F.
   (a) Find the change in length of the boiler tubes if they expand freely.
   (b) Assume that the steel tubes are fully restrained so they cannot expand or contract freely. Determine the thermal stress that develops in the tubes when the boiler is at its operating temperature.

9.9 A steel member is fully restrained against tension and compression. At room temperature (70°F), the member has a tensile stress of 11 830 lb/in^2.
   (a) Determine the stress experienced by the member if it is cooled from room temperature to 50°F.
   (b) Determine the stress experienced by the member if it is heated from room temperature to 100°F.

(c) Determine the stress experienced by the member if it is heated from room temperature to 150°F.

9.10 An aluminum bar is clamped (fully restrained in compression only) between two immovable supports. Initially, at room temperature (22°C), it has a compressive stress of 20 MPa. The temperature of the bar is slowly decreased. Determine the temperature at which the bar will begin to slip from the clamp.

9.11 As shown, a curtain wall is constructed of brick masonry ($\alpha = 0.000\ 005$ in/in-°F). The bottom of the masonry wall bears on the leg of a steel angle that is welded to and extends from a steel spandrel beam at each floor. The wall extends to the bottom side of the leg of a steel angle on the upper floor. The wall measures 12.0 ft and is fully restrained between the angles. It is installed at a temperature of 55°F. On a hot summer day, the temperature of the wall can rise to 90°F. In the winter, the temperature of the wall can fall to 20°F. Stress due to weight of the wall is 15 lb/in² at the base of the wall (at the bearing point on the angle).

(a) Determine the maximum thermal stress that develops in the curtain wall.

(b) Determine the maximum stress that develops in the curtain wall with this temperature change.

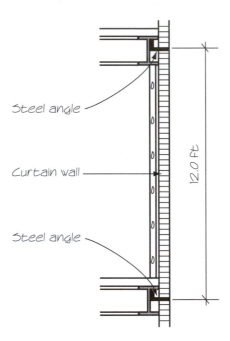

Chapter 10
MATERIAL TESTING AND SIMPLE DESIGN

With the presentation of properties of materials thus far in this book, it may appear that materials under stress behave in an easily understood manner. However, the behavior of materials is complex and properties associated with these behaviors vary significantly. For example, most steels have nearly equal strengths in tension and compression, but strength and other properties can vary significantly with the addition of a small fraction of alloying elements. Steel tends to be 15 to 20 times stiffer than concrete and wood. The mechanical properties of wood vary considerably with species, grain structure and orientation, and imperfections such as knots, splits, and checks; shear and bearing strengths vary significantly when applied with or against the grain. Generally, however, wood is slightly stronger in tension than in compression. Concrete, on the other hand, is about 10 times stronger in compression than in tension. Strength properties of concrete vary for a particular type of cement. Type and water-to-cement ratio will affect the strength of concrete.

With such variations in properties, it is helpful for the technician to gain an understanding of the patterns of behavior of a material under various magnitudes of stress. In this chapter we review select methods of identifying behaviors of materials and determining their mechanical properties. Simple design methods related to these behaviors are also introduced.

10.1   MATERIAL TESTING

Effective selection and use of engineering materials requires a good understanding of the materials themselves and their properties. Many standardized tests have been developed and are used to ascertain the properties of materials. Simple tensile and compression tests are discussed in this chapter. A flexure test determines the bending (tension) stresses at which a material fails. Toughness, the capacity of a material to absorb energy from impact, can be determined through the Charpy impact test. This test involves swinging a heavy pendulum through a specimen to determine the amount of energy required to break the specimen; a "tough" material withstands greater impact. Hardness, the capacity of a material to resist indentation or abrasion, is determined through the Rockwell or Brinell hardness tests. A hardness test involves indenting a specimen with a weighted indenter and measuring the depth or width of penetration.

Rigorous testing of a vast number of specimens is performed. Observations from these tests help the designer understand the differences in behavior of the different materials. Test results are used by a group of professionals to determine mechanical properties such as strength and elasticity. These properties are then used by engineers as acceptable design values in material specification.

10.2 TENSILE TEST

The *tensile test* involves clamping a specimen of a material in a testing apparatus that is designed to apply a tensile force that slowly pulls the specimen apart, as shown in Figure 10.1. The apparatus applies a controlled and gradually increasing tensile force to the specimen. As the material is pulled apart, an extensometer that can measure a change in length as little as 0.000 01 in is used to measure longitudinal deformation (ΔL), the elongation of the specimen as it is in tension. The force applied to the specimen at any time throughout the test and the resulting elongation are chronicled by an electronic data recorder.

Once the test is complete, stress and strain are calculated from data observed during the test. Observed data and calculation results of a tensile test of a 0.509-in-diameter mild steel bar are shown in Table 10.1. From these tensile test observations, apparent stresses and longitudinal strains are calculated. For example, a force (F) of 1.0 000 lb results in an elongation (ΔL) of 0.000 20 in a 1.000-in measurement length of bar (L_o). The area of the bar is calculated by $a = \pi D^2/4 = \pi(0.509 \text{ in})^2/4 = 0.2035$ in^2. Stress is found by $\sigma = F/a = 1000 \text{ lb}/0.2035 \text{ in}^2 = 4910 \text{ lb/in}^2$. Longitudinal strain at this stress is found by $\epsilon = \Delta L/L_o = 0.000 20 \text{ in}/1.000 \text{ in} = 0.000 20 \text{ in/in}$.

Once calculated, values of stress and the associated longitudinal strain are then plotted on a graph known as a *stress–strain curve*, a mathematical curve that relates apparent stress (read from the vertical axis of graph) to a specific longitudinal strain (read from the horizontal axis of the graph). The stress–strain curve for the mild steel bar tested in tension as discussed above is illustrated in Figure 10.2. An enhanced illustration of the left side of this curve is provided in Figure 10.3.

A Mathematical Curve

In mathematics, a *curve* is a path traced out by a series of related points. Although it can take the shape of a curved line, a curve can also be a straight line. A curve can have limited or infinite length and can be open such that its ends never close or can be closed, such as a circle or ellipse.

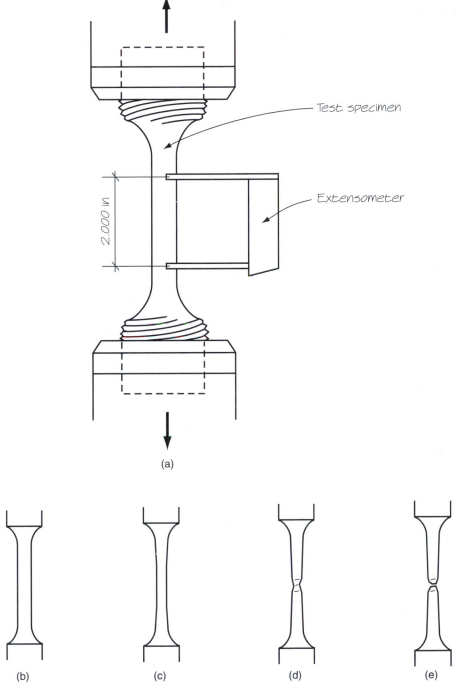

Test specimen

Extensometer

2.000 in

(a)

(b)   (c)   (d)   (e)

FIGURE 10.1 (a) In a tensile test of mild steel, the test specimen is placed in tensile testing apparatus. An extensometer is attached to the specimen to measure change in length (b) as a controlled and gradually increasing tensile force is applied to the specimen. The specimen deforms longitudinally (elongates) under the applied force (c). As the applied force increases, a reduction in cross section called necking (d) becomes evident. Eventually, the specimen fails as it is pulled apart (e).

Table 10.1 OBSERVED DATA AND RELATED STRESS AND STRAIN CALCULATION
RESULTS OF A TENSILE TEST OF MILD STEEL [a]

Observed data		Calculation results	
Force, F (lb)	Elongation, ΔL (in)	Stress, σ (lb/in^2)	Strain, ϵ (in/in)
0	0.000	—	0.000
1 000	0.000 20	4 910	0.000 20
2 000	0.000 40	9 830	0.000 40
3 000	0.000 55	14 740	0.000 55
4 000	0.000 75	19 660	0.000 75
5 000	0.000 95	24 570	0.000 95
6 000	0.001 05	29 490	0.001 05
7 000	0.001 20	34 400	0.001 20
8 000	0.001 45	39 320	0.001 45
9 000	0.001 75	44 230	0.001 75
10 000	0.002 05	49 140	0.002 05
9 900	0.005 00	48 650	0.005 00
9 840	0.010 00	48 360	0.010 00
9 860	0.015 00	48 460	0.015 00
9 860	0.020 00	48 460	0.020 00
9 800	0.025 00	48 160	0.025 00
10 180	0.030 00	50 030	0.030 00
10 540	0.035 00	51 800	0.035 00
10 880	0.040 00	53 470	0.040 00
11 200	0.045 00	55 040	0.045 00
11 500	0.050 00	56 520	0.050 00
11 600	0.050 00	57 010	0.050 00
13 180	0.100 00	64 770	0.100 00
13 740	0.150 00	67 520	0.150 00
13 960	0.200 00	68 610	0.200 00
14 000	0.250 00	68 800	0.250 00
13 940	0.300 00	68 510	0.300 00
13 780	0.350 00	67 720	0.350 00
13 400	0.400 00	65 850	0.400 00
12 800	0.445 00	62 910	0.445 00

[a]Specimen was a 0.509-in-diameter steel rod.

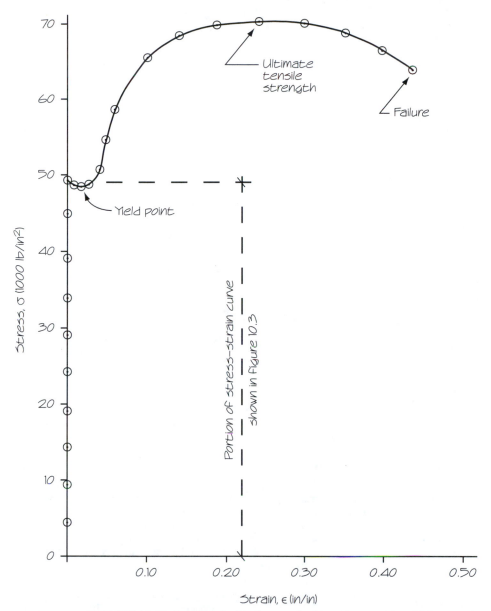

FIGURE 10.2 Results of a tensile test of mild steel are graphically illustrated on a stress–strain curve. The yield point and ultimate tensile strength of the steel is clearly evident. Observed data related to this test and calculation results used in developing this stress–strain curve are provided in Table 10.1.

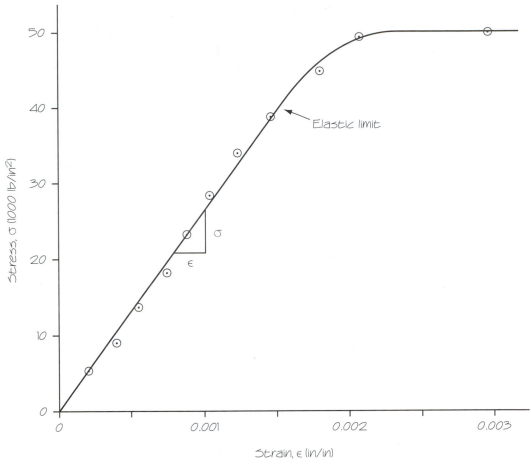

FIGURE 10.3 Enhanced illustration of the left side of the stress–strain curve of a tensile test of mild steel. In this illustration, plotted values of strain have been expanded from Figure 10.2 to illustrate the linear characteristics of the elastic region of the stress–strain curve. Linear behavior of the stress–strain curve for mild steel is evident between 0 to 48 000 lb/in². Such linear behavior represents proportionality between stress and strain; in this range, increases or decreases in stress are followed by instantaneous and proportional increases or decreases in longitudinal strain.

10.3 COMPRESSION TEST

Under a heavy compressive force, a brittle material such as concrete will fail by crushing. *Crushing* is a true compressive stress failure due to an excessive compressive force applied to the material, causing the molecules of a material to slide past one another. A test specimen carries a gradually increasing compressive force until rapid failure of the specimen results. The *ultimate compressive strength* is the stress at compressive stress failure.

■ EXAMPLE 10.1

A 6-in-diameter concrete specimen failed at a force of 127.5 kips in a compression test exactly four weeks after it was mixed. Determine the ultimate compressive strength (the stress at failure 28 days after mixing) for this concrete mix.

Solution

$$A = \frac{\pi D^2}{4} = \frac{\pi (6.0 \text{ in})^2}{4} = 28.3 \text{ in}^2$$

$$\sigma = \frac{F}{A} = \frac{127\,500 \text{ lb}}{28.3 \text{ in}^2} = 4505 \text{ lb/in}^2$$

A *compression test* is used to determine the compressive strength of a specimen. It involves a test setup and procedure similar to the tensile test except that the specimen is placed under compression rather than tension. The specimen is placed in a testing apparatus as shown in Figure 10.4. Care is taken to ensure that the applied force is evenly distributed during the test. For example, with a test of concrete, sulfur caps or steel caps are placed at the ends of the concrete cylinder before placing the concrete specimen in the test apparatus. The sulfur evenly

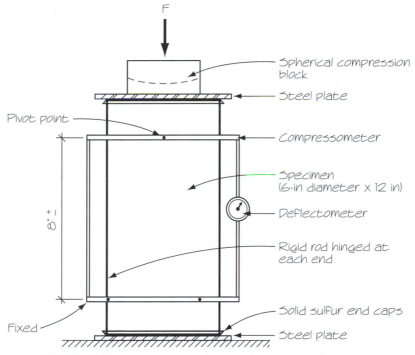

FIGURE 10.4   Compression test setup of a concrete cylinder.

10.3 Compression Test

distributes the force from the apparatus to the specimen. Under a gradually increasing compressive force the specimen is compressed until failure of the specimen results. The force applied to the specimen at any time throughout the test and the resulting deformation are chronicled by an electronic data recorder. Once the compression test is complete, stress and strain are calculated from data observed during the test. Observed data and calculation results of a compression test of dry-cured concrete on the 28th day after mixing are provided in Table 10.2. The related stress–strain diagram of the concrete specimen is provided in Figure 10.5.

Table 10.2 OBSERVED DATA AND RELATED STRESS AND STRAIN CALCULATION RESULTS OF A COMPRESSION TEST OF DRY-CURED CONCRETE ON THE 28TH DAY AFTER MIXING.[a]

Observed data		Calculation results	
Force, F (lb)	Elongation, ΔL (in)	Stress, σ (lb/in^2)	Strain, ϵ (in/in)
0	—	—	—
5 000	0.000 25	193	0.000 03
10 000	0.000 60	385	0.000 07
15 000	0.001 10	578	0.000 14
20 000	0.001 50	770	0.000 18
25 000	0.001 55	963	0.000 19
30 000	0.002 30	1155	0.000 28
35 000	0.002 65	1348	0.000 33
40 000	0.003 05	1540	0.000 38
45 000	0.003 35	1733	0.000 41
50 000	0.003 70	1926	0.000 46
55 000	0.004 45	2118	0.000 55
60 000	0.005 05	2311	0.000 62
65 000	0.005 65	2503	0.000 70
70 000	0.006 15	2696	0.000 76
75 000	0.006 70	2888	0.000 83
80 000	0.007 30	3081	0.000 90
85 000	0.007 90	3273	0.000 97
90 000	0.008 60	3466	0.001 06
95 000	0.009 35	3658	0.001 15
100 000	0.010 15	3851	0.001 25
105 000	0.011 00	4044	0.001 35
110 000	0.012 05	4236	0.001 48
115 000	0.013 25	4429	0.001 63
120 000	0.014 90	4621	0.001 83
124 500	Failure	4795	—

[a]Specimen was a concrete cylinder with a diameter of 5.75 in.

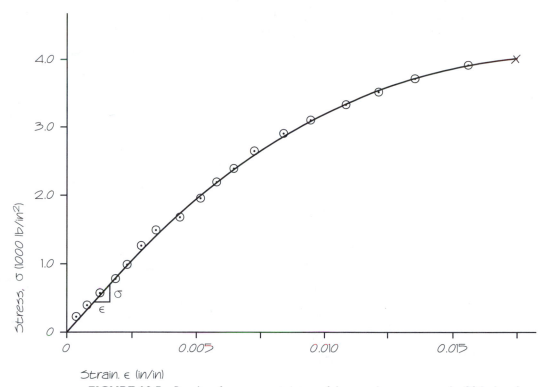

FIGURE 10.5 Results of a compression test of dry-cured concrete on the 28th day after mixing are graphically illustrated on the stress–strain curve. Observed data related to this test and calculation results used in developing this stress–strain curve are provided in Table 10.2.

A second mode of failure by compression is called buckling. *Buckling* is a sudden, lateral bending of a long, slender specimen caused by an excessive compressive force. A specimen that buckles bends sideways. Buckling is due to a combination of excessive end loading and an instability in elastic behavior of the material and the specimen cross section. Buckling failure typically occurs in members that are long and slender in cross section.

If the specimen is immediately unloaded when buckling is first evident, it will realign itself elastically. However, if the compressive force continues after buckling is evident, only smaller compressive forces are required to continue the buckling effect; once the specimen buckles, it can no longer carry the maximum force. Ductile materials such as mild steel can easily fail by buckling. Observed data and related stress and strain calculation results of a compression test of mild steel (the same steel used in the tensile test) are shown in Table 10.3. The related stress–strain diagram is provided in Figure 10.6.

Table 10.3 OBSERVED DATA AND RELATED STRESS AND STRAIN CALCULATION RESULTS OF A COMPRESSION TEST OF A LONG, SLENDER MILD STEEL SPECIMEN[a]

Observed data		Calculation results	
Force, F (lb)	Elongation, ΔL (in)	Stress, σ (lb/in^2)	Strain, ϵ (in/in)
0	0.000	—	0.0000
1000	0.010	4 910	0.0016
2000	0.015	9 830	0.0024
3000	0.020	14 740	0.0032
4000	0.027	19 660	0.0043
5000	0.033	24 570	0.0052
6000	0.040	29 490	0.0063
7000	0.049	34 400	0.0078
8000	0.066	39 320	0.0105
9000	0.087	44 230	0.0138
9100	0.095	44 720	0.0150
9000	0.100	44 230	0.0158
8000	0.129	39 320	0.0204
7000	0.150	34 400	0.0238
6000	0.165	29 490	0.0261
5000	0.179	24 570	0.0284
4000	0.195	19 660	0.0309
3000	0.231	14 740	0.0366
2000	0.317	9 830	0.0502
1450	0.430	4 910	0.0681

[a]Specimen was a 0.509-in-diameter steel rod with a length of 6.31 in.

10.4 INTERPRETING MATERIAL TEST RESULTS

The mechanical properties of materials reported in technical standards come from interpretation of the results of countless tests of materials. The stress–strain behavior of mild steel in tension traditionally serves as the theoretical basis upon which properties of materials are established. The results of the tensile test for mild steel as reported in Table 10.1 and shown in the stress–strain curves in Figures 10.2 and 10.3 are examined below.

A review of the stress–strain curve for mild steel shows that under low stresses ($<48\ 000$ lb/in^2), a straight, steeply sloped (steep diagonal) line exists. This straight curve is better illustrated in the enhancement of the stress–strain curve shown in Figure 10.3. A straight line on a stress–strain curve indicates linear behavior. *Linear* behavior of a material on a stress–strain curve indicates proportion-

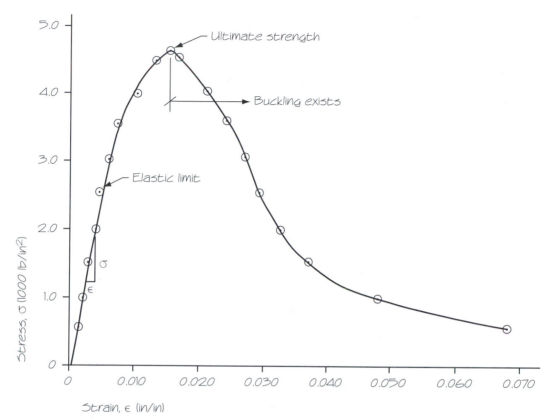

FIGURE 10.6 Results of a compression test of mild steel are graphically illustrated on the stress–strain curve. Observed data related to this test and calculation results used in developing this stress–strain curve are provided in Table 10.3. The mode of failure of the specimen was by buckling (bending sideways) due to a combination of the length (6.31 in) and cross-sectional slenderness (0.509 in) of the specimen. Observe that once the material achieved its ultimate compressive strength (just below 45 000 lb/in²) a decreasing stress is required to continue the buckling effect; a member generally fails once buckling occurs.

ality between stress and strain. Thus, in this range, increases or decreases in stress are followed by instantaneous and proportional increases or decreases in longitudinal strain; that is, twice the stress produces twice the longitudinal strain, three times the stress causes three times the longitudinal strain, and so on. In this case, linear behavior demonstrates Hooke's law.

Linear behavior of the stress–strain curve for mild steel is evident between 0 and 48 000 lb/in². In this region of the curve, steel behaves so that stress is proportional to longitudinal strain. The slope (rise over run) of this portion of the stress–strain curve represents the modulus of elasticity of mild steel; the modulus of elasticity *(E)* was introduced in Chapter 7 as a proportionality constant relating stress and longitudinal strain within the elastic range of a material.

10.4 Interpreting Material Test Results

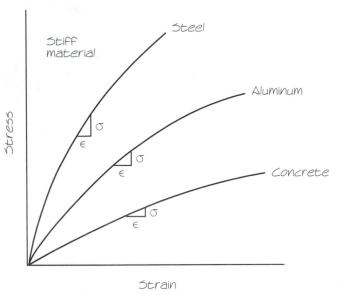

FIGURE 10.7 A material with a steep slope on the stress–strain curve indicates a high modulus of elasticity and a stiff material. A shallow slope indicates a material that has a small modulus of elasticity.

A steep slope on the stress–strain curve indicates a material with a high modulus of elasticity and a stiff material; a shallow slope indicates a material that is not as stiff and therefore has a small modulus of elasticity (Figure 10.7).

From 0 to about 48 000 lb/in^2 on the stress–strain curve for mild steel, stress and longitudinal strain are proportional. So, for this material, the *proportional limit* or *elastic limit* occurs at about 48 000 lb/in^2. If the material was subjected to stresses beyond this value, the test specimen would yield and no longer return to its original length after removal of the load. If the material was stressed above about 48 000 lb/in^2, it would become permanently deformed and gain *permanent set*. As shown by the dashed line in Figure 10.8, the elastic strain is recovered but permanent deformation remains. Stressing a material beyond the elastic limit is similar to stretching a spring too far; it never returns to its original shape when unloaded.

Shortly after the elastic limit had been reached by the mild steel specimen (at approximately 48 000 lb/in^2), the steel suddenly yields. A rapid sudden elongation takes place while the force applied to the specimen actually decreases. This point is called the *yield point*. The material has become permanently deformed. It will not elastically return to its original shape when the load is removed.

Under increasing load, the specimen again picks up the load, but longitudinal strains increase at a much faster rate than the stresses; the curve is no longer a straight line. The stress–strain curve becomes nonlinear. *Nonlinear* behavior on a

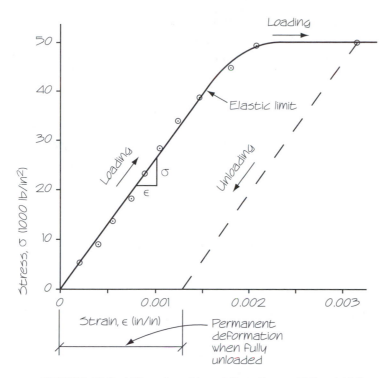

FIGURE 10.8 When the mild steel (from Figures 10.2 and 10.3) is stressed beyond its elastic limit, it becomes permanently deformed. Upon gradual removal of the load, the elastic strain is recovered as the material returns on a path parallel with the slope of the modulus of elasticity. This path is represented by the dashed line. When fully unloaded, the material remains permanently deformed. It has gained permanent set.

stress–strain curve means that strains and stresses are not proportional: twice the stress *does not* produce twice the strain. The stress–strain curve shows that beyond stresses of about 48 000 lb/in^2, longitudinal strain occurs at a faster rate with asymmetrical behavior. The stress–strain curve behaves in a nonlinear manner beyond the yield point.

Shortly after the yield point has been reached, decreases in cross-sectional area of the specimen become quite evident and are easily measured. Beyond the steel's ultimate strength, an obvious reduction in cross-sectional area called necking is evident near the midpoint of the specimen (Figure 10.1c). At the point of maximum stress on the stress–strain curve for mild steel (about 69 000 lb/in^2) the material's *ultimate tensile strength* has been reached. Beyond this point, longitudinal strain will result from what appear to be gradually decreasing stresses until failure occurs at the breaking point. A more rapid failure appears to occur beyond the ultimate strength.

By convention, values of stress that are plotted on the stress–strain curve are calculated with a cross-sectional area based on the original diameter of the specimen.

The cross-sectional area of the specimen is assumed to be constant throughout a test even though longitudinal elongation precipitates a reduction in specimen diameter. Diameter and cross-sectional area decrease as force applied to the specimen increases.

The necking that occurs beyond the ultimate strength of mild steel significantly reduces the cross-sectional area of the specimen. Therefore, values used for stresses are not *true* stresses based on the decreasing diameter of the specimen. Instead, *apparent* stresses based on the original diameter of the specimen are used. The differences in apparent and true stresses are demonstrated in Example 10.2.

■ EXAMPLE 10.2

A 0.505-in-diameter steel specimen failed at a force of 14.35 kips in a tensile test.

(a) Determine the apparent stress of the steel specimen at failure based on the original cross-sectional area.

(b) Determine the true ultimate strength of the steel specimen. Assume that necking caused a portion of the specimen to be reduced to about 0.4 in. in diameter at failure.

Solution

(a)
$$A = \frac{\pi D^2}{4} = \frac{\pi (0.505 \text{ in})^2}{4} = 0.200 \text{ in}^2$$

$$\sigma = \frac{F}{A} = \frac{14\,350 \text{ lb}}{0.200 \text{ in}^2} = 71\,750 \text{ lb/in}^2$$

(b)
$$A = \frac{\pi D^2}{4} = \frac{\pi (0.4 \text{ in})^2}{4} = 0.126 \text{ in}^2$$

$$\sigma = \frac{F}{A} = \frac{14\,350 \text{ lb}}{0.126 \text{ in}^2} = 113\,900 \text{ lb/in}^2$$

It is standard practice to calculate and plot the stress–strain curve based on apparent stress rather than true stresses. If true stresses were used in the analysis (by using the true cross-sectional area as the specimen narrows and necks), the stress–strain curve would climb above the values for ultimate strength and stress at failure. At failure, the true stress would be much greater than the apparent stress; see Example 10.2. If calculated on true stress, the slope of a stress–strain curve would not decrease beyond the ultimate strength. The curve would climb upward to higher values of true stress, as approximated in Figure 10.9. Use of apparent stress is common practice, however, since it is more convenient and involves less measurement.

Although the stress–strain behavior of mild steel in tension traditionally serves as the theoretical basis on which properties of materials are established, this well-defined behavior is only representative of certain materials. Common types of stress–strain curves are illustrated in Figure 10.10. The stress–strain behavior of

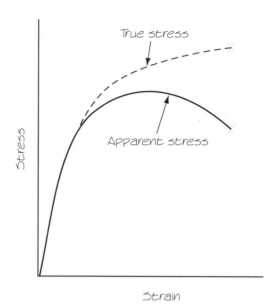

FIGURE 10.9 On a stress–strain curve of mild steel, the true stresses beyond the yield point would be much greater than the apparent stresses shown here. This behavior is due to a phenomenon called necking, which is visibly evident in the specimen as it attains its ultimate strength.

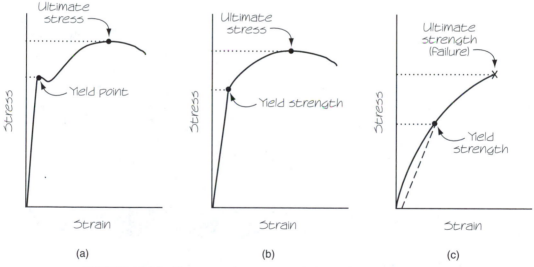

FIGURE 10.10 Variations of stress–strain behavior are illustrated in common stress–strain curves. Curve (a) is characteristic of a material with a well-defined yield point and modulus of elasticity. Curve (b) represents a material where the transition from elastic and plastic deformation is somewhat evident, so that yield stress can be approximated. Curve (c) is representative of a material exhibiting gradual plastic deformation so it has a poorly defined yield point. In this case, the yield strength is defined as that point on the stress–strain curve where permanent set is offset at a strain of 0.2% (0.002 in/in).

mild steel in tension is similar to the stress–strain curve shown in Figure 10.10a. Mild steel (as shown in Figures 10.2 and 10.3) exhibits a well-defined yield point and modulus of elasticity. Mild steel is a very ductile material; it deforms greatly before it fractures. Other materials have significantly different behaviors that are usually less defined than those of this type of steel. Their stress–strain proportionality (the modulus of elasticity) and yield point are not well defined, as shown in Figure 10.10b and c.

Typically, the stress–strain curves for aluminum, concrete, copper, cast iron, and high-carbon steel are a curved line on which there is no true proportionality between stress and strain; the line representing stress–strain is nonlinear. This lack of definition in the stress–strain curve complicates identification of design values such as the yield stress and modulus of elasticity. Since the stress–strain line is only slightly curved, stress–strain proportionality is assumed to exist. Design values such as the elastic limit, modulus of elasticity, and yield strength are determined through standardized approximation techniques. For example, for materials with poorly defined yield points, an arbitrary stress called the *yield strength* is established to approximate where near-elastic behavior ends, as shown in Figure 10.10b. On curves such as the one shown in Figure 10.10c, the yield strength is defined as that point on the stress–strain curve where permanent set is offset at a strain of 0.2% (0.002 in/in).

Behavior of a material tends to change with type of stress. A review of the stress–strain diagram of the compression test of mild steel (Figure 10.5) shows that a well-defined modulus of elasticity exists. It does, however, exhibit a less defined yield point than it did in tension. At about 45 000 lb/in^2 the steel specimen buckled. Smaller compressive forces were required to continue the buckling effect at strains beyond that point.

The stress–strain curve for the compression test of dry-cured concrete (Figure 10.6) shows a poorly defined yield point. There is no true proportionality between stress and strain, so a poorly defined modulus of elasticity exists. The specimen clearly failed at about 4800 lb/in^2 and a strain of 0.001 83 in/in (from Table 10.2). The concrete tested is a nonductile material in comparison to the mild steel. Mild steel (Figure 10.2) failed at a strain of 0.445 00 in/in, a strain that is 243 times greater than the strain of the concrete at compression failure.

10.5   BASIC DESIGN METHODS

The results from material testing are used by the engineer in design of a member. The designer selects a material based on standard test results, estimates the maximum probable load the member will support during its service life, and through analysis, computes the required size of the member such that stresses are limited to safe levels. In many cases, structural members are designed to perform below the yield strength and thus within the elastic range of the stress–strain curve. This practice ensures that the material does not permanently deform under the load for which it is designed. This design technique is known as *elastic design. Plastic design* involves design such that when its service load is applied the member or component can be stressed beyond its elastic limit but never above its ultimate strength.

There are two widely accepted approaches used in design of structural members: the *allowable* or *working stress design* method and the *strength design* method. Both methods are presented below. The allowable or working stress method was once the most widely accepted method used in the structural engineering profession. Research over the past few decades has resulted in accepted use of the strength design method, mainly because it has been proven to provide more uniform reliability and better economy.

The allowable stress or working stress design method is presented in this book because its design approach is less complicated, so it serves as an easier method of acquainting the technician with basic structural engineering design principles. The technician interested in strength design is referred to the many references and textbooks on *load and resistance factor design* (LRFD).

10.6 ALLOWABLE STRESS OR WORKING STRESS DESIGN

Allowable stress design or *working stress design* limits the stress to which the material is subjected to an allowable or working stress. It ensures that a specified stress that is assigned to the specified material is not exceeded under the maximum probable load. *Allowable stress* ($\sigma_{allowable}$), sometimes called *working stress,* is the maximum permissible stress that a material can develop under maximum probable load. For example, a member composed of steel with an allowable stress of 24 000 lb/in^2 should never exceed a stress of 24 000 lb/in^2 when in service.

The proportional limit, yield strength, and ultimate strength of many materials are difficult to determine accurately or does not truly exist with many materials. Therefore, a *factor of safety* (FS) is frequently applied to a material's yield or ultimate strength, which limits stress to a fraction of the yield or ultimate strength:

$$\sigma_{allowable} = \frac{\sigma_{ultimate\ strength}}{FS} \quad or \quad \frac{\sigma_{yield\ strength}}{FS}$$

The factor of safety reduces the yield strength or ultimate strength to the allowable stress. For example, laboratory testing of a certain concrete specimen may show that the concrete fails at a stress of 4000 lb/in^2; that is, the concrete exhibits an ultimate strength of 4000 lb/in^2. In design of a concrete column, a designer may use an allowable stress of 800 lb/in^2. Thus the designer is working with a factor of safety of 5: 4000 lb/in^2/5 = 800 lb/in^2. A factor of safety of 5 against ultimate strength means that the maximum anticipated stress applied to the concrete column will not exceed one-fifth or 20% of the ultimate strength.

■ EXAMPLE 10.3

The ultimate compressive strength determined through laboratory testing of a certain mix of concrete was found to be 35.0 MPa. Determine the allowable stress for this concrete mix at a factor of safety of 5 against ultimate compressive strength.

Solution

$$\sigma_{allowable} = \frac{\sigma_{ultimate\ strength}}{FS} = \frac{35.0\ MPa}{5} = 7.0\ MPa$$

■ EXAMPLE 10.4

The minimum yield strength of a low-carbon steel is specified as 36 000 lb/in^2. Determine the allowable stress for this steel at a factor of safety of 1.5 against yield strength.

Solution

$$\sigma_{allowable} = \frac{\sigma_{yield\ strength}}{FS} = \frac{36\ 000\ lb/in^2}{1.5} = 24\ 000\ lb/in^2$$

Values of allowable stress or factors of safety applied to an ultimate or yield strength used to compute the allowable stress are determined after careful consideration of results of many standardized tests. Standard values of allowable stress or factors of safety are reported in various codes (such as the Uniform Building Code) and specifications (such as from American Society for Testing and Materials specifications). Typically, factors of safety are not specified directly. Instead, allowable stresses are generally established for different materials under different conditions of use. In many cases the allowable or working stress is computed using a prescribed formula; for example, the working stress of wood, steel, concrete, and masonry columns is determined by a prescribed set of formulas that are based on properties such as material strength, cross-section characteristics, and height. Values for allowable stresses of common materials are provided in Tables A.3 and A.4.

■ EXAMPLE 10.5

A steel hanger must carry a load of 20 000 lb. The minimum yield strength of a low-carbon steel in tension is 36 000 lb/in^2. Determine the minimum cross-sectional area of the hanger assuming a factor of safety of 1.5 against yield strength.

Solution

$$\sigma_{allowable} = \frac{\sigma_{yield\ strength}}{FS} = \frac{36\ 000\ lb/in^2}{1.5} = 24\ 000\ lb/in^2$$

$$\sigma = \frac{F}{A} \quad \text{therefore} \quad A \geq \frac{F}{\sigma}$$

$$A \geq \frac{F}{\sigma}$$

$$\geq \frac{20\ 000\ \text{lb}}{24\ 000\ \text{lb/in}^2}$$

$$\geq 0.833\ \text{in}^2$$

In Example 10.5 the cross-sectional area of the steel hanger must be at least 0.833 in². The designer could safely specify a 1-in² bar, since its cross-sectional area is 1 in². This would result in an actual stress for the 1-in² bar of

$$\sigma = \frac{F}{A} = \frac{20\ 000\ \text{lb}}{1\ \text{in}^2} = 20\ 000\ \text{lb/in}^2$$

The actual stress for the 1-in² bar of 20 000 lb/in² is below the allowable stress of 24 000 lb/in² and thus is acceptable.

10.7 STRENGTH DESIGN

Strength design, referred to as *load and resistance factor design* (LRFD) in structural engineering, involves designing a member on the basis of its ultimate load-carrying capacity. In this design method a *strength reduction factor* (ϕ) is used to reduce the load-carrying capacity of the member to a permissible strength. The strength reduction factor is a multiplier that is less than 1; it usually ranges from 0.5 to 0.9. The strength reduction factor is a type of factor of safety (a multiplier) that accounts for imperfections in material and variations in construction and load.

■ EXAMPLE 10.6

A 1.00-in² steel bar will be used to hang a load. In tension, the specified steel has a minimum yield strength of 36 000 lb/in² and a minimum ultimate strength of 58 000 lb/in².

(a) Determine the maximum permissible load that the bar can carry using the allowable strength design method. The appropriate standard calls for a factor of safety of 1.5 against yield strength.

(b) Determine the maximum permissible load that the bar can carry using the strength design method. Since the load carried by the bar is an accurately determined dead load based on known weights of materials, use a strength reduction factor of 0.90 against yield strength.

(c) Determine the maximum permissible load that the bar can carry using the plastic design method. Use a factor of safety of 2.0 against ultimate strength.

Solution (a) Determining the allowable stress:

$$\sigma_{allowable} = \frac{\sigma_{yield\ strength}}{FS} = \frac{36\ 000\ lb/in^2}{1.5} = 24\ 000\ lb/in^2$$

Determining the maximum permissible load:

$$F \leq \sigma A$$
$$\leq 24\ 000\ lb/in^2 \cdot (1.00\ in \cdot 1.00\ in)$$
$$\leq 24\ 000\ lb$$

(b) Beginning with the stress formula, $\sigma = F/A$, the ultimate load is found by

$$F \leq \sigma A$$
$$\leq 36\ 000\ lb/in^2 \cdot (1.00\ in \cdot 1.00\ in)$$
$$\leq 36\ 000\ lb$$

The maximum permissible load is found by multiplying the ultimate load by the strength reduction factor:

$$\phi F = (0.90)(36\ 000lb) = 32\ 400\ lb$$

(c) Determining the allowable stress:

$$\sigma_{allowable} = \frac{\sigma_{yield\ strength}}{FS} = \frac{58\ 000\ lb/in^2}{2} = 29\ 000\ lb/in^2$$

Determining the maximum permissible load:

$$F \leq \sigma A$$
$$\leq 29\ 000\ lb/in^2 \cdot (1.00\ in \cdot 1.00\ in)$$
$$\leq 29\ 000\ lb$$

Strength design differs from allowable and working stress design in that a factor of safety is applied to the load rather than to the stress. It typically leads to more efficient use of materials because a smaller factor of safety is applied to known loads such as weights of building materials.

EXERCISES

Material Testing

10.1. A 0.505-in-diameter mild steel specimen was tested to failure in a tensile testing machine by slowly pulling it apart with a steadily increasing force. Deformation (elongation) of the specimen was measured at force intervals of 1000 lb until the specimen failed. The initial gauge length of the specimen was 2.00 in. Observed data are provided in Table E10.1.

(a) Calculate the stress and strain at each force interval.
(b) Plot a graph of the stress–strain curve.
(c) Estimate the yield point of the steel. Note its location on the curve.
(d) Estimate the ultimate strength of the steel. Note its location on the curve.

Table E10.1   TENSILE TEST OF MILD STEEL: OBSERVED DATA

Force (lb)	Deformation (in)	Force (lb)	Deformation (in)
0	—	10 000	0.1170
1 000	0.0004	11 000	0.1555
2 000	0.0007	12 000	0.1705
3 000	0.0011	13 000	0.1875
4 000	0.0014	14 000	0.2058
5 000	0.0017	15 000	0.2380
6 000	0.0021	16 000	0.2710
7 000	0.0024	16 200	0.2982
8 000	0.0032	15 000	0.3570
9 000	0.0748	14 100	Fracture

10.2 A 0.505-in-diameter aluminum specimen was tested to failure in a tensile testing machine by slowly pulling it apart with a steadily increasing force. Deformation (elongation) of the specimen was measured at 500-lb-force intervals until the specimen failed. The initial gauge length of the specimen was 2.00 in. Observed data are provided in Table E10.2.
(a) Calculate the stress and strain at each force interval.
(b) Plot a graph of the stress–strain curve.
(c) Estimate the yield point. Note its location on the curve.
(d) Estimate the ultimate strength. Note its location on the curve.

Table E10.2   TENSILE TEST OF ALUMINUM: OBSERVED DATA

Force (lb)	Deformation (in)	Force (lb)	Deformation (in)
0	—	9 000	0.0085
1 000	0.0009	10 000	0.0095
2 000	0.0019	11 000	0.0104
3 000	0.0028	12 000	0.0143
4 000	0.0038	13 000	0.0197
5 000	0.0047	14 000	0.0325
6 000	0.0057	14 550	0.0710
7 000	0.0066	14 000	0.1320
8 000	0.0076	13 700	Fracture

10.3 A $5\frac{3}{4}$-in-diameter concrete specimen was compression tested to failure 7 days after the concrete was mixed. Deformation of the specimen was measured at force intervals of

5000 lb. The initial deflectometer gauge length was 4.0625 in initially. Observed data are provided in Table E10.3.

(a) Calculate the stress and strain at each force interval.
(b) Plot a graph of the stress–strain curve.
(c) Estimate the ultimate strength. Note its location on the curve.

Table E10.3 COMPRESSION TEST OF CONCRETE: OBSERVED DATA

Force (lb)	Deformation (in)	Force (lb)	Deformation (in)
0	—	60 000	0.0054
5 000	0.0001	65 000	0.0059
10 000	0.0003	70 000	0.0065
15 000	0.0007	75 000	0.0071
20 000	0.0012	80 000	0.0078
25 000	0.0016	85 000	0.0085
30 000	0.0021	90 000	0.0090
35 000	0.0026	95 000	0.0100
40 000	0.0032	100 000	0.0110
45 000	0.0037	105 000	0.0122
50 000	0.0042	110 000	0.0140
55 000	0.0048	113 000	Fracture

10.4 A 1×2 ($\frac{3}{4}$ in $\times$ $1\frac{5}{8}$ in) block of wood was tested in a compression test. The original length of the block was 4.50 in. Deformation of the specimen was measured at the force intervals noted in Table E10.4.

(a) Calculate the stress and strain at each force interval.
(b) Plot a graph of the stress–strain curve.
(c) Estimate the ultimate strength. Note its location on the curve.

Table E10.4 TENSILE TEST OF WOOD: OBSERVED DATA

Force (lb)	Deformation (in)	Force (lb)	Deformation (in)
0	—	8000	0.0432
225	0.0034	8500	0.0468
550	0.0066	9000	0.0507
1000	0.0095	9500	0.0595
1500	0.0123	9570	0.0641
2000	0.0125	9570	0.0701
2600	0.0175	9300	0.0808
3000	0.0193	9000	0.0975
4000	0.0236	8950	0.1096
5000	0.0275	9000	0.1215
6000	0.0322	9100	0.1293
7000	0.0373	8700	0.1423
7500	0.0397	Specimen was removed from test setup	

10.5   With the results in Exercises 10.1 through 10.4, plot the stress–strain curves for each material on a single stress–strain diagram. Compare ultimate strengths and moduli of elasticity.

10.6   With the results in Exercise 10.1, approximate the strain that remains if:
   (a) The specimen is initially tensioned to a stress of 15 000 lb/in^2 and then unloaded.
   (b) The specimen is initially tensioned to a stress of 30 000 lb/in^2 and then unloaded.
   (c) The specimen is initially tensioned to a stress of 45 000 lb/in^2 and then unloaded.
   (d) The specimen is initially tensioned to a stress of 60 000 lb/in^2 and then unloaded.

10.7   As shown in Tables A.3 and A.4, aluminum, concrete, copper, steel, and wood (Douglas fir) have unique moduli of elasticity.

   (a) Plot the slope of the modulus of elasticity for each material on a single stress–strain diagram (similar to that shown in Figure 10.7).

   (b) Identify the stiffest material of those examined.

Simple Design

10.8   On the 28th day after concrete was mixed, a 6.00-in-diameter concrete cylinder failed in a compressive testing at a load of 145 000 lb.
   (a) Determine the ultimate strength of the concrete, in lb/in^2.
   (b) Determine the allowable compressive stress with a factor of safety of 5.0 against ultimate strength.
   (c) Determine the maximum permissible load that the concrete can carry using the strength design method. Use a strength reduction factor of 0.60 against ultimate strength.

10.9   A 25-mm-diameter round bar will be used to hang a load. In tension, the material has a yield strength of 170 MPa and an ultimate strength of 380 MPa.

   (a) Determine the maximum permissible load that the bar can carry using the allowable strength design method. Use a factor of safety of 2.0 against yield strength.

   (b) Determine the maximum permissible load that the bar can carry using the strength design method. Use a strength reduction factor of 0.75 against ultimate strength.

10.10 A 0.50-in^2 steel bar will be used to hang a load. In tension, the steel has a minimum yield strength of 36 000 lb/in^2 and an ultimate strength of 58 000 lb/in^2.

   (a) Determine the maximum permissible load that the bar can carry using the allowable strength design method. Use a factor of safety of 1.5 against yield strength.

   (b) Determine the maximum permissible load that the bar can carry using the strength design method. Use a strength reduction factor of 0.85 against yield strength.

   (c) Determine the maximum permissible load that the bar can carry using the strength design method. Use a strength reduction factor of 0.6 against ultimate strength.

Chapter 11
SHEAR IN BEAMS

A beam must be designed to withstand the bending and shearing stresses that are induced by the effects of the applied loads and resulting reactions. Such design involves determining the shear forces and bending moments across the beam length. Once these are determined a beam material and cross section can be selected to withstand these stresses. In this chapter, shear forces that act on and within a beam will be examined. Bending moment and related bending stresses in beams are examined in Chapter 12. Each investigation will result in diagrams that graphically illustrate the shear forces acting on the beam at any point across its length.

11.1   SHEAR FORCES IN BEAMS

Shear stress occurs when two forces with parallel but offset lines of action act in opposite directions on a body. In a beam under load, there are two major tendencies for it to fail due to shear: vertical and horizontal shear. *Vertical shear* is the shearing force that tends to cause a member to fail by a cutting action perpendicular to the beam's longitudinal axis, as shown in Figure 11.1a. Consider a beam suspended between two supports. As a force or load is applied, the member *tends* to drop between the two supports but is restrained by the resistance or strength of the material of the beam. The ends of the beam tend to remain on the supports. If sufficient force or load is applied, the member may break by shearing at or near the beam supports. This type of failure is often a concern in short beams carrying heavy loads.

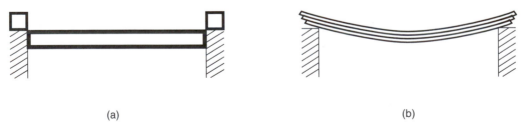

(a)   (b)

FIGURE 11.1   Exaggerated examples of the effects of vertical (a) and horizontal (b) shear in simple beams under load. Vertical shear (a) occurs when the beam tends to drop at its supports. Horizontal shear (b) is the sliding of beam fibers in a horizontal manner.

Horizontal shear refers to the tendency of theoretical layers in a member to slide horizontally. This is best illustrated by loading stacked boards that span between two supports, as shown in Figure 11.1b. When a load or force is applied, the boards bend and tend to slide horizontally; they become uneven at the ends. The surfaces in contact encounter a sliding or horizontal shearing effect. The molecules of the material of any beam under load experiences this horizontal shearing effect.

11.2   SHEAR CALCULATIONS AND DIAGRAMS

Vertical shear is commonly designated by the symbol V. It is computed by moving an imaginary section along the length of the beam and summing only those forces (loads and reactions) acting to the left of that section. The section is placed perpendicular to the longitudinal axis of the beam and is assumed to have negligible thickness. Vertical shear at any point along a beam is calculated by summing any forces acting upward (F_{up}) and subtracting any forces acting downward (F_{down}) to the left of the section under consideration. External forces to the right of the section are neglected. This approach requires the assumption that the beam is in static equilibrium; that is, the external forces and moments acting at any point along the beam counteract each another ($\Sigma F = 0$ and $\Sigma M = 0$).

The expression for calculating a vertical shear force (V) in a beam is noted below and is stated as: *The vertical shear force at the section under consideration is equal to the sum of the forces acting upward minus the sum of the forces acting downward calculated to the left of the section:*

$$V @ x = \Sigma F_{up} - \Sigma F_{down} \text{ to the left of section under consideration}$$

The section under consideration is placed at a convenient location on the free body. It can be positioned at any location across the beam: at 2 ft to the right of R_1, 2.5 ft to the left of R_2, 4 m from the left end of the beam, and so on. The designer selects where this section is located and looks only to the left of that section when making computations of vertical shear. Computations are made by summing the forces on the left side of the section, that is, adding those forces acting upward and subtracting those forces acting downward. The resulting value is the vertical shear force at that section location.

It is customary to cite the location of the section by expressing a measurement from the left end of the beam: for example $V @ 2$ ft indicates that the computation is made with the section placed 2 ft away from the left end of the beam. In this text, the location of the vertical shear force will always be identified by its distance from the left end of the beam.

There is no vertical shear force that exists directly below a reaction or concentrated force because a shear force requires offsetting forces. There are no offsetting forces at a reaction or concentrated force, so vertical shear is zero at these locations. Vertical shear forces do exist on either side of a concentrated force or re-action, however. Consequently, calculation of the shear force must be made a small distance to the right and left of a concentrated force or reaction. This distance is so small that it is not worthy of precise measurement other than to say that it is just to the left or just to the right of the concentrated force or reaction.

In shear force calculations, the location of the section (i.e., V @ 2 ft, V @ 3 m, etc.) followed by a negative $(-)$ or positive $(+)$ symbol denotes placement *just to the left* $(-)$ or *just to the right* $(+)$ of the concentrated force or reaction. The negative sign in V @ $x-$ indicates section placement a very small distance to the left of the x mark on the beam. V @ $x+$ indicates a very small distance to the right of the concentrated force or reaction. So, V @ 2 ft$-$ indicates a small distance to the left of the 2-ft position on the beam, V @ 3 m$+$ means a small distance to the right of the 3-m mark on the beam, V @ R_1+ indicates a small distance to the right of R_1, and so on. Again, the exact distance designated by $+$ and $-$ is inconsequential and not worthy of exact measurement.

■ EXAMPLE 11.1

A free-body diagram of a simple beam is shown in Figure 11.2. The 4-m beam carries a concentrated load of 40 kN at its center. Reactions R_1 and R_2 each equal 20 kN. Calculate the vertical shear forces at 1-m intervals across the beam.

Solution Beginning at just to the right of R_1 (R_1+) and ending at just to the left $(-)$ of R_2 at intervals of 1 ft along the length of the beam, calculations are

$$V @ x = \Sigma F_{up} - \Sigma F_{down} \text{ to the left of the section under consideration}$$
$$V @ R_1+ = 20 \text{ kN} - 0 = 20 \text{ kN}$$
$$V @ 1 \text{ m} = 20 \text{ kN} - 0 = 20 \text{ kN}$$
$$V @ 2 \text{ m}- = 20 \text{ kN} - 0 = 20 \text{ kN}$$
$$V @ 2 \text{ m}+ = 20 \text{ kN} - 40 \text{ kN} = -20 \text{ kN}$$
$$V @ 3 \text{ m} = 20 \text{ kN} - 40 \text{ kN} = -20 \text{ kN}$$
$$V @ R_2- = 20 \text{ kN} - 40 \text{ kN} = -20 \text{ kN}$$

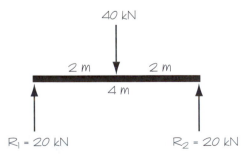

FIGURE 11.2 Free-body diagram of the beam for Example 11.1

■ EXAMPLE 11.2

A free-body diagram of a simple beam is shown in Figure 11.3. The 4-m beam carries a uniformly distributed load of 10 kN/m across its entire length. Reactions R_1 and R_2 are equal to 20 kN each. Calculate the vertical shear forces at 1-m intervals across the beam.

Solution The uniformly distributed load is 10 kN/m (w). It extends across the entire 4-m length of the beam. Beginning at just to the right (+) of R_1 and ending at just to the left (−) of R_2 at intervals of 1 ft along the length of the beam, calculations are

$$V @ x = \Sigma F_{up} - \Sigma F_{down} \text{ to the left of the section under consideration}$$
$$V @ R_1 + = 20 \text{ kN} - 0 = 20 \text{ kN}$$
$$V @ 1 \text{ m} = 20 \text{ kN} - (10 \text{ kN/m} \cdot 1 \text{ m}) = 10 \text{ kN}$$
$$V @ 2 \text{ m} = 20 \text{ kN} - (10 \text{ kN/m} \cdot 2 \text{ m}) = 0$$
$$V @ 3 \text{ m} = 20 \text{ kN} - (10 \text{ kN/m} \cdot 3 \text{ m}) = -10 \text{ kN}$$
$$V @ R_2 - = 20 \text{ kN} - (10 \text{ kN/m} \cdot 4 \text{ m}) = -20 \text{ kN}$$

The vertical shear calculations in Examples 11.1 and 11.2 provide values of vertical shear force at 1-ft intervals along the length of the beam. As demonstrated in Example 11.1, the shear force must be calculated on the sides of the 40-kN concentrated force and on the side of each reaction. The positive and negative symbols at locations R_1+, 2 m−, 2 m+, and R_2- serve as an indication that the shear force is computed a small distance to the left (−) or to the right (+) of the position of the concentrated force or reaction under consideration. Calculation of the shear force on the sides of a concentrated force or reaction is important, since the shear force is different at these locations. Computation of the shear force on the side of a reaction or concentrated force where there is not any beam is not necessary because a vertical shear force cannot exist if the beam does not exist. In Examples 11.1 and 11.2, $V @ R_1-$ and $V @ R_2+$ do not exist.

Shear diagrams are created from vertical shear force calculation results by plotting the values of vertical shear force in the form of a curve, at an appropriate

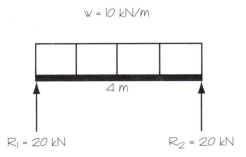

FIGURE 11.3 Free-body diagram of the beam for Example 11.2.

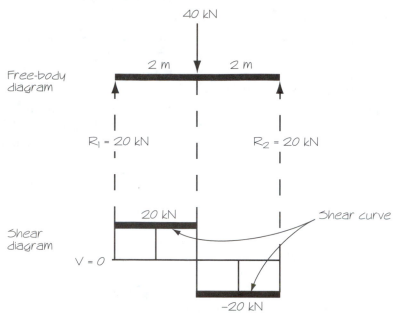

FIGURE 11.4 Vertical shear diagram for Example 11.1.

A Mathematical Curve

In mathematics, a *curve* is a path traced out by a series of related points. Although it can take the shape of a curved line, a curve can also be a straight line. A curve can have limited or infinite length and can be open, such that its ends never close or can be closed such as a circle or ellipse.

---

scale, below the free body diagram. Figures 11.4 and 11.5 illustrate the shear force curve plotted from calculation results in Examples 11.1 and 11.2, respectively.

The shear force diagram for Example 11.1 illustrates that a 40-kN concentrated force creates a significant change in shear force (from 20 kN at 2 m− to −20 kN at 2m+). Also, where there is no force or reaction, there is no change in magnitude of shear force; the vertical shear force curve is horizontal. A vertical shear force does not exist directly below the 40 kN concentrated force. At this location, the vertical shear force is said to *pass through zero* from 20 kN at 2 m− to −20 kN at 2 m+.

The shear force diagram for Example 11.2 illustrates that uniformly distributed forces result in a sloping (diagonal) shear force curve. The curve is linear (straight) because the force is uniformly distributed. Also, note that the vertical shear force passes through zero at the center of the member (at 2 m).

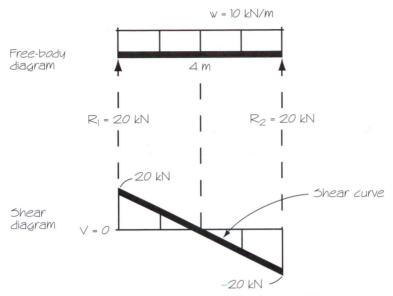

FIGURE 11.5 Vertical shear diagram for Example 11.2.

Note that in both Examples 11.1 and 11.2 the vertical shear force switches sign from positive to negative at the 2-m mark; the shear force is positive on the left side of the 2-m mark and negative on the right side. The shear force at any point across a beam length is the absolute value because a force cannot have a negative sign. The value of the vertical shear force at 3 m from R_1 (V @ 3 m) is expressed as -20 kN in Example 11.1 and as -10 kN in Example 11.2, but the vertical shear forces are actually 20 kN and 10 kN, respectively.

The positive and negative signs for the vertical shear force are useful in determining the directions of the shear forces. By sign convention, a positive $(+)$ value for a vertical shear force signifies that a shearing force on the left side of the section acts upward and on the right side acts downward ($\uparrow\downarrow$). A negative sign $(-)$ indicates that the shear forces act down on the left side and up on the right ($\downarrow\uparrow$). This sign convention and force action holds true only when the vertical shear force is calculated to the *left of the specified section.*

- Positive values of shear force $(+V)$ act $\uparrow\downarrow$ at the section under consideration.
- Negative values of shear force $(-V)$ act $\downarrow\uparrow$ at the section under consideration.

Consider the effects of the uniformly distributed load on the simple beam in Example 11.2: The load has the effect of pushing the beam downward, so it tends to shear at and drop between the reactions. The strength of the beam material resists failure, but these vertical shear forces do develop in the beam. At the left reaction (R_1), the material above the reaction remains in place and the material to the right of the reaction tends to drop downward. The molecules on either side of this thin section (at R_1+) undergo an upward force on the left side and a downward force on the right. With the calculation technique presented (calculating the shear force to the left of the specified section), the shear force has a positive sign at this section (20 kN). By sign convention, a positive value of a vertical shear force represents a shearing force on the left side of the section that acts upward and a right side that acts downward ($\uparrow\downarrow$). The opposite effect holds true on the other side of the beam. The molecules along a section at R_2- undergo a downward force on the left side and an upward force at the right ($\downarrow\uparrow$). By convention, the sign of the shear force at this point is negative.

This sign convention described above holds true only when the shear force is calculated to the left of the specified section. Calculations based on forces acting to the right of the section result in the same absolute value but the signs are reversed. In Example 11.1:

$$V @ x = \Sigma F_{up} - \Sigma F_{down} \text{ to the } left \text{ of the section under consideration}$$
$$V @ 1 \text{ m} = 20 \text{ kN} - 0 = 20 \text{ kN}$$

$$V @ x = \Sigma F_{up} - \Sigma F_{down} \text{ to the } right \text{ of the section under consideration}$$
$$V @ 1\text{m} = 20 \text{ kN} - 40 \text{ kN} = -20 \text{ kN}$$

A shear force calculation can be performed by looking on the right side of the section. It may be helpful to compute to the right of the section to simplify work or compare the absolute value on the right with the value on the left. However, by convention, vertical shear is always computed and reported as if it were calculated to the left of the section specified.

11.3   THE MAXIMUM VERTICAL SHEARING FORCE

Maximum vertical shearing force (V_{max}) is the greatest absolute value of shear across the entire length of the beam. The shear diagrams in Examples 11.1 and 11.2 have maximum vertical shearing forces that (coincidentally) have a magnitude of 20 kN. In Example 11.1, maximum vertical shear occurs across the entire beam length (between R_1+ and R_2-) except below the concentrated force. In Example 11.2, the maximum vertical shear force occurs just to the right of R_1 (at R_1+) and just to the left of R_2 (at R_2-).

■ EXAMPLE 11.3

A free-body diagram of a simple beam is shown in Figure 11.6. The 8-ft-long beam carries a 1000-lb concentrated load at the center of its span and a uni-

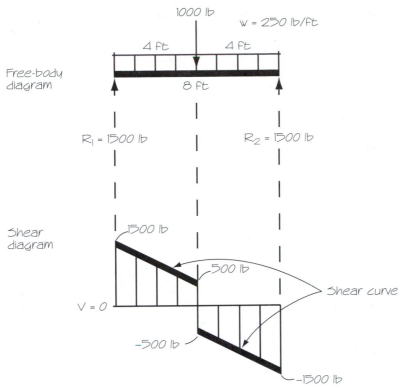

FIGURE 11.6 Free-body and vertical shear diagrams of the beam for Example 11.3.

formly distributed load of 250 lb/ft across its entire length. Calculate vertical shear forces at 1-ft intervals across the beam length.

Solution Beginning at just to the right (+) of R_1 and ending at just to the left (−) of R_2 at intervals of 1 ft along the length of the beam, calculations are:

$$V @ x = \Sigma F_{up} - \Sigma F_{down} \text{ to the left of the section under consideration}$$
$$V @ R_1+ = 1500 \text{ lb} - 0 = 1500 \text{ lb}$$
$$V @ 1 \text{ ft} = 1500 \text{ lb} - (250 \text{ lb/ft} \cdot 1 \text{ ft}) = 1250 \text{ lb}$$
$$V @ 2 \text{ ft} = 1500 \text{ lb} - (250 \text{ lb/ft} \cdot 2 \text{ ft}) = 1000 \text{ lb}$$
$$V @ 3 \text{ ft} = 1500 \text{ lb} - (250 \text{ lb/ft} \cdot 3 \text{ ft}) = 750 \text{ lb}$$
$$V @ 4 \text{ ft}- = 1500 \text{ lb} - (250 \text{ lb/ft} \cdot 4 \text{ ft}) = 500 \text{ lb}$$
$$V @ 4 \text{ ft}+ = 1500 \text{ lb} - (250 \text{ lb/ft} \cdot 4 \text{ ft}) - 1000 \text{ lb} = -500 \text{ lb}$$
$$V @ 5 \text{ ft} = 1500 \text{ lb} - (250 \text{ lb/ft} \cdot 5 \text{ ft}) - 1000 \text{ lb} = -750 \text{ lb}$$
$$V @ 6 \text{ ft} = 1500 \text{ lb} - (250 \text{ lb/ft} \cdot 6 \text{ ft}) - 1000 \text{ lb} = -1000 \text{ lb}$$
$$V @ 7 \text{ ft} = 1500 \text{ lb} - (250 \text{ lb/ft} \cdot 7 \text{ ft}) - 1000 \text{ lb} = -1250 \text{ lb}$$
$$V @ R_2- = 1500 \text{ lb} - (250 \text{ lb/ft} \cdot 8 \text{ ft}) - 1000 \text{ lb} = -1500 \text{ lb}$$

In Example 11.3 the maximum vertical shear force occurs just to the right of R_1 (at R_1+) and just to the left of R_2 (at R_2-). It has a magnitude of 1500 lb. On either side of the concentrated load, the vertical shear force changes sign. By sign convention the positive sign of the vertical shear forces on the left side of the concentrated load indicates that the shearing forces act ↑↓. The opposite effect holds true just on the right side of the concentrated load. The negative sign of the vertical shearing forces between 4 ft+ and R_2- indicates that the shearing forces act ↓↑. Also, observe that the slope of the shear curve is identical across the length of the beam because the uniformly distributed load is consistent across the entire beam length; like uniformly distributed loads have like slopes.

■ EXAMPLE 11.4

A free-body diagram of a simple beam is shown in Figure 11.7. The 4-ft beam carries a 100-kip concentrated load at the center of the beam. Uniformly distributed loads of 10 kip/ft on the left side of the beam and 5 kip/ft on the right side also rest on the beam. Reaction R_1 is 67.5 kip and reaction R_2 is 62.5 kip. Calculate the vertical shear forces at 1-m intervals across the beam.

Solution   Beginning at just to the right $(+)$ of R_1 and ending at just to the left $(-)$ of R_2 at intervals of 1 ft along the length of the beam, calculations are

$$V @ x = \Sigma F_{up} - \Sigma F_{down} \text{ to the left of the section under consideration}$$
$$V @ R_1+ = 67.5 \text{ kip} - 0 = 67.5 \text{ kip}$$
$$V @ 1 \text{ ft} = 67.5 \text{ kip} - (10 \text{ kip/ft} \cdot 1 \text{ ft}) = 57.5 \text{ kip}$$
$$V @ 2 \text{ ft}- = 67.5 \text{ kip} - (10 \text{ kip/ft} \cdot 2 \text{ ft}) = 47.5 \text{ kip}$$
$$V @ 2 \text{ ft}+ = 67.5 \text{ kip} - (10 \text{ kip/ft} \cdot 2 \text{ ft}) - 100 \text{ kip} = -52.5 \text{ kip}$$
$$V @ 3 \text{ ft} = 67.5 \text{ kip} - (10 \text{ kip/ft} \cdot 2 \text{ ft}) - 100 \text{ kip} - (5 \text{ kip/ft} \cdot 1 \text{ ft})$$
$$= -57.5 \text{ kip}$$
$$V @ R_2- = 67.5 \text{ kip} - (10 \text{ kip/ft} \cdot 2 \text{ ft}) - 100 \text{ kip} - (5 \text{ kip/ft} \cdot 2 \text{ ft})$$
$$= -62.5 \text{ kip}$$

In Example 11.4 the uniformly distributed load of 10 kip/ft extends across the left 2 ft of the beam. A second uniformly distributed load of 5 kip/ft covers the right 2 ft of the beam. As uniformly distributed loads, they result in shear curves that are linear (straight) and sloping (diagonal). Because they are loads of different magnitudes, the slope of each curve is different. The curve is steep for the 10 kip/ft load and gradual for the 5 kip/ft load. The slope of a uniformly distributed load on a shear curve is characteristic of the load; similar uniformly distributed loads have shear curves with like slopes and different uniformly distributed loads have shear curves with different slopes.

Additionally, in Example 11.4, the maximum vertical shear force occurs just to the left of R_2 (at R_2-) with a magnitude of 62.5 kip. The maximum shear force is the maximum absolute value of shear. Shear passes through zero below the 100-kip load (at 2 ft on the beam).

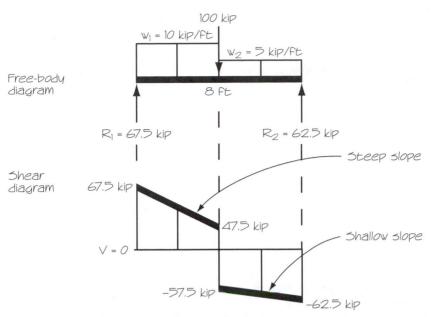

FIGURE 11.7 Free-body and vertical shear diagrams of the beam for Example 11.4.

11.4 WHERE SHEAR PASSES THROUGH ZERO

The location or locations along a beam where shear passes through zero are the points where the beam will have the greatest tendency to fail due to bending. As we discuss in Chapter 12, these points must be examined in more detail to determine the beam cross section accurately so that the beam will safely resist bending stresses. Thus in shear analysis it is necessary to locate the points where shear passes through zero ($V = 0$).

Frequently, a vertical shear force will pass through zero at a known location. In Example 11.4 the shear force passes through zero below the 100-kip concentrated force because shear is a positive value at the $x = 2$ ft− location and it is negative at $x = 2$ ft+. By a *mathematical pinch play,* shear must pass through zero at $x = 2$ ft because shear is a positive value on one side of the section and negative on the other side. The pinch play method is an acceptable method of finding the point or points in the beam where shear passes through zero.

■ EXAMPLE 11.5

A free-body diagram of a single overhang beam is shown in Figure 11.8. The 3-m beam carries a concentrated load of 100 kN at the right end of the 1-m overhang. Reaction R_1 equals 50 kN acting downward, and reaction R_2 equals 150 kN acting upward. Calculate the vertical shear forces at 1-m intervals across the beam.

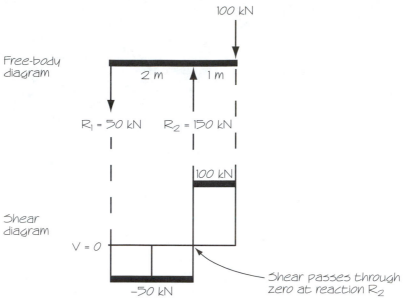

FIGURE 11.8 Free-body and vertical shear diagrams of the beam for Example 11.5.

Solution Beginning at just to the right $(+)$ of R_1 and ending at $x = 3$ m at intervals of 1 ft along the length of the beam, calculations are

$$V @ x = \Sigma F_{\text{up}} - \Sigma F_{\text{down}} \text{ to the left of the section under consideration}$$
$$V @ R_1+ = + 0 - 50 \text{ kN} = -50 \text{ kN}$$
$$V @ 1 \text{ m} = + 0 - 50 \text{ kN} = -50 \text{ kN}$$
$$V @ R_2- = + 0 - 50 \text{ kN} = -50 \text{ kN}$$
$$V @ R_2+ = + 150 \text{ kN} - 50 \text{ kN} = 100 \text{ kN}$$
$$V @ 3 \text{ m}- = + 150 \text{ kN} - 50 \text{ kN} = 100 \text{ kN}$$

In Example 11.5 the vertical shear force is negative at R_2- and positive at R_2+. By the mathematical pinch play method, the vertical shear force was found to pass through zero at R_2. Also, observe that the shear curve is straight and horizontal where there are no forces acting on the beam, from R_1+ to R_2- and from R_2+ to 3 m$-$. To simplify computations, it is only necessary to calculate the value of shear in segments of the beam where loads or reactions exist; the shear curve will be straight and horizontal in this segment of the free body. Finally, notice that the maximum vertical shear force occurs just to the right of R_2 (at R_2+) with a magnitude of 100 kN.

Under many loading conditions involving uniformly distributed loads, shear will pass through zero between the convenient 1-ft or 1-m intervals used in the calculation technique that was just presented. This behavior may be attributed to the sloping nature of the shear curve that is associated with uniformly distributed

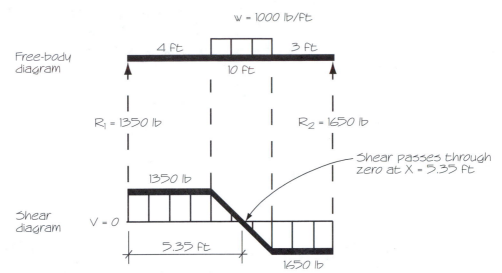

FIGURE 11.9 Free-body and vertical shear diagrams of the beam for Example 11.6.

forces. In such cases the designer may locate the point where shear passes through zero through the use of linear algebra and the relationship between similar triangles. The *similar triangle technique* of finding where shear passes through zero is demonstrated in Example 11.6.

■ EXAMPLE 11.6

A free-body diagram of a simple beam is shown in Figure 11.9. The 10-ft-long beam carries a uniformly distributed load of 1000 lb/ft for 3 ft beginning 4 ft to the right of R_1. Reaction R_1 equals 1350 lb and reaction R_2 equals 1650 lb. Calculate the vertical shear forces at 1-ft intervals across the beam and find where shear passes through zero.

Solution Shear is calculated at select sections. To simplify computations, the position of the section is to the side of the reactions and load and at 1-ft intervals where the uniformly distributed load is present. Shear is not calculated where a load or reaction is not present.

$$V @ x = \Sigma F_{up} - \Sigma F_{down} \text{ to the left of the section under consideration}$$
$$V @ R_1+ = 1350 \text{ lb} - 0 = 1350 \text{ lb}$$
$$V @ 4 \text{ ft} = 1350 \text{ lb} - 0 = 1350 \text{ lb}$$
$$V @ 5 \text{ ft} = 1350 \text{ lb} - (1000 \text{ lb/ft} \cdot 1 \text{ ft}) = 350 \text{ lb}$$
$$V @ 6 \text{ ft} = 1350 \text{ lb} - (1000 \text{ lb/ft} \cdot 2 \text{ ft}) = -650 \text{ lb}$$
$$V @ 7 \text{ ft} = 1350 \text{ lb} - (1000 \text{ lb/ft} \cdot 3 \text{ ft}) = -1650 \text{ lb}$$
$$V @ R_2- = 1350 \text{ lb} - (1000 \text{ lb/ft} \cdot 3 \text{ ft}) = -1650 \text{ lb}$$

11.4 Where Shear Passes Through Zero   **199**

Vertical shear passes through zero between the 5- and 6-ft locations. The uniformly distributed load begins at $x = 4$ ft. So by algebraically equating the unknown position (x) to $V = 0$, the location where shear passes through zero is found:

$$V = 0 = 1350 \text{ lb} - (1000 \text{ lb/ft})(x - 4 \text{ ft})$$
$$0 = 1350 \text{ lb} - (1000 \text{ lb/ft})(x - 4 \text{ ft})$$
$$= 1350 - 1000x + 4000$$
$$= 5350 - 1000x$$
$$1000x = 5350$$
$$x = \frac{5350}{1000}$$
$$= 5.35 \text{ ft}$$

In Example 11.6, shear passes through zero at 11.35 ft from R_1. The symbol x represents the length of the beam carrying the uniformly distributed force required to set $V = 0$. A similar algebraic approach may be used to solve for the location where shear passes through zero.

When behavior associated with uniformly distributed forces causes shear to pass through zero at an inconvenient location, a second approach is to recognize that this behavior is due to the shear curve's slope. The slope of the shear curve is the value of the uniformly distributed force (w). Therefore, the point where shear passes through zero may be found by relating the closest known value of shear (V_{CKV}) and the uniformly distributed force (w):

$$x' = \frac{V_{CKV}}{w}$$

In this case, shear passes through zero at a point that is x' away from the location of the closest known value of shear. *This expression applies only when the sloping curve of a uniformly distributed force is causing shear to pass through zero* at an odd location, as illustrated in Examples 11.7 and 11.8.

■ EXAMPLE 11.7

In Example 11.6 it was found that shear passes through zero between the 6- and 7-ft marks on the beam. Use the $x' = V_{CKV}/w$ equation to find where shear passes through zero.

Solution   The closest known values of shear (V_{CKV}) are found as

$$V \text{ @ } 5 \text{ ft} = 1350 \text{ lb} - (1000 \text{ lb/ft} \cdot 1 \text{ ft}) = 350 \text{ lb}$$
$$V \text{ @ } 6 \text{ ft} = 1350 \text{ lb} - (1000 \text{ lb/ft} \cdot 2 \text{ ft}) = -650 \text{ lb}$$

The uniformly distributed load (w) at these points on the curve is 1000 lb/ft. The location where shear passes through zero can be found by using the vertical shear force at 5 ft (350 lb) as the closest known value of shear:

$$x' = \frac{V_{CKV}}{w} = \frac{350 \text{ lb}}{1000 \text{ lb/ft}} = 0.35 \text{ ft}$$

Shear passes through zero at 0.35 ft to the right of the 5-ft mark. By convention, this location is at $x = 5.35$ ft, 5.35 ft from the left end of the beam.

The location where shear passes through zero can also be found by using shear at 6 ft (-650 lb) as the closest known value of shear:

$$x' = \frac{V_{CKV}}{w} = \frac{-650 \text{ lb}}{1000 \text{ lb/ft}} = -0.65 \text{ ft}$$

Shear passes through zero at 0.65 ft to the left of the 6-ft mark; the negative sign implies to the left of the known location (x'). By convention, this location is at $x = 5.35$ ft, 5.35 ft from the left end of the beam.

Use of the $x' = V_{CKV}/w$ equation makes finding the location where shear passes through zero easier, but only when *the sloping curve of a uniformly distributed force is causing shear to pass through zero.*

■ EXAMPLE 11.8

A free-body diagram of a beam with a double overhang is shown in Figure 11.10. The 8-ft-long beam carries a uniformly distributed load of 1 kip/ft over 6 ft of its length. The load does not cover a 2-ft portion of the beam just to the left of R_2. Reaction R_1 equals 3.5 kip and reaction R_2 equals 2.5 kip. Calculate the vertical shear forces at 1-ft intervals across the beam.

Solution   To simplify computations, the shear force is calculated at select sections. The position of the section is to the side of the reactions and at the beginning and ending points of the uniformly distributed load:

$V @ x = \Sigma F_{up} - \Sigma F_{down}$ to the left of the section under consideration
$V @ 0 \text{ ft} = 0 - 0 = 0$
$V @ R_1- = 0 - (1 \text{ kip/ft} \cdot 2 \text{ ft}) = -2 \text{ kip}$
$V @ R_1+ = 3.5 \text{ kip} - (1 \text{ kip/ft} \cdot 2 \text{ ft}) = 1.5 \text{ kip}$
$V @ 4 \text{ ft} = 3.5 \text{ kip} - (1 \text{ kip/ft} \cdot 4 \text{ ft}) = -0.5 \text{ kip}$
$V @ R_2- = 3.5 \text{ kip} - (1 \text{ kip/ft} \cdot 4 \text{ ft}) = -0.5 \text{ kip}$
$V @ R_2+ = 3.5 \text{ kip} + 2.5 \text{ kip} - (1 \text{ kip/ft} \cdot 4 \text{ ft}) = 2 \text{ kip}$
$V @ 8 \text{ ft} = 3.5 \text{ kip} + 2.5 \text{ kip} - (1 \text{ kip/ft} \cdot 4 \text{ ft}) - (1 \text{ kip/ft} \cdot 2 \text{ ft}) = 0$

Vertical shear passes through zero at R_1, at R_2, and at an unknown location between 5 and 6 ft. The closest known values of shear (V_{CKV}) that are near the unknown location and that are associated with the slope of the uniformly distributed load are at R_1+ and at 4 ft. The uniformly distributed load (w) between these points on the curve is 1 kip/ft.

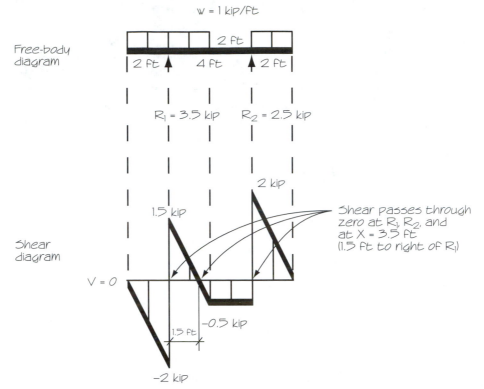

FIGURE 11.10   Free-body and vertical shear diagrams of the beam for Example 11.8.

The location where shear passes through zero can be found by using the vertical shear force at $R_1 +$ (1.5 kip) as the closest known value of shear:

$$x' = \frac{V_{\mathrm{CKV}}}{w} = \frac{1.5 \text{ kip}}{1 \text{ kip/ft}} = 1.5 \text{ ft}$$

Shear passes through zero at 1.5 ft to the right of $R_1 +$. By convention, this location is at $x = 3.5$ ft, 3.5 ft from the left end of the beam.

The location where shear passes through zero can also be found by using shear at 4 ft (-0.5 kip) as the closest known value of shear:

$$x' = \frac{V_{\mathrm{CKV}}}{w} = \frac{-0.5 \text{ kip}}{1 \text{ kip/ft}} = -0.5 \text{ ft}$$

Shear passes through zero at 0.5 ft to the left of the 4-ft mark; the negative sign implies to the left of the known location (x'). By convention, this location is at $x = 3.5$ ft, 3.5 ft from the left end of the beam.

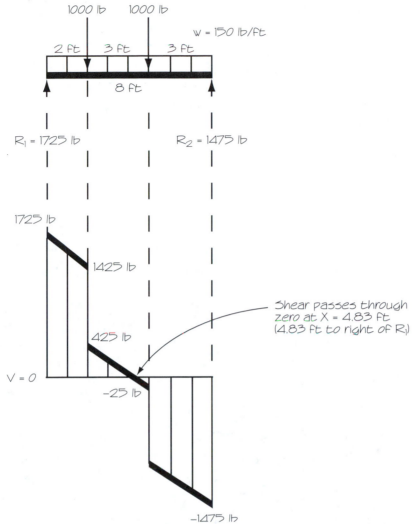

FIGURE 11.11 Free-body and vertical shear diagrams of the beam for Example 11.9.

■ EXAMPLE 11.9

A free-body diagram of a simple beam is shown in Figure 11.11. The 8-ft-long beam carries a uniformly distributed load of 150 lb/ft over its entire length. Two concentrated loads of 1000 lb each are at the 2- and the 5-ft marks. Reaction R_1 equals 1725 lb and reaction R_2 equals 1475 lb. Determine where vertical shear passes through zero.

Solution   Shear passes through zero between the 4- and 5-ft− calculation intervals as follows.

V @ $x = \Sigma F_{up} - \Sigma F_{down}$ to the left of the section under consideration
V @ 4 ft $= 1725$ lb $- 1000$ lb $- (4$ ft $\cdot 150$ lb/ft$) = 125$ lb
V @ 5 ft$- = 1725$ lb $- 1000$ lb $- (5$ ft $\cdot 150$ lb/ft$) = -25$ lb

Review of the shear diagram shows that the 150-lb/ft uniformly distributed force is the only force applied between the 4- and 5-ft– marks. Therefore, using the known value for V @ 4 ft (125 lb) and the 150-lb/ft uniformly distributed force (w) yields

$$x' = \frac{V_{CKV}}{w} = \frac{125 \text{ lb}}{150 \text{ lb/ft}} = 0.83 \text{ ft from 4 ft} \quad \text{or} \quad x = 4.83 \text{ ft}$$

Or, by using the known value for V @ 5 ft– (-25 lb) and the 150-lb/ft uniformly distributed force (w), we obtain

$$x' = \frac{V_{CKV}}{w} = \frac{-25 \text{ lb}}{150 \text{ lb/ft}} = -0.67 \text{ ft from 5 ft} \quad \text{or} \quad x = 4.83 \text{ ft}$$

Shear passes through zero 4.83 ft to the right of R_1, at $x = 4.83$ ft.

Calculation Approach for Shear

In summary, shear calculations and development of shear diagrams involve the following series of steps:

1. Neatly draw the free-body diagram of the beam showing forces and reactions. Solving for unknown reactions may be necessary.

2. Solve for shear at convenient (1-ft) intervals along the beam length:

   V @ $x = \Sigma F_{up} - \Sigma F_{down}$ to the left of the section under consideration

3. Using a convenient scale, neatly plot the calculated values of shear along the beam directly below the beam body diagram.

4. Determine the maximum and minimum shear forces. Indicate them on a shear diagram.

5. Determine where shear passes through zero using any of the techniques noted below, and indicate the location(s) where shear passes through zero on the shear diagram.

   (a) Mathematical pinch play (by observation on the shear diagram, such as at a location where there is a change in sign on either side of a concentrated load or reaction)

   (b) Algebraic solution

   (c) $x' = V_{CKV}/w$ for locations where shear passes through zero due to a uniformly distributed load only

---

Eleven sets of illustrations and the corresponding free-body diagrams of building-related beams were provided in Chapter 4 as Exercises 4.1 through 4.11. These free-body diagrams constitute exercises that will be used here and again in Chapter 12. *Follow these instructions carefully to save time later:*

(a) On each sheet of engineering paper reserved for analysis, calculate the vertical shear specific to that member at appropriate intervals across the length of the free-body diagram. Be sure to solve for locations where shear passes through zero.

(b) On each sheet of engineering paper containing the free-body diagram, neatly draw the related shear diagram directly below the associated free-body diagram. The shear diagram should be located in the middle one-third of the page. Reserve the bottom one-third of the sheet for the bending moment diagram introduced in Chapter 12. Be sure to show locations where shear passes through zero.

11.1 The free-body diagram in Exercise 4.1.

11.2 The free-body diagram in Exercise 4.2.

11.3 The free-body diagram in Exercise 4.3.

11.4 The free-body diagram in Exercise 4.4.

11.5 The free-body diagram in Exercise 4.5.

11.6 The free-body diagram in Exercise 4.6.

11.7 The free-body diagram in Exercise 4.7.

11.8 The free-body diagram in Exercise 4.8.

11.9 The free-body diagram in Exercise 4.9.

11.10 The free-body diagram in Exercise 4.10.

11.11 The free-body diagram in Exercise 4.11.

Chapter 12
BENDING IN BEAMS

A beam must be designed to withstand the bending stresses that are induced by the forces or loads it supports. In this chapter the method of computing the bending and resisting moments in a beam is examined. This approach requires that the beam be in static equilibrium. Each investigation will result in a bending moment diagram that is characteristic of that distinct loading condition. The bending moment diagram serves to facilitate a clear understanding of bending moments and the stresses resulting from resisting moments acting on a beam.

12.1   BENDING IN BEAMS

A beam will have a tendency to bend when forces are applied to it. When forces act on a beam, bending stresses develop and strain is produced. Strain associated with bending reveals itself as deflection. *Deflection* is a form of strain caused by stresses that develop in the beam material when the beam bends. The forces that cause deflection in a beam are the external forces, that is, the loads and the reactions. The resulting bending stresses that develop internally in the beam are compressive and tensile in nature. The action of these stresses is examined by intuitively considering the behavior of a hypothetical beam.

The hypothetical beam is assumed to be a simple beam that is carrying a single concentrated load acting downward at its center. As a simple beam it has reactions at its ends. Being composed of a *homogeneous material,* the beam material has properties that are uniform and flawless throughout such that it has properties of equal strength, elasticity, density, and so on. To examine strain from bending, it is also assumed that the beam is composed of an infinite number of very thin, interconnected layers called *fibers*. These fibers lay flat and parallel to the length of the beam, as shown in Figure 12.1a.

When a force is applied to the hypothetical beam, the beam tends to bend, as illustrated in Figure 12.1b. With bending of the beam, the fiber on top of the beam gets shorter and the bottom fiber gets longer. Change in length of the top and bottom fibers is equal but opposite; the top fiber shortens an amount equal to the elongation of the bottom fiber. Additionally, there exists a central fiber that does not

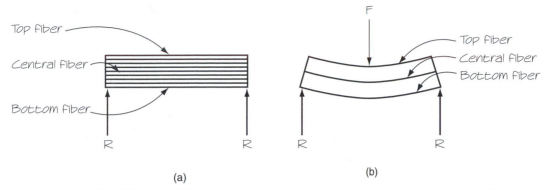

FIGURE 12.1 A hypothetical beam is composed of an infinite number of very thin, interconnected layers called fibers (a). In a hypothetical beam under load (b), the top fiber experiences compression (shortening) and the bottom fiber experiences tension (elongation).

experience shortening or elongation. Several conclusions may be drawn from the behavior of the hypothetical beam:

- Since strain is proportional to stress, the top fiber that shortens experiences compressive stress equal to the tensile stress that develops due to the elongation of the bottom fiber.
- Since there is no strain experienced by the central fiber, it experiences neither tensile nor compressive stresses.
- Stresses are compressive on one side of the central fiber and tensile on the other side of the central fiber.
- Stresses must increase uniformly beginning with no stress at the central fiber to a maximum stress at the outermost fibers.

Engineers refer to the central fiber as the neutral plane. The *neutral plane* is that fiber in a beam that does not experience tensile nor compressive stresses caused by bending. When discussing a beam cross section, this fiber is referred to as the *neutral axis of bending* or simply the *neutral axis* (Figure 12.2). The neutral plane or neutral axis is always perpendicular to the applied forces. The neutral plane or neutral axis will, however, not always be located in the geometric center of the beam. Location is related to symmetry of the beam cross section and composition of the beam material.

A central location of the neutral plane or neutral axis only occurs in a beam that is symmetrical in cross section and composed of a homogeneous material. A rectangular joist and an H-shaped steel beam have sections that are symmetrical in cross section and assumed to be composed of a homogeneous material. Thus the neutral axis and neutral plane are assumed to be located in the geometric center of these beams. Angles and tees are asymmetrical (not symmetrical) in cross section,

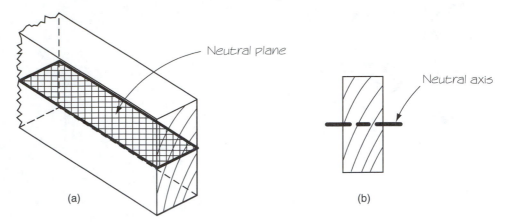

FIGURE 12.2 The neutral plane (a) is that fiber in a beam under load that does not experience compressive nor tensile stresses. This plane is referred to as the neutral axis (b) when discussing a beam cross section.

so the neutral plane or neutral axis does not occur in the geometric center of these members. Reinforced concrete structural members are composed of a composite of concrete and steel reinforcement; they are not homogeneous, so they do not typically have the neutral plane or neutral axis in the center of the member cross section.

Regardless of the symmetry of cross section or composition of a beam material, there is always a neutral plane in a beam experiencing bending. Material along this plane does not undergo tensile or compressive stresses associated with bending. As shown in Figure 12.3, compressive stresses always occur on one side of the neutral plane and tensile stresses on the other side of this plane. Tensile stresses occur on that side of the neutral plane where the fibers are experiencing elongation and compressive stresses exist in the material where the fibers are compressed.

Bending stresses increase as the distance of a fiber increases from the neutral plane or neutral axis. The *extreme fibers* are the outermost fibers of a beam experiencing bending. As a result, bending stresses are greatest at the extreme fibers. The *most extreme fiber* is located at that surface that is the farthest away from the neutral plane. Beams with symmetrical cross sections that are composed of a homogeneous material have two most extreme fibers, so they experience the greatest stresses at both the top and bottom fibers. Asymmetrical beam cross sections (i.e., L-shaped angles and T-shaped structural tees) and beams composed of nonhomogeneous or composite materials (i.e., beams of reinforced concrete) will typically have a single most extreme fiber. Because the greatest bending stresses occur at a beam's extreme fibers, the beam cross section and material must be investigated to ensure that the beam will resist the tensile and compressive stresses that occur at the beam's extreme fibers.

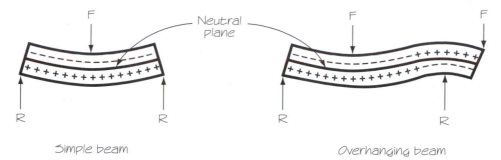

Simple beam   **Overhanging beam**

FIGURE 12.3   In a beam under load, the compressive and tensile stresses occur on opposite sides of the neutral plane. Tensile stresses (+) occur on that side of the neutral plane where the fibers are experiencing elongation and compressive stresses (−) occur in the material where the fibers are compressed.

12.2   STRESSES CAUSED BY BENDING

A fundamental assumption in beam calculations is that the beam is in static equilibrium; that is, the sum of the external forces acting on that beam equals zero ($\Sigma F = 0$) and the sum of the external moments of force acting on that beam equals zero ($\Sigma M = 0$). All external forces acting on the full length of the beam cause the beam to remain at rest. Simply, a beam under load is in static equilibrium under the action of the external forces (the loads and reactions). Furthermore, any segment of the beam when isolated must also be in static equilibrium because the entire beam as a rigid body is at or assumed to be at rest. However, *external* forces alone do not maintain a state of static equilibrium in a beam.

Consider the free-body diagram of the simple beam as shown in Figure 12.4a. Assume that the sides of the beam are divided at section *x*. Observe that external forces R_1 and P acting on the left side of the beam are creating a net clockwise moment about section *x*. On the right side of the beam, external force R_2 is creating a counterclockwise moment about section *x*. The sum of the moments of each side of the beam about section *x* is not zero. The beam itself must induce additional forces and moments to keep each segment in equilibrium.

As discussed previously and as shown graphically in Figure 12.4b, internal compressive and tensile forces also act at the isolated section. These forces act perpendicular to the plane of section *x* and increase as the distance from the neutral axis increases. These forces combine to develop an internal moment couple that resists the moment created by the external forces. The moment couple is equal in magnitude but opposite in rotation to the moment created by the external forces.

Only forces R_1 and P act externally on the left segment of the beam. Since the upward-acting reaction R_1 is less than downward-acting force P, shear forces develop

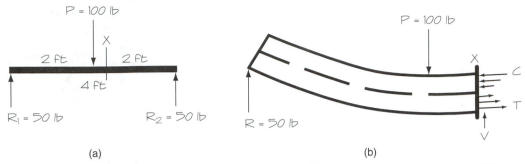

FIGURE 12.4 (a) Free-body diagram of a 4-ft-long simple beam supporting a 100-lb concentrated load at its center. (b) When the left side of the beam is isolated at section x, internal forces keep the beam segment in equilibrium. Since the upward-acting reaction R_1 is less than downward-acting force P, a shear force (V) develops along the plane of section x. Additionally, compressive (C) and tensile (T) forces acting perpendicular to the plane of section x combine to create a resisting moment couple that counteracts the moment created by external forces P and R_1.

along the plane of section x. These internal shearing forces develop within the material to resist the shearing effect of external forces. It is these forces that relate to the vertical shear (V) introduced in Chapter 11. At a point in a beam where shear passes through zero, it must follow that internal moment alone must provide an internal resistance to bending. Hence, it is at the point(s) in a beam where shear passes through zero ($V = 0$) that the moment created internally is the greatest; it is at this point that the beam will have the greatest tendency to fail due to tensile and/or compressive bending stresses.

Based on these deductions, it can be concluded that an isolated segment of a beam at rest is in equilibrium under the action of external forces *and* internal forces and moments acting at the section. This internal moment must counter the moment of the external forces (reaction R_1 and force P) that are created about any section. Similarly, a shear force must develop at any beam section that counters the difference in upward and downward acting forces acting on either beam segment.

12.3   BENDING MOMENT AND RESISTING MOMENT

A *bending moment* is created by the external forces (the reactions and loads) acting on the beam. It is bending moment that causes a beam to deflect. In a simple beam under load, bending moment varies across the length of the beam. Bending moment at a section in a simple beam increases with an increase in load and decreases with a reduction in load.

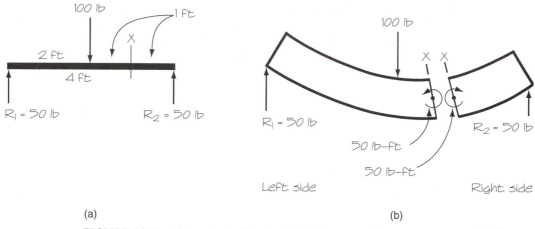

FIGURE 12.5 (a) Free-body diagram of a 4-ft-long sample beam supporting a 100-lb concentrated load at its center. (b) To keep the beam shown above in static equilibrium, the material at the 3-ft mark must produce and apply a moment on each side of that section; that is, the beam material on the right side of the 3-ft mark must induce a moment to the beam material on the left side of the 3-ft mark and the beam material on left side must induce a moment on the beam material on the right side.

At all sections of a beam under load, compressive stresses and tensile stresses are found in the fibers on opposite sides of the neutral plane. These stresses produce a couple known as *resisting moment* that acts about the neutral axis of the beam at the section under consideration. Resisting moment is a couple created by the bending forces that act internally in the material. These forces are actually tension and compression stresses that develop to counter deflection caused by bending moment. The bending forces that combine to create resisting moment are produced by bending moment, so resisting moment reacts to and counters bending moment.

Consider a simple beam that is 4 ft long and carrying a 100-lb concentrated load at its center as shown in Figure 12.5a. Reaction R_1 is a pinned support and reaction R_2 is a bearing support. Under the conditions of static equilibrium, reactions, R_1 and R_2, are equal to 50 lb. Assume that the beam is divided into two segments at the 3-ft mark on the beam as shown in 12.5b. This creates a left segment that is 3 ft long with forces R_1 and the 100-lb concentrated load acting on it and a right segment that is 1 ft long with only force R_2 acting on it. Moments of the external forces acting on each section about the 3-ft mark are calculated as follows:

$\circlearrowright\ \Sigma M_{\text{left side about 3 ft}} = -(50 \text{ lb} \cdot 3 \text{ ft}) + (100 \text{ lb} \cdot 1 \text{ ft}) = -50 \text{ lb-ft} = 50 \text{ lb-ft}$
$$\text{(a clockwise moment)}$$

$\circlearrowleft\ \Sigma M_{\text{right side about 3 ft}} = (50 \text{ lb} \cdot 1 \text{ ft}) = 50 \text{ lb-ft} = 50 \text{ lb-ft}$
$$\text{(a counterclockwise moment)}$$

12.3 Bending Moment and Resisting Moment **211**

The external forces acting on the left segment of the beam create a bending moment of 50 lb-ft that tends to cause a clockwise rotation about the 3-ft mark. The lone external force acting on the right beam segment also causes a 50-lb-ft moment but the tendency is to cause counterclockwise rotation about the 3-ft mark. The magnitudes of bending moment are equal, but their directions of rotation are opposite; one is clockwise and the other is counterclockwise.

To maintain static equilibrium at the 3-ft mark in the beam, the right side of the beam must apply a resisting moment equivalent in magnitude but opposite in rotation to the bending moment created by the external forces acting on the left side of the beam. At the 3-ft mark, this resisting moment must be a 50-lb-ft moment with clockwise rotation that counters the counterclockwise bending moment of 50 lb-ft. The bending moment is created by the external forces acting on the left segment of the beam. Similarly, the left side of the beam must exert a 50-lb-ft clockwise moment to counter the bending moment on the right side of the beam.

At any section in the beam, the left segment of the beam is in static equilibrium due to a resisting moment applied by the right segment. Similarly, the right segment of the beam is in static equilibrium due to a resisting moment applied by the left segment. The resisting moment is applied by the bending stresses of tension and compression. At any section in a beam in static equilibrium, resisting moment counters bending moment; that is, resisting moment is equal in magnitude but opposite in direction of rotation to bending moment.

If the beam section and material cannot develop a resisting moment that counters bending moment, failure of the beam results. A beam that fails is not in static equilibrium; it is not a rigid body, nor is it stationary. The capability of a beam to develop a resisting moment that reacts to a bending moment is related to the properties of the beam material and section. Properties of materials and sections are covered elsewhere in the book.

When a beam meets the conditions of static equilibrium, its resisting moment equals the magnitude of (but counters the direction of rotation of) bending moment. Therefore, resisting moment is easily determined by calculating the bending moment created by the external forces acting about the section. For now, it is helpful to understand that the principal reason for making bending moment calculations is to observe how bending moment, and thus bending stresses, occur across the length of the beam.

12.4   BENDING MOMENT CALCULATIONS AND DIAGRAMS

Bending moment, M, is a moment found by summing the moments of the external loads and reactions about a selected section. As with vertical shear force calculations, the section is placed at any location on the free body. The section is assumed

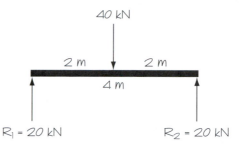

FIGURE 12.6 Free-body diagram of the beam for Example 12.1.

to contain the axis of rotation about which moments of forces on *one side of the beam* are calculated.

Bending moment may be calculated about any section in a beam by the following rule: *The sum of the moments of the upward forces to the left of the section under consideration minus the moments of the downward forces to the left of the section.* This may be expressed as

$$M = \Sigma M_{up} - \Sigma M_{down} \text{ to the left of the section under consideration}$$

Like vertical shear force computations, the designer selects where to position the section. Since bending moment is a force multiplied by distance, it is expressed in units such as lb-ft, kip-ft, and N-m.

■ EXAMPLE 12.1

A free-body diagram of a simple beam is shown in Figure 12.6. The 4-m beam carries a concentrated load of 40 kN at its center. Reactions R_1 and R_2 each equal 20 kN. Calculate the bending moment at 1-m intervals across the beam.
Solution   Beginning at R_1 and ending at R_2 at intervals of 1 ft along the length of the beam, bending moment calculations are:

$$M = \Sigma M_{up} - \Sigma M_{down} \text{ to the left of the section under consideration}$$
$$M \text{ @ } R_1 = (20 \text{ kN} \cdot 0) - 0 = 0$$
$$M \text{ @ } 1 \text{ m} = (20 \text{ kN} \cdot 1 \text{ m}) - 0 = 20 \text{ kN-m}$$
$$M \text{ @ } 2 \text{ m} = (20 \text{ kN} \cdot 2 \text{ m}) - (40 \text{ kN} \cdot 0) = 40 \text{ kN-m}$$
$$M \text{ @ } 3 \text{ m} = (20 \text{ kN} \cdot 3 \text{ m}) - (40 \text{ kN} \cdot 1 \text{ m}) = 20 \text{ kN-m}$$
$$M \text{ @ } R_2 = (20 \text{ kN} \cdot 4 \text{ m}) - (40 \text{ kN} \cdot 2 \text{ m}) = 0$$

■ EXAMPLE 12.2

A free-body diagram of a simple beam is shown in Figure 12.7. The 4-m beam carries a uniformly distributed load of 10 kN/m across its entire length. Reactions R_1 and R_2 are equal to 20 kN each. Calculate the bending moment at 1-m intervals across the beam.

12.4   Bending Moment Calculations and Diagrams

213

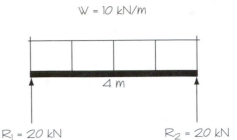

FIGURE 12.7 Free-body diagram of the beam for Example 12.2.

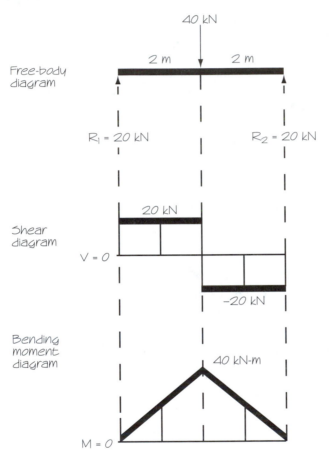

FIGURE 12.8 Free-body, vertical shear, and bending moment diagrams for Example 12.1. Reactions were calculated in Example 5.1. Vertical shear calculations were made in Example 11.1. Bending moment analysis is shown in Example 12.1.

Solution   Beginning at R_1 and ending at R_2 at intervals of 1 ft along the length of the beam, bending moment calculations are:

$$M = \Sigma M_{up} - \Sigma M_{down} \text{ to the left of the section under consideration}$$
$$M @ R_1 = (20 \text{ kN} \cdot 0) - 0 = 0$$

$$M@ \ 1 \ m = (20 \ kN \cdot 1 \ m) - [(10 \ kN/m \cdot 1 \ m) \ (0.5 \ m)] = 15 \ kN\text{-}m$$
$$M @ \ 2 \ m = (20 \ kN \cdot 2 \ m) - [(10 \ kN/m \cdot 2 \ m) \ (1 \ m)] = 20 \ kN\text{-}m$$
$$M @ \ 3m = (20 \ kN \cdot 3 \ m) - [(10 \ kN/m \cdot 3 \ m) \ (1.5 \ m)] = 15 \ kN\text{-}m$$
$$M @ \ R_2 = (20 \ kN \cdot 4 \ m) - [(10 \ kN/m \cdot 4 \ m) \ (2 \ m)] = 0 \ kN\text{-}m$$

Results of bending moment calculations are plotted below the free-body and vertical shear force diagrams. The free-body, shear, and bending moment diagrams for the results in Examples 12.1 and 12.2 are shown in Figures 12.8 and 12.9, respectively.

As shown in Example 12.2, bending moments created by uniformly distributed forces are found by considering only that portion of the load that is on the beam segment under consideration: first, finding the total distributed load (w) on the segment, and then by multiplying this partial load by a distance measured from center of gravity of the load to the axis of rotation (the section under consideration). For example, the uniformly distributed load portion of the calculation at M @ 3 m in Example 12.2 was found by multiplying $[(10 \ kN/m \cdot 3 \ m) \ (1.5 \ m)]$. The section is located at the 3-m mark in the beam, so only the portion of the uniformly distributed load to the left of that mark is considered. The "$(10 \ kN/m \cdot 3 \ m)$" segment of the computation is the uniformly distributed force under consideration, that portion of the uniformly distributed load to the left of the 3-m mark in the beam. The "1.5 m" segment of the computation is the distance from the center of gravity of the uniformly distributed load to the axis of rotation at the 3-m mark on the beam; that is, the center of gravity of that portion of the uniformly distributed load under consideration is 1.5 m away from the 3-m mark on the beam.

■ EXAMPLE 12.3

A free-body diagram of a simple beam is shown in Figure 12.10. The 4-ft beam carries a 100-kip concentrated load at the center of the beam. Uniformly distributed loads of 10 kip/ft on the left side of the beam and 5 kip/ft on the right side also rest on the beam. Reaction R_1 is 67.5 kip and reaction R_2 is 62.5 kip. Calculate the bending moment at 1-ft intervals across the beam.

Solution Beginning at R_1 and ending at R_2 at intervals of 1 ft along the length of the beam, bending moment calculations are:

$$M = \Sigma M_{up} - \Sigma M_{down} \text{ to the left of the section under consideration}$$
$$M @ \ R_1 = (67.5 \ kip \cdot 0 \ ft) - 0 = 0$$
$$M @ \ 1 \ ft = (67.5 \ kip \cdot 1 \ ft) - [(10 \ kip/ft \cdot 1 \ ft) \ (0.5 \ ft)] = 62.5 \ kip\text{-}ft$$
$$M @ \ 2 \ ft = (67.5 \ kip \cdot 2 \ ft) - [(10 \ kip/ft \cdot 2 \ ft) \ (1 \ ft)] - (100 \ kip \cdot 0 \ ft)$$
$$= 115.0 \ kip\text{-}ft$$
$$M @ \ 3 \ ft = (67.5 \ kip \cdot 3 \ ft) - [(10 \ kip/ft \cdot 2 \ ft) \ (2 \ ft)] - (100 \ kip \cdot 1 \ ft)$$
$$- [(5 \ kip/ft \cdot 1 \ ft) \ (0.5 \ ft)] = 60.0 \ kip\text{-}ft$$
$$M @ \ R_2 = (67.5 \ kip \cdot 4 \ ft) - [(10 \ kip/ft \cdot 2 \ ft) \ (3 \ ft)] - (100 \ kip \cdot 2 \ ft)$$
$$- [(5 \ kip/ft \cdot 2 \ ft) \ (1 \ ft)] = 0$$

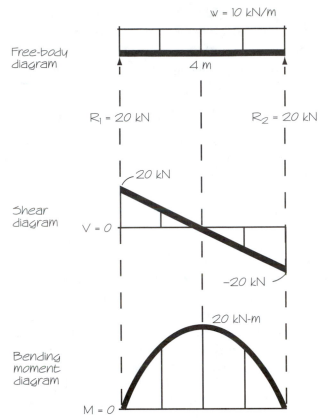

FIGURE 12.9 Free-body, vertical shear, and bending moment diagrams for Example 12.2. Reactions were calculated in Example 5.2. Vertical shear calculations were made in Example 11.2. Bending moment analysis is shown in Example 12.2.

By convention, bending moment is calculated by analyzing the forces to the *left* of the section under consideration. Bending moment calculation results may be verified by calculating it to the right of the section. For example, bending moment at the 3-ft mark on the beam in Example 12.3 can be found by analyzing the forces to the *right* of the same section:

$$M = \Sigma M_{up} - \Sigma M_{down} \text{ to the } right \text{ of the section under consideration}$$
$$M \text{ @ 3 ft} = (62.5 \text{ kip} \cdot 1 \text{ ft}) - [(5 \text{ kip/ft} \cdot 1 \text{ ft}) (0.5 \text{ ft})] = 60.0 \text{ kip-ft}$$

Results are equal in magnitude and sign to those found when bending moment is calculated to the left of the section. As demonstrated, calculating to the right of the section under consideration sometimes makes computation less complicated. However, it can be confusing to switch from the left section to the right section, which may lead to errors in analysis. Therefore, it is recommended that the student grow accustomed to solving bending moment problems regularly by computing to one

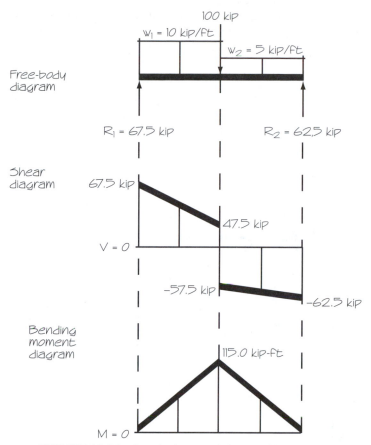

FIGURE 12.10 Free-body, vertical shear, and bending moment diagrams for Example 12.3. Reactions were calculated in Example 5.4. Vertical shear calculations were made in Example 11.4. Bending moment analysis is shown in Example 12.3.

side (left or right) of the section. Computation on the other side should be reserved for verifying results; that is, always working with forces on the left segment of the section in initial computations and then computing to the right to verify results.

12.5 MAXIMUM BENDING MOMENT

Maximum bending moment (M_x or $M_{\max}$) is the largest absolute value of bending moment; the largest magnitude regardless of sign. The point at which maximum bending moment occurs is the section that experiences the greatest bending stresses. When a beam cross section is to be uniform throughout the length of the beam, such as a wood joist or an I-shaped steel member, maximum bending moment governs the cross-sectional size of the member. Thus it is customary to identify the location and magnitude of maximum bending moment.

■ EXAMPLE 12.4

A free-body diagram of a simple beam is shown in Figure 12.11. The 8-ft-long beam carries a 1000-lb concentrated load at the center of its span and a uniformly distributed load of 250 lb/ft across its entire length. Calculate the bending moment at 1-ft intervals across the beam.

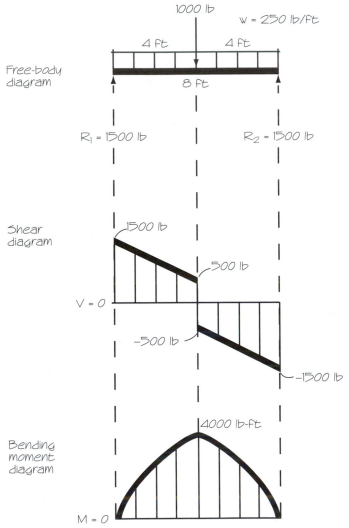

FIGURE 12.11 Free body, vertical shear, and bending moment diagrams for Example 12.4. Vertical shear calculations were made in Example 11.3. Bending moment analysis is shown in Example 12.4.

Solution Beginning at R_1 and ending at R_2 at intervals of 1 ft along the length of the beam, bending moment calculations are

$$M = \Sigma M_{\text{up}} - \Sigma M_{\text{down}} \text{ to the left of the section under consideration}$$

M @ R_1 = (1500 lb · 0 ft) − 0 = 0

M @ 1 ft = (1500 lb · 1 ft) − [(250 lb/ft · 1 ft) (0.5 ft)] = 1375 lb-ft

M @ 2 ft = (1500 lb · 2 ft) − [(250 lb/ft · 2 ft) (1 ft)] = 2500 lb-ft

M @ 3 ft = (1500 lb · 3 ft) − [(250 lb/ft · 3 ft) (1.5 ft)] = 3375 lb-ft

M @ 4 ft = (1500 lb · 4 ft) − [(250 lb/ft · 4 ft) (2 ft)] − (1000 lb · 0 ft)
        = 4000 lb-ft

M @ 5 ft = (1500 lb · 5 ft) − [(250 lb/ft · 5 ft) (2.5 ft)] − (1000 lb · 1 ft)
        = 3375 lb-ft

M @ 6 ft = (1500 lb · 6 ft) − [(250 lb/ft · 6 ft) (3 ft)] − (1000 lb · 2 ft)
        = 2500 lb-ft

M @ 7 ft = (1500 lb · 7 ft) − [(250 lb/ft · 7 ft) (3.5 ft)] − (1000 lb · 3 ft)
        = 1375 lb-ft

M @ R_2 = (1500 lb · 8 ft) − [(250 lb/ft · 8 ft) (4 ft)] − (1000 lb · 4 ft) = 0

Review of the calculations in Example 12.4 indicates that the maximum magnitude of bending moment occurs at the beam center (the 4-ft mark). Maximum bending moment is 4000 lb-ft because this is the maximum absolute value of bending moment on the beam. The designer must specify a cross-sectional size and material that can safely handle an internal resisting moment of 4000 lb-ft to ensure that the beam will withstand failure due to bending.

Upon review of previous examples on bending moment, it appears that maximum bending moment occurs where vertical shear was found to pass through zero. In beams supporting only forces (no external moments induced on the beam), maximum bending moment will always occur at a point where the shear force (V) was found to pass through zero. Therefore, it is necessary for the designer to calculate bending moment where the shear force passes through zero.

■ EXAMPLE 12.5

A free-body diagram of a simple beam is shown in Figure 12.12. The 10-ft-long beam carries a uniformly distributed load of 1000 lb/ft for 3 ft beginning 4 ft to the right of R_1. Reaction R_1 equals 1350 lb and reaction R_2 equals 1650 lb. Calculate the bending moment at 1-ft intervals across the beam. Identify the maximum bending moment.

Solution Beginning at R_1 and ending at R_2 at intervals of 1 ft along the length of the beam, bending moment calculations are

$$M = \Sigma M_{\text{up}} - \Sigma M_{\text{down}} \text{ to the left of the section under consideration}$$

M @ R_1 = (1350 lb · 0 ft) − 0 = 0

M @ 1 ft = (1350 lb · 1 ft) − 0 = 1350 lb-ft

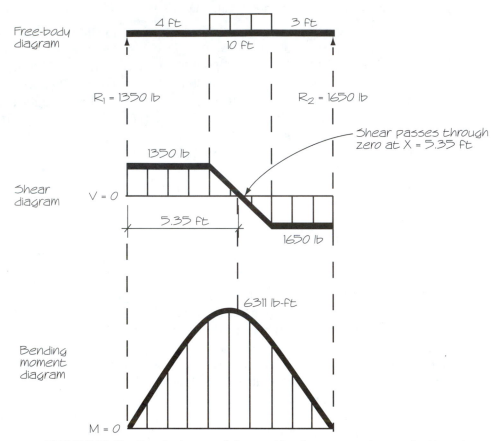

FIGURE 12.12 Free-body, vertical shear, and bending moment diagrams for Example 12.5. Reactions were calculated in Example 5.6. Vertical shear calculations were made in Example 11.6. Bending moment analysis is shown in Example 12.5.

M @ 2 ft = (1350 lb · 2 ft) − 0 = 2700 lb-ft
M @ 3 ft = (1350 lb · 3 ft) − 0 = 4050 lb-ft
M @ 4 ft = (1350 lb · 4 ft) − 0 = 5400 lb-ft
M @ 5 ft = (1350 lb · 5 ft) − [(1000 lb/ft · 1 ft) (0.5 ft)] = 6250 lb-ft
M @ 6 ft = (1350 lb · 6 ft) − [(1000 lb/ft · 2 ft) (1 ft)] = 6100 lb-ft
M @ 7 ft = (1350 lb · 7 ft) − [(1000 lb/ft · 3 ft) (1.5 ft)] = 4950 lb-ft
M @ 8 ft = (1350 lb · 8 ft) − [(1000 lb/ft · 3 ft) (2.5 ft)] = 3300 lb-ft
M @ 9 ft = (1350 lb · 9 ft) − [(1000 lb/ft · 3 ft) (3.5 ft)] = 1650 lb-ft
M @ R_1 = (1350 lb · 10 ft) − [(1000 lb/ft · 3 ft) (4.5 ft)] = 0

In Examples 11.6 and 11.7 it was found that the shear force passes through zero at the 5.35-ft mark on the beam. Bending moment is calculated at that location:

$$M \text{ @ } 5.35 \text{ ft} = (1350 \text{ lb} \cdot 5.35 \text{ ft}) - [(1000 \text{ lb/ft} \cdot 1.35 \text{ ft})(1.35 \text{ ft}/2)]$$
$$= 6311 \text{ lb-ft}$$

The maximum bending moment is 6311 lb-ft and occurs at the 5.35 ft mark in the beam.

In Example 12.5 it appears initially that maximum bending moment occurs at the 5-ft mark on the beam. As shown, however, bending moment is greatest at the point where the shear force passes through zero, at the 5.35-ft mark in the beam.

■ EXAMPLE 12.6

Free-body and shear diagrams of a simple beam are shown in Figure 12.13. The 8-ft-long beam carries a uniformly distributed load of 150 lb/ft over its entire length. Two concentrated loads of 1000 lb each are at the 2- and 5-ft marks. Reaction R_1 equals 1725 lb and reaction R_2 equals 1475 lb. Calculate the maximum bending moment.

Solution   As in Example 11.9, the shear force passes through zero at the 4.83-ft mark on the beam. The bending moment at that mark in the beam is found by

$$M = \Sigma M_{\text{up}} - \Sigma M_{\text{down}} \text{ to the left of the section under consideration}$$

$$M \text{ @ } 4.83 \text{ ft} = (1725 \text{ lb} \cdot 4.83 \text{ ft}) - \left[(150 \text{ lb/ft} \cdot 4.83 \text{ ft}) \frac{(4.83 \text{ ft})}{2} \right]$$
$$- (1000 \text{ lb} \cdot 2.83 \text{ ft}) = 3752 \text{ lb-ft}$$

The maximum bending moment is 3752 lb-ft.

12.6   BENDING CURVATURE

In a beam under load, compressive stresses and tensile stresses occur in fibers on opposite sides of the neutral plane. Internal forces develop from these stresses. The effect of these forces causes a beam to deflect. Deflection results in curvature of the beam.

Concavity of curvature indicates the location of compressive and tensile stresses acting at that point in the beam. In a simple beam under gravity load, curvature is always concave up (i.e., bent so that the beam can hold water). In this beam, tensile stresses occur below the neutral plane and compressive stresses occur above it. At points in a beam where bending is concave up, compressive stresses always occur above the neutral plane and tensile stresses occur below it. In cases where bending is concave down (i.e., bent so that the beam drains water), tensile stresses occur above the neutral plane and compressive stresses occur below it.

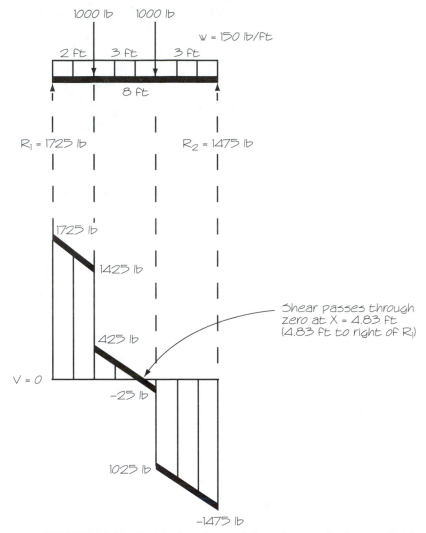

FIGURE 12.13 Free-body and vertical shear diagrams for Example 12.6. In vertical shear calculations made in Example 11.9 it was found that shear passed through zero at 4.83 ft from R_1. Maximum bending moment occurs where shear passed through zero. Analysis is shown in Example 12.6.

By computation convention, a positive bending moment (+) indicates that at that point in the beam the curvature is concave up. Bending moment with a negative sign (−) denotes that bending is concave down.

- $+M$ bends concave up ⌣.
- $-M$ bends concave down ⌢.

Beams experiencing certain loading conditions, such as a beam with a single over-hang carrying a uniformly distributed load, will have curvature that changes from concave up to concave down across the beam length. In that segment of a beam where bending moment is negative, curvature is concave down (⌒). Conversely, in that segment of the beam where bending moment is positive, curvature is concave up (⌣).

■ EXAMPLE 12.7

A free-body diagram of a single overhang beam is shown in Figure 12.14. The 3-m beam carries a concentrated load of 100 kN at the right end of the 1-m overhang. Reaction R_1 equals 50 kN acting downward and reaction R_2 is 150 kN acting upward. Calculate the bending moment at 1-m intervals across the beam.

Solution Beginning at R_1 and ending at R_2 at intervals of 1 ft along the length of the beam, bending moment calculations are

$$M = \Sigma M_{up} - \Sigma M_{down} \text{ to the left of the section under consideration}$$
$$M \text{ @ } R_1 = 0 - (50 \text{ kN} \cdot 0 \text{ m}) = 0$$
$$M \text{ @ } 1 \text{ m} = 0 - (50 \text{ kN} \cdot 1 \text{ m}) = -50 \text{ kN-m}$$
$$M \text{ @ } R_2 = (150 \text{ kN} \cdot 0 \text{ m}) - (50 \text{ kN} \cdot 2 \text{ m}) = -100 \text{ kN-m}$$
$$M \text{ @ } 3 \text{ m} = (150 \text{ kN} \cdot 1 \text{ m}) - (50 \text{ kN} \cdot 3 \text{ m}) = 0$$

The beam in Example 12.7 has bending moments with a negative sign ($-$) across its entire length. This denotes that bending is concave down across its entire length; throughout the beam, compressive stresses occur above the neutral plane and tensile stresses occur below it.

■ EXAMPLE 12.8

A free-body diagram of a beam with a double overhang is shown in Figure 12.15. The 8-ft-long beam carries a uniformly distributed load of 1 kip/ft over 6 ft of its length. The load does not cover a 2-ft portion of the beam just to the left of R_2. Reaction R_1 equals 3.5 kip and reaction R_2 equals 2.5 kip. Calculate the bending moment at 1-ft intervals across the beam. Identify the maximum bending moment.

Solution Beginning at R_1 and ending at R_2 at intervals of 1 ft along the length of the beam, bending moment calculations are

$$M = \Sigma M_{up} - \Sigma M_{down} \text{ to the left of the section under consideration}$$
$$M \text{ @ } 0 \text{ ft} = 0$$
$$M \text{ @ } 1 \text{ ft} = 0 - [(1 \text{ kip/ft} \cdot 1 \text{ ft}) (0.5 \text{ ft})] = -0.5 \text{ kip-ft}$$
$$M \text{ @ } R_1 = (3.5 \text{ kip} \cdot 0 \text{ ft}) - [(1 \text{ kip/ft} \cdot 2 \text{ ft}) (1 \text{ ft})] = -2.0 \text{ kip-ft}$$
$$M \text{ @ } 3 \text{ ft} = (3.5 \text{ kip} \cdot 1 \text{ ft}) - [(1 \text{ kip/ft} \cdot 3 \text{ ft}) (1.5 \text{ ft})] = -1.0 \text{ kip-ft}$$
$$M \text{ @ } 4 \text{ ft} = (3.5 \text{ kip} \cdot 2 \text{ ft}) - [(1 \text{ kip/ft} \cdot 4 \text{ ft}) (2 \text{ ft})] = -1.0 \text{ kip-ft}$$

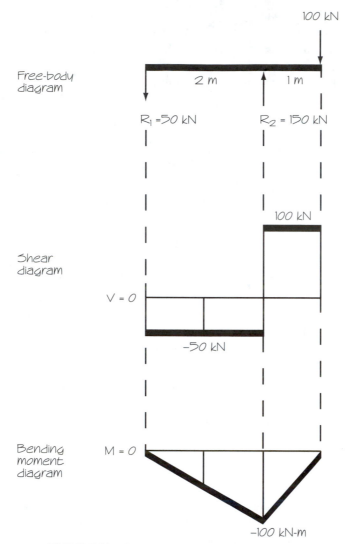

FIGURE 12.14 Free-body, vertical shear, and bending moment diagrams for Example 12.7. Reactions were calculated in Example 5.5. Vertical shear calculations were made in Example 11.5. Bending moment analysis is shown in Example 12.7.

$$M \text{ @ } 5 \text{ ft} = (3.5 \text{ kip} \cdot 3 \text{ ft}) - [(1 \text{ kip/ft} \cdot 4 \text{ ft}) (3 \text{ ft})] = -1.5 \text{ kip-ft}$$
$$M \text{ @ } R_2 = (3.5 \text{ kip} \cdot 4 \text{ ft}) - [(1 \text{ kip/ft} \cdot 4 \text{ ft}) (4 \text{ ft})] - (2.5 \text{ kip} \cdot 0 \text{ ft})$$
$$= -2.0 \text{ kip-ft}$$
$$M \text{ @ } 7 \text{ ft} = (3.5 \text{ kip} \cdot 5 \text{ ft}) + (2.5 \text{ kip} \cdot 1 \text{ ft}) - [(1 \text{ kip/ft} \cdot 4 \text{ ft}) (5 \text{ ft})]$$
$$- [(1 \text{ kip/ft} \cdot 1 \text{ ft}) (0.5 \text{ ft})] = -0.5 \text{ kip-ft}$$
$$M \text{ @ } 8 \text{ ft} = (3.5 \text{ kip} \cdot 6 \text{ ft}) + (2.5 \text{ kip} \cdot 2 \text{ ft}) - [(1 \text{ kip/ft} \cdot 4 \text{ ft}) (6 \text{ ft})]$$
$$- [(1 \text{ kip/ft} \cdot 2 \text{ ft}) (1 \text{ ft})] = 0$$

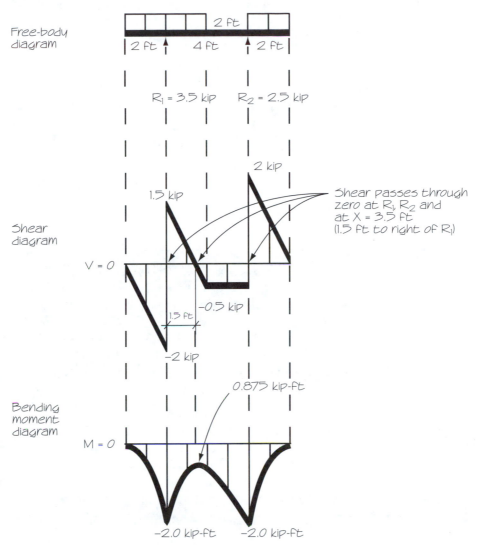

FIGURE 12.15 Free-body, vertical shear, and bending moment diagrams for Example 12.8. Reactions were calculated in Example 5.7. Vertical shear calculations were made in Example 11.8. Bending moment analysis is shown in Example 12.7.

In Example 11.8 it was found that shear passes through zero at R_1, R_2, and at the 3.5-ft mark on the beam. The bending moment must be calculated at the 3.5-ft mark:

$$M \text{ @ } 3.5 \text{ ft} = (3.5 \text{ kip} \cdot 1.5 \text{ ft}) - \left[(1 \text{ kip/ft} \cdot 3.5 \text{ ft}) \frac{(3.5 \text{ ft})}{2} \right] = -0.875 \text{ kip-ft}$$

The maximum bending moment is 2.0 kip-ft and occurs at R_1 and R_2.

12.7   INTERPRETING BEAM ANALYSIS

The tedious task of developing shear and bending moment diagrams is sometimes neglected in design work. However, shear and bending moment diagrams serve as an excellent illustration of the forces and moments experienced by a beam under load. Example 12.9 shows a typical longhand approach to beam analysis. Free-body, shear force, and bending moment diagrams are shown with pertinent information in Figure 12.16. The designer typically follows a similar approach in beam analysis.

■ EXAMPLE 12.9

A free-body diagram of a beam with a double overhang is shown in Figure 12.16. The 19-ft-long beam carries a uniformly distributed load of 1000 lb/ft over the first 15 ft of its length. The load does not cover the 4-ft overhang to

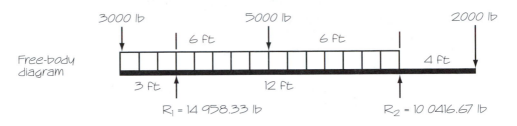

Free-body
diagram

w = 1000 lb/ft

3000 lb   5000 lb   2000 lb

6 ft   6 ft

4 ft

3 ft   12 ft

R_1 = 14 958.33 lb R_2 = 10 0416.67 lb

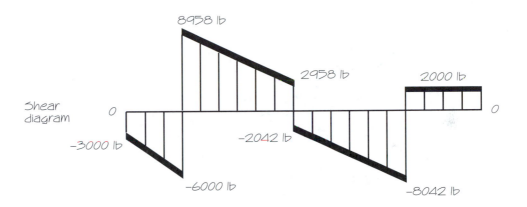

Shear
diagram

8958 lb

2958 lb

2000 lb

0   0

−3000 lb

−2042 lb

−6000 lb

−8042 lb

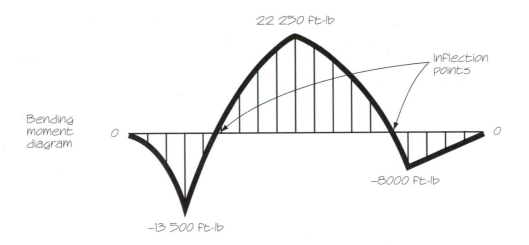

Bending
moment
diagram

22 250 ft-lb

Inflection
points

0   0

−8000 ft-lb

−13 500 ft-lb

Beam
flexure

FIGURE 12.16 Free-body, vertical shear, and bending moment diagrams for Example 12.9.

227

the right of R_2. The following concentrated loads rest on the beam: 3000 lb at the left end, 5000 lb at the 9-ft mark, and 2000 lb at the right end. Reaction R_1 is a pinned support and reaction R_2 is a bearing support.
(a) Solve for the unknown reactions R_1 and R_2.
(b) Calculate the vertical shear forces at 1-ft intervals across the beam and find where shear passes through zero.
(c) Calculate bending moment at 1-ft intervals across the beam. Identify the maximum bending moment.

Solution

(a) $\circlearrowright \Sigma M_{R1} = 0$

$\circlearrowright \Sigma M_{R1} = 0 = (R_2 \cdot 12 \text{ ft}) + (3000 \text{ lb} \cdot 3 \text{ ft}) + (1000 \text{ lb/ft} \cdot 15 \text{ ft} \cdot 4\frac{1}{2} \text{ ft})$
$$- (5000 \text{ lb} \cdot 6 \text{ ft}) - (2000 \text{ lb} \cdot 16 \text{ ft})$$
$$0 = 12R_2 + 9000 \text{ lb-ft} + 67\,500 \text{ lb-ft} - 30\,000 \text{ lb-ft} - 32\,000 \text{ lb-ft}$$
$$R_2 = 10\,041.67 \text{ lb}$$

$\overset{+}{\rightarrow} \Sigma F_x = 0$

$\overset{+}{\uparrow} \Sigma F_y = 0$

$\overset{+}{\uparrow} \Sigma F_y = 0 = R_1 + 10\,041.67 \text{ lb} - 3000 \text{ lb} - 5000 \text{ lb} - 2000 \text{ lb}$
$$- (1000 \text{ lb/ft} \cdot 15 \text{ ft})$$
$$R_1 = 14\,958.33 \text{ lb}$$

(b) $V @ x = \Sigma F_{up} - \Sigma F_{down}$ to the left of section x

$V @ 0 \text{ ft} = 0 - 3000 \text{ lb} = -3000 \text{ lb}$

$V @ R_1- = 0 - 3000 \text{ lb} - (3 \text{ ft} \cdot 1000 \text{ lb/ft}) = -6000 \text{ lb}$

$V @ R_1+ = 14\,958.33 \text{ lb} - 3000 \text{ lb} - (3 \text{ ft} \cdot 1000 \text{ lb/ft}) = 8958.33 \text{ lb}$

$V @ 9 \text{ ft}- = 14\,958.33 \text{ lb} - 3000 \text{ lb} - (9 \text{ ft} \cdot 1000 \text{ lb/ft}) = 2958.33 \text{ lb}$

$V @ 9 \text{ ft}+ = 14\,958.33 \text{ lb} - 3000 \text{ lb} - (9 \text{ ft} \cdot 1000 \text{ lb/ft}) - 5000 \text{ lb}$
$$= -2041.67 \text{ lb}$$

$V @ R_2- = 14\,958.33 \text{ lb} - 3000 \text{ lb} - (15 \text{ ft} \cdot 1000 \text{ lb/ft}) - 5000 \text{ lb}$
$$= -8041.67 \text{ lb}$$

$V @ R_2+ = 14\,958.33 \text{ lb} + 10\,041.67 \text{ lb} - 3000 \text{ lb} - (15 \text{ ft} \cdot 1000 \text{ lb/ft})$
$$- 5000 \text{ lb} = 2000 \text{ lb}$$

$V @ 19 \text{ ft}- = 14\,958.33 \text{ lb} + 10\,041.67 \text{ lb} - 3000 \text{ lb} - (15 \text{ ft} \cdot 1000 \text{ lb/ft})$
$$- 5000 \text{ lb} = 2000 \text{ lb}$$

$V @ 19 \text{ ft} = 14\,958.33 \text{ lb} + 10\,041.67 \text{ lb} - 3000 \text{ lb} - (15 \text{ ft} \cdot 1000 \text{ lb/ft})$
$$- 5000 \text{ lb} - 2000 = 0$$

Shear passes through zero at R_1, R_2, and below the 5000-lb load.

(c) $M = \Sigma M_{up} - \Sigma M_{down}$ to the left of the section under consideration

$M @ 0 \text{ ft} = 0$

$M @ 1 \text{ ft} = -(3000 \text{ lb} \cdot 1 \text{ ft}) - (1000 \text{ lb/ft} \cdot 1 \text{ ft} \cdot \frac{1}{2} \text{ ft}) = -3500 \text{ lb-ft}$

$M @ 2 \text{ ft} = -(3000 \text{ lb} \cdot 2 \text{ ft}) - (1000 \text{ lb/ft} \cdot 2 \text{ ft} \cdot 1 \text{ ft}) = -8000 \text{ lb-ft}$

$M @ R_1 = -(3000 \text{ lb} \cdot 3 \text{ ft}) - (1000 \text{ lb/ft} \cdot 3 \text{ ft} \cdot 1\frac{1}{2} \text{ ft}) = -13\,500 \text{ lb-ft}$

M @ 4 ft = (14 958.33 lb · 1 ft) − (3000 lb · 4 ft) − (1000 lb/ft · 4 ft · 2 ft)
$\qquad$ = −5041.67 lb-ft

M @ 5 ft = (14 958.33 lb · 2 ft) − (3000 lb · 5 ft) − (1000 lb/ft · 5 ft · $2\frac{1}{2}$ ft)
$\qquad$ = 2416.67 lb-ft

M @ 6 ft = (14 958.33 lb · 3 ft) − (3000 lb · 6 ft) − (1000 lb/ft · 6 ft · 3 ft)
$\qquad$ = 8875 lb-ft

M @ 7 ft = (14 958.33 lb · 4 ft) − (3000 lb · 7 ft) − (1000 lb/ft · 7 ft · $3\frac{1}{2}$ ft)
$\qquad$ = 14 333.33 lb-ft

M @ 8 ft = (14 958.33 lb · 5 ft) − (3000 lb · 8 ft) − (1000 lb/ft · 8 ft · 4 ft)
$\qquad$ = 18 791.67 lb-ft

M @ 9 ft = (14 958.33 lb · 6 ft) − (3000 lb · 9 ft) − (1000 lb/ft · 9 ft · $4\frac{1}{2}$ ft)
$\qquad$ = 22 250 lb-ft

M @ 10 ft = (14 958.33 lb · 7 ft) − (3000 lb · 10 ft) − (1000 lb/ft · 10 ft · 5 ft)
$\qquad$ − (5000 lb · 1 ft) = 19 708 lb-ft

M @ 11 ft = (14 958.33 lb · 8 ft) − (3000 lb · 11 ft) − (1000 lb/ft · 11 ft · $5\frac{1}{2}$ ft)
$\qquad$ − (5000 lb · 2 ft) = 16 167 lb-ft

M @ 12 ft = (14 958.33 lb · 9 ft) − (3000 lb · 12 ft) − (1000 lb/ft · 12 ft · 6 ft)
$\qquad$ − (5000 lb · 3 ft) = 11 625 lb-ft

M @ 13 ft = (14 958.33 lb · 10 ft) − (3000 lb · 13 ft) − (1000 lb/ft · 13 ft · $6\frac{1}{2}$ ft)
$\qquad$ − (5000 lb · 4 ft) = 6083 lb-ft

M @ 14 ft = (14 958.33 lb · 11 ft) − (3000 lb · 14 ft) − (1000 lb/ft · 14 ft · 7 ft)
$\qquad$ − (5000 lb · 5 ft) = −458 lb-ft

M @ R_2 = (14 958.33 lb · 12 ft) − (3000 lb · 15 ft) − (1000 lb/ft · 15 ft · $7\frac{1}{2}$ ft)
$\qquad$ − (5000 lb · 6 ft) = −8000 lb-ft

M @ 16 ft = (14 958.33 lb · 13 ft) + (10 041.67 lb · 1 ft) − (3000 lb · 16 ft)
$\qquad$ − (1000 lb/ft · 15 ft · $8\frac{1}{2}$ ft) − (5000 lb · 7 ft) = −6000 lb-ft

M @ 17 ft = (14 958.33 lb · 14 ft) + (10 041.67 lb · 2 ft) − (3000 lb · 17 ft)
$\qquad$ − (1000 lb/ft · 15 ft · $9\frac{1}{2}$ ft) − (5000 lb · 8 ft) = −4000 lb-ft

M @ 18 ft = (14 958.33 lb · 15 ft) + (10 041.67 lb · 3 ft) −(3000 lb ·18 ft)
$\qquad$ − (1000 lb/ft · 15 ft · $10\frac{1}{2}$ ft) − (5000 lb · 9 ft) = −2000 lb-ft

M @ 19 ft = (14 958.33 lb · 16 ft) + (10 041.67 lb · 4 ft) − (3000 lb · 19 ft)
$\qquad$ − (1000 lb/ft · 15 ft · $11\frac{1}{2}$ ft) − (5000 lb · 10 ft) = 0

The maximum bending moment is 22 250 lb-ft, as this is the greatest absolute value of bending moment where shear passes through zero.

Following is an interpretation of the results from Example 12.9.

Reactions R_1 is significantly greater than R_2 as a result of a greater percentage of total load being positioned closer to R_1. The bearing point of the beam at R_1 must be designed to carry, or react to 14 958 lb. At R_2, the bearing point must be designed to react to 10 042 lb. Note that the total load resting on the beam equals the sum of the reactions:

$$3000 \text{ lb} + 5000 \text{ lb} + 2000 \text{ lb} + (1000 \text{ lb/ft} \cdot 15 \text{ ft}) = 14\,958 \text{ lb} + 10\,042 \text{ lb}$$
$$25\,000 \text{ lb} = 25\,000 \text{ lb}$$

Shear   As with most vertical shear force calculations involving two reactions, both positive and negative values of shear exist. At a specified section on a beam, a positive value indicates that the shear force will develop as forces to the left of the section act upward and forces to the right act downward ($\uparrow\downarrow$); this occurs to the right of R_1 through the 9-ft mark and to the right of R_2 through the 19-ft mark (right end) of the beam. Conversely, a negative value indicates that shear forces to the left of the section act downward, with forces to the right acting upward ($\downarrow\uparrow$); this occurs from the 0-ft mark (left end) through R_1+ and again from the 9-ft mark through just to the right of R_2. In all cases the opposing forces attempt to slice through the member creating a shear stress.

Maximum vertical shear force occurs just to the right of R_1 and is equal to a force of 8958 lb. Since the shear force is a positive (+) value, the right side of the section is experiencing a downward force; the beam tends to shear at its support. Sloping shear force curves occur directly below uniformly distributed forces. A horizontal shear curve exists along lengths of the beam where no force is present. Significant changes in shear occur on the sides of the reactions and concentrated forces.

Bending moment   Both positive and negative values for bending moment result from this loading condition. A positive value for bending moment indicates that curvature or bending at that point in the beam is concave up; that is, the beam bends such that it holds water. A negative value indicates that curvature is concave down. Additionally, maximum bending moment occurs below the 5000-lb force and is equal to 22 500 lb-ft. This is the largest absolute value for bending moment where shear passes through zero.

12.8   INFLECTION POINTS

Points in the beam where bending moment equals zero are called *inflection points*. On either side of an inflection point, tensile and compressive stresses are in an opposite position with respect to the neutral plane; that is, compressive and tensile stresses switch position with respect to the neutral axis as the section under consideration moves through the inflection point. With certain beams, such as those in Figures 12.11 through 12.15, there are no inflection points. In these examples, bending moment never equals zero so compressive and tensile stresses do not switch position. In Figure 12.16, however, there are two locations where bending moment equals zero and two inflection points.

Determining the exact position where an inflection point occurs is necessary in placement of reinforcing bars in a concrete beam; reinforcement must be located on the tensile side of the neutral plane. An inflection point occurs at the point where bending moment equals zero ($M = 0$). The process used to ascertain the exact position of an inflection point is demonstrated in Example 12.10.

■ EXAMPLE 12.10

Find the exact location of the inflection points in the beam shown in Figure 12.16.

Solution *First inflection point.* From Example 12.9, M @ 4 ft = -5041.67 lb-ft and M @ 5 ft = 2416.67 lb-ft, so bending moment must pass through zero between these known points. Assume that $M = 0$ at x, which is somewhere between the 4- and 5-ft marks on the beam. Computing bending moment to the left of the unknown section x:

$$M @ x = 0 = [14\,958.33 \text{ lb} \cdot (x - 3)\text{ft}] - (3000 \text{ lb} \cdot x \text{ ft}) - \left(1000 \text{ lb/ft} \cdot x \text{ ft} \cdot \frac{x}{2 \text{ ft}}\right)$$

$$= 14\,958.33(x - 3) - 3000x - 500x^2$$
$$= 14\,958.33x - 44\,875.0 - 3000x - 500x^2$$

This expression simplifies to

$$0 = -500x^2 + 11\,958.33x - 44\,875.0$$

At this point values between 4 and 5 can be substituted in for x until the right side of the equation equals zero, or the quadratic formula can be used to solve for x:

$$x = \frac{-b \pm \sqrt{b^2 - 4ac}}{2a}$$

From $0 = -500x^2 + 11\,958.33x - 44\,875.0$, where $a = -500$, $b = 11\,958.33$, and $c = -44\,875.0$:

$$x_1 = \frac{-11\,958.33 + \sqrt{(11\,958.33)^2 - 4(-500)(-44\,875.0)}}{2(-500)}$$

$$x_1 = 4.661 \text{ ft}$$

$$x_2 = \frac{-11\,958.33 - \sqrt{(11\,958.33)^2 - 4(-500)(-44\,875.0)}}{2(-500)}$$

$$x_2 = 19.26 \text{ ft}$$

The value $x_2 = 19.26$ ft cannot exist because the beam is not that long, so the inflection point occurs at $x_1 = 4.66$ ft. The results can be checked by substituting $x_1 = 4.66$ ft into the original expression:

$$0 = -500x^2 + 11\,958.33x - 44\,875.0$$
$$-500(4.661)^2 + 11\,958.33(4.661) - 44\,875.0 = 0$$

12.8   Inflection Points

Second inflection point. From Example 12.9, M @ 13 ft = 6083 lb-ft and M @ 14 ft = −458 lb-ft, so the second inflection point occurs between the 13- and 14-ft marks on the beam. Assume that $M = 0$ at x, which is somewhere between the 13- and 14-ft marks on the beam. Computing bending moment to the left of the unknown section x:

$$M @ 14 \text{ ft} = 0 = [14\,958.33 \text{ lb} \cdot (x - 3)\text{ft}] - (3000 \text{ lb} \cdot x \text{ ft})$$
$$- \left(1000 \text{ lb/ft} \cdot x \text{ ft} \cdot \frac{x}{2 \text{ ft}}\right) - [5000 \text{ lb} \cdot (x - 9) \text{ ft}]$$
$$= 14\,958.33(x - 3) - 3000x - 500x^2 - 5000(x - 9)$$
$$= 14\,958.33x - 44\,875.0 - 3000x - 500x^2 - 5000x + 45\,000$$

This expression simplifies to

$$0 = -500x^2 + 6958.33x + 125$$

Using the quadratic formula to solve for x where $a = -500$, $b = 6958.33$, and $c = 125$, we have

$$x = \frac{-b \pm \sqrt{b^2 - 4ac}}{2a}$$
$$x = \frac{-6958.33 + \sqrt{(6958.33)^2 - 4(-500)(125)}}{2(-500)}$$
$$x_1 = -0.018 \text{ ft}$$
$$x = \frac{-6958.33 - \sqrt{(6958.33)^2 - 4(-500)(125)}}{2(-500)}$$
$$x_2 = 13.93 \text{ ft}$$

A negative dimension ($x_1 = -0.018$ ft) cannot exist, so the inflection point occurs at $x_2 = 13.93$ ft. The results can be checked by substituting 13.93 for x into the original expression.

12.9   LIMITATIONS OF THIS ANALYSIS APPROACH

In this chapter, analysis of shear and bending moment was limited to forces and reactions acting along the same plane and acting perpendicular to the beam length. In actual practice in the profession, designers will encounter instances where this is not the case; that is, forces may act diagonally and still cause the member to bend. Furthermore, moments may be applied to a beam; for example, a shaft may intersect the member and induce a twisting moment, causing the member to bend. Discussion of analysis techniques used in solving these problems is beyond the scope of this book. The reader should be aware of the limitations of the analysis approach presented.

The methods for computing reactions, shear, and bending moment enables one to find these values under a wide variety of loading conditions and at various points along a beam length. However, the designer is typically most interested in maximum values. Certain beam conditions occur so frequently, it is convenient and customary to use formulas that have been derived to give the maximum values directly. Table A.5 in Appendix A has a condensed collection of formulas that are frequently used when beam loading conditions permit. Designers use these formulas *if and only if* they fully describe a free-body diagram for a particular loading condition.

An example derivation of the formulas for a beam with a concentrated load at its center follows. Consider a simple beam carrying a concentrated force at its center, as described in Figure 12.17. Intuitively, one can conclude that since beam loading is symmetrical, $R_1 = R_2$ and the reactions equal one-half the concentrated force (P). Therefore,

$$R_1 = R_2 = \frac{P}{2}$$

or simply,

$$R = \frac{P}{2}$$

Maximum vertical shear (V) occurs to the left of the force P, because

$$V @ x = \Sigma F_{up} - \Sigma F_{down} \text{ to the left of section under consideration}$$

In this case, maximum shear occurs when $\Sigma F_{down} = 0$; therefore,

$$V = R$$

But $R = P/2$; therefore,

$$R = V = \frac{P}{2}$$

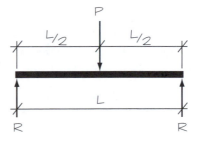

FIGURE 12.17 Simple beam with span L carrying a concentrated force P at the center of the span.

Maximum bending moment (M_{max}) is found by

$$M = \Sigma M_{up} - \Sigma M_{down}$$

M_{max} occurs at the beam center, as this is where the ΣM_{up} is at its greatest with the ΣM_{down} equal to zero; therefore,

$$M = \Sigma M_{up} \text{ (at beam center)} = R\frac{L}{2}$$

But $R = P/2$; therefore,

$$M = \frac{P}{2} \cdot \frac{L}{2} = \frac{PL}{4}$$

Results of this analysis yields the following derived formulas:

$$R = V = \frac{P}{2}$$

$$M = \frac{PL}{4}$$

These formulas may be used *only* for a simple beam with a concentrated force at its center.

Included in Table A.5 in Appendix A is an abridged set of formulas for a variety of common loading conditions. A more complete set of formulas may be found in other sources, such as the American Institute of Steel Construction (AISC) *Steel Construction Manual*. The designer is cautioned to use correct units when using any derived formula. For example, w in this book refers to lb/ft or kip/ft, whereas in the AISC manual, w is expressed in kip/in. Use of improper units usually results in significant error in results.

EXERCISES

Eleven sets of illustrations and the corresponding free-body diagrams of building-related beams were provided in Chapter 4 as Exercises 4.1 through 4.11. Shear diagrams were developed in Chapter 11 as Exercises 11.1 through 11.11. These diagrams constitute exercises that will be used again here in Chapter 12. *Follow these instructions carefully to save time later:*

(a) On each sheet of engineering paper reserved for analysis, calculate bending moment specific to that member at appropriate intervals across the length of the free-body diagram. Be sure to solve for bending moment at locations where shear passes through zero.

(b) On each sheet of engineering paper containing the free-body and shear diagrams, neatly draw the related bending moment diagram directly below the associated free-body and shear diagrams. Label the maximum bending values. The bending moment diagram should be located on the bottom one-third of the page.

(c) On bending moment diagrams having inflection point(s), determine the exact position of the inflection point(s).

12.1 The free-body diagram in Exercise 4.1 and shear diagram from Exercise 11.1.

12.2 The free-body diagram in Exercise 4.2 and shear diagram from Exercise 11.2.

12.3 The free-body diagram in Exercise 4.3 and shear diagram from Exercise 11.3.

12.4 The free-body diagram in Exercise 4.4 and shear diagram from Exercise 11.4.

12.5 The free-body diagram in Exercise 4.5 and shear diagram from Exercise 11.5.

12.6 The free-body diagram in Exercise 4.6 and shear diagram from Exercise 11.6.

12.7 The free-body diagram in Exercise 4.7 and shear diagram from Exercise 11.7.

12.8 The free-body diagram in Exercise 4.8 and shear diagram from Exercise 11.8.

12.9 The free-body diagram in Exercise 4.9 and shear diagram from Exercise 11.9.

12.10 The free-body diagram in Exercise 4.10 and shear diagram from Exercise 11.10.

12.11 The free-body diagram in Exercise 4.11 and shear diagram from Exercise 11.11.

Chapter 13
CENTER OF GRAVITY AND CENTROID OF A SHAPE

It is frequently necessary for a designer to establish the center of gravity of a body to simplify computations involving forces. Such simplification was evident in Chapter 3, where uniformly and evenly increasing distributed forces were treated as a single concentrated force to simplify static equilibrium computations. In this chapter, concepts associated with the center of gravity are presented. An understanding of the concept of the centroid is fundamental to developing an understanding of the properties of beam and column sections that are introduced in Chapter 14. Additionally, the concept of the centroid of a shape is more easily understood when presented with the concept of the center of gravity.

13.1 THE CENTER OF GRAVITY

At the earth's surface, the force of gravity acts on a body, giving it weight. In many instances in engineering analysis, forces under consideration involve the weight of a relatively large body being analyzed. In such analysis, it is helpful to assume that all the weight of a body is concentrated at a single point called the center of gravity. The weight of a body is composed of a system of small gravitational forces each acting on the individual molecules of the material making up the body. At the earth's surface, these forces act downward toward the center of the earth. The *center of gravity* of a body is that point where the total weight of the body may be assumed to act. It is that point at which the combined effect of the small gravitational forces acting on each molecule may be theoretically replaced with a single larger resultant force.

Consider a 10-lb steel plate: When it is said that the plate weighs 10 lb, it is meant that there is a force of 10 lb acting on the plate that is directed toward the center of the earth. Actually, the force of gravitational attraction is exerted on the plate as a whole; that is, the force of gravity is exerted on each molecule in the plate. The resultant or sum of these small forces is a single force of 10 lb directed downward from the center of gravity of the plate toward the center of the earth.

The center of gravity of a body is that point where a body can be suspended or balanced in any orientation without tending to rotate. When a yard or meter stick is balanced on a finger such that it does not rotate, the center of gravity of the yard or meter stick is located directly above the contact point of the finger. The yard or meter stick may be balanced such that it is standing vertically or balanced such that it is laying horizontally. In both cases the center of gravity of the yard or meter stick lies directly above the contact point of the finger.

The location of the center of gravity of a body occurs at its geometric center if the body is symmetrical about all axes and composed of a homogeneous material. Thus, for spheres, cubes, and cylinders composed of a homogeneous material, the center of gravity is situated at an easily identified geometric center. In such cases the exact location of the center of gravity can be found through observation; analysis is not necessary.

The location of the center of gravity of an irregularly shaped body is not as easily identified. Many times a body's center of gravity lies outside the body, such as in the case of a boomerang, steel angle, or channel section. Additionally, a symmetrical body may be composed of materials of different densities, thereby causing the center of gravity of the body to be located away from its geometric center.

In engineering analysis, it is frequently necessary to determine the exact location of the center of gravity of a body. Analysis involves dividing the body into constituent bodies, each with a known weight and geometric center, and placing the body on a Cartesian (x–y–z) coordinate system. A *constituent body* is an easily identified part of the whole body that has a form that is geometrically simple with a known geometric center. The approach is described below.

The following technique may be used to approximate the center of gravity of an asymmetrical or composite body:

1. Divide the body into symmetrical (or near symmetrical) constituent bodies each having a known weight (w) and known location of the geometric center. If the body is composed of materials having different densities, it is necessary to consider these as separate constituent bodies. The weight of each constituent body (i.e., w_1, w_2, w_3, etc.) may be found by multiplying the density of the material (i.e., ρ_1, ρ_2, ρ_3, etc.) by the volume (i.e., V_1, V_2, V_3, etc.) of each constituent body:

$$w = \rho V$$

The density (ρ) of a solid material is the mass of the material divided by its volume. Densities for common material are given in Tables A.3 and A.4 in Appendix A.

2. Position the body within a Cartesian (x–y–z) coordinate system origin and determine the x, y, and z coordinate location of the centers of gravity of each constituent body.

3. Solve for the overall center of gravity coordinate location ($\bar{x}$, $\bar{y}$, and $\bar{z}$) using the following formulas. All formulas are needed for three-dimensional bodies. Formulas for $\bar{x}$ and $\bar{y}$ are needed for bodies of a constant thickness, such as an odd-shaped plate or panel.

$$\bar{x} = \frac{x_1 w_1 + x_2 w_2 + x_3 w_3 + \cdots}{w_1 + w_2 + w_3 + \cdots}$$

$$\bar{y} = \frac{y_1 w_1 + y_2 w_2 + y_3 w_3 + \cdots}{w_1 + w_2 + w_3 + \cdots}$$

$$\bar{z} = \frac{z_1 w_1 + z_2 w_2 + z_3 w_3 + \cdots}{w_1 + w_2 + w_3 + \cdots}$$

The resulting $\bar{x}$, $\bar{y}$, and $\bar{z}$ values are the coordinates of the center of gravity of the body based on the initially positioned coordinate system origin.

■ EXAMPLE 13.1

The channel shown in Figure 13.1a is made from aluminum. Determine the location of its center of gravity.

Solution The channel is divided into constituent bodies 1, 2, and 3 that have known geometric centers as shown in Figure 13.1b. A coordinate system origin is placed on the channel as shown in Figure 13.1c. Coordinate values for the centers of gravity of each constituent body are shown in Figure 13.1d.

In Table A.3 in Appendix A, the density of aluminum is found to be 0.0955 lb/in^3, and the weight of each constituent body is found by multiplying the volume of each constituent body by the density (ρ) of aluminum:

$$w_1 = \rho V = (0.0955 \text{ lb/in}^3)(3 \text{ in} \cdot 1 \text{ in} \cdot 4 \text{ in}) = 1.146 \text{ lb}$$
$$w_2 = \rho V = (0.0955)(1 \text{ in} \cdot 4 \text{ in} \cdot 4 \text{ in}) = 1.528 \text{ lb}$$
$$w_3 = \rho V = (0.0955)(3 \text{ in} \cdot 1 \text{ in} \cdot 4 \text{ in}) = 1.146 \text{ lb}$$

The $\bar{x}$, $\bar{y}$, and $\bar{z}$ coordinate locations of the center of gravity of the channel are found by first summing the products of the weight of each constituent body (w) and the x, y, and z coordinates of the center of gravity of each constituent body and then dividing it by the total weight of the body:

$$\bar{x} = \frac{1.5 \text{ in}(1.146 \text{ lb}) + 0.5 \text{ in}(1.528 \text{ lb}) + 1.5 \text{ in}(1.146 \text{ lb})}{1.146 \text{ lb} + 1.528 \text{ lb} + 1.146 \text{ lb}} = 1.10 \text{ in}$$

$$\bar{y} = \frac{5.5(1.146) + 3(1.528) + 0.5(1.146)}{1.146 + 1.528 + 1.146} = 3.0 \text{ in}$$

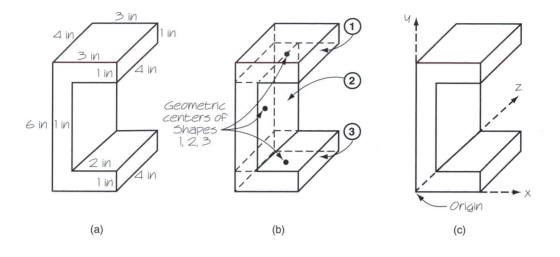

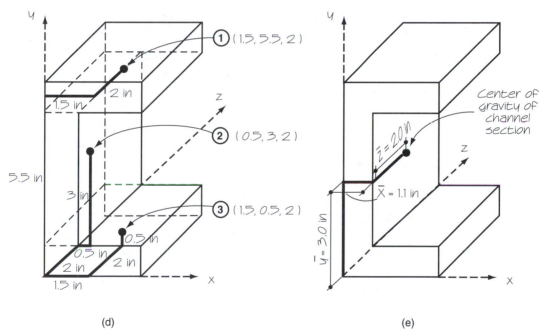

FIGURE 13.1 Aluminum channel for Example 13.1: (a) dimensions of the channel; (b) the channel divided into simple geometric shapes; (c) positioning of the coordinate system; (d) coordinate location of the centers of gravity of each geometric shape; and (e) center of gravity of the channel.

$$\bar{z} = \frac{2(1.146) + 2(1.528) + 2(1.146)}{1.146 + 1.528 + 1.146} = 2.0 \text{ in}$$

The center of gravity of the channel in Example 13.1 has coordinates of 1.10 in, 3.0 in, and 2.0 in. based on the origin selected. It is shown in Figure 13.1e. This center of gravity is outside of the channel itself.

■ EXAMPLE 13.2

A 50-mm-thick rubber disk with a diameter of 150 mm is attached to a 15-mm-thick steel disk of the same diameter, as shown in Figure 13.2. The disks are joined to form a composite cylindrical member that will be used as a large pad. Determine the location of the center of gravity of the composite cylindrical member.

Solution The composite cylindrical member is divided by material into two disks having known geometric centers in the center of each disk. A coordinate system origin is placed at the center point of the outside surface of the steel portion of the composite cylindrical member. The y axis falls along the longitudinal axis of the cylinder.

In Table A.4 in Appendix A the density of rubber is 1500 kg/m^3 and the density of steel is 7840 kg/m^3. The volume of a disk or cylinder is $h(\pi D^2/4)$, where D is the diameter of the cylinder and h is the cylinder height. The weight of each disk is found by multiplying the volume of each disk by the density of the material (ρ).

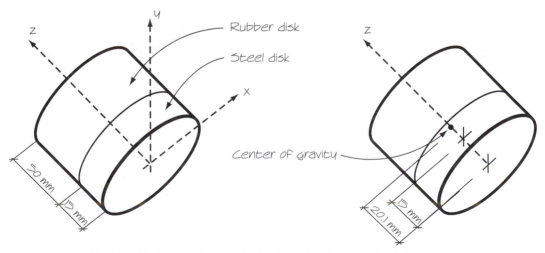

FIGURE 13.2 Composite cylindrical member for Example 13.2.

Steel disk:

$$w_1 = \rho h \left(\frac{\pi D^2}{4}\right) = (7840 \text{ kg/m}^3)\left[(0.015 \text{ m})\frac{\pi (0.150 \text{ m})^2}{4}\right] = 2.08 \text{ kg}$$

Rubber disk:

$$w_2 = \rho h \left(\frac{\pi D^2}{4}\right) = (1500 \text{ kg/m}^3)\left[(0.050 \text{ m})\frac{\pi (0.150 \text{ m})^2}{4}\right] = 1.325 \text{ kg}$$

The $\bar{x}$, $\bar{y}$, and $\bar{z}$ coordinates are found by first summing the products of the weight of each disk (w) and the x, y, and z coordinates of each center of gravity and then dividing by the total weight of the composite disk:

$$\bar{x} = \frac{0 \,(2.08 \text{ kg}) + 0 \,(1.325 \text{ kg})}{2.08 \text{ kg} + 1.325 \text{ kg}} = 0$$

$$\bar{y} = \frac{0 \,(2.08 \text{ kg}) + 0 \,(1.325 \text{ kg})}{2.08 \text{ kg} + 1.325 \text{ kg}} = 0$$

$$\bar{z} = \frac{7.5 \text{ mm} \,(2.08 \text{ kg}) + 40 \text{ mm} \,(1.325 \text{ kg})}{2.08 \text{ kg} + 1.325 \text{ kg}} = 20.1 \text{ mm}$$

In Example 13.2, two disks form a single composite shape that is a simple geometric shape, a cylinder. As shown in Example 13.2, the composite cylinder has a center of gravity that is located 20.1 mm from the center point of the outside surface of the steel portion of the member. It falls along the longitudinal axis of the cylinder. The center of gravity is located in the rubber disk portion of the composite cylinder but not in the geometric center of the cylinder. Steel is more dense than rubber, so the center of gravity of the composite member is closer to the side of the steel disk. Because the cylinder is a composite of two materials having different densities, it is necessary to treat each disk as a separate constituent body when finding the center of gravity.

■ **EXAMPLE 13.3**

As shown in Figure 13.3, a steel beam spans 32 ft and supports a 10-ft-high masonry wall across its entire length. For the first 2 ft from the left side of the beam and for 12 ft on the right side of the beam, the masonry wall is composed of solid face brick with hollow concrete masonry backing. This composite wall of brick and concrete masonry weighs 70 lb/ft² of wall surface area. The remaining 18 ft of masonry wall is composed of hollow concrete masonry that weighs 55 lb/ft² of wall surface area. The designer desires to place a column at the midpoint of the beam. The column must be located within 18 in. of the center of gravity of the wall. Is the desired column placement acceptable?

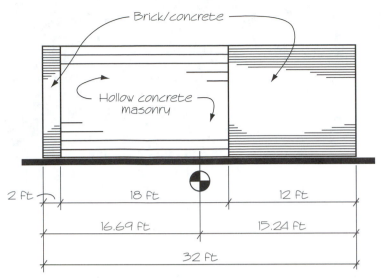

FIGURE 13.3 Masonry wall for Example 13.3.

Solution   The wall is divided by material into three assemblies. The weight of each wall assembly is found by multiplying the area of each wall assembly by the lb/ft^2 of wall surface area:

2-ft brick/concrete assembly:	70 lb/ft^2 · (2 ft · 10 ft) = 1400 lb
18-ft hollow concrete assembly:	55 lb/ft^2 · (18 ft · 10 ft) = 9900 lb
12-ft brick/concrete assembly:	70 lb/ft^2 · (12 ft · 10 ft) = 8400 lb

A coordinate system origin is placed at the left end of the beam. The x axis of the wall falls along the longitudinal axis of the beam, so only the x coordinate of the center of gravity of the wall is needed. The center of gravity along the x axis of each wall assembly exists at the geometric center of each assembly. The x coordinate of the geometric center of each assembly is found by:

2-ft brick/concrete assembly:
$$x = \frac{2 \text{ ft}}{2} = 1 \text{ ft}$$

18-ft hollow concrete assembly:
$$x = 2 \text{ ft} + \frac{18 \text{ ft}}{2} = 11 \text{ ft}$$

12-ft brick/concrete assembly:
$$x = 2 \text{ ft} + 18 \text{ ft} + \frac{12 \text{ ft}}{2} = 26 \text{ ft}$$

The center of gravity of the wall is found by first summing the products of the weight of each wall assembly by the $\bar{x}$ coordinate of the center of gravity of each assembly and then dividing the sum by the weight of the total wall:

$$\bar{x} = \frac{1 \text{ ft}(1400 \text{ lb}) + 11 \text{ ft}(9900 \text{ lb}) + 26 \text{ ft}(8400 \text{ lb})}{1400 \text{ lb} + 9900 \text{ lb} + 8400 \text{ lb}} = 16.69 \text{ ft}$$

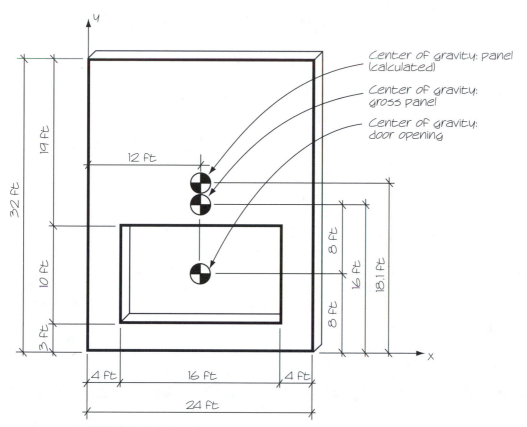

FIGURE 13.4 Reinforced concrete wall panel for Example 13.4.

The midpoint of the beam is 18 ft from the left end of the beam. The center of gravity of the wall is located 16.69 ft (about 16 ft 8¼ in) from the left end of the beam; 1 ft 3¾ in. from the midpoint of the beam. Column placement at the midpoint of the beam is within 18 in. and acceptable.

■ **EXAMPLE 13.4**

An 8-in-thick reinforced concrete wall panel is site cast. The panel is formed to be 32 ft high and 24 ft wide. The panel has a 10-ft-high × 16-ft-wide door opening located in the center of the panel width and 3 ft from the bottom of the panel, as shown in Figure 13.4. The density of reinforced concrete is 150 lb/ft³. Determine the location of the center of gravity of the panel.

Solution   The panel is divided into the gross panel body and the door opening that have known geometric centers, as shown in Figure 13.4. A coordinate system origin is placed at the lower left corner of the panel. The weight of each constituent body is found by multiplying the volume of each constituent body by the density of reinforced concrete (150 lb/ft³). Note that 8 in. is 0.667 ft.

Gross panel: $w_1 = (150 \text{ lb/ft}^3)(32 \text{ ft} \cdot 24 \text{ ft} \cdot 0.667 \text{ ft}) = 76\,838 \text{ lb}$
Door opening: $w_2 = (150 \text{ lb/ft}^3)(10 \text{ ft} \cdot 16 \text{ ft} \cdot 0.667 \text{ ft}) = 16\,008 \text{ lb}$

The $\bar{x}$ and $\bar{y}$ coordinate locations of the center of gravity are found by first summing the products of the weight of each constituent body (w) and the x and y coordinates of the center of gravity of each constituent body and then dividing it by the total weight of the body.

$$\bar{x} = \frac{12.0 \text{ ft } (76\,838 \text{ lb}) - 12.0 \text{ ft } (16\,008 \text{ lb})}{76\,838 \text{ lb} - 16\,008 \text{ lb}} = 12.0 \text{ ft}$$

$$\bar{y} = \frac{16.0 \text{ ft } (76\,838 \text{ lb}) - 8.0 \text{ ft } (16\,008 \text{ lb})}{76\,838 \text{ lb} - 16\,008 \text{ lb}} = 18.1 \text{ ft}$$

Because the concrete wall panel in Example 13.4 is a uniform thickness, only formulas for x and y coordinate values are used in analysis. Since the door opening is a void, it is necessary to subtract the door opening weight in the expression. Also note that while the density of concrete is listed as 144 lb/ft^3 in Table A.3 in Appendix A, a density of 150 lb/ft^3 for reinforced concrete was specified in the statement of the problem. The added weight accounts for use of steel reinforcement in the concrete; steel is much more dense (489 lb/ft^3) than the concrete (144 lb/ft^3) it displaces.

13.3   THE CENTROID

A *centroid* is the point on a one- or two-dimensional shape that may be considered to be the center of that shape. The centroid of a straight line occurs at the midpoint of the length of the line. For simple geometric shapes such as triangles, rectangles, circles, and arcs, the centroid is found at the geometric center of the shape: It is at the center point of a circle, the intersecting point of diagonal lines drawn from opposite corners of a rectangle or any parallelogram, and at the intersection of the medians of a triangle. The centroid of symmetrical structural steel shapes such as wide-flange beam sections is found at the midpoint of the web.

13.4   FINDING THE CENTROID

The centroid of a shape is situated at the geometric center of the shape. Thus if the shape is symmetrical about both axes, such as a rectangle or circle, the centroid is found at the easily identified geometric center of the shape. In such cases the exact location of the centroid can be attained through observation; analysis is not necessary.

The centroid of an irregular or complex two-dimensional shape may be found by first assuming that the shape is composed of a homogeneous material of uniform thickness and then placing the centroid at the center of gravity of the shape as the assumed thickness of the shape decreases to zero and it becomes weightless. The centroid of an irregular or complex two-dimensional shape may be found by the following steps:

1. Divide the shape into symmetrical (or near symmetrical) constituent areas, each having a known constituent area (a) and known location of the geometric center of the area. A *constituent area* is an easily identified part of the whole body that has a shape that is geometrically simple with a known geometric center.

2. Position the section profile within a Cartesian (x–y) coordinate system and determine the x and y coordinate location of the centroid of each constituent area.

3. Solve for the overall centroid coordinate location ($\bar{x}$ and $\bar{y}$) by using the following formulas:

$$\bar{x} = \frac{x_1 a_1 + x_2 a_2 + x_3 a_3 + \cdots}{a_1 + a_2 + a_3 + \cdots}$$

$$\bar{y} = \frac{y_1 a_1 + y_2 a_2 + y_3 a_3 + \cdots}{a_1 + a_2 + a_3 + \cdots}$$

The $\bar{x}$ and $\bar{y}$ are the coordinates of the centroid based on the initially positioned coordinate system origin.

■ EXAMPLE 13.5

An irregular tee shape is illustrated in Figure 13.5. Determine the location of its centroid.

Solution The shape is divided into two constituent areas each having the following areas:

$$30 \text{ mm} \cdot 150 \text{ mm} = 4500 \text{ mm}^2$$
$$30 \text{ mm} \cdot 90 \text{ mm} = 2700 \text{ mm}^2$$

The origin of the coordinate system is placed at the upper left corner of the tee. The $\bar{x}$ and $\bar{y}$ coordinates of the centroid are found by first summing the products of the constituent areas and the x and y coordinates of the centroid of each constituent area and then dividing by the total area of the section profile:

$$\bar{x} = \frac{75 \text{ mm} (4500 \text{ mm}^2) + 75 \text{ mm} (2700 \text{ mm}^2)}{4500 \text{ mm}^2 + 2700 \text{ mm}^2} = 75 \text{ mm}$$

$$\bar{y} = \frac{-15 \text{ mm} (4500 \text{ mm}^2) + -75 \text{ mm} (2700 \text{ mm}^2)}{4500 \text{ mm}^2 + 2700 \text{ mm}^2} = -37.5 \text{ mm}$$

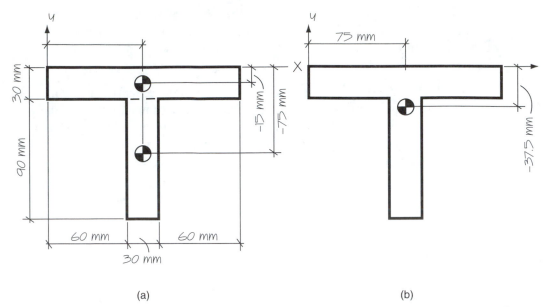

FIGURE 13.5 (a) Tee shape for Example 13.5. (b) The centroid of the tee is located 37.5 mm from the top side of the tee.

The centroid is located at the intersection of a vertical axis that is at the center of the width of the tee and a horizontal axis that is 37.5 mm below the top edge.

Although the tee in Example 13.5 is 120 mm deep, its centroid is located along an *x*-axis that is only 37.5 mm below the top edge of the section profile. The constituent area along the top of the tee draws the centroid closer to the top side of the shape.

A centroid is considered to be the *first moment of an area;* when the area is divided up into infinitesimal (very small) constituent areas and moments of these constituent areas are taken about an axis that passes through the centroid, the moments cancel one another. Consider a rectangle: The rectangle can be divided into long, thin constituent areas that are parallel to the axis of the rectangle. The center of each constituent area are series of points that form a line that runs through the centroid of the rectangle. When moments are calculated from the centroid of each constituent area about the centroid of the rectangle, the moments cancel one another.

The position of the centroid and the loading condition of a beam influence the location of the neutral axis of bending on a beam section. (The neutral axis of bending on a beam section was presented in Chapter 12.) When the beam is composed of a homogeneous material, the neutral axis of the beam section passes through the centroid of the section. As shown in Figure 13.6, the neutral axis of bending is always perpendicular to the applied forces (the reactions and loads).

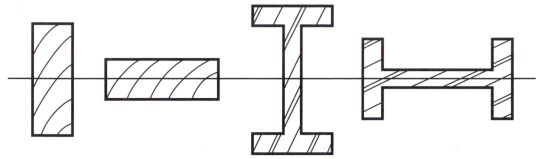

FIGURE 13.6 The neutral axis of bending passes through the centroid of homogeneous sections and is perpendicular to the applied force.

EXERCISES

13.1 Calculate the center of gravity of the object shown. Be sure to use the coordinate origin placement shown in expressing your answer.

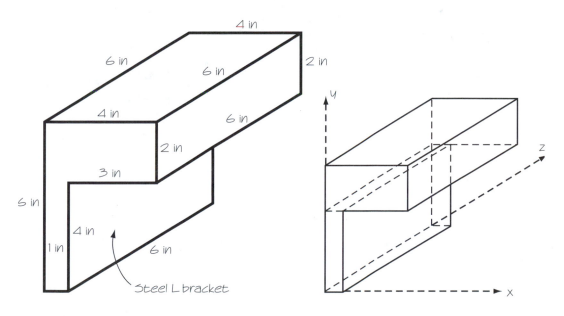

13.2 Calculate the center of gravity of the object shown. Be sure to use the coordinate origin placement shown in expressing your answer.

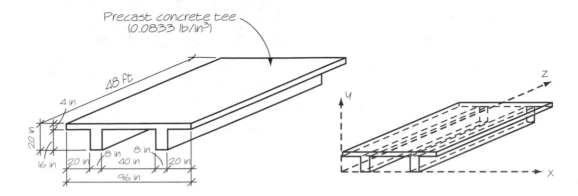

13.3 Calculate the center of gravity of the object shown. Be sure to use the coordinate origin placement shown in expressing your answer.

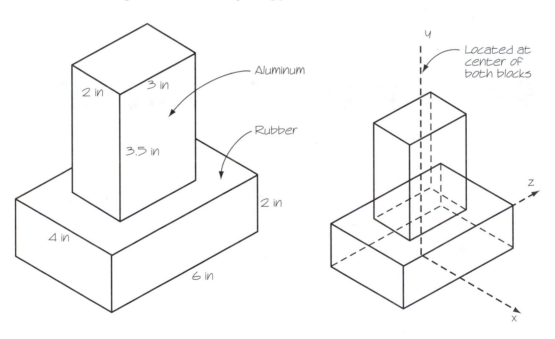

13.4 Calculate the center of gravity of the sledge hammer shown. Be sure to use the coordinate origin placement shown in expressing your answer. The *sledge hammer specifications* are: 75 mm × 75 mm × 200 mm steel head and a 40-mm-diameter wood (Douglas Fir) handle centrally located in the steel head.

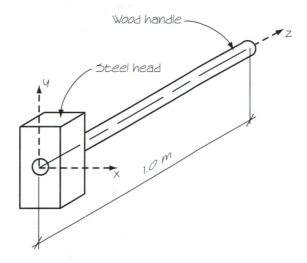

13.5 Calculate the center of gravity of the object shown. Be sure to use the coordinate origin placement shown in expressing your answer.

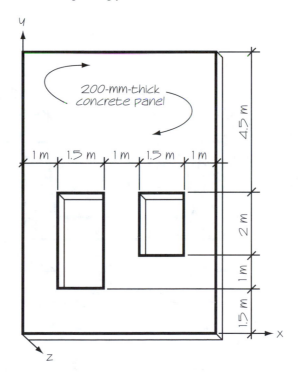

13.6 Calculate the centroid of the shape shown. Be sure to use the coordinate origin placement shown in expressing your answer. Note that the members are positioned symmetrically about the *y* axis.

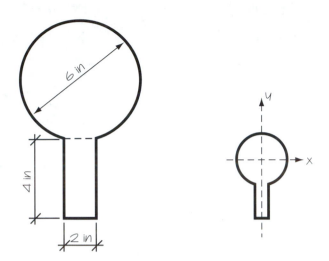

13.7 Calculate the centroid of the shape shown. Be sure to use the coordinate origin placement shown in expressing your answer.

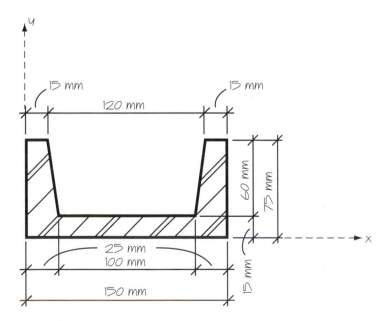

13.8 Calculate the centroid of the 2 in × 2 in × $\frac{1}{2}$ in steel angle shown. Be sure to use the coordinate origin placement shown in expressing your answer.

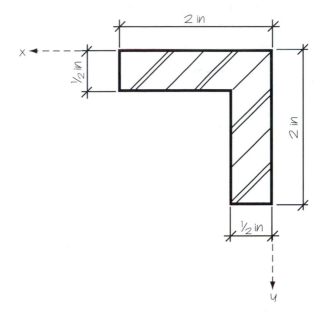

13.9 Calculate the centroid of the shape shown. Be sure to use the coordinate origin placement shown in expressing your answer. Note that the members are positioned symmetrically about the *y* axis.

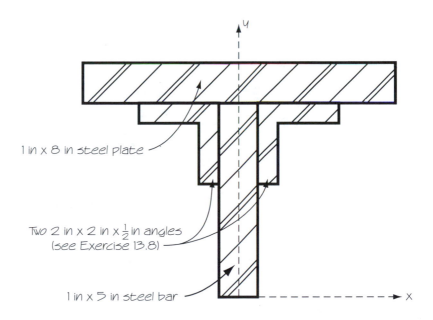

1 in x 8 in steel plate

Two 2 in x 2 in x $\frac{1}{2}$ in angles
(see Exercise 13.8)

1 in x 5 in steel bar

13.10 Calculate the centroid of the shape shown. Be sure to use the coordinate origin placement shown in expressing your answer. Note that the members are positioned symmetrically about the y axis.

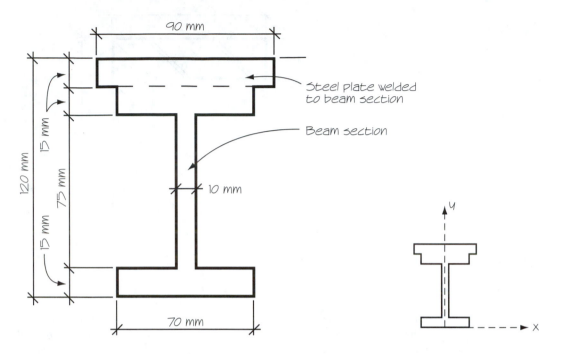

90 mm

Steel plate welded to beam section

Beam section

120 mm

15 mm

75 mm

15 mm

15 mm

10 mm

70 mm

y

x

Chap. 13   Center of Gravity and Centroid of a Shape

Chapter 14
PROPERTIES OF SECTION PROFILES

In this chapter, concepts associated with properties of beam and column section profiles will be explored. These properties relate to the capacity of a beam or column section profile to resist bending and buckling independent of the material's elastic behavior. Section profile properties relate to why a rectangular joist section profile is stiffer and stronger under load when positioned on edge rather than flat in a floor system or why a column with a rectangular section profile tends to buckle under load in a direction that is perpendicular to its wider surface.

14.1   THE MOMENT OF INERTIA

The *moment of inertia (I)* is the second moment of an area with respect to an axis in the plane of the area. In analysis of beams and columns, the most commonly used moment of inertia is computed about the centroidal axis of the shape. In this book, use of the term *moment of inertia* will refer to this property of the shape. A second type of moment of inertia, called the *polar moment of inertia,* is computed about an axis of a twisting area and is introduced separately later in this chapter.

14.2   MOMENT OF INERTIA ABOUT THE CENTROIDAL AXIS

The most commonly used moment of inertia is computed about the centroidal axis of the shape. In computing the moment of inertia about the centroidal axis of the shape, the area under consideration is assumed to be composed of an infinite number of very small constituent areas; these constituent areas are so small that there is an infinite number of them. Each constituent area is a specific distance from the axis under consideration. In the case of a beam section profile, the axis under consideration is the neutral axis of bending. The moment of inertia of an area is found mathematically by finding the sum of the products obtained by multiplying these constituent areas (a) by the square of the distance of each constituent area to the axis under consideration (d) (Figure 14.1):

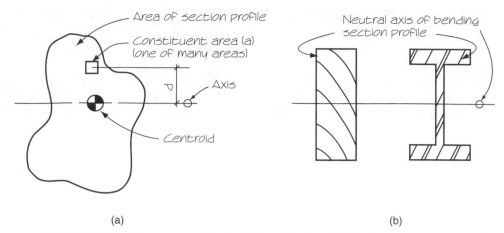

Area of section profile

Constituent area (a) (one of many areas)

Axis

Centroid

Neutral axis of bending section profile

(a)

(b)

FIGURE 14.1 The *moment of inertia, I*, is found by first dividing the area or section profile into an infinite number of constituent areas and then finding the sum of the products obtained by multiplying these constituent areas (*a*) by the square of the distance of each constituent area to an axis that passes through the centroid of the shape (*b*). In the case of a beam-section profile, the centroidal axis is the neutral axis of bending.

$$\text{moment of inertia, } I = \Sigma a_n d_n^2 = a_1 d_1^2 + a_2 d_2^2 + a_3 d_3^2 + \cdots$$

The moment of inertia is expressed in in^4, mm^4, and m^4. The fourth power of these units is somewhat confusing; it has no physical meaning other than serving as units that describe the moment of inertia.

In building structure analysis, the moment of inertia represents the relative capacity of a beam section profile to provide stiffness against bending in a direction perpendicular to the neutral axis. A beam section profile with a high value of moment of inertia is stiffer against bending in contrast to a section profile with a low moment of inertia.

Table 14.1 provides moment of inertia data for common two-by solid wood lumber section profiles. A 2×10 member resting on edge, like a floor or ceiling joist ($I = 98.93$ in^4), is a much stiffer section profile against bending than is a 2×8 member resting on edge ($I = 47.63$ in^4) because the 2×10 member has a greater moment of inertia: 98.93 in^4 / 47.63 in^4 = 2.077 times stiffer against bending. Orientation of a section profile with respect to loading also influences the moment of inertia of a section. A 2×10 member resting on edge, such as a floor or ceiling joist ($I = 98.93$ in^4), is a substantially stiffer section profile than is a 2×10 member laying flat ($I = 2.60$ in^4). The 2×10 member resting on edge is about 38 times stiffer: 98.93 in^4/2.60 in^4 = 38.05.

Simply stated, the moment of inertia is a measure of the relative resistance to bending of a beam section. Do not confuse the property of moment of inertia with the property of modulus of elasticity; they both describe stiffness. However, while moment of inertia describes stiffness of a section profile against bending, the mod-

Table 14.1 SECTION PROPERTIES OF COMMON TWO-BY SOLID WOOD LUMBER SECTIONS

Nominal beam size (in)		Surfaced (actual) size (in)		Section modulus S (in^3)		Moment of inertia, I (in^4)		Radius of gyration, r (in)	
Breadth, b	Depth, d	Breadth, b	Depth, d	Resting on edge	Laying flat	Resting on edge	Laying flat	Breadth	Depth
	2	1.5	1.5	0.563	0.563	0.422	0.422		0.433
	3	1.5	2.5	1.56	0.938	1.95	0.703		0.722
	4	1.5	3.5	3.06	1.31	5.36	0.981		1.01
2	6	1.5	5.5	7.56	2.06	20.80	1.55	0.433	1.59
	8	1.5	7.25	13.14	2.72	47.63	2.04		2.09
	10	1.5	9.25	21.39	3.47	98.93	2.60		2.67
	12	1.5	11.25	31.64	4.22	177.98	3.16		3.25
	14	1.5	13.25	43.89	4.97	290.78	3.73		3.82

ulus of elasticity (E) describes the stiffness of a material. A stiff material such as steel, when combined with a beam section profile having a high moment of inertia, results in a very stiff structural member, a member that can carry a beam load with minimal deflection.

14.3 APPROXIMATING THE MOMENT OF INERTIA

By definition, the moment of inertia (I) of a section profile is determined by dividing the section profile into an infinite number of constituent areas (a) and summing the products of each constituent area by the square of its distance to the centroidal axis (d). The centroidal axis is the neutral axis of a beam section profile. The moment of inertia is the sum of each constituent area multiplied by a distance; that is, $I = \Sigma a_n d_n^2$ or $I = a_1 d_1^2 + a_2 d_2^2 + a_3 d_3^2 + \cdots$

Nevertheless, the moment of inertia may be *approximated* by first dividing a beam section profile into a reasonable number of mathematically manageable constituent areas (i.e., 10 mm $\times$ 10 mm or $\frac{1}{2}$ in $\times$ $\frac{1}{2}$ in constituent areas). Each constituent area is then multiplied by the square of the distance from the centroid of the constituent area to the neutral axis of the entire section profile. The *approximation technique* of determining the moment of inertia is demonstrated in Example 14.1.

■ EXAMPLE 14.1

Use the approximation technique to estimate the moment of inertia of a 2 in $\times$ 6 in section profile (not a 1.5 in $\times$ 5.5 in lumber section but a true 2 in $\times$ 6 in section) resting flat, as shown in Figure 14.2. Divide the section profile into the constituent areas noted below and use these constituent areas in the analysis.

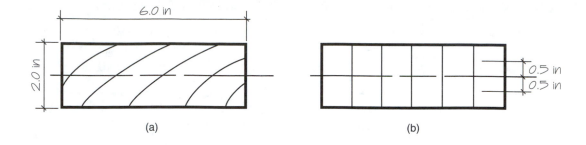

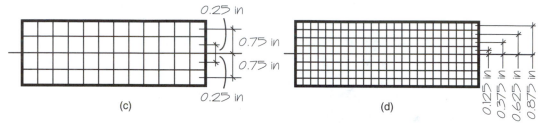

FIGURE 14.2 The 2 in by 6 in section for Example 14.1.

(a) 1.0 in × 1.0 in constituent areas
(b) 0.5 in × 0.5 in constituent areas
(c) 0.25 in × 0.25 in constituent areas

Solution (a) The 2 in × 6 in section profile is divided into 1.0 in × 1.0 in constituent areas, as illustrated in Figure 14.2b. Note that on each side of the neutral axis there are six 1.0 in × 1.0 in constituent areas in two rows, with each row having the same distance to the neutral axis. The moment of inertia is calculated beginning with constituent areas on the top side of the section:

$$I = a_1d_1^2 + a_2d_2^2 + a_3d_3^2 + \cdots$$
$$= 6(1.0 \text{ in} \cdot 1.0 \text{ in}) (0.5 \text{ in})^2 + 6(1.0 \text{ in} \cdot 1.0 \text{ in}) (0.50 \text{ in})^2$$
$$= 3.00 \text{ in}^4$$

(b) The 2 in × 6 in section profile is divided into 0.5 in × 0.5 in constituent areas as illustrated in Figure 14.2c. Note that there are 12 constituent areas in four rows, with each row having the same distance to the neutral axis. The moment of inertia is calculated beginning with constituent areas on the top side of the section.

$$I = a_1d_1^2 + a_2d_2^2 + a_3d_3^2 + \cdots$$
$$= 12(0.5 \text{ in} \cdot 0.5 \text{ in}) (0.75 \text{ in})^2 + 12(0.5 \text{ in} \cdot 0.5 \text{ in}) (0.25 \text{ in})^2$$
$$+ 12(0.5 \text{ in} \cdot 0.5 \text{ in}) (0.25 \text{ in})^2 + 12(0.5 \text{ in} \cdot 0.5 \text{ in}) (0.75 \text{ in})^2$$
$$= 3.75 \text{ in}^4$$

(c) The 2 in × 6 in section profile is divided into 0.25 in × 0.25 in constituent areas as illustrated in Figure 14.2d. Note that there are 24 constituent areas in eight rows, with each row having the same distance to the neutral axis. The moment of inertia is calculated beginning with constituent areas on the top side of the section.

$$
\begin{aligned}
I &= a_1 d_1^2 + a_2 d_2^2 + a_3 d_3^2 + \cdots \\
&= 24(0.25 \text{ in} \cdot 0.25 \text{ in})\,(0.875 \text{ in})^2 + 24(0.25 \text{ in} \cdot 0.25 \text{ in})\,(0.625 \text{ in})^2 \\
&\quad + 24(0.25 \text{ in} \cdot 0.25 \text{ in})\,(0.375 \text{ in})^2 + 24(0.25 \text{ in} \cdot 0.25 \text{ in})\,(0.125 \text{ in})^2 \\
&\quad + 24(0.25 \text{ in} \cdot 0.25 \text{ in})\,(0.125 \text{ in})^2 + 24(0.25 \text{ in} \cdot 0.25 \text{ in})\,(0.375 \text{ in})^2 \\
&\quad + 24(0.25 \text{ in} \cdot 0.25 \text{ in})\,(0.625 \text{ in})^2 + 24(0.25 \text{ in} \cdot 0.25 \text{ in})\,(0.875 \text{ in})^2 \\
&= 3.94 \text{ in}^4
\end{aligned}
$$

As shown in Example 14.1, the approximation method provides different values of the moment of inertia when the section profile is divided into different-size constituent areas. The actual moment of inertia of a 2 in × 6 in section resting flat is 4.00 in^4, as shown later in Example 14.3. In the approximation method, use of a large number of small constituent areas achieves more accurate results in comparison to use of a small number of large constituent areas. Use of a large number of small constituent areas also involves more tedious computation. As evident in Example 14.1, analysis using the 0.25 in × 0.25 in constituent areas results in a reasonably accurate approximation; the approximated moment of inertia of 3.94 in^4 is within 2% of the actual moment of inertia of 4.00 in^4.

With use of small constituent areas in analysis, the approximation method of determining the moment of inertia is usually sufficiently accurate for most engineering work, especially when the need arises to compute the moment of inertia of complex or irregular-shaped section profiles. Widely available CAD software packages make determining the moment of inertia of a complex shape relatively simple, as it usually involves drawing the shape and running a command.

14.4   FINDING THE ACTUAL MOMENT OF INERTIA

Simple Shapes

Higher-level mathematics (calculus) provides the capability of determining the *actual* moment of inertia by first dividing the section profile into an infinite number of constituent areas and then summing the products of these areas by the square of their distances to the neutral axis. Formulas can be derived and used to determine the actual moment of inertia of simple section profiles, much like the formulas $A = \pi R^2$ and $A = \pi D^2/4$ have been developed to easily determine the area of a circle. These formulas are used to determine the moment of inertia of simple shapes

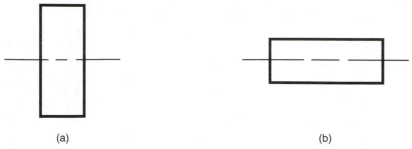

(a)   (b)

FIGURE 14.3   The 2 in by 6 in section for Example 14.2.

such as rectangles, triangles, and circles from the physical dimensions of the section profile, such as depth, width, or diameter.

Table A.6 in Appendix A contains formulas for determining the actual moments of inertia of other common sections. Manufacturers' technical literature is another good source for the moments of inertia of other structural sections. Some of the common formulas include:

- The actual moment of inertia for rectangular section profiles may be found by

$$I_{\text{rectangular}} = \frac{bd^3}{12}$$

where beam section profile breadth (width), b, and depth, d, are measured in inches, millimeters, or meters.

- The actual moment of inertia for circular section profiles may be found by

$$I_{\text{circular}} = \frac{\pi D^4}{64}$$

where the circular section profile diameter, D, is measured in inches, millimeters, or meters.

■ EXAMPLE 14.2

Calculate the actual moment of inertia for the 2 in × 6 in beam section profile about an axis passing through the centroid of the section (the neutral axis).
(a) Such that it is resting on edge, as shown in Figure 14.3a
(b) Such that it is resting flat, as shown in Figure 14.3b

Solution

(a) $$I_{\text{rectangular}} = \frac{bd^3}{12} = \frac{(2 \text{ in})(6 \text{ in})^3}{12} = 36.00 \text{ in}^4$$

(b) $$I_{\text{rectangular}} = \frac{bd^3}{12} = \frac{(6 \text{ in})(2 \text{ in})^3}{12} = 4.00 \text{ in}^4$$

In Example 14.2 it is evident that the 2 in × 6 in beam section profile resting on edge is a substantially stiffer section than the same section profile resting flat. A 2 in × 6 in section profile is nine times stiffer against bending resting on edge versus laying flat: $36.00\ \text{in}^4/4.00\ \text{in}^4 = 9.0$.

Some hollow shapes and I- and H-section profiles are composed of simple solid and vacant shapes having a common axis. For example, a pipe is composed of a solid circle with a vacant circular center sharing common axes, square structural tubing is a vacant square overlying a solid square with common axes, and the H shape of a wide-flange beam is composed of a solid rectangle with two vacant rectangles, each having a common axis.

The moment of inertia of a section profile comprised of geometric shapes that have a common axis may be computed by adding or subtracting the constituent moments of inertia that make up the shape. For example, the moment of inertia of square structural tubing can be found by subtracting the moment of inertia of the vacant square from the moment of inertia of the solid square.

■ EXAMPLE 14.3

A section of pipe has a 50-mm outside diameter and a 40-mm inside diameter. Calculate the actual moment of inertia about an axis passing through the centroid of the section (the neutral axis).

Solution From Table A.6 in Appendix A, the moment of inertia of a circle may be found by the equation $I_{\text{circle}} = \pi d^4/64$. The moment of inertia of a circle created by the outer wall is found by

$$I_{\text{outer wall}} = \frac{\pi D^4}{64} = \frac{\pi (50\ \text{mm})^4}{64} = 306\ 796\ \text{mm}^4$$

The moment of inertia of a circle created by the inner wall is found by

$$I_{\text{inner wall}} = \frac{\pi D^4}{64} = \frac{\pi (40\ \text{mm})^4}{64} = 125\ 664\ \text{mm}^4$$

By subtracting the moment of inertia of the vacant circle formed by the inner wall from the moment of inertia of the circle formed by the outer wall, the moment of inertia of the pipe section is found:

$$I_{\text{pipe}} = I_{\text{outer wall}} - I_{\text{inner wall}} = 306\ 796\ \text{mm}^4 - 125\ 664\ \text{mm}^4$$
$$= 181\ 132\ \text{mm}^4$$

In Example 14.3 the pipe section profile consists of two concentric circles; circles with a common center point. Because the centroids of these circles fall along the same neutral axis, the moments of inertia of these shapes can be treated arithmetically.

■ EXAMPLE 14.4

An I-shaped beam section profile is shown in Figure 14.4. It is 15.0 in deep and 11.0 in wide and has flanges that are 2 in thick and a web that is 1.0 in

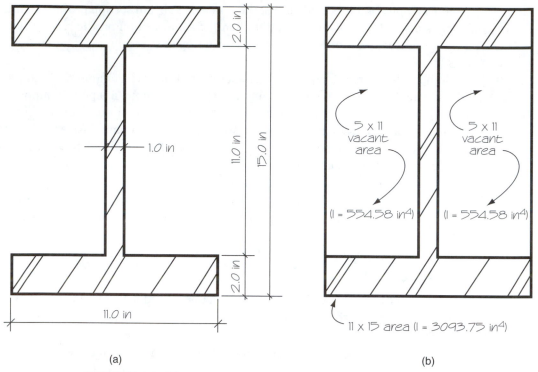

(a)   (b)

FIGURE 14.4 The I-shaped beam section for Example 14.4.

thick. Calculate the actual moment of inertia of the section profile about the centroidal axis (neutral axis).

Solution From above or from Table A.6 in Appendix A, the actual moment of inertia for a rectangular section profile may be found by the formula, $I_{\text{rectangular}} = bd^3/12$. The moment of inertia of the 11.0 in × 15.0 in rectangle is

$$I_{11\times 15} = \frac{bd^3}{12} = \frac{(11.0 \text{ in})(15.0 \text{ in})^3}{12} = 3093.75 \text{ in}^4$$

Two vacant 5.0 in × 11.0 in rectangles are created by the web and the inner faces of the upper and lower flanges. The moment of inertia of a 5.0 in × 11.0 in rectangle is

$$I_{5\times 11} = \frac{bd^3}{12} = \frac{(5.0 \text{ in})(11.0 \text{ in})^3}{12} = 554.58 \text{ in}^4$$

By subtracting the moment of inertia of the two vacant rectangles from the moment of inertia of the 11.0 in × 15.0 in rectangle, the moment of inertia of the I-shaped beam section profile (I_{beam}) is found:

$$I_{\text{beam}} = I_{11\times 15} - 2(I_{5\times 11}) = 3093.75 \text{ in}^4 - 2(554.58 \text{ in}^4) = 1984.59 \text{ in}^4$$

Irregular and Built-up Shapes

Frequently, it is necessary to find the moment of inertia of irregular or built-up structural section profiles. The moment of inertia of such a section profile may be found by transferring the moment of inertia of constituent sections to a common axis that is parallel to the axes of the constituent sections. The moment of inertia of a two-dimensional irregular or built-up structural shape may be found by the following steps:

1. Identify the location and orientation of an axis about which the moment of inertia of the entire section profile will be calculated. This is typically an axis passing through the centroid of the entire section, that is the neutral axis of the entire section.

2. Divide the entire section profile into constituent sections that each have a moment of inertia with an axis that is parallel to but a known distance away from the axis of the entire section profile.

3. Determine the moment of inertia of each constituent section ($I_1, I_2, I_3, \ldots, I_n$) about the axis of the entire section profile by the following formula, where the variables are moment of inertia about the centroidal axis of the constituent section ($I_{constituent}$), the area of the constituent section (a), and the distance between the centroidal axis of the constituent section and the centroidal axis of the entire section profile (d):

$$I_n = I_{constituent} + ad^2$$

4. Determine the moment of inertia of the entire section profile (I) by summing the moments of inertia of the constituent sections ($I_1, I_2, I_3, \ldots, I_n$) about the centroidal axis of the entire section profile:

$$I = \Sigma I_n = I_1 + I_2, + I_3 + \cdots + I_n$$

■ EXAMPLE 14.5

An I-shaped beam section profile is shown in Figure 14.5. It is 15.0 in deep and 11.0 in wide and has flanges that are 2 in thick and a web that is 1.0 in thick. Calculate the actual moment of inertia of the section about the centroidal axis (neutral axis).

Solution   The section profile is divided into three constituent sections: two 11 in × 2 in flanges laying horizontally and a 1 in × 11 in web that is oriented vertically. The axis of the entire section profile is positioned in the center of the section profile, 7.5 in away from the outer edge of the outer flanges. The axes of the entire section profile and the constituent sections are parallel with the flanges.

The moment of inertia of the 11 in wide × 2 in deep flange about an axis that is parallel with the axis of the entire section profile is found by

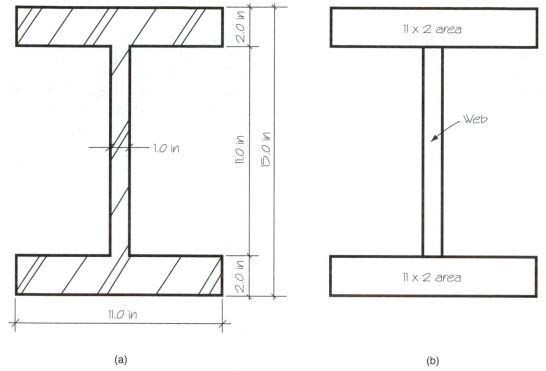

(a)   (b)

FIGURE 14.5   The I-shaped beam section for Example 14.5.

$$I_{11 \times 2} = \frac{bd^3}{12} = \frac{(11.0 \text{ in})(2.0 \text{ in})^3}{12} = 7.33 \text{ in}^4$$

Each flange has an axis that is 6.5 in from the axis of the entire section profile (d). The moment of inertia of each 11 in × 2 in flange section about the axis of the entire section profile (I_{flange}) is found by

$$I_{n \text{ flange}} = I_{\text{constituent}} + ad^2 = 7.33 \text{ in}^4 + (11 \text{ in} \cdot 2 \text{ in})(6.5 \text{ in})^2 = 936.83 \text{ in}^4$$

Since the axis of the 1 in × 11 in web is the same as the axis of the entire section profile, the moment of inertia of the web as a constituent section ($I_{n \text{ web}}$) is found by

$$I_{n \text{ web}} = \frac{bd^3}{12} = \frac{(1.0 \text{ in})(11.0 \text{ in})^3}{12} = 110.92 \text{ in}^4$$

The moment of inertia of the entire section profile (I) is found by

$$I = 2(I_{n \text{ flange}}) + I_{n \text{ web}} = 2(936.83 \text{ in}^4) + 110.92 \text{ in}^4 = 1984.58 \text{ in}^4$$

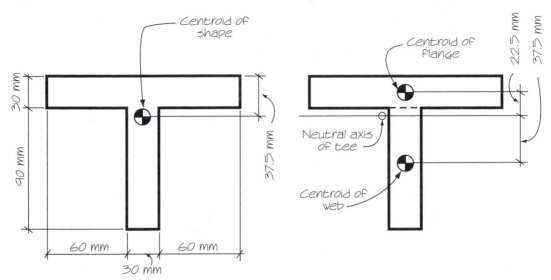

FIGURE 14.6 The tee section for Example 14.6.

■ EXAMPLE 14.6

The tee section profile illustrated in Figure 14.6 was found to have a centroid that is 37.5 mm below the top surface of the flange (see Example 13.4). Calculate the actual moment of inertia of the section about the centroidal axis of the flange (the neutral axis).

Solution The section profile is divided into two constituent sections: a 150 mm × 30 mm flange and a 30 mm × 90 mm web. The moment of inertia of the 150 mm wide × 30 mm deep flange about an axis that is parallel with the axis of the entire section profile is found by

$$I_{\text{flange}} = \frac{bd^3}{12} = \frac{(150 \text{ mm})(30 \text{ mm})^3}{12} = 337\ 500 \text{ mm}^4$$

The flange has an axis that is 22.5 mm from the axis of the entire section profile (d). The moment of inertia of the flange section about the axis of the entire section profile (I_{flange}) is found by

$$I_{n \text{ flange}} = I_{\text{constituent}} + ad^2 = 337\ 500 \text{ mm}^4 + (150 \text{ mm} \cdot 30 \text{ mm})(22.5 \text{ mm})^2$$
$$= 2\ 615\ 625 \text{ mm}^4$$

The moment of inertia of the 30 mm wide × 90 mm deep web about an axis that is parallel with the axis of the entire section profile is found by

$$I_{\text{web}} = \frac{bd^3}{12} = \frac{(30 \text{ mm})(90 \text{ mm})^3}{12} = 1\ 822\ 500 \text{ mm}^4$$

The web has an axis that is coincidentally 37.5 mm from the axis of the entire section profile (d). The moment of inertia of the web section about the axis of the entire section profile (I_{flange}) is found by

$$I_{n \text{ web}} = I_{\text{constituent}} + ad^2 = 1\ 822\ 500 \text{ mm}^4 + (30 \text{ mm} \cdot 90 \text{ mm})(37.5 \text{ mm})^2$$
$$= 5\ 619\ 375 \text{ mm}^4$$

The moment of inertia of the entire tee section profile (I) is found by

$$I = I_{n \text{ flange}} + I_{n \text{ web}} = 1\ 822\ 500 \text{ mm}^4 + 5\ 619\ 375 \text{ mm}^4 = 7\ 441\ 875 \text{ mm}^4$$

14.5 MOMENT OF INERTIA AND A BEAM SECTION PROFILE

In examining the mathematical definition of the moment of inertia, $I = \Sigma a_n d_n^2$, observe that a constituent area (a) that is located a large distance (d) from the neutral axis has a great influence on the section profile's moment of inertia. With $a_n d_n^2$, a constituent area (a) located a large distance (d) away from the neutral axis has a great influence on the moment of inertia of the section profile because the distance is squared. Thus, when $d = 2$, $d^2 = 4$; when $d = 3$, $d^2 = 9$; when $d = 4$, $d^2 = 16$; and so on. A section profile with constituent areas far away from the neutral axis has a much larger moment inertia than a section profile that is close to the neutral axis and a greater capacity to resist bending.

The position of a beam section profile with respect to bending influences its stiffness against bending. Compare the results of the 2 in × 6 in section profile resting flat versus resting on edge, as shown in Example 14.2. The large contrast in the moments of inertia for the section profile resting on edge and resting flat is clearly evident. A 2 in × 6 in section profile of a member resting on edge is nine times stiffer than the same 2 in × 6 in section profile resting flat: 36.00 in^4/4.00 in^4 = 9.

Example 14.7 is an *approximation* of the moment of inertia of the 2 in × 6 in section profile resting on edge. Again, since the 2 in × 6 in section profile is not divided into an infinite number of constituent areas as required by definition but only 0.5 in × 0.5 in constituent areas, the technique shown produces only approximate results. The actual moment of inertia of a 2 in × 6 in member resting on edge was found to be 36.00 in^4 in Example 14.2. The approximation results in Example 14.11 are a close estimate that will be used to demonstrate the influence that location of a constituent area has on the magnitude of moment of inertia.

■ EXAMPLE 14.7

Use the approximation technique to determine the moment of inertia of a 2 in × 6 in rectangular section profile resting on edge, as shown in Figure 14.7.

Solution The 2 in × 6 in section profile is divided into the 0.5 in × 0.5 in constituent areas, as illustrated in Figure 14.7b. Note that there are four con-

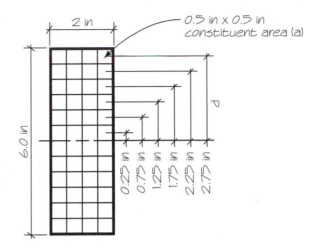

FIGURE 14.7 A 2 in by 6 in rectangular cross section resting on edge that is divided into rows of equal constituent areas for the approximation of the moment of inertia as shown in Example 14.7.

stituent areas in 12 rows, with each constituent area in a row having the same distance to the neutral axis. The moment of inertia is calculated beginning with constituent areas on the top side of the section profile.

$$
\begin{aligned}
I &= a_1 d_1^2 + a_2 d_2^2 + a_3 d_3^2 + \cdots \\
&= 4(0.5 \text{ in} \cdot 0.5 \text{ in})(2.75 \text{ in})^2 &&= 7.5625 \text{ in}^4 \\
&+ 4(0.5 \text{ in} \cdot 0.5 \text{ in})(2.25 \text{ in})^2 &&= +5.0625 \text{ in}^4 \\
&+ 4(0.5 \text{ in} \cdot 0.5 \text{ in})(1.75 \text{ in})^2 &&= +3.0625 \text{ in}^4 \\
&+ 4(0.5 \text{ in} \cdot 0.5 \text{ in})(1.25 \text{ in})^2 &&= +1.5625 \text{ in}^4 \\
&+ 4(0.5 \text{ in} \cdot 0.5 \text{ in})(0.75 \text{ in})^2 &&= +0.5625 \text{ in}^4 \\
&+ 4(0.5 \text{ in} \cdot 0.5 \text{ in})(0.25 \text{ in})^2 &&= +0.0625 \text{ in}^4 \\
&+ 4(0.5 \text{ in} \cdot 0.5 \text{ in})(0.25 \text{ in})^2 &&= +0.0625 \text{ in}^4 \\
&+ 4(0.5 \text{ in} \cdot 0.5 \text{ in})(0.75 \text{ in})^2 &&= +0.5625 \text{ in}^4 \\
&+ 4(0.5 \text{ in} \cdot 0.5 \text{ in})(1.25 \text{ in})^2 &&= +1.5625 \text{ in}^4 \\
&+ 4(0.5 \text{ in} \cdot 0.5 \text{ in})(1.75 \text{ in})^2 &&= +3.0625 \text{ in}^4 \\
&+ 4(0.5 \text{ in} \cdot 0.5 \text{ in})(2.25 \text{ in})^2 &&= +5.0625 \text{ in}^4 \\
&+ 4(0.5 \text{ in} \cdot 0.5 \text{ in})(2.75 \text{ in})^2 &&= \underline{+7.5625 \text{ in}^4} \\
& && \ 35.750 \text{ in}^4
\end{aligned}
$$

Clearly evident in Example 14.7 is the influence that a constituent area has on the magnitude of a section's moment of inertia as its distance away from the neutral axis increases. The two constituent areas that are farthest away from the neutral axis (one on the top surface and one on the bottom) contribute 15.250 in^4 to the moment of inertia of the 2 in $\times$ 6 in section profile (7.5625 in^4 + 7.5625 in^4), over 42% of the moment of inertia. Thus, almost half of the moment of inertia of the section profile under consideration is contributed by these two outermost constituent areas alone, even though they comprise less than 17% of the section profile's total area.

14.5 Moment of Inertia and a Beam Section Profile **265**

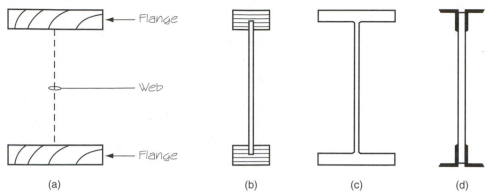

FIGURE 14.8 (a) A theoretical beam section with flanges separated by a web of negligible thickness. Actual cross sections that are similar to this theoretical member include the plywood I joist (b), wide-flange steel beam (c), and open-web steel joist (d).

Additionally, whereas a 2 in × 6 in section profile has an approximate moment of inertia of 35.750 in^4, reducing it to a 2 in × 4 in section profile offers an approximate moment of inertia of only 20.500 in^4: 35.750 in^4 − 15.250 in^4 = 20.500 in^4. The 2 in × 6 in section profile is 1.74 times (35.750/20.500) stiffer against bending in comparison to the 2 in × 4 in section profile, even though it only has 1.5 times the material.

Situating constituent areas in a section profile at a large distance away from the neutral axis produces a greater capacity for that section profile to resist bending. Beam section profiles with wide flanges and narrow webs such as the shapes shown in Figure 14.8 are designed with this theory in mind. These shapes result in more efficient use of material because less material is required to resist bending.

14.6 THE POLAR MOMENT OF INERTIA

The *polar moment of inertia* (J) is a moment of inertia of an area with respect to an axis perpendicular to the plane of the area under consideration. It is a property of a rotating section profile that is used in the design of members subjected to twisting or rotating about an axis that is perpendicular to the plane under consideration.

The polar moment of inertia is found mathematically by first dividing the area or section profile into an infinite number of constituent areas. Again, each constituent area is infinitesimal in size and a specific distance from the axis under consideration. The distance is computed from an axis that is perpendicular to the plane of the area under consideration; that is, the axes of all constituent areas made up a point rather than a line. In engineering analysis, this point typically represents the axis of rotation about which a shaftlike member twists.

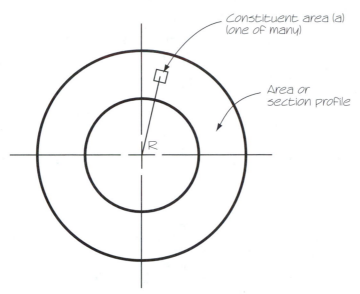

Constituent area (a)
(one of many)

Area or
section profile

R

FIGURE 14.9 The polar moment of inertia is found mathematically by first dividing the area or section profile into an infinite number of constituent areas and then finding the sum of the products obtained by multiplying these constituent areas (*a*) by the square of the distance of each constituent area to an axis that is perpendicular to the constituent area (*R*).

The polar moment of inertia is found by summing the products obtained by multiplying the constituent areas (*a*) by the square of the distance of each constituent area to the axis (*R*) (Figure 14.9):

$$\text{polar moment of inertia, } J = \Sigma a_n R_n^2 = a_1 R_1^2 + a_2 R_2^2 + a_3 R_3^2 + \cdots$$

Like the moment of inertia about the centroidal axis of the shape, the polar moment of inertia can be approximated by dividing the shape into small manageable constituent areas and then summing the products obtained by multiplying these constituent areas by the square of the distance of each constituent area to the axis. For common shapes, formulas have been derived; for example, the polar moment of inertia for a solid circular shaft may be found by $J = \pi D^4/32$, where D is the shaft diameter.

14.7 THE SECTION MODULUS

The *section modulus* (*S*) is a property of a beam section profile that is a measure of a beam section profile's strength against bending stresses (Figure 14.10). It relates the moment of inertia (*I*) to the distance of the *most* extreme fiber of the beam section to the neutral axis (*c*). The section modulus is defined as

$$\text{section modulus, } S = \frac{I}{c}$$

14.7 The Section Modulus

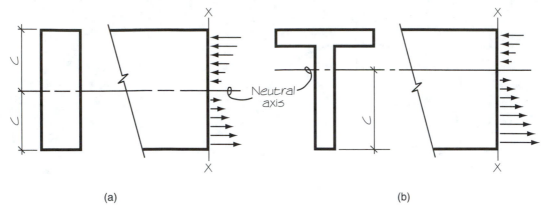

(a)   (b)

FIGURE 14.10   The section modulus (S) is a property of a beam section profile that is a measure of a beam section's strength against bending stresses. It relates the moment of inertia (I) to the distance of the *most* extreme fiber of the beam section to the neutral axis (c): $S = I/c$. Bending stresses are greatest at the most extreme fiber. A symmetrical section profile has two most extreme fibers (a) whereas an asymmetrical section profile has one most extreme fiber (b).

The most extreme fiber (c) under consideration is the fiber that is farthest away from the neutral axis of the section profile. Irregular-shaped section profiles may have a single most extreme fiber, while symmetrical sections may have two most extreme fibers. The section modulus is commonly expressed in units of in^3, mm^3, and m^3. These units do not express a volume. Instead, they describe the section modulus as a measure of a section profile's strength against bending stresses.

■ EXAMPLE 14.8

The tee-section profile illustrated in Figure 14.6 was found to have a centroid that is 37.5 mm below the top surface of the flange (see Example 13.4) and a moment of inertia of 7 441 875 mm^4 (see Example 14.6). Calculate the section modulus of the tee section about the neutral axis parallel with the top edge of the flange.

Solution   The tee is 120 mm deep. The neutral axis of the section profile passes through the centroid. Thus the neutral axis of the tee is 37.5 mm below the top surface of the flange and 82.5 mm above the bottom surface of the web. The distance of the most extreme fiber of the beam section profile to the neutral axis (c) is 82.5 mm. The moment of inertia of 7 441 875 mm^4 is provided in the statement of the problem. The section modulus of the tee is found by

$$S = \frac{I}{c} = \frac{7\ 441\ 875\ mm^4}{82.5\ mm} = 90\ 206\ mm^3$$

Example 14.8 demonstrates that the extreme fibers of an irregular section profile can be different distances away from the neutral axis. It is the most extreme fiber (c) that experiences the greatest bending stress.

The section modulus serves as a measure of a beam section profile's strength against bending stresses. Table 14.1 lists section modulus data for common two-by solid wood lumber sections. A 2 × 12 ($S = 31.64$ in^3) member resting on edge and carrying a load such as a floor or ceiling joist is substantially stronger than a 2 × 10 ($S = 21.39$ in^3). Orientation of a section profile with respect to loading influences the section modulus of a section profile and its strength against bending. A comparison of the section modulii of a 2 × 12 resting on edge and laying flat shows that the 2 × 12 resting on edge is 7.5 times stronger in resisting bending stresses than if it were laying flat: 31.64 in^3/4.22 in^3 = 7.50. As shown in Example 14.9, the 2 in × 6 in section profile is three times stronger against bending when resting on edge than it is laying flat.

■ EXAMPLE 14.9

Calculate the section modulus of the 2 in × 6 in beam section profile that is oriented so that it is:

(a) Resting on edge
(b) Resting flat

Solution (a) As calculated in Example 14.2, the moment of inertia of a 2 in × 6 in beam section profile resting on edge is 36.00 in^4. The neutral axis of bending is at the centroid of the section profile: 6.00 in/2 = 3.00 in from the edges of the section profile. The most extreme fibers are 3.00 in from the neutral axis. The section modulus is found by $S = I/c$:

$$S = \frac{I}{c} = \frac{36.00 \text{ in}^4}{3.00 \text{ in}} = 12.00 \text{ in}^3$$

(b) As calculated in Example 14.2, the moment of inertia of a 2 in × 6 in beam section profile resting on edge is 4.00 in^4. The neutral axis of bending is at the centroid of the section profile: 2 in/2 = 1.00 in. from the flat surfaces of the section profile. The most extreme fibers are 1 in. from the neutral axis. The section modulus is found by $S = I/c$:

$$S = \frac{I}{c} = \frac{4.00 \text{ in}^4}{1.00 \text{ in}} = 4.00 \text{ in}^3$$

14.8 FINDING THE SECTION MODULUS

A section modulus for a common geometric shape may be easily calculated by using the derived formulas found in Table A.6 in Appendix A. These formulas are derived from the $S = I/c$ expression. For example, a formula to find the section modulus of rectangular beam section profiles may be derived by relating the moment of inertia of a rectangular section profile (I) to the distance of the most extreme fiber of the rectangular section profile to the neutral axis (c):

$$I_{\text{rectangular}} = \frac{bd^3}{12}$$

In the case of a rectangle, the distance of the most extreme fiber of the rectangular section profile to the neutral axis (c) is half the depth (d) of the rectangle:

$$c = \frac{d}{2}$$

The section modulus of a rectangular beam section profile is found by

$$S = \frac{I}{c} = \frac{bd^3/12}{d/2}$$

This expression simplifies to

$$S_{\text{rectangular}} = \frac{bd^2}{6}$$

where b is the actual beam breadth (width) and d is the beam depth in inches, meters, or millimeters.

■ **EXAMPLE 14.10**

Calculate the section modulus of the 2 in $\times$ 6 in beam section profile that is oriented so that it is:
(a) Resting on edge
(b) Resting flat

Solution

(a) $$S_{\text{rectangular}} = \frac{bd^2}{6} = \frac{(2 \text{ in})(6 \text{ in})^2}{6} = 12.00 \text{ in}^3$$

(b) $$S_{\text{rectangular}} = \frac{bd^2}{6} = \frac{(6 \text{ in})(2 \text{ in})^2}{6} = 4.00 \text{ in}^3$$

The position of a beam section profile with respect to bending affects its strength against bending. Based on the difference in section moduli for a 2 in $\times$ 6 in as found in Example 14.10, a 2 in $\times$ 6 in resting on edge is three times stronger than the same 2 in $\times$ 6 in section profile resting flat ($12.00/4.00 = 3$).

The *radius of gyration* (*r*) of a section profile is the distance from the neutral axis of bending at which the entire area of the section profile could be concentrated without changing the moment of inertia. The radius of gyration is a section profile property that serves as a measure of the section capacity to resist column buckling independent of the material's elastic behavior. A section profile with a high value of radius of gyration is more resistant to buckling in contrast to a section profile with a low radius of gyration.

The radius of gyration of a section profile can easily be determined with the moment of inertia (*I*) and the cross-sectional area (*A*) of the section profile:

$$r = \sqrt{\frac{I}{A}}$$

Section profiles that are asymmetrical typically have moments of inertia that are not alike in the *x* and *y* axes, such as the 2 in × 6 in in Example 14.10. As a result, the radius of gyration for one axis of an asymmetrical section profile is less than the radius of gyration for the other.

■ **EXAMPLE 14.11**

Calculate the radii of gyration of a 2 in × 6 in column section profile parallel to each surface:

(a) With buckling in the direction of its wide dimensional axis
(b) With buckling in the direction of its narrow dimensional axis

Solution   (a) From Example 14.2, the moment of inertia for the 2 in × 6 in beam section profile perpendicular to its edge is 36.00 in^4 and the area of a 2 in × 6 in is 12.00 in^2. Solving for the radius of gyration yields

$$r = \sqrt{\frac{I}{A}} = \sqrt{\frac{36.00 \text{ in}^4}{12.00 \text{ in}^2}} = 1.732 \text{ in}$$

(b) From Example 14.2, the moment of inertia for the 2 in × 6 in beam section profile parallel to its edge is 4.00 in^4 and the area of a 2 in × 6 in is 12.00 in^2. Solving for the radius of gyration gives

$$r = \sqrt{\frac{I}{A}} = \sqrt{\frac{4.00 \text{ in}^4}{12.00 \text{ in}^2}} = 0.577 \text{ in}$$

Columns, piers, posts, and walls are long, slender members; their width is much less than their height. A long, slender member that carries a heavy compressive load at its end has a tendency to fail suddenly by buckling rather than by compression. Buckling is lateral bending that results from an excessive compressive force that is applied through the column's longitudinal axis. It is caused by a combination of excessive end loading, an instability in elastic behavior of the material, and inconsistencies in the column cross-section profile. Consider a thin yard (meter) stick that stands vertically on the ground while it is held at the top end. When an excessive load is applied downward at the top end, the stick buckles and fails.

A long, slender compression member with an irregular section profile (i.e., I, H, C, L, and T shapes) has a tendency to buckle in the direction of the least radius of gyration of the section. Again, consider the yard (meter) stick: When an excessive load is applied downward at the top end, the stick always buckles and fails in the direction of its least dimension, not its largest dimension. It buckles in the direction of its least radius of gyration.

The 2 in × 6 in section profile in Example 14.11 has different radii of gyration related to the x and y axes. The section is much more resistant to buckling against its widest dimension, in contrast to buckling in the direction of its slender dimension. Accordingly, when analyzing a long, slender compression member for potential failure by buckling, the designer typically investigates the least radius of gyration of the section profile.

Computations for masonry columns and walls are based on the radius of gyration of the rectangular horizontal face of the specified masonry unit or assembly. For solid masonry units that have a rectangular horizontal face, $I = lw^3/12$ and $A = lw$, based on the specified unit length (l) and width (w) of the masonry unit.

$$r = \sqrt{\frac{I}{A}}$$

$$= \sqrt{\frac{lw^3/12}{lw}}$$

$$= \sqrt{\frac{w^2}{12}}$$

$$= \sqrt{\frac{w}{3.464}}$$

This simplifies to

$$r = 0.289w$$

The radius of gyration of solid masonry units having a rectangular horizontal face may be approximated by this formula based on the specified width (w) of the unit or wall. The formula applies only to solid sections, so it does not apply to hollow masonry assemblies.

14.10 TABULAR STRUCTURAL SECTION PROFILE DATA

Although properties of structural section profiles may be computed using the techniques outlined above, the moment of inertia (I), section modulus (S), and radius of gyration (r) for commonly available beam and column section profiles are generally reported in tabular data available through the manufacturer or a professional association. For example, a good source of tabular data for steel structural section profiles is the *Manual of Steel Construction* published by the American Institute of Steel Construction (AISC). Section profile properties for selected section profiles of steel and wood members are provided in Appendix B and Appendix C, respectively. Designers generally rely on available tabulated data when specifying a member.

Members with irregular section profiles (i.e., rectangular, I, H, L, C, and T shapes) have different section properties about their different axes. In tables, these properties are expressed in values where the neutral axis of bending is about the x–x axis or y–y axis. In the AISC *Manual of Steel Construction,* properties of structural section profiles are expressed as I_x, S_x, and r_x for properties related to a neutral axis of bending computed about the x–x axis and I_y, S_y, and r_y for properties related to a neutral axis about the y–y axis.

Section properties about each axis can vary considerably. For example, a W16 × 100 steel member has an $I_x = 1490$ in^4 and an $I_y = 186$ in^4; the section is eight times stiffer against bending about the x–x axis! When relying on tabular data, the designer must use caution in selecting values about the appropriate axis.

14.11 BUILT-UP SECTIONS AND PLATE GIRDERS

Strength and stiffness of a standard beam and column sections can be enhanced by welding heavy plates to the outside surface of the standard beam section flanges. The *built-up sections* improve the structural properties of the section profile. Additionally, a *plate girder* can be fabricated from heavy plates when a standard beam section is inadequate. Plate girders are used when spans are too long, loads are too heavy, or lateral stability is a problem with standard beam sections. Plate girders are usually fabricated in the form of an I, but C, Z, and box shapes are not uncommon. The section properties of built-up beam sections and plate girders are determined using the analysis techniques outlined in this chapter.

EXERCISES

14.1. For the section profile shown:
   (a) Using the approximation technique, calculate the moment of inertia about the axis shown. Use $\frac{1}{2}$ in $\times$ $\frac{1}{2}$ in areas for the approximation.

(b) Calculate the *exact* moment of inertia about the axis shown.

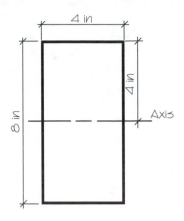

14.2. For the section profile shown:

(a) Using the approximation technique, calculate the moment of inertia about the axis shown. Use $\frac{1}{2}$ in $\times$ $\frac{1}{2}$ in areas for the approximation.

(b) Calculate the *exact* moment of inertia about the axis shown.

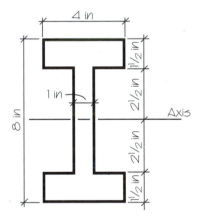

14.3. Calculate the *exact* moment of inertia of the section profile shown, about the axis shown.

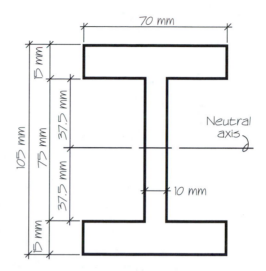

14.4. Calculate the *exact* moment of inertia of the section profile shown, about the axis shown.

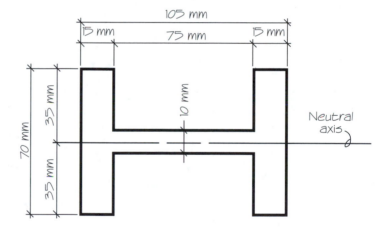

14.5. Calculate the *exact* moment of inertia of the section profile shown, about the axis ·shown.

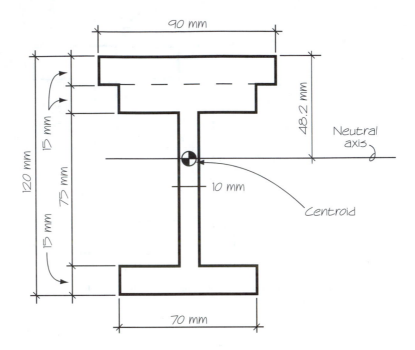

14.6. Calculate the *exact* moment of inertia of the section profile shown, about the axis shown.

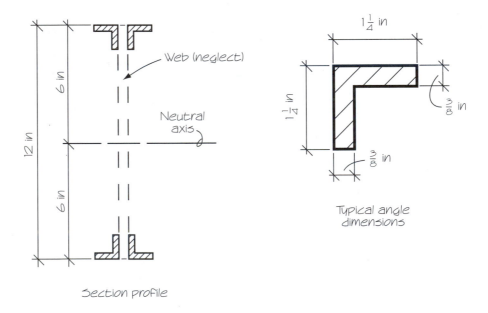

Section profile

Typical angle dimensions

14.7. For the following section profiles, calculate the *exact* moment of inertia, in^4, about the narrow dimensional axis (i.e., $b = 3\frac{1}{2}$ in). Compare the moment of inertia of each section profile to the moment of inertia of the 4×4 by percentage difference. Actual dimensions of the section profiles are shown in parentheses.

(a) 4×4 ($3\frac{1}{2}$ in $\times$ $3\frac{1}{2}$ in)
(b) 4×6 ($3\frac{1}{2}$ in $\times$ $5\frac{1}{2}$ in)
(c) 4×8 ($3\frac{1}{2}$ in $\times$ $7\frac{1}{4}$ in)
(d) 4×10 ($3\frac{1}{2}$ in $\times$ $9\frac{1}{4}$ in)
(e) 4×12 ($3\frac{1}{2}$ in $\times$ $11\frac{1}{4}$ in)

14.8. For the following section profiles, calculate the *exact* moment of inertia, in^4, about the wide dimensional axis (i.e., $b = 3\frac{1}{2}$ in, $5\frac{1}{2}$ in, $7\frac{1}{4}$, etc.). Compare the moment of inertia of each section profile to the moment of inertia of the 4×4 by percentage difference. Actual dimensions of the section profiles are shown in parentheses.

(a) 4×4 ($3\frac{1}{2}$ in $\times$ $3\frac{1}{2}$ in)
(b) 4×6 ($3\frac{1}{2}$ in $\times$ $5\frac{1}{2}$ in)
(c) 4×8 ($3\frac{1}{2}$ in $\times$ $7\frac{1}{4}$ in)
(d) 4×10 ($3\frac{1}{2}$ in $\times$ $9\frac{1}{4}$ in)
(e) 4×12 ($3\frac{1}{2}$ in $\times$ $11\frac{1}{4}$ in)

14.9. For the following section profiles, calculate the section modulus, in^3, about the narrow dimensional axis (i.e., $b = 3\frac{1}{2}$ in). Compare the section modulus of each section profile to the section modulus of the 4×4 by percentage difference. Actual dimensions of the section profiles are shown in parentheses.

(a) 4×4 ($3\frac{1}{2}$ in $\times$ $3\frac{1}{2}$ in)
(b) 4×6 ($3\frac{1}{2}$ in $\times$ $5\frac{1}{2}$ in)
(c) 4×8 ($3\frac{1}{2}$ in $\times$ $7\frac{1}{4}$ in)
(d) 4×10 ($3\frac{1}{2}$ in $\times$ $9\frac{1}{4}$ in)
(e) 4×12 ($3\frac{1}{2}$ in $\times$ $11\frac{1}{4}$ in)

14.10. For the following section profiles, calculate the section modulus, in^3, about the wide dimensional axis (i.e., $b = 3\frac{1}{2}$ in, $5\frac{1}{2}$ in, $7\frac{1}{4}$, etc.). Compare the section modulus of each section profile to the section modulus of the 4×4 by percentage difference. Actual dimensions of the section profiles are shown in parentheses.

(a) 4×4 ($3\frac{1}{2}$ in $\times$ $3\frac{1}{2}$ in)
(b) 4×6 ($3\frac{1}{2}$ in $\times$ $5\frac{1}{2}$ in)
(c) 4×8 ($3\frac{1}{2}$ in $\times$ $7\frac{1}{4}$ in)
(d) 4×10 ($3\frac{1}{2}$ in $\times$ $9\frac{1}{4}$ in)
(e) 4×12 ($3\frac{1}{2}$ in $\times$ $11\frac{1}{4}$ in)

14.11. For the following section profiles, calculate the radius of gyration, in., about the narrow dimensional axis (i.e., $b = 3\frac{1}{2}$ in). Compare the radius of gyration of each section profile to the radius of gyration of the 4×4 by percentage difference. Actual dimensions of the section profiles are shown in parentheses.

(a) 4×4 ($3\frac{1}{2}$ in $\times$ $3\frac{1}{2}$ in)
(b) 4×6 ($3\frac{1}{2}$ in $\times$ $5\frac{1}{2}$ in)
(c) 4×8 ($3\frac{1}{2}$ in $\times$ $7\frac{1}{4}$ in)
(d) 4×10 ($3\frac{1}{2}$ in $\times$ $9\frac{1}{4}$ in)
(e) 4×12 ($3\frac{1}{2}$ in $\times$ $11\frac{1}{4}$ in)

14.12. For the following section profiles, calculate the radius of gyration, in., about the wide dimensional axis (i.e., $b = 3\frac{1}{2}$ in, $5\frac{1}{2}$ in, $7\frac{1}{4}$, etc.). Compare the radius of gyration of each section profile to the radius of gyration of the 4×4 by percentage difference. Actual dimensions of the section profiles are shown in parentheses.

(a) 4×4 ($3\frac{1}{2}$ in $\times$ $3\frac{1}{2}$ in)

(b) 4×6 ($3\frac{1}{2}$ in $\times$ $5\frac{1}{2}$ in)

(c) 4×8 ($3\frac{1}{2}$ in $\times$ $7\frac{1}{4}$ in)

(d) 4×10 ($3\frac{1}{2}$ in $\times$ $9\frac{1}{4}$ in)

(e) 4×12 ($3\frac{1}{2}$ in $\times$ $11\frac{1}{4}$ in)

14.13. For the following section profiles, calculate the radius of gyration, mm. Compare the radius of gyration of each section profile to the radius of gyration of the 20 mm $\times$ 20 mm section by percentage difference.

(a) 20 mm $\times$ 20 mm

(b) 40 mm $\times$ 40 mm

(c) 60 mm $\times$ 60 mm

(d) 80 mm $\times$ 80 mm

(e) 100 mm $\times$ 100 mm

14.14. For the following section profiles, calculate the polar moment of inertia. Compare the polar moment of inertia of each section profile to the polar moment of inertia of the solid 150-mm circular shape by percentage difference.

(a) A solid 150-mm circular shape.

(b) A hollow 150-mm circular shape with a 50-mm hole that is concentrically located.

(c) A hollow 150-mm circular shape with a 75-mm hole that is concentrically located.

(d) A hollow 150-mm circular shape with a 100-mm hole that is concentrically located.

(e) A hollow 150-mm circular shape with a 125-mm hole that is concentrically located.

14.15. Included in Table B.1 in Appendix B are the properties of wide-flange (W) steel beam sections. For a W8 $\times$ 31 section profile, identify the moment of inertia (I), section modulus (S), and radius of gyration (r) about the x–x axis and the y–y axis.

(a) Compare each property about the x–x axis to the like property about the y–y axis by percentage difference.

(b) Which axis offers better stiffness against bending?

(c) Which axis offers better strength against bending stresses?

(d) Which axis offers better resistance against buckling?

14.16. Included in Table B.1 in Appendix B are the properties of wide-flange (W) steel beam sections. For a W12$\times$ 120 section profile, identify the moment of inertia (I), section modulus (S), and radius of gyration (r) about the x–x axis and the y–y axis.

(a) Compare each property about the x–x axis to the like property about the y–y axis by percentage difference.

(b) Which axis offers better stiffness against bending?

(c) Which axis offers better strength against bending stresses?

(d) Which axis offers better resistance against buckling?

Chapter 15
BUILDING LOADS

A building structural member must be strong enough to safely support the loads imposed on it but no larger than necessary to prevent waste of material. Centuries ago, the master builders of the great European cathedrals relied on experiences from past successes and failures and a perceptive eye for impending failure as the new structure was being built. Today, design and selection of building structural members is based on sophisticated analysis that is founded upon sound engineering theory. Many of these basic engineering principles are introduced in this and the remaining chapters.

In previous chapters a free-body diagram was used to represent the loading condition on a beam. The free-body diagram served as the core of shear and bending moment computations that ultimately lead to selection of a member size and material. In this chapter, types of loads and the process used to establish free-body diagrams of beams and transfer of loads are introduced.

15.1 TYPES OF BUILDING LOADS

In the construction industry, external forces applied to beams and other structural members are called *loads*. Loads are caused by occupants and weight of building materials (Table 15.1), furniture, and equipment. Loads can also be caused by snow, wind, earthquake, and other natural forces. Forces at the beam supports that react to the applied loads are called *reacting forces* or simply *reactions*. Loads and reactions combine to cause a beam to bend and cause a column to compress or buckle.

There are many types of loads that act on a building structure. Generally, building loads are classified as live or dead loads. *Dead loads* are vertical loads that result from the weight of the building itself and any fixed equipment in the building. *Live loads* result from the weight of occupants, furniture, and any piece of equipment or installation that will probably be relocated over the life of the building. Additional types of loads are created by wind, water (snow and ice), earthquakes, and other forces. Types of building loads are described below:

Dead loads   These loads are vertical gravity loads that result from the weight of the building itself and any fixed equipment in the building. Dead loads include

Table 15.1 WEIGHTS OF COMMON BUILDING MATERIALS

System/material description	lb/ft^2	kN/m^2 (kPa)
Ceilings		
Acoustical ceiling tile	1	0.05
Gypsum board		
$\frac{1}{2}$ in thick	2	0.10
$\frac{5}{8}$ in thick	2.5	0.12
Gypsum plaster, per in thickness	10	0.48
Metal lath	0.5	0.02
Suspended steel channels	1	0.05
Floors		
Carpet with pad	3	0.14
Ceramic tile	10	0.48
Concrete deck, per in thickness	12.5	0.60
Hardwood, $\frac{1}{2}$ in	4	0.19
Resilient tile or sheet (linoleum)	1	0.05
Softwood, $\frac{1}{2}$ in	2.5	0.12
Terrazzo	15	0.72
Wood deck, per in thickness	3	0.14
Wood joists		
2 × 8 @ 16 in on center (O.C.)	2.2	0.11
2 × 10 @ 16 in O.C.	2.7	0.13
2 × 12 @ 16 in O.C.	3.2	0.15
Partitions/walls		
Solid or fully-grouted hollow masonry assemblies		
4 in thick	40	1.92
6 in thick	60	2.88
8 in thick	80	3.84
10 in thick	100	4.80
12 in thick	120	5.76
Hollow (ungrouted) masonry assemblies		
4 in thick, normal weight	30	1.44
4 in thick, lightweight	21	1.01
6 in thick, normal weight	42	2.02
6 in thick, lightweight	30	1.44
8 in thick, normal weight	55	2.64
8 in thick, lightweight	40	1.82
10 in thick, normal weight	67	3.22
10 in thick, lightweight	47	2.26
12 in thick, normal weight	80	3.84
12 in thick, lightweight	55	2.64
Curtain wall (varies by manufacturer)	12–20	0.58–0.96
Glass block	18	0.86
Gypsum board		
$\frac{1}{2}$ in thick	2	0.10
$\frac{5}{8}$ in thick	2.5	0.12
Gypsum plaster, per in thickness	10	0.48
Metal lath	0.5	0.02
Steel framing	3	0.14
Stone, per in, per in thickness	15	0.72
Wood studs, 2 × 4 @ 16 in O.C.	2	0.10

Table 15.1 WEIGHTS OF COMMON BUILDING MATERIALS *(continued)*

System/material description	lb/ft²	kN/m² (kPa)
Roofs		
Coverings		
Built-up, composition, three-ply with aggregate surfacing	5	0.24
Built-up, composition, four-ply with aggregate surfacing	5.5	0.26
Built-up, composition, five-ply with aggregate surfacing	6	0.29
Shingles, composition, three tab or t-lock, per layer	3	0.14
Shingles, slate	10	0.48
Shingles, wood shakes	3	0.14
Single-ply membrane	1.5	0.07
Insulation		
Fiberglass batts or loose fill, per in thickness	0.5	0.02
Insulating concrete, per in thickness	3	0.14
Rigid boards, per in thickness	1.5	0.07
Decking		
Metal, thin gauge	1.5–3	0.07–0.14
Wood, per in, per in thickness	3	0.14

the weight of structural members, building materials, and any permanently fixed installation or piece of equipment, such as heating, cooling, and ventilation equipment, piping, lighting, ductwork, escalators, and elevators. Dead loads are computed based on the actual unit weights of the materials. Temporary supports, bracing, and other construction loads must also be considered part of the dead load. Dead loads can change over the life of a building, such as when the building is renovated or type of occupancy is changed.

■ EXAMPLE 15.1

A 48'-0" × 104'-0" low-slope (flat) roof deck system is composed of the following materials:

> Roof membrane and gravel (5 lb/ft²)
> Insulation (2.1 lb/ft²)
> Steel decking (1.4 lb/ft²)

Determine the total dead load of the roof deck system. Neglect the roof structure and any mechanical and electrical equipment supported by the roof.

Solution   The total area of the roof is found by

$$48 \text{ ft} \cdot 104 \text{ ft} = 4992 \text{ ft}^2$$

The unit weight of the roof deck system in lb/ft² is found by

$$5 \text{ lb/ft}^2 + 2.1 \text{ lb/ft}^2 + 1.4 \text{ lb/ft}^2 = 8.5 \text{ lb/ft}^2$$

15.1 Types of Building Loads   **281**

The dead load of the roof system is

$$8.5 \text{ lb/ft}^2 \cdot 4992 \text{ ft}^2 = 42\,432 \text{ lb} = 42.4 \text{ kip}$$

Partition loads Partitions (interior walls) within a building are treated as a dead load and are based on the unit weight of the material in the partition. In buildings, such as offices, having interior partitions with locations that are subject to change, building codes require application of a uniformly distributed load in addition to live loads. Typically, this load is 20 lb/ft^2 (0.96 kN/m^2).

Floor live loads These are loads that result from the weight of occupants, furniture, and any piece of equipment or installation that will probably be relocated over the life of the building. As shown in Table 15.2, floor loads are related to the expected use of a space. Minimum floor loads are generally stipulated by the building code. Because they are minimum requirements, the designer can increase the live load when a loading condition on a building element is anticipated to be greater. *Uniform floor loads* are treated as uniformly distributed loads that have units expressed in lb/ft^2 or kN/m^2 and differ by use and occupancy. Most floor systems must also be designed to support a *concentrated load* specified by the building code. The floor system must be capable of supporting this concentrated load over a $2\frac{1}{2}$-ft^2 area anywhere on the floor.

Roof live loads To accommodate maintenance and construction traffic, roofs are required to support a uniformly distributed live load. These loads typically range from 12 to 20 lb/ft^2 for low-sloped (below a 4/12 slope) roofs. These loads are generally with respect to the horizontal projection.

Snow loads Snow loads are dependent on the climate at the location of a building, the exposure of the building to winds and blowing snow, roof slope, roof configuration, and tendency of a roof to drift snow. Because a snow load is not permanent, it can be considered a live load. Snow loads are generally based on a 100-year occurrence; the heaviest snow load that will statistically occur only once during a 100-year period. Local building code typically stipulates a minimum snow load that is treated as a uniformly distributed load expressed in units of lb/ft^2 or kN/m^2. Depending on the geographic location of and climatic conditions at the building site, snow loads usually range from 5 to 80 lb/ft^2 (0.24 to 3.8 kN/m^2) in the United States.

A sloped roof will not carry as much snow as a flat roof, and codes generally allow a reduction in snow load for sloped roofs. On the other hand, drifting snow can result in loads that are greater than the minimum snow load required by local code. For example, a building with different elevation roofs can have a large accumulation of drifting snow on the single-story roof when it is on the leeward side of a second story. The designer must anticipate the location of and account for any snow drifts that will apply a heavier than normal load.

Table 15.2 EXAMPLES OF UNIFORM FLOOR LOADS BY USE OR OCCUPANCY[a]

Use or occupancy		Floor live load			
		U.S. Customary units		Metric (SI) units	
Type	Description of load	Uniformly distributed (lb/ft^2)	Concentrated (lb)	Uniformly distributed (kN/m^2)	Concentrated (kN)
Garages	General storage	100	—	4.8	—
	Motor vehicle storage	50	—	2.4	—
Hospitals	Wards and rooms	40	1000	1.8	4.5
Libraries	Reading rooms	60	1000	2.9	4.5
	Stack rooms	125	1500	6.0	6.7
Manufacturing	Light	75	2000	3.6	9.0
	Heavy	125	3000	6.0	13.4
Offices	—	50+ partition[b]	2000	2.4	9.0
Printing plants	Press rooms	150	2500	7.2	11.2
	Composing rooms	100	2000	4.8	9.0
Residential	Basic floor area	40	300 for	1.9	1.3 for
	Exterior balconies	60	stair treads	2.9	stair treads
	Decks	40		1.9	
Schools	Classroom	40+ partition[b]	1000	1.9	4.5
Storage	Light	125	—	6.0	—
	Heavy	250	—	12.0	—
Stores	—	100	3000	4.8	13.4

Source: 1994 Uniform Building Code.

[a]Specific design requirements are found in the building code.

[b]In buildings, such as offices and schools, having interior partitions with locations that are subject to change, building codes require application of a uniformly distributed load in addition to live loads. Typically, this load is 20 lb/ft^2 (0.96 kN/m^2).

Rain/ice loads Rain and ice loads relate to water or ice accumulation on a roof system due to a drainage system that is partially or fully clogged. Water exerts a pressure of 5.2 lb/ft^2 per inch of depth; water ponding at a depth of 6 in exerts 31.2 lb/ft^2 pressure on the roof (6 in · 5.2 lb/ft per inch = 31.2 lb/ft^2). Ice is less dense. The designer must anticipate rain and ice loads and make provisions in design to support these loads.

Dynamic loads A load that is gradually applied to a structural member is called a *static load.* Most loading conditions in buildings involve static loading. A rapid loading condition is referred to as a *dynamic load.* An *impact load* is a sudden dynamic load, such as a load being dropped on a structural member. An impact load can cause immediate failure of a structural member at a magnitude significantly less than the static load that would cause failure. Impact loads can occur at elevators, escalators, equipment and machinery supports, dance floors, auditoriums, gymnasiums, conference halls, manufacturing areas, loading docks, and helipads. A *moving load* is a constant dynamic load that moves across the building structure. Examples include the load from automobiles and trucks in a parking garage or the load from a crane moving across rails hung from structural roof members in a manufacturing plant.

Design for impact and moving loads can involve evaluating the dynamic loads and translating them into a static load, that is, increasing the static load by a factor known as the *equivalent static load factor*. The equivalent load factor is a multiplier greater than one that increases a static load to account for the dynamic effect of the load. Further discussion of dynamic loading is beyond the intended breadth of this book.

■ EXAMPLE 15.2

An office building has an overall floor system that covers an open area measuring 48′-0″ × 104′-0″. Determine the live load carried by the floor system. Assume a minimum uniformly distributed design live load of 125 lb/ft^2.

Solution   The total area of the floor system is

$$48 \text{ ft} \cdot 104 \text{ ft} = 4992 \text{ ft}^2$$

The live load on the floor system is

$$125 \text{ lb/ft}^2 \cdot 4992 \text{ ft}^2 = 624\ 000 \text{ lb} = 624 \text{ kip}$$

■ EXAMPLE 15.3

A 48′-0″ × 104′-0″ low-slope (flat) roof deck system carries a uniformly distributed snow load of 30 lb/ft^2. There is little drifting of snow from wind. Determine the total snow load on the roof system.

Solution   The total area of the roof from above is 4992 ft^2. The snow load on the roof system is

$$30 \text{ lb/ft}^2 \cdot 4992 \text{ ft}^2 = 149\ 760 \text{ lb} = 149.8 \text{ kip}$$

Wind loads   Wind loads result from a pressure or suction that develops on the exterior surface of a building due to blowing wind. Nearly all areas of the United States and Canada are susceptible to wind velocities that exceed 100 miles per hour (161 km/h). Strong hurricanes can produce wind velocities of up to 200 miles per hour (322 km/h). Winds from tornados have been estimated to reach as high as 600 miles per hour (966 km/h).

Winds that strike a wall or roof surface produce air pressure differences. On most sloped roofs, wind develops a direct velocity pressure on the windward side of a building and a suction on the leeward side. On most flat roofs, wind tends to develop a suction pressure. The theoretical *wind stagnation pressure* ($W_{\text{theoretical}}$) that develops on a flat surface that is perpendicular to the wind direction is found by the following formula where v is in mile/hr and $W_{\text{theoretical}}$ is in lb/ft^2:

$$W_{\text{theoretical}} = 0.002558v^2$$

In SI units, the following formula for theoretical wind stagnation pressure applies where v is in km/hr and $W_{\text{theoretical}}$ is in kN/m^2:

$$W_{\text{theoretical}} = 0.0473v^2$$

In practice, the basic formula for the wind stagnation pressure is expanded to include design factors for type of wind exposure, building height above ground, and building importance, a factor that is dependent on the perceived importance of the building (i.e., hospital or police/fire/rescue stations versus agricultural or storage buildings). In extreme conditions, these adjustment factors could combine to increase the theoretical wind stagnation pressure by a factor of 3.

Wind forces can create an uplift force on a roof that can lift a light roof structure. Wind can also cause a low-frequency harmonic or rocking effect in a building that can lead to structural failure. The complex nature of wind loading extends any further discussion of wind loads beyond the intended breadth of this book.

■ EXAMPLE 15.4

Determine the theoretical wind velocity pressure and load that develops on a wall surface that is perpendicular to the wind direction:
(a) With a sustained wind velocity of 100 mile/h.
(b) With a hurricane wind gust velocity of 200 mile/h.
(c) Approximate the total wind load that develops on a 36'-6" high × 104'-0" wide wall surface that is perpendicular to a 100-mile/h wind.
(d) Approximate the total wind load that develops on a 36'-6" high × 104'-0" wide wall surface that is perpendicular to a 200-mile/h wind.

Solution

(a) $W_{\text{theoretical}} = 0.002558v^2 = 0.002558(100)^2 = 25.6$ lb/ft^2
(b) $W_{\text{theoretical}} = 0.002558v^2 = 0.002558(200)^2 = 102.3$ lb/ft^2

(c) The area of the wall is

$$(36'\text{-}6'' \cdot 104'\text{-}0'') = (36.5 \text{ ft} \cdot 104.0 \text{ ft}) = 3796 \text{ ft}^2$$

The total wind load is found by

$$25.6 \text{ lb/ft}^2 \cdot 3796 \text{ ft}^2 = 97\ 178 \text{ lb} = 97 \text{ kip}$$

(d) The area of the wall is

$$(36'\text{-}6'' \cdot 104'\text{-}0'') = (36.5 \text{ ft} \cdot 104.0 \text{ ft}) = 3796 \text{ ft}^2$$

The total wind load is found by

$$102.3 \text{ lb/ft}^2 \cdot 3796 \text{ ft}^2 = 388\ 331 \text{ lb} = 388 \text{ kip}$$

Seismic loads   An earthquake is a sudden seismic movement of the tectonic plates that make up the earth's crust. Motion is caused by the release of stress accumulated along geologic faults or by volcanic activity. The sudden release of energy results in vibrational waves that travel along the surface of the earth's crust and cause damage near the epicenter of the earthquake. Although vibrational waves oscillate both horizontally and vertically, it is the horizontal movement that causes the greatest potential for damage of a building structure. A building structure is principally designed to react to gravity loads, loads that act downward and not horizontally. The horizontal thrust of the moving earth coupled with the inertia of the building structure causes the development of a horizontal shearing effect at the base of the structure; simply, as the earth moves the building foundation horizontally, the building superstructure above ground attempts to remain in place.

Building codes typically classify the severity of this horizontal shearing effect with seismic zones having numerical designations (0 through 4) that serve as a factor indicating peak ground acceleration. For example, much of California (seismic zone 4) is likely to have an earthquake with a peak acceleration force that is four times greater than what may occur in the Front Range and Western Slope of Colorado (seismic zone 1) and twice as great as the peak force in much of the New England area (seismic zone 2). Building codes in severe seismic zones require significant bracing and reinforcement for lateral movement.

A specific frequency of horizontal ground vibration can cause the structure of a tall building to oscillate (swing back and forth) with potentially catastrophic results. Due to the complexity of seismic analysis, further discussion is beyond the intended breadth of this book.

Construction loads   Construction loads are short-duration loads that the building structure supports during construction. These loads relate to temporary placement of equipment and installations used to construct the building and storage of construction materials and equipment on the structure before use in construction. A tower crane used in erection of a skyscraper is supported by the building structure. It induces large dead, dynamic, and wind loads on the structure that will not exist once the building is completed. Placement of stacks of roofing materials during installation of a roof system creates temporary concentrated loads at the location where the materials are stacked temporarily. The designer must anticipate how and where these construction loads will temporarily occur and make provisions in design to support these loads.

Earth loads   Soils and soil water can exert vertical (upward) and lateral forces on concrete slabs, building foundations, and retaining walls. Compacted backfill soil and hydrostatic forces from groundwater produce significant lateral pressure on walls below grade.

Expansive soils, soils containing a significant fraction of certain types of clay, heave when water is introduced to them and consolidate when moisture is removed. Soil shrinking and swelling are related to the fraction of clay in the soil, the type of clay mineral found, the denseness of the soil, the amount of moisture change, and the structural load transmitted to the soil. Upon being moistened, an expansive soil can exert a lateral pressure on a foundation wall and an upward pressure on a concrete slab that may easily reach 20 000 lb/ft^2 (3.1 MN/m^2). Lateral forces from swelling expansive soil can cause bulging in building foundations and retailing walls. Upward-acting forces from expansive soil heave can lift foundation and retailing walls, concrete slabs, and foundation piers. The designer must anticipate how and where expansive soil creates a load and make provisions in design to counter these loads or isolate movement to prevent damage to the structure.

Live, wind, seismic, and construction loads tend to be much more unpredictable than dead loads. They are generally based on an anticipated 50- or 100-year occurrence; the maximum load that will statistically occur only once during the period of occurrence. Unfortunately, a load greater than the 50- or 100-year load can occur during that period, sometimes with catastrophic damage.

15.2   SERVICE AND DESIGN LOADS

In designing a structural member, the designer must first quantify the maximum total load that must be supported by the member under consideration. The *service load* ($L_{service}$) is the maximum load that *will* likely be supported by the structural member or system while it is in service. The *design load* (L_{design}) is the largest load that a member *can* carry. It is used by the designer when selecting a structural member. The design load of a member should always be greater than the service load. When establishing service loads, the designer uses experience, quantity-take-off estimation techniques, building code requirements, and the application of statistical and analytical methods.

15.3   FACTORED LOADS AND LOAD COMBINATIONS

A building structural member must be strong enough to safely support the maximum anticipated load but no larger than is necessary to prevent waste of material. The design load on a roof system may be determined by combining the effects of anticipated occupant, snow, wind, dynamic, seismic, and dead loads. It is, however, not probable that maximum snow, wind, and seismic loads will occur at the same time. Therefore, in determining the minimum acceptable design load, the designer must examine a number of loading scenarios to ascertain the maximum effect of snow, wind, and seismic loads.

Load combinations are a series of prescribed loading scenarios that are used to calculate and determine the design load. Building codes regulate the method of ascertaining the design load by specifying standard load combinations. In allowable stress design, the Uniform Building Code (UBC) specifies the following combinations of dead loads (D_{load}), live floor loads (L_{load}), roof loads ($L_{roof\ load}$), snow loads (S_{load}), rain/ice load (R_{load}), wind load (W_{load}), and earthquake load (E_{load}):

$$L_{design-a} = D_{load} + L_{load} + L_{roof\ load}\ or\ S_{load}$$
$$L_{design-b} = D_{load} + L_{load} + W_{load}$$
$$L_{design-c} = D_{load} + L_{load} + W_{load} + 0.5S_{load}$$
$$L_{design-d} = D_{load} + L_{load} + S_{load} + 0.5W_{load}$$
$$L_{design-e} = D_{load} + L_{load} + S_{load} + E_{load}$$

The designer calculates each prescribed load combination. The largest load of the combinations becomes the minimum design load.

A *factored load* is the service load multiplied by an appropriate load factor; it determines the design load that will be applied to or carried by a structural member. *Load factors* are factors by which the anticipated or specified load is multiplied to ascertain the factored design load. Load factors are greater than 1 and are based on research, experience, and the degree of accuracy by which a load can be determined. For example, a dead load is more reliably determined than a live load; the factor may be 1.4 for a dead load and 1.7 for a live load.

Factored load combinations are a prescribed series of factored loads that are used to calculate and determine the factored design load. Use of load combinations serves as an analytical method of determining a design load. In its strength design method, the American Concrete Institute, Inc. (ACI) specifies that the strength of a member must be at least equal to the factored load calculated on the basis of dead loads (D_{load}) and live loads (L_{load}) according to the following equation:

$$L_{design} = 1.4D_{load} + 1.7L_{load}$$

■ EXAMPLE 15.5

A concrete column will carry an 80-kip dead load and a 120-kip live load. Determine the factored (design) load.

Solution

$$L_{design} = 1.4D_{load} + 1.7L_{load}$$
$$= 1.4(80\ kip) + 1.7(120\ kip) = 316\ kip$$

Through their joint specification, *ANSI/ASCE 7-95 Minimum Design Loads for Buildings and Other Structures,* the American National Standards Institute

(ANSI) and the American Society of Civil Engineers, Inc. (ASCE) prescribe use of the largest load found by any of the following load combinations of dead loads (D_{load}), live floor loads (L_{load}), roof loads ($L_{roof\ load}$), snow loads (S_{load}), rain/ice load (R_{load}), wind load (W_{load}), and earthquake load (E_{load}).

$$L_{design\ a} = 1.4D_{load}$$
$$L_{design\ b} = 1.2D_{load} + 1.6L_{load} + 0.5(L_{roof\ load}\ or\ S_{load}\ or\ R_{load})$$
$$L_{design\ c} = 1.2D_{load} + 1.6(L_{roof\ load}\ or\ S_{load}\ or\ R_{load}) + (0.5L_{load}\ or\ 0.8W_{load})$$
$$L_{design\ d} = 1.2D_{load} + 1.3W_{load} + 0.5L_{load} + 0.5(L_{roof\ load}\ or\ S_{load}\ or\ R_{load})$$
$$L_{design\ e} = 1.2D_{load} \pm E_{load} + 0.5L_{load} + 0.2S_{load}$$
$$L_{design\ f} = 0.9D_{load} \pm (1.3W_{load}\ or\ E_{load})$$

■ EXAMPLE 15.6

Determine the factored (design) load of a roof system, in lb/ft^2, based on ANSI/ASCE 7-95. The roof system is supporting the following uniformly distributed loads. Assume that the building is in an area with no seismic activity ($E_{load} = 0$) and that there are no additional loads on the roof (i.e., movable signs, cranes, etc.).

> Service dead load of 30 lb/ft^2
> Service roof load of 20 lb/ft^2
> Snow load of 30 lb/ft^2
> Rain/ice load of 20.8 lb/ft^2 (4 in · 5.2 lb/ft^2 per inch)
> Wind load of 12 lb/ft^2 acting downward

Solution Each load combination is calculated as follows:

$$L_{design\ a} = 1.4D_{load}$$
$$= 1.4(30\ lb/ft^2) = 42\ lb/ft^2$$

$$L_{design\ b} = 1.2D_{load} + 1.6L_{load} + 0.5(L_{roof\ load}\ or\ S_{load}\ or\ R_{load})$$
$$= 1.2(30\ lb/ft^2) + 1.6(20\ lb/ft^2) + 0.5(30\ lb/ft^2) = 83\ lb/ft^2$$

$$L_{design\ c} = 1.2D_{load} + 1.6(L_{roof\ load}\ or\ S_{load}\ or\ R_{load}) + (0.5L_{load}\ or\ 0.8W_{load})$$
$$= 1.2(30\ lb/ft^2) + 1.6(30\ lb/ft^2) + 0.5(20\ lb/ft^2) = 88\ lb/ft^2$$

$$L_{design\ d} = 1.2D_{load} + 1.3W_{load} + 0.5L_{load} + 0.5(L_{roof\ load}\ or\ S_{load}\ or\ R_{load})$$
$$= 1.2(30\ lb/ft^2) + 1.3(12\ lb/ft^2) + 0.5(20\ lb/ft^2) + 0.5(30\ lb/ft^2)$$
$$= 76.6\ lb/ft^2$$

$$L_{design\ e} = 1.2D_{load} \pm E_{load} + 0.5L_{load} + 0.2S_{load}$$
$$= 1.2(30\ lb/ft^2) + 0.5(20\ lb/ft^2) + 0.2(30\ lb/ft^2) = 52\ lb/ft^2$$

$$L_{design\ f} = 0.9D_{load} \pm (1.3W_{load}\ or\ E_{load})$$
$$= 0.9(30\ lb/ft^2) + 1.3(12\ lb/ft^2) = 42.6\ lb/ft^2$$

The factored load is 88 lb/ft^2 of roof area; the largest result ($L_{design\ c}$) of the load combinations.

In Example 15.6, the factored load is 88 lb/ft^2. The designer would use this load (or a greater load) as the design load when designing roof structure elements (i.e., joists, beams, and girders) through investigations of shear, moment, and deflection. In addition, establishing the design loading condition on a beam or column requires good reasoning and a basic understanding of building systems on the part of the designer.

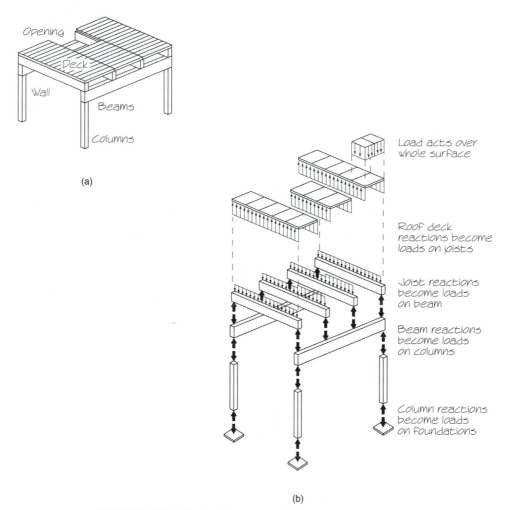

FIGURE 15.1 The load-carrying members of a building structural system supports a portion of the load acting on the building: (a) structure; (b) free-body diagrams. Each load-carrying member transfers the load to a second member, the second member transfers it to a third, and so on until the load is transferred to the earth. Reproduced from Daniel L. Schodek, *Structures,* © 1980, p. 114, by permission of Prentice Hall, Inc.

In a building structural system, each member supports a portion of the load acting on the building. A load-carrying member transfers the load to a second member, the second member transfers it to a third, and so on, until the load is transferred through the building foundation to the earth (Figure 15.1). Joists or beams resting on a girder support the load of roof deck and floor deck systems. A portion of the roof or floor load carried by the beam or joist is transferred to a girder at the reaction of each joist or beam. The load on the girder is equivalent in magnitude to the reactions of the joists or beams it supports. The girder then transfers the load it supports to a column. The load transferred by the girder to the column is equivalent in magnitude to the girder reaction. The column supports the loads from the beams and girders connected to it. The load carried by a column is equivalent to the sum of the reactions of the girders connected to the column. Finally, the column transfers the load to the building foundation, where it is distributed to the foundation bed in soil or bedrock.

A residential floor system during construction. Horizontally positioned wood joists spaced repetitively below the floor deck make up the structure of the floor platform. A horizontally positioned steel beam spans between a vertically positioned column and the foundation walls. The beam supports the floor load and transfers it to the column and walls.

A part of the roof truss system during construction. The smaller, repetitively spaced steel joists will support a roof platform.

Horizontally positioned steel joists bear on joist girders that span between columns. The joists transfer the roof load to joist girders which, in turn, transfer their load to vertically positioned columns. Columns transfer their load to individual column foundations.

Rooftop HVAC units impose a concentrated dead load on top of the columns.

A structural system of a precast concrete structure during construction. Concrete double tees make up the structure of the floor and roof platforms. These horizontally positioned double tees span between girders and transfer the floor and roof loads to the girders. Girders span between vertically positioned columns and transfer their end loads to the columns. Columns support the girder ends and transfer the building loads from the girders to individual foundations located at the base of each column.

A horizontally positioned structural member such as a joist or beam supports a specific portion of the floor or roof area. The roof or floor area that contributes to the load on a horizontally positioned structural member is known as the *tributary area.* With repetitively spaced members such as floor or roof joists carrying a uniformly distributed load across the entire span of the member, the tributary area is found by

$$\text{tributary area} = \text{span} \cdot \text{member spacing}$$

The calculation technique used to determine the total tributary areas on floor joist systems with common repetitive spacings is demonstrated in Example 15.7. A similar approach is used to determine the total tributary areas of other types of horizontally positioned members that are repetitively spaced.

■ EXAMPLE 15.7

Floor joists span 12'-6". Determine the tributary area of each joist at the following repetitive spacings:
(a) At 12 in on center (O.C.)
(b) At 16 in O.C.
(c) At 19.2 in O.C.
(d) At 24 in O.C.

Solution (a) The spacing of the joist is 1.0 ft O.C.: 12 in/(12 in/ft) = 1.0 ft.

$$\text{tributary area} = \text{span} \cdot \text{member spacing} = 12.5 \text{ ft} \cdot 1.0 \text{ ft} = 12.5 \text{ ft}^2$$

(b) The spacing of the joist is 1.33 ft O.C.: 16 in/(12 in/ft) = 1.33 ft.

$$\text{tributary area} = \text{span} \cdot \text{member spacing} = 12.5 \text{ ft} \cdot 1.33 \text{ ft} = 16.7 \text{ ft}^2$$

(c) The spacing of the joist is 1.60 ft O.C.: 19.2 in/(12 in/ft) = 1.60 ft.

$$\text{tributary area} = \text{span} \cdot \text{member spacing} = 12.5 \text{ ft} \cdot 1.60 \text{ ft} = 20 \text{ ft}^2$$

(d) The spacing of the joist is 2.0 ft O.C.: 24 in/(12 in/ft) = 2.0 ft.

$$\text{tributary area} = \text{span} \cdot \text{member spacing} = 12.5 \text{ ft} \cdot 2.0 \text{ ft} = 25.0 \text{ ft}^2$$

Joists and beams typically connect to a girder that carries the load of these members. The tributary area of a girder includes a portion of the tributary areas carried by the members it supports. The calculation technique used to determine the total tributary area of a girder is demonstrated in Example 15.8.

■ EXAMPLE 15.8

A steel girder and foundation wall support floor joists as shown in Figure 15.2. The joists are placed perpendicular to the girder and foundation wall and

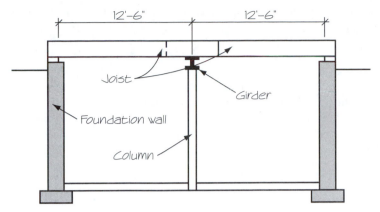

Transverse section

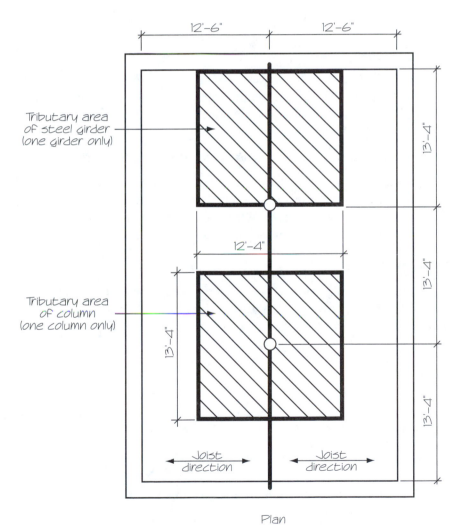

Plan

FIGURE 15.2 Illustration for Example 15.8.

span 12'-6". The girder runs continuously down the center of the building and is supported by columns spaced at 13'-4". Assume that the floor system is carrying a uniformly distributed load. Determine the tributary area of the girder.

Solution Each joist carries a uniformly distributed load across its entire length. Therefore, half of the tributary area on the joist is transferred to the foundation wall and half is transferred to the girder:

$$\frac{12'\text{-}6''}{2} = 6'\text{-}3''$$

The girder spans 13'-4" between columns and carries joists on both sides, so

$$13'\text{-}4'' = 13.33 \text{ ft}$$
$$6'\text{-}3'' + 6'\text{-}3'' = 12'\text{-}6'' = 12.5 \text{ ft}$$

The tributary area is found by

$$13.33 \text{ ft} \cdot 12.5 \text{ ft} = 166.6 \text{ ft}^2$$

Any loads applied to the tributary area are known as *tributary loads.* When loads are uniformly distributed over the entire tributary area of a member such as with uniformly distributed live loads on a floor deck or snow loads on a roof deck, the *total tributary load* is found by

$$\text{total tributary load} = \text{uniformly distributed load} \cdot \text{tributary area}$$

The calculation technique used to determine the total tributary loads on floor joist systems with common repetitive spacings is demonstrated in Example 15.9. A similar approach is used to determine the total tributary load on other types of repetitively spaced members such as roof joists and floor beams.

■ EXAMPLE 15.9

In Example 15.7, floor joists span 12'-6" at different repetitive spacings. Determine the tributary load on each joist at the following joist spacings. Assume that the joist will be supporting a design load of 50 lb/ft^2.
(a) At 12 in O.C.
(b) At 16 in O.C.
(c) At 19.2 in O.C.
(d) At 24 in O.C.

Solution (a) From Example 15.7, the tributary area for this spacing is 12.5 ft^2. The tributary load is found by

$$\text{tributary load} = \text{uniformly distributed load} \cdot \text{tributary area}$$
$$= 50 \text{ lb/ft}^2 \cdot 12.5 \text{ ft}^2 = 625 \text{ lb}$$

(b) From Example 15.7, the tributary area for this spacing is 16.7 ft^2. The tributary load is found by

$$\text{tributary load} = \text{uniformly distributed load} \cdot \text{tributary area}$$
$$= 50 \text{ lb/ft}^2 \cdot 16.7 \text{ ft}^2 = 833 \text{ lb}$$

(c) From Example 15.7, the tributary area for this spacing is 20 ft^2. The tributary load is found by

$$\text{tributary load} = \text{uniformly distributed load} \cdot \text{tributary area}$$
$$= 50 \text{ lb/ft}^2 \cdot 20 \text{ ft}^2 = 1000 \text{ lb}$$

(d) From Example 15.7, the tributary area for this spacing is 25 ft^2. The tributary load is found by

$$\text{tributary load} = \text{uniformly distributed load} \cdot \text{tributary area}$$
$$= 50 \text{ lb/ft}^2 \cdot 25 \text{ ft}^2 = 1250 \text{ lb}$$

The total tributary load supported by a horizontally positioned structural member carrying a uniformly distributed load such as a joist, beam, or girder is similar to the total uniformly distributed force (W) introduced in Chapter 3: The total uniformly distributed force (W) exerted on a rigid body is the product of the uniformly distributed force per unit length (w) multiplied by the member length (L). In the case of a beam, this relationship is expressed as

total tributary load = uniformly distributed load per unit length · member length

or

$$W = wL$$

The uniformly distributed load per unit length is expressed in force per unit length, such as lb/ft, kip/ft, N/m, kN/m, and so on. It can also be identified by the symbol w. The total tributary load supported by a horizontally positioned structural member carrying a uniformly distributed load can be identified by the symbol W.

When loads are uniformly distributed over the entire tributary area of a member, the uniformly distributed load per unit length supported by a horizontally positioned member is found by dividing the total tributary load or total uniformly distributed force (W) on the member by the length of the member (L):

$$\text{uniformly distributed load per unit} = \frac{\text{total tributary load}}{\text{member length}}$$

or

$$w = \frac{W}{L}$$

■ EXAMPLE 15.10

In Example 15.9, the total tributary loads for floor joists spanning 12'-6" at different repetitive spacings were found. Determine the uniformly distributed design load per unit length on each joist at the following joist spacings.

15.5 Tributary Loads on Beams

(a) At 12 in O.C.

(b) At 16 in O.C.

(c) At 19.2 in O.C.

(d) At 24 in O.C.

Solution (a) From Example 15.9 the total tributary load for this spacing is 625 lb and the member length is 12.5 ft. The uniformly distributed load per unit length is found by

$$w = \frac{W}{L} = \frac{625 \text{ lb}}{12.5 \text{ ft}} = 50 \text{ lb/ft}$$

(b) From Example 15.9 the total tributary load for this spacing is 833 lb and the member length is 12.5 ft. The uniformly distributed load per unit length is found by

$$w = \frac{W}{L} = \frac{833 \text{ lb}}{12.5 \text{ ft}} = 67 \text{ lb/ft}$$

(c) From Example 15.9 the total tributary load for this spacing is 1000 lb and the member length is 12.5 ft. The uniformly distributed load per unit length is found by

$$w = \frac{W}{L} = \frac{1000 \text{ lb}}{12.5 \text{ ft}} = 80 \text{ lb/ft}$$

(d) From Example 15.9 the total tributary load for this spacing is 1250 lb and the member length is 12.5 ft. The uniformly distributed load per unit length is found by

$$w = \frac{W}{L} = \frac{1250 \text{ lb}}{12.5 \text{ ft}} = 100 \text{ lb/ft}$$

■ EXAMPLE 15.11

A low-slope (flat) roof system is supported by 20K5 steel joists spaced at 4'-0" O.C. that span 24'-0" as shown in Figure 15.3. The 20K5 steel joists weigh approximately 8.2 lb/ft. Assume a uniformly distributed service live load of 20 lb/ft², a rain/ice load of 20.8 lb/ft², a wind load of 8 lb/ft² acting downward, and a snow load of 30 lb/ft². Neglect seismic (earthquake) loads. The roof system carried by the joists is composed of the following materials and equipment:

Roof membrane and gravel (5 lb/ft²)

Insulation (2.1 lb/ft²)

Steel decking (1.4 lb/ft²)

Mechanical (HVAC ducts) and electrical (lights) equipment (20 lb/ft²)

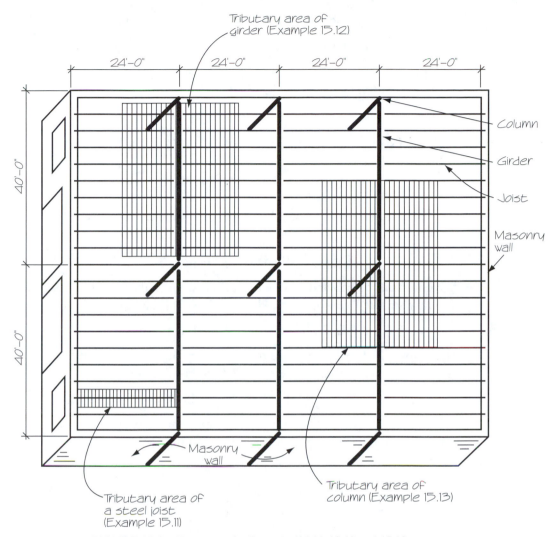

FIGURE 15.3 Illustration for Examples 15.11, 15.12, and 15.13.

(a) Determine the tributary area of each joist.

(b) Determine the loads supported by a single joist.

(c) Determine the design (factored service) load on each joist based on ANSI/ASCE 7-95.

(d) Determine the uniformly distributed design (factored service) load per linear foot of joist.

Solution (a) The tributary area carried by each joist is found by multiplying the joist span by the joist spacing:

$$24 \text{ ft} \cdot 4 \text{ ft} = 96 \text{ ft}^2$$

15.5 Tributary Loads on Beams

299

(b) The uniformly distributed dead load carried by the roof in lb/ft^2 is found by

$$5 \text{ lb/ft}^2 + 2.1 \text{ lb/ft}^2 + 1.4 \text{ lb/ft}^2 + 20 \text{ lb/ft}^2 = 28.5 \text{ lb/ft}^2$$

The tributary area carried by each joist was found to be 96 ft^2. The tributary dead load carried by each joist is found by

$$96 \text{ ft}^2 \cdot 28.5 \text{ lb/ft}^2 = 2736 \text{ lb}$$

The 20K5 steel joists weigh approximately 8.2 lb/ft. The dead load (weight) of the joist alone is found by

$$24 \text{ ft} \cdot 8.2 \text{ lb/ft} = 196.8 \text{ lb}$$

The tributary service dead load of each joist and the dead load it supports is found by

$$2736 \text{ lb} + 196.8 \text{ lb} = 2932.8 \text{ lb} = 2933 \text{ lb}$$

The uniformly distributed service live load resting on the roof is 20 lb/ft^2. The tributary service live load carried by each joist is found by

$$96 \text{ ft}^2 \cdot 20 \text{ lb/ft}^2 = 1920 \text{ lb}$$

The uniformly distributed snow load resting on the roof is 30 lb/ft^2. The tributary snow load carried by each joist is found by

$$96 \text{ ft}^2 \cdot 30 \text{ lb/ft}^2 = 2880 \text{ lb}$$

The uniformly distributed rain/ice load resting on the roof is 20.8 lb/ft^2. The tributary rain/ice load carried by each joist is found by

$$96 \text{ ft}^2 \cdot 20.8 \text{ lb/ft}^2 = 1996.8 \text{ lb} = 1997 \text{ lb}$$

The uniformly distributed wind load acting downward on the roof is 8 lb/ft^2. The wind load carried by each joist is found by

$$96 \text{ ft}^2 \cdot 8 \text{ lb/ft}^2 = 768 \text{ lb}$$

(c) Each load combination is calculated as follows:

$$
\begin{aligned}
L_{\text{design a}} &= 1.4D_{\text{load}} \\
&= 1.4(2933 \text{ lb}) = 4106 \text{ lb} \\
L_{\text{design b}} &= 1.2D_{\text{load}} + 1.6L_{\text{load}} + 0.5(L_{\text{roof load}} \text{ or } S_{\text{load}} \text{ or } R_{\text{load}}) \\
&= 1.2(2933 \text{ lb}) + 1.6(1920 \text{ lb}) + 0.5(2880 \text{ lb}) = 8032 \text{ lb} \\
L_{\text{design c}} &= 1.2D_{\text{load}} + 1.6(L_{\text{roof load}} \text{ or } S_{\text{load}} \text{ or } R_{\text{load}}) + (0.5\,L_{\text{load}} \text{ or } 0.8\,W_{\text{load}}) \\
&= 1.2(2933 \text{ lb}) + 1.6(2880 \text{ lb}) + 0.5(1920 \text{ lb}) = 9088 \text{ lb} \\
L_{\text{design d}} &= 1.2D_{\text{load}} + 1.3W_{\text{load}} + 0.5L_{\text{load}} + 0.5(L_{\text{roof load}} \text{ or } S_{\text{load}} \text{ or } R_{\text{load}}) \\
&= 1.2(2933 \text{ lb}) + 1.3(768 \text{ lb}) + 0.5(1920 \text{ lb}) + 0.5(2880 \text{ lb}) = 6918 \text{ lb}
\end{aligned}
$$

$$L_{\text{design e}} = 1.2D_{\text{load}} \pm E_{\text{load}} + 0.5L_{\text{load}} + 0.2S_{\text{load}}$$
$$= 1.2(2933 \text{ lb}) + 0.5(1920 \text{ lb}) + 0.2(2880 \text{ lb}) = 5056 \text{ lb}$$
$$L_{\text{design f}} = 0.9D_{\text{load}} \pm (1.3W_{\text{load}} \text{ or } E_{\text{load}})$$
$$= 0.9(2933 \text{ lb}) + 1.3(768 \text{ lb}) = 3638 \text{ lb}$$

The factored load is 9088 lb of roof area, the largest result ($L_{\text{design c}}$) of the load combinations.

(d) The load resting on each linear foot of joist is found by dividing the total tributary load on the joist by the length of the joist:

$$w = \frac{W}{L} = \frac{9088 \text{ lb}}{24 \text{ft}} = 379 \text{ lb/ft}$$

The tributary load is used to determine the loading condition on a beam. The designer uses the tributary load to develop the free-body diagram that serves as a foundation for analysis of a beam. It is frequently necessary to determine the tributary loads on subordinate members before ascertaining the tributary load on the member under consideration. As shown in Examples 15.10 and 15.11, loads on the joists supported by a girder must be determined before determining the total tributary load on the girder.

■ EXAMPLE 15.12

As shown in Figure 15.3 and as described in Example 15.11, a low-slope (flat) roof system is supported by 20K5 steel joists spaced at 4′-0″ on center. The total tributary load supported by each joist, including its own weight, was found to be 9088 lb. Assume that a group of nine repetitively spaced joists are supported on each side of a 40′-0″ steel girder. Determine the total tributary load resting on the girder.

Solution Each joist carries a total tributary load of 9088 lb that is uniformly distributed. The reaction (R) at the end of each joist is found by the appropriate derived formula in Table A.5:

$$R = \frac{W}{2} = \frac{9088 \text{ lb}}{2} = 4544 \text{ lb} = 4.54 \text{ kip}$$

Nine joists are resting on the girder. The number of joists was computed by

$$40 \text{ ft/4 ft joist spacing} = 10 \text{ joist spacings}$$
$$10 \text{ joist spacings} - 1 = 9 \text{ joists}$$

The total tributary load resting on the girder is found by

$$2 \text{ sides} \cdot 9 \text{ joists} \cdot 4.54 \text{ kip} = 81.7 \text{ kip}$$

As discussed above, building loads from the roof and floor decks are transferred to a system of horizontally positioned structural members called *joists, beams,* or *girders.* This system of horizontally positioned beams and girders is connected to a system of vertically positioned structural members called *columns.* Gravity loads on the building are transferred from the horizontally positioned structural members to the column at the beam or girder and column connections.

Beams or girders connected to a column transfer a portion of the load they are supporting to the column. The column reacts to this load. This transferred load is equivalent to the reaction of the beam or girder at the column connection. The total gravity load carried by a column is equivalent to the sum of the reactions of all beams or girders connected to the column.

■ EXAMPLE 15.13

As shown in Figure 15.3 and as described in Examples 15.11 and 15.12, a column supports the two W18×76 steel girders and two 20K5 steel joists. Determine the total tributary load resting on the column.

Solution From Example 15.12, each girder is carrying a total tributary load of 81.7 kip that is uniformly distributed across its length. The reaction of each girder is found by using the appropriate derived formula in Table A.5:

$$R = \frac{W}{2} = \frac{81.7 \text{ kip}}{2} = 40.9 \text{ kip}$$

From Example 15.11, each joist carries a total tributary load of 9088 lb, including its own weight that is uniformly distributed. The reaction (R) at the end of each joist is found by the appropriate derived formula in Table A.5:

$$R = \frac{W}{2} = \frac{9088 \text{ lb}}{2} = 4544 \text{ lb} = 4.5 \text{ kip}$$

With two girders and two joists connected to the column, the total tributary load resting on the column is found by summing the reactions of the connected members:

$$40.9 \text{ kip} + 40.9 \text{ kip} + 4.5 \text{ kip} + 4.5 \text{ kip} = 90.8 \text{ kip}$$

15.7 LATERAL LOADS

The technique above that was used to determine tributary loads presumes that loading is applied downward on the structure. Downward loading is frequently called *gravity loading* because the forces exerted on the structure are due to the force of gravity. *Lateral loads* from sources such as wind, soil and hydrostatic pressure, and seismic activity are also applied to the building superstructure. Seismic movement and soil pressure can produce a lateral force that is exerted on the building foundation. Lateral forces can produce excessive deflection, swaying, harmonic, or rocking, vibrational, and torsional effects on the building structure that could not be caused by gravity loads. Treatment of the effects of lateral forces in building structures generally involves use of diagonal bracing, shear walls, and proper anchoring and connection of structural members. Analysis of the effects of lateral loads is detailed and beyond the intended breadth of this book.

EXERCISES

15.1 Calculate the wind stagnation pressure ($W_{theoretical}$) that develops on a 1.0-ft^2 flat surface from 0 to 200 miles/hr at wind velocity intervals of 10 miles/hr. Assume that the surface is perpendicular to the wind direction. Plot a graph of wind stagnation pressure versus wind velocity.

15.2 Approximate the total theoretical wind load that develops on a 200-m-high × 45-m-wide wall surface. Assume that the wind velocity is 210 km/hr.

15.3 A 30 m × 80 m low-slope (nearly flat) roof system is supported by steel joists. The protected-membrane roof system consists of concrete pavers (0.50 kN/m^2), a roof membrane (0.06 kN/m^2), thermal insulation (0.04 kN/m^2), and steel decking (0.05 kN/m^2). The roof system carries a snow load of 1.5 kN/m^2 and a uniformly distributed live load of 1.0 kN/m^2.
   (a) Determine the total dead load of the roof deck system. Neglect the steel joists and any mechanical and electrical equipment on the roof.
   (b) Determine the total roof live load on the roof system.
   (c) Determine the total snow load on the roof system with the assumption that there is little drifting.

15.4 A floor system is supported by 18K7 steel joists spaced at 2 ft O.C. that span 28′-3″. The floor system carries a uniformly distributed live load of 100 lb/ft^2. The floor system consists of a 3-in-thick concrete slab (144 lb/ft^3) cast on steel decking (1.0 lb/ft^2). 18K7 steel joists weigh approximately 9.0 lb/ft of joist length.
   (a) Determine the tributary area of a single joist.
   (b) Determine the total service live load carried by the joist.
   (c) Determine the total service dead load carried by the joist.
   (d) Determine the factored (design) load of a single joist using the ACI prescribed load combination.
   (e) Determine the factored (design) load of a single joist using the ANSI/ASCE load combinations.

15.5 Determine the factored (design) load of a roof system, in lb/ft^2, based on allowable stress design under the requirements of the Uniform Building Code (UBC). The roof system is supporting the following uniformly distributed loads. Assume that the building is in an area with no seismic activity ($E_{load} = 0$) and that there are no additional loads on the roof (i.e., movable signs, cranes, etc.).

   Service dead load of 35 lb/ft^2
   Service roof load of 20 lb/ft^2
   Snow load of 30 lb/ft^2

Rain/ice load (6 in of ponding water)

Wind load of 15 lb/ft^2 acting downward

15.6 Wood joists will be used to support a floor deck. The joists will span 4 m. The joists will be repetitively spaced at 400-mm O.C. intervals. The deck will be designed to carry a total uniformly distributed service load of 3.5 kN/m^2. Neglecting the weight of the structural members, draw a free-body diagram of the loading condition and compute the reactions of the joist.

(a) Draw a free-body diagram of the loading condition.

(b) Determine the tributary load of a single joist.

(c) Compute the joist reactions.

15.7 A floor deck will be constructed of repetitively spaced floor joists. The deck will have a total uniformly distributed service load of 65 lb/ft^2, including the weight of the joists. Joists will have a simple span of 16 ft between bearing points. Joist spacings under consideration include 12 in, 16 in, and 19.2 in O.C.

(a) Determine the tributary area on each joist at each joist spacing.

(b) Determine the loads supported by a single joist at each joist spacing.

(c) Compute the reactions of a single joist at each joist spacing.

15.8 As in Exercise 15.7, a floor deck will be constructed of repetitively spaced floor joists and will have a total uniformly distributed service load of 65 lb/ft^2, including the weight of the joists. Again, joist spacings under consideration include 12 in, 16 in and 19.2 in O.C. In this exercise, however, the 16-ft-long floor joists will span 12 ft between bearing points and have a single 4-ft overhang.

(a) Determine the tributary area on each joist at each joist spacing.

(b) Determine the loads supported by a single joist at each joist spacing.

(c) Compute the reactions of a single joist at each joist spacing.

15.9 A partial framing plan of a simple building structure is shown. The beams support the floor load carried by the deck. The beams are connected to girders marked G1, so the load carried by the beams is transferred to these girders. The total uniformly distributed service load carried by the floor system is 10 kN/m^2. The weight of the floor is supported by 8-m-long beams marked B1. Neglecting the weight of the structural members:

(a) Draw a free-body diagram of the loading condition.

(b) Compute the reactions of beams B1 and girders G1.

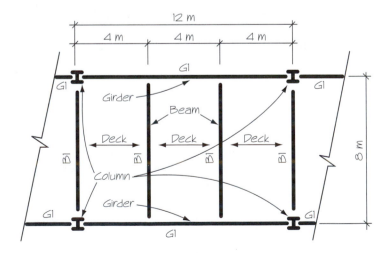

A framing plan of a second-story floor and a roof framing plan of a two-story structure are shown. The total uniformly distributed service load of the floor is 200 lb/ft². The total uniformly distributed service load of the roof, including the joists, is 60 lb/ft². On the floor structure, the load of a 12-ft-high section of exterior curtain wall is supported by spandrel beams SB1 and spandrel girders SG1 and SG3. On the roof structure, the weight of a 6-ft-high section of exterior curtain wall is supported by spandrel beams SB2 and spandrel girder SG5. The service load of the curtain wall is 20 lb/ft². Beams are estimated to weigh 50 lb/ft and girders are estimated to weigh 100 lb/ft. Columns support the second-story floor and roof and transfer their loads to individual column foundations at the first-floor level. All columns are 24 ft tall (12 ft between floors) and are estimated to weigh 100 lb/ft.

(a) Draw a free-body diagram of the loading condition and compute the reactions of the following members.

(1) Beam B1	(2) Beam B2	(3) Beam B3	(4) Beam B4
(5) Beam SB1	(6) Beam SB2	(7) Girder G1	(8) Girder G2
(9) Girder SG3	(10) Girder G4	(11) Girder SG5	(12) Girder G6

(b) Determine the load at the base of each column.

(1) Column C1	(2) Column C2	(3) Column C3	(4) Column C4
(5) Column C5	(6) Column C6	(7) Column C7	

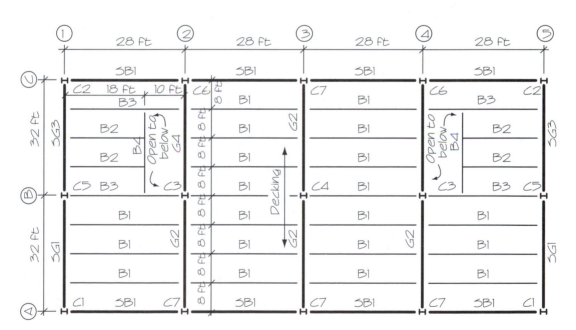

Floor framing plan

B   beam
C   column
G   girder
SB spandrel beam

FIGURE SP 15.1

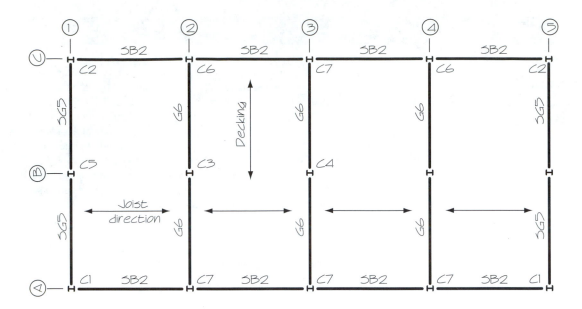

Roof framing plan

FIGURE SP 15.1   (continued)

Chapter 16
FUNDAMENTAL BEAM DESIGN PRINCIPLES

A beam is a structural member that rests between two or more bearing points and is subjected to loads or forces acting normal (perpendicular) to its longitudinal axis thereby causing it to bend. Types of beams were discussed in Chapter 4. In the building construction industry, the term *beam* generally relates to a horizontally positioned structural member that supports a load. Joists, purlins, girders, headers, and lintels act as beams.

A beam can fail by a variety of different modes, including failure by flexure, shear, deflection, bearing, lateral buckling, torsional buckling, and rotation. The theories behind these modes of failure are introduced in this chapter. Select procedures used in design of steel, wood, and concrete members are presented in later chapters. The presentation in this book does not provide full coverage of all design principles involved with design of building structures. Such coverage would require an extensive presentation requiring several volumes and references. This book is intended to acquaint the technician with basic design principles used by an engineer. Additional coverage is beyond the intended breadth of this book.

A steel beam supporting a masonry wall.

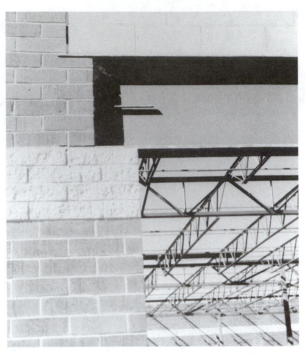

The end of a steel beam transferring its load to a masonry wall. The masonry wall reacts to the end load.

16.1 FLEXURE

As a beam bends under load, a resisting moment is developed internally to counteract the bending moment caused by the external forces (i.e., the loads and reactions). *Flexure* is caused by resistance of the beam material to bending. Bending results in the development of compressive and tensile stresses on opposite sides of the neutral axis. With bending moment and thus internal resisting moment varying across the beam length, stresses caused by flexure vary significantly across the length of the beam. Flexure failure usually results in the extreme (outermost) fiber being torn apart at the fiber, experiencing tensile stresses (Figure 16.1a).

16.2 THE FLEXURE FORMULA

The *flexure formula* physically relates the important influences on the bending strength of a beam: the beam material, the beam section, and the beam loading condition. Bending moment (M), the moment of inertia (I) of the beam section,

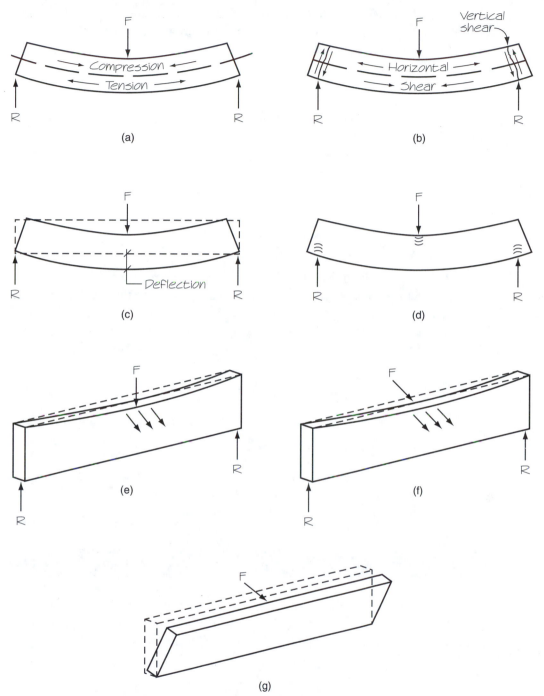

FIGURE 16.1 Examples of modes of failure in beams: (a) flexure; (b) shear; (c) deflection; (d) bearing; (e) lateral buckling; (f) torsional buckling; (g) rotation.

distance of the most extreme fiber to the neutral axis (c), and bending stress at the extreme fiber ($\sigma_{bending}$) are related by the flexure formula as expressed in its original conceptual form:

$$\sigma_{bending} = \frac{M}{I/c}$$

Frequently, the term I/c is substituted with the section modulus (S) of the beam section profile. The modified expression becomes

$$\sigma_{bending} = \frac{M}{S}$$

The bending stress at the extreme fiber ($\sigma_{bending}$) is the stress experienced by the extreme fiber at the location in the beam length where this moment exists.

■ EXAMPLE 16.1

When loaded, the bending moment at a certain point in a beam is 100 lb-ft. The beam is composed of a 1-in^2 metal bar. Determine the bending stress at the extreme fiber at that point in the beam.

Solution   Since $\sigma_{bending} = M/S$, it is necessary to find the section modulus (S) of a 1-in^2 bar:

$$S_{rectangular} = \frac{bd^2}{6} = \frac{1 \text{ in}(1 \text{ in})^2}{6} = 0.1667 \text{ in}^3$$

Solving for bending stress at the extreme fiber ($\sigma_{bending}$), we obtain

$$\sigma_{bending} = \frac{M}{S} = \frac{100 \text{ lb-ft} \cdot 12 \text{ in/ft}}{0.1667 \text{ in}^3} = 7199 \text{ lb/in}^2$$

Note: To maintain constant units, 100 lb-ft was converted to in-lb by multiplying by 12.

16.3   FLEXURE INVESTIGATION

The flexure formula is applied when sizing a beam section against bending (flexural) stresses at the extreme fiber. By algebraically manipulating the formula and substituting I/c with the section modulus (S), the flexure formula takes the following form:

$$S = \frac{M}{\sigma_{bending}}$$

With this revised formula, the designer selects a required section for the bending moment (M) of a specific loading condition and the allowable bending stress ($\sigma_{bending}$) for the material specified for the member.

■ EXAMPLE 16.2

A proposed beam has a maximum bending moment of 3000 lb-ft. The designer has selected wood as the beam material. Allowable stress in bending ($\sigma_{bending}$) for the wood species selected is 1200 lb/in^2.
(a) Find the minimum required section modulus under these conditions.
(b) Select a two-by member from Table 16.1 that will not experience flexure failure under these conditions.

Solution

(a)
$$S = \frac{M}{\sigma_{bending}} = \frac{3000 \text{ lb-ft} \cdot 12 \text{ in/ft}}{1200 \text{ lb/in}^2} = 30.0 \text{ in}^3$$

Note: To maintain constant units, 3000 lb-ft was converted to in-lb by multiplying by 12.
(b) From Table 16.1 the 2 × 12 section has a section modulus of 31.64 in^3. It is the only common two-by section that is equal to or greater than 30.0 in^3 and is acceptable.

With the results in Example 16.2, the designer must select a beam section that has a minimum section modulus of 30 in^3 to ensure that the allowable bending stresses are not exceeded. Since a 2 in × 12 in has a section modulus of 31.64 in^3, it can safely resist these maximum bending stresses as long as the species of wood has an allowable stress in bending of at least 1200 lb/in^2. A 2 in × 10 in section with a section modulus of 21.39 in^3 is too small for the loading condition described in Example 16.2; if it were used, the allowable bending stresses for the wood species (1200 lb/in^2) would be exceeded:

$$\sigma_{bending} = \frac{M}{S} = \frac{3000 \text{ lb-ft} \cdot 12 \text{ in/ft}}{21.39 \text{ in}^3} = 1683 \text{ lb/in}^2$$

Table 16.1 SECTION PROPERTIES OF COMMON TWO-BY SOLID WOOD LUMBER SECTIONS

Nominal beam size (in)		Surfaced (actual) size (in)		Section modulus S (in^3)		Moment of inertia, I (in^4)		Radius of gyration, r (in)	
Breadth, b	Depth, d	Breadth, b	Depth, d	Resting on edge	Laying flat	Resting on edge	Laying flat	Breadth	Depth
2	2	1.5	1.5	0.563	0.563	0.422	0.422	0.433	0.433
	3	1.5	2.5	1.56	0.938	1.95	0.703		0.722
	4	1.5	3.5	3.06	1.31	5.36	0.981		1.01
	6	1.5	5.5	7.56	2.06	20.80	1.55		1.59
	8	1.5	7.25	13.14	2.72	47.63	2.04		2.09
	10	1.5	9.25	21.39	3.47	98.93	2.60		2.67
	12	1.5	11.25	31.64	4.22	177.98	3.16		3.25
	14	1.5	13.25	43.89	4.97	290.78	3.73		3.82

16.3 Flexure Investigation

■ EXAMPLE 16.3

A proposed beam has a maximum bending moment of 3000 lb-ft. The designer has selected structural steel as the beam material. Allowable stress in bending ($\sigma_{bending}$) for the steel selected is 24 000 lb/in^2. Find the minimum required section modulus under these conditions.

Solution

$$S = \frac{M}{\sigma_{bending}} = \frac{3000 \text{ lb-ft} \cdot 12 \text{ in/ft}}{24\,000 \text{ lb/in}^2} = 1.50 \text{ in}^3$$

Note: To maintain constant units, 24 000 lb-ft was converted to in-lb by multiplying by 12.

Example 16.3 substitutes structural steel ($\sigma_{bending} = 24\,000$ lb/in^2) for wood ($\sigma_{bending} = 1200$ lb/in^2) in the same loading condition. As the stronger material, steel will require a much smaller section because of its greater allowable bending stress. For example, a section of TS4 $\times$ 2 $\times \frac{3}{16}$ rectangular structural tubing with dimensions of 4 in $\times$ 2 in with a $\frac{3}{16}$-in wall thickness is acceptable if it is positioned so that it is resting on its slender edge; it has a section modulus of 1.93 in^3. An S3 $\times$ 5.7 standard I-beam section is also acceptable if it is oriented with the load applied to either of its flanges; it has a section modulus of 1.68 in^3.

As shown in Examples 16.2 and 16.3, the flexure formula is typically used in its revised form of $S = M/\sigma_{bending}$ to select a beam cross section once a beam material is selected. The flexure formula serves as a powerful tool used in design against bending stresses.

16.4   SHEAR

Shear forces vary across the beam length. *Vertical shear* develops internally within a beam under load, due to the opposing forces created by the loads and the reactions. The vertical shearing effect has a tendency to cause the beam to drop between its supports (reactions). The vertical shear force is reacted to by an equal horizontal shearing effect. The *horizontal shear* force tends to cause theoretical longitudinal fibers in the beam to slide and shear horizontally (Figure 16.1b).

16.5   SHEAR INVESTIGATION

Maximum shear stress in a beam with a uniform cross section across its entire length will occur where the shear force is the largest absolute value. The maximum shear stress in common rectangular C-, T-, and I-shaped sections occurs at the neutral axis. The formula for calculating *maximum shear stress* ($\sigma_{max\ shear}$) *in a rectangular section* of a beam is based on the shear force (V) at that location in the beam and the breadth (b) and depth (d) of the rectangular section:

$$\sigma_{max\ shear} = \frac{3V}{2bd}$$

■ EXAMPLE 16.4

Calculate the maximum shear stress in a 4 × 12 rectangular beam section with actual dimensions of $3\frac{1}{2}$ in × $11\frac{1}{4}$ in. The shear force at that location in the beam is 2080 lb.

Solution

$$\sigma_{max\ shear} = \frac{3V}{2bd} = \frac{3(2080\ lb)}{2(3\frac{1}{2}\ in \times 11\frac{1}{4}\ in)} = 79.2\ lb/in^2$$

Web shear stress ($\sigma_{web\ shear}$) in C-, T-, and I-shaped beam sections approximates maximum shear stress in these types of sections. The formula for calculating *web shear stress* ($\sigma_{web\ shear}$) of a beam is based on the shear force (V) at that location in the beam and the beam section depth (d) and beam section web thickness (t_w):

$$\sigma_{web\ shear} = \frac{V}{dt_w}$$

The web shear stress formula assumes that the web resists all the shearing stress. It does not account for the flange(s) beyond the thickness of the web. This approach provides results that are approximate, typically lower than the actual maximum shear stress by 10 to 35%.

■ EXAMPLE 16.5

Calculate the web shear stress in a W8×31 steel beam section with web thickness of 0.285 in and a depth of 8.00 in. The shear force at that location in the beam is 12 000 lb.

Solution

$$\sigma_{web\ shear} = \frac{V}{dt_w} = \frac{12\ 000\ lb}{8.00\ in \cdot 0.285\ in} = 5263\ lb/in^2$$

16.6   DEFLECTION

The action of external forces (i.e., loads and reactions) acting on a beam causes a downward deformation that is commonly known as *bending*. When elastic in nature, this deformation is called *deflection*. *Deflection* (Δ) is a linear measure of the longitudinal deformation in a beam as a result of the action of external forces that are causing it to bend elastically. Deflection is measured as the distance between a point on the beam when unloaded and that same point when the beam is loaded (Figure 16.1c).

Deflection is an important design concern in buildings; a beam may have sufficient strength but deflects severely under the same loading condition. A diving board safely supports the weight of a person who slowly walks across it. It is, however, very springy and deflects significantly as a person walks to its end. A diving board would be unsuitable as a structural member in a building because it deflects excessively under load. A beam section profile and material may be suitable to resist anticipated shear and bending stresses, but deflect excessively.

In buildings, moderate deflection may result in the uncomfortable feeling of the structure not being stiff enough (i.e., bouncy floors). A floor that is too bouncy may cause dishes to clatter and the occupant to feel unsafe. Excessive deflection will create cracks in drywall ceiling finishes that could result in an unsafe condition as the ceiling material loosens and collapses with movement. Vertical structural members such as framing (studs) carrying a wind load must also be investigated for deflection. Wall cladding material often dictates the maximum deflection permitted. For this reason, prudent design practice and the building code limit the amount of deflection a beam may experience under design conditions.

16.7 DEFLECTION INVESTIGATION

A designer must investigate deflection to ascertain that a beam does not deflect beyond acceptable limits. Deflection varies across the length of the beam. The designer is typically most interested in the maximum amount of deflection. *Maximum deflection* (Δ_{max}) is the largest measured or anticipated deflection across the entire beam length when the member is under a design loading condition. A deflection investigation involves comparing maximum deflection under design conditions to the *maximum deflection permitted* (D_{code}). In design of beams, maximum deflection (Δ_{max}) must never exceed the maximum deflection permitted (D_{code}):

$$D_{code} \geq \Delta_{max}$$

Maximum deflection permitted is generally expressed as a fraction of the span length (l) in inches, such as $l/360$ or $l/240$. The design engineer may establish a deflection tolerance simply based on intuition and experience or by relying on specific code requirements. The most common deflection limitations required by building codes are:

Floors:	$l/240$ for total load	$l/360$ for live loads
Roofs:	$l/180$ for total load	$l/240$ for live loads

Long-term deflection under sustained loading (creep) must also be investigated, especially for some horizontal structural members, such as wide garage door headers and long beams. In such cases, more restrictive deflection limits than those noted above are recommended.

■ EXAMPLE 16.6

A floor beam is estimated to deflect 0.412 in under design conditions. Building code limits deflection to $l/360$ of span. Assuming a span of 15 ft, investigate deflection.

Solution

Converting beam length from feet to inches: $l = 15 \text{ ft} \cdot 12 \text{ in/ft} = 180 \text{ in}$

For maximum deflection permitted: $D_{code} = \dfrac{l}{360} = \dfrac{180 \text{ in}}{360} = 0.500 \text{ in}$

Maximum deflection (provided): $\Delta_{max} = 0.412 \text{ in}$

$$0.412 \, (\Delta_{max}) \leq 0.500 \, (D_{code})$$

Since Δ_{max} does not exceed D_{code}, the beam meets the deflection requirement established by code.

The beam in Example 16.6 passes the deflection investigation because maximum deflection under design conditions (Δ_{max}) is less than or equal to maximum deflection permitted (D_{code}). In the event that maximum deflection would exceed the deflection permitted, the beam would fail the deflection investigation. The designer would have to select and investigate a stiffer beam section (a larger I) or material (a larger E).

Maximum deflection under simple loading conditions may be estimated using the formulas available for specific loading conditions in Table A.5. Additional formulas may be found in technical resources such as the AISC *Manual of Steel Construction*. These were derived using higher-level mathematics. They relate deflection to load, loading condition, and material and cross-sectional stiffness. The actual location of maximum deflection on the beam length is dependent on loading condition; it does not always occur at the beam center.

■ EXAMPLE 16.7

A repetitive wood floor joist with a span of 16 ft will carry a uniformly distributed load of 75 lb/ft. A 2 in × 10 in beam section passes the flexure and shear investigation. The beam materials have properties of $E = 1\,600\,000$ lb/in^2. Building code limits deflection to $l/360$ of span. Investigate deflection.

Solution   Since the beam is carrying a uniform load, the following formula from Table A.5 applies. The moment of inertia of a 2 in × 10 in joist section is found in Table 16.1 to be 98.93 in^4. Maximum deflection of a 2 in × 10 in joist section under design conditions (Δ_{max}) is found by

$$\Delta_{max} = \frac{5Wl^3}{384EI}$$

$$= \frac{5(75 \text{ lb/ft} \cdot 16 \text{ ft})(16 \text{ ft} \cdot 12 \text{ in/ft})^3}{384 \cdot 1\ 600\ 000 \text{ lb/in}^2 \cdot 98.93 \text{ in}^4}$$

$$= 0.699 \text{ in.}$$

Maximum deflection permitted (D_{code}) is $l/360$. It is found by

$$D_{code} = \frac{l}{360} = \frac{16 \text{ ft} \cdot 12 \text{ in/ft}}{360} = 0.533 \text{ in}$$

Since Δ_{max} is greater than D_{code}, the 2 in $\times$ 10 in joist section fails the deflection investigation for this loading condition. Maximum deflection under design conditions is too great.

The 2 in $\times$ 10 in joist section in Example 16.7 fails the deflection investigation. The designer would need to select a larger section (greater I) or choose a stiffer material (greater E) to ensure that deflection is not exceeded. In Example 16.8, a 2 in $\times$ 12 in section is considered and passes the deflection investigation.

■ EXAMPLE 16.8

With failure by deflection of the 2 in $\times$ 10 in joist section in Example 16.7, the designer may choose to substitute a larger joist section. For the condition described in Example 16.7, investigate deflection of the 2 in $\times$ 12 in joist section:

Solution The moment of inertia of a 2 in $\times$ 12 in joist section is found in Table 16.1 to be 177.98 in^4. Maximum deflection under design conditions (Δ_{max}) of a 2 in $\times$ 12 in joist section is found by

$$\Delta_{max} = \frac{5Wl^3}{384EI}$$

$$= \frac{5(75 \text{ lb/ft} \cdot 16 \text{ ft})(16 \text{ ft} \cdot 12 \text{ in/ft})^3}{384 \cdot 1\ 600\ 000 \text{ lb/in}^2 \cdot 177.98 \text{ in}^4}$$

$$= 0.388 \text{ in.}$$

Maximum deflection permitted (D_{code}) remains at 0.533 in because the span remains the same. Since Δ_{max} does not exceed D_{code}, the 2 in $\times$ 12 in joist section passes the deflection investigation.

The 2 in $\times$ 12 in joist section passes the deflection investigation for the conditions described in Example 16.8; maximum deflection under design conditions does not exceed maximum deflection permitted. An increase from a 10-in to a 12-in joist depth limited maximum deflection to an acceptable level.

A designer usually cannot alter a loading condition (the Wl in Example 16.8) without changing the layout and design of the structure. The designer can, however, select the moment of inertia (I) of the section and the modulus of elasticity (E) of the beam material. The product IE is a measure of relative stiffness of a beam; a greater IE provides a stiffer beam. By increasing IE, the designer decreases maximum deflection under design conditions.

16.8 BEARING STRESSES

Bearing stress is a type of compressive stress that affects the external surface of a body when it contacts another body, as shown in Figure 16.1d. Bearing stress develops when the beam material experiences a crushing effect where loads and reactions contact the beam. It is a contact stress between two bodies or members.

16.9 BEARING STRESS INVESTIGATION

The designer must ascertain that the beam material at the bearing points does not become crushed excessively. Typically, this analysis is accomplished by evaluating bearing stresses ($\sigma_{bearing}$) at the areas where loads (F) and reactions (R) contact the beam and ensuring that these stresses do not exceed allowable design values. The contact area (A) under consideration in bearing stress analysis is that area where the force is transferred from one body to the next, usually where both bodies are in contact with one another. The basic bearing stress formula is

$$\sigma_{bearing} = \frac{F}{A} = \frac{R}{A}$$

■ EXAMPLE 16.9

A W8×15 steel beam supports three 7-kip loads equally spaced at fourth points of a simple span. The loads are transferred to the steel beam by laminated veneer lumber (LVL) wood members that directly contact the upper surface of the steel beam flange. The flange breadth (width) of a W8×15 steel section is 4.00 in wide. The ends of the beam extend 8.00 in into and rest on solid concrete walls.

(a) Determine the bearing stress that exists at the beam reactions and ascertain that it does not exceed the allowable bearing stresses of 24 000 lb/in² for steel and 400 lb/in² for concrete.

(b) Determine the minimum required contact thickness of each LVL member if they are resting across the entire 4.00-in-wide flange of the steel member. Allowable bearing stress for the LVL member is 2460 lb/in².

(c) Investigate bearing stress in the single $1\frac{3}{4}$-in-wide LVL member that is transferring the 7-kip load to the steel beam. Allowable bearing stress for the LVL member is 2460 lb/in².

Solution

(a) The reactions are 10.5 kip each: $\dfrac{3 \text{ loads} \cdot 7 \text{ kip}}{2} = 10.5 \text{ kip}$

The contact area at each reaction is 32.00 in²: 4.00 in · 8.00 in = 32.00 in.
The bearing stress is found by

$$\sigma_{\text{bearing}} = \frac{R}{A} = \frac{10.5 \text{ kip}}{32.00 \text{ in}^2} = 0.328 \text{ kip/in}^2 = 328 \text{ lb/in}^2$$

The actual bearing stress at the contact point of steel and concrete members is compared to the allowable bearing stresses of each material. The actual bearing stress at each reaction is 328 lb/in². It is acceptable because it does not exceed the allowable bearing stresses of either material.

(b)

$$\sigma_{\text{bearing}} = \frac{F}{A}$$

$$\geq \frac{F}{A}$$

$$A \geq \frac{F}{\sigma_{\text{bearing}}}$$

The contact area (A) of the two members is the product of beam width (4.00 in) multiplied by minimum contact thickness (t) of the LVL member; $A = (4.00 \text{ in})t$:

$$(4.00 \text{ in})t \geq \frac{F}{\sigma_{\text{bearing}}}$$

$$t \geq \frac{F}{\sigma_{\text{bearing}} \cdot 4.00 \text{ in}}$$

$$\geq \frac{7000 \text{ lb}}{2460 \text{ lb/in}^2 \cdot 4.00 \text{ in}}$$

$$\geq 0.71 \text{ in}$$

The minimum contact thickness (t) of the LVL member should be at least 0.71 in.

(c) The bearing stress is

$$\sigma_{\text{bearing}} = \frac{F}{A} = \frac{7000 \text{ lb}}{1\frac{3}{4} \text{ in} \cdot 4.00 \text{ in}} = 1000 \text{ lb/in}^2$$

Since the contact width of an LVL member is $1\frac{3}{4}$ in wide, the LVL member can transfer its load to the steel beam without exceeding allowable bearing stresses.

If a beam is not stiff enough to resist lateral forces created when it bends and it is not braced laterally, it will buckle to its side (Figure 16.1e). Under load, the fibers of the beam that develop compressive stresses (the top fibers of a simple beam) tend to buckle laterally like a column. This resulting buckling effect is referred to as *lateral buckling*. The lateral buckling phenomenon is best demonstrated by placing a yard or meter stick on edge while it spans across two tables or chairs and applying a load downward at its center; under load the top edge (edge under compression) tends to buckle.

To minimize lateral buckling, deep, narrow beams are typically secured to a rigid deck and/or secured with bridging. Properly secured plywood floor panels adequately keep joists in a floor deck system from buckling laterally. In a system of open web joists, decking is typically inadequate in resisting lateral buckling. Horizontal and sometimes diagonal bridging is required in an open web joist system.

Eccentric or misaligned loads may cause the beam to twist across a portion of its length, thereby causing a *torsional buckling* effect (Figure 16.1f). Again, the beam is typically secured to a rigid deck and/or secured with horizontal or diagonal bridging that minimizes twisting caused by torsion of the beam. If the entire beam tends to roll (but not twist or buckle) when it is subjected to a lateral force, it experiences *rotation* (Figure 16.1g). The tendency of rotation is usually counteracted with adequate lateral blocking at the beam ends and/or bridging in the center or at center points across the beam length. The joists in a floor deck system are kept from rotating by securing their ends in a *joist header*, a board that runs perpendicular to the joists.

The dynamics of loading also influence a beam's tendency to fail by bending or shearing. A rapid increase in load, such as a weight being dropped on the beam, can cause rapid failure. A rapid loading condition is referred to as *impact loading*. In comparison, a slow, gradual load buildup typically results in the beam supporting a greater load before eventually bending and failing. Impact loads are frequently treated by applying an adjustment factor known as the *equivalent static load factor*, as discussed in Chapter 15.

Flexure

16.1 A beam experiences a maximum bending moment of 32 000 lb-ft. Use the flexure formula to determine the required section modulus of the beam section for the following materials. Maximum allowable bending stress for each material is shown in parentheses.
   (a) Steel (24 000 lb/in^2)
   (b) Aluminum (16 000 lb/in^2)
   (c) Wood (1200 lb/in^2)

16.2 A 4-m-long simple beam supports a 40-kN load at its center. Determine the required section modulus of the beam section for the following materials. Maximum allowable bending stress for each material is shown in parentheses.
   (a) Steel (150 MPa)
   (b) Aluminum (100 MPa)
   (c) Wood (4.0 MPa)

16.3 A 16-ft-long beam carries a uniformly distributed load of 260 lb/ft across its entire length. Reaction R_1 is a pinned support, and reaction R_2 is a bearing support. Both reactions are equal to 2080 lb. The beam has a section with a constant breadth of $3\frac{1}{2}$ in. The beam consists of a material with a maximum allowable bending stress of 1500 lb/in^2.
   (a) Calculate bending moment at 1-ft intervals across the beam length (from 0 to 16 ft).
   (b) Use the flexure formula to determine the required section modulus of the beam section at 1-ft intervals across the beam length.
   (c) From Chapter 14, a section modulus of a rectangular section is related to breadth (b) and depth (d) of the section by the formula $S_{rectangular} = bd^2/6$. Determine the required depth of the section at 1-ft intervals across the beam length.
   (d) Draw a plot of required depth of the section versus beam length.

16.4 Free-body, shear, and bending moment diagrams are shown. Maximum allowable bending stress for the wood species and grade selected for the wood beam is 1200 lb/in^2. Determine the minimum required section modulus of the beam section at the following locations:
   (a) At reaction R_1
   (b) Below the 5000-lb load
   (c) At reaction R_2

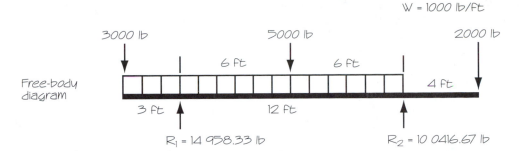

Free-body diagram

3000 lb   5000 lb   W = 1000 lb/ft   2000 lb

6 ft   6 ft   4 ft

3 ft   12 ft

R_1 = 14 958.33 lb R_2 = 10 0416.67 lb

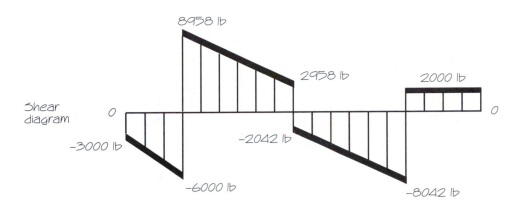

Shear diagram

8958 lb

2958 lb   2000 lb

0   0

−3000 lb   −2042 lb

−6000 lb   −8042 lb

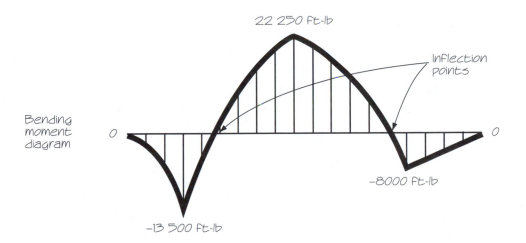

Bending moment diagram

22 250 ft-lb

Inflection points

0   0

−8000 ft-lb

−13 500 ft-lb

16.5 Free-body, shear, and bending moment diagrams are shown. Maximum allowable bending stress for the steel beam is 24 000 lb/in². Determine the minimum required section modulus of the beam section at the following locations:

(a) At reaction R_1
(b) 1.5 ft to the right of R_1
(c) At reaction R_2

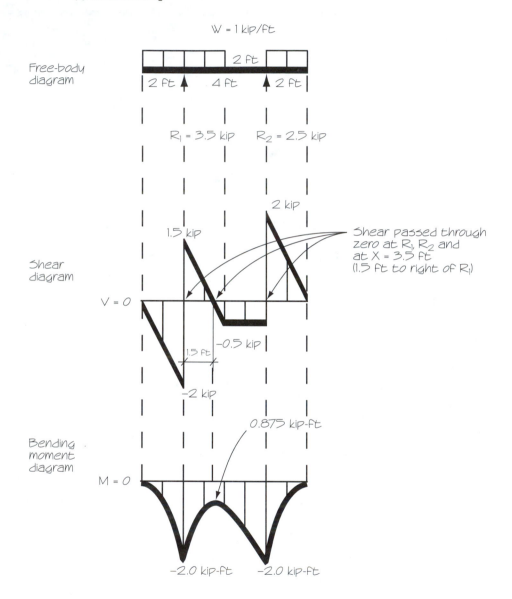

Shear

16.6 Calculate the maximum shear stress in a 50 mm × 300 mm rectangular beam section. The shear force at that location in the beam is 1.6 kN.

16.7 Calculate the web shear stress in a steel beam section with a web thickness of 12.5 mm and a depth of 325.0 mm. The shear force at that location in the beam is 62.5 kN.

16.8 A 16-ft-long beam carries a uniformly distributed load of 260 lb/ft across its entire length like the beam in Exercise 16.2. Reaction R_1 is a pinned support and reaction R_2 is a bearing support. Both reactions are equal to 2080 lb. The beam has a section with a constant breadth of $3\frac{1}{2}$ in. The beam consists of a material with a maximum allowable shear stress of 90 lb/in^2.

    (a) Calculate the shear force (V) at 1-ft intervals across the beam length (from 0 ft to 16 ft).

    (b) Shear stress ($\sigma_{\text{max-shear}}$) of a rectangular section is related to breadth (b) and depth (d) of the section by the formula $\sigma_{\text{max-shear}} = 3V/2bd$. Determine the required depth of the section at 1-ft intervals across the beam length.

    (c) Draw a plot of required depth of the section versus beam length.

Deflection

16.9 A 20-ft-long beam is estimated to exhibit a maximum deflection of 0.62 in under design conditions. Code limits deflection to $l/360$ of span. Investigate deflection to determine whether maximum deflection is acceptable.

16.10 A 3.5-m-long beam is estimated to exhibit a maximum deflection of 10.2 mm under design conditions. Code limits deflection to $l/360$ of span. Investigate deflection to determine whether maximum deflection is acceptable.

16.11 A wood beam with a span of 16 ft is carrying a uniformly distributed load of 800 lb/ft across its entire length. A 6 in × 14 in beam section passes the flexure and shear investigation. The beam section and material have properties of $I = 1\ 127.67$ in^4 and $E = 1\ 400\ 000$ lb/in^2. Code limits deflection to $l/360$ of span. Investigate deflection to determine whether maximum deflection is acceptable.

Bearing Stresses

16.12 A 2 × 12 floor joist spans 16'-6" and rests on its $1\frac{1}{2}$ in edge. It carries a uniformly distributed load of 67 lb/ft across its entire length.

    (a) One end of the joist rests on the $3\frac{1}{2}$-in surface of a 2 × 4 plate. Determine the bearing stress that develops on the plate.

    (b) The other end of the joist is attached to a joist header and this assembly rests on a sill plate. The assembly is constructed so that only 2 in of the joist end is bearing on the sill plate. Determine the bearing stress that develops on the sill plate.

16.13 A 150 mm × 150 mm steel plate at the end of a steel joist rests on a smooth concrete surface. The plate transfers a load of 50 kN to the concrete. Determine the theoretical bearing stress.

Chapter 17
FUNDAMENTAL COLUMN DESIGN PRINCIPLES

A *column* is a structural member that resists compression forces at its ends. Since the load carried by a column tends to compress the column material, columns are referred to as compression members. In building construction, posts, piers, pillars, and studs typically behave like columns. A simple column must be investigated for

A column is a vertically positioned structural member that supports an axial load.

failure by buckling and crushing. The theory behind these modes of failure is explored in this chapter. Presentation of basic design procedures for steel, wood, and concrete columns is reserved for later chapters. Again, the reader is advised that the following presentation does not provide full coverage of all principles involved with design of building structures. Such coverage would require several texts and references and is beyond the intended breadth of this book.

17.1   COLUMN CRUSHING

A compressive stress is produced when the molecular structure of the material resists being compressed as a force is applied. *Crushing* is failure of the material due to excessive compressive stress. Crushing of a material results when the molecular structure of the material can no longer resist a compressive stress; the molecular structure collapses when the molecular bond holding the molecules fails and the molecules slide past one another.

A *perfect column* is a theoretical column that is completely straight and axially loaded (the line of action of the load passes through the longitudinal center of the column). It is made from a homogeneous material and has a uniform cross section without flaws. Under excessive loading, a perfect column would fail only by crushing of the column material and not by bowing or bending sideways. Realistically, all columns have flaws in material and inconsistencies in cross section. Additionally, columns are rarely, if ever, perfectly axially loaded. The perfect column described above is hypothetical and does not exist. However, a short column that is axially loaded tends to act like a perfect column; it does not tend to bow or bend sideways under excessive load. It fails by crushing only.

In theory, the maximum theoretical load a *short* column can carry without failure is the force at which the ultimate compressive strength (compressive stress at failure) of the material is produced. Stress (σ) is defined as the external force (F) acting on the body divided by the cross-sectional area (A) of the body: $\sigma = F/A$. So, the maximum theoretical load a short column can carry without true compression failure is found by algebraically rearranging terms in the $\sigma = F/A$ formula to

$$F = (\sigma_{\text{ultimate strength}})A$$

Stressing a column material to its ultimate compressive strength will result in substantial plastic deformation of the column length. To limit deformation to the elastic range, the stress used in the analysis of a short column could be the yield stress of the material:

$$F = (\sigma_{\text{yield strength}})A$$

In allowable stress design, the allowable stress (the working stress) of a column material should be used to determine the maximum safe load of a short column:

$$F = (\sigma_{\text{allowable strength}})A$$

■ EXAMPLE 17.1

A steel W10×60 section is used as a very short, axially loaded column. The steel used in the column has an ultimate compressive strength of 76 000 lb/in² (kips/in²), a yield stress of 36 000 lb/in² , and an allowable stress of 21 600 lb/in². The cross-sectional area of the W10×60 section is 17.6 in².
(a) Determine the ultimate load the column can support.
(b) Determine the maximum load the column can support with only elastic deformation.
(c) Determine the maximum allowable load the column can support based on allowable stress design.

Solution

$$(a) \ \ F = (\sigma_{\text{ultimate strength}})A = 76\ 000 \ \text{lb/in}^2 \cdot 17.6 \ \text{in}^2 = 1338 \ \text{kip}$$
$$(b) \ \ F = (\sigma_{\text{yield strength}})A = 36\ 000 \ \text{lb/in}^2 \cdot 17.6 \ \text{in}^2 = 634 \ \text{kip}$$
$$(c) \ \ F = (\sigma_{\text{allowable strength}})A = 21\ 600 \ \text{lb/in}^2 \cdot 17.6 \ \text{in}^2 = 380 \ \text{kip}$$

The W10×60 column in Example 17.1 will fail by crushing if it is subjected to an axial end load above 1338 kip. In allowable stress design, a steel W10×60 section used as a very short column can safely carry a load of 380 kip.

■ EXAMPLE 17.2

A very short, unreinforced concrete column is axially loaded. The circular column section has a diameter of 150 mm. The concrete has an ultimate compressive strength of 27.5 MN/m². Determine the ultimate load the column can support before it fails by crushing.

Solution

$$27.5 \ \text{MN/m}^2 = 27\ 500\ 000 \ \text{N/m}^2 \quad \text{and} \quad 150 \ \text{mm} = 0.150 \ \text{m}$$
$$A = \frac{\pi D^2}{4} = \frac{\pi (0.150 \ \text{m})^2}{4} = 0.0177 \ \text{m}^2$$
$$F = (\sigma_{\text{ultimate strength}})A = 27\ 500\ 000 \ \text{N/m}^2 \cdot 0.0177 \ \text{m}^2 = 486\ 750 \ \text{N}$$
$$= 487 \ \text{kN}$$

For a short, axially loaded, composite column such as a steel-reinforced concrete column, the theoretical load before compression failure is found by

$$F_{\text{composite}} = (\sigma A)_a + (\sigma A)_b$$

where $(\sigma A)_a$ is the product of the cross-sectional area and permissible stress of material a, $(\sigma A)_b$ is the product of the cross-sectional area and permissible stress of material b, and $F_{\text{composite}}$ is the maximum theoretical axial load before compression failure. Although this theoretical composite column formula appears logical, it is significantly limited in application. This limitation is due to the combined effect of differences in the properties of strength and elasticity of the materials used. These limitations are introduced below.

In Chapter 7 it was found that the modulus of elasticity (E) of a material is a relationship of stress (σ) and longitudinal strain (ϵ) below the proportional limit: $E = \sigma/\epsilon$. By rearranging terms algebraically, the relationship $\sigma = \epsilon E$ is found. In the form of $\sigma = \epsilon E$, it is evident that for a specific strain, the stress that develops is dependent on the modulus of elasticity of the material. Simply, under a specific strain, a material with a large modulus of elasticity develops a larger stress in comparison to a material having a small modulus of elasticity.

■ EXAMPLE 17.3

A very short, 10 in × 10 in concrete column is axially loaded. The concrete has an ultimate compressive strength of 3000 lb/in^2.

(a) Determine the ultimate load the unreinforced concrete column can theoretically support before it fails by crushing.

(b) Assume that the column is reinforced with four No. 9 reinforcing bars (cross-sectional area = 1.00 in^2 per bar) placed vertically. Steel has an ultimate strength of 80 000 lb/in^2. Determine the ultimate load the reinforced concrete column can theoretically support before it fails by crushing.

Solution

(a) $\quad A = 10 \text{ in} \cdot 10 \text{ in} = 100 \text{ in}^2$

$\quad\quad F = (\sigma_{\text{ultimate strength}})A = 3000 \text{ lb/in}^2 \cdot 100 \text{ in}^2 = 300\,000 \text{ lb}$

(b) $\quad A_{\text{steel}} = 4 \text{ bars} \cdot 1.00 \text{ in}^2/\text{bar} = 4.00 \text{ in}^2$

$\quad\quad A_{\text{concrete}} = 100 \text{ in}^2 - 4.00 \text{ in}^2 = 96 \text{ in}^2$

$\quad F_{\text{composite}} = (\sigma A)_{\text{steel}} + (\sigma A)_{\text{concrete}}$

$\quad\quad\quad = (80\,000 \text{ lb/in}^2 \cdot 4.00 \text{ in}^2) + (3000 \text{ lb/in}^2 \cdot 96 \text{ in}^2) = 608\,000 \text{ lb}$

First, consider a composite column comprised of materials having similar ultimate strengths but significantly different moduli of elasticity: As the composite column deforms under load, longitudinal deformation (ΔL) and longitudinal strain (ϵ) of the different materials are identical. Therefore, by the $\sigma = \epsilon E$ relationship, the column material having the larger modulus of elasticity develops larger stresses than the material having the smaller modulus of elasticity. Under increasing load, the stiffer composite column material develops greater stresses and eventually fails before the material having a smaller modulus of elasticity if the materials have similar ultimate strengths. The load the column can support is limited to the properties of the stiffer material if the materials have similar ultimate strengths. On the other hand, a stiffer material with a higher ultimate strength could develop higher stresses without failure; the composite column could carry a greater load without failure.

Materials used in composite columns have significantly different moduli of elasticity and ultimate strengths. With the proper match of materials (i.e., steel and concrete), composite column strength more closely represents the theoretical composite column formula. In practice, the theoretical composite column formula is modified with multipliers so that it can account for differences in material properties and behavior under load.

17.2 COLUMN BUCKLING

Columns, especially long columns, can fail laterally due to a phenomenon called buckling. *Buckling* is a sudden, lateral deflection of a column caused by an excessive end load. Standing a yard or meter stick upright and pushing down on its end with increasing force eventually produces buckling in the yard or meter stick. Continued end loading when buckling is present produces buckling failure as the yard or meter stick buckles excessively and breaks.

Buckling is due to a combination of excessive end loading, an instability in elastic behavior of the material, and flaws in the column cross section. If a column is unloaded immediately when buckling first becomes evident, it will realign itself elastically. However, if loading continues after buckling is evident, only lighter end loads are required to continue the buckling effect. So, once a column buckles, it can no longer carry a substantial load. In a building structure, a column generally fails once it has buckled. The maximum load that a long column can carry is a load slightly less than the load at which the column just begins to buckle.

17.3 EULER'S COLUMN FORMULA

In 1757, a Swiss mathematician named Leonard Euler (pronounced "oiler") developed what has become known as *Euler's column formula:*

$$\text{Euler's column formula, } \frac{F}{A} = \frac{\pi^2 E}{(l/r)^2}$$

Euler related the modulus of elasticity of the column material (E), the column length (l), and the radius of gyration of the column cross section (r) to the stress (F/A) that a column with ends free to rotate can support without buckling. His formula applies only to columns that are stressed below the elastic limit of the column material.

Derivation and application of Euler's column formula is complex and beyond the intended breadth of this book. However, by reviewing Euler's column formula it can be observed that for a specific column section and material (i.e., a constant E and r and thus A) an increase in column length results in a decrease in the load

(F) that the column can carry without buckling; that is, a long column fails more easily by buckling. The formulas used to determine the safe load a long slender column can carry without buckling are generally founded on Euler's work.

■ EXAMPLE 17.4

A 150 mm × 150 mm wood column has a radius of gyration (r) of 43.3 mm. The wood species has a modulus of elasticity of 11 000 MPa (11 000 MN/m^2). The column has ends that are free to rotate.

(a) At a column length of 3 m, compute the maximum stress the wood column can theoretically develop and the maximum load the column can support without failure by buckling.

(b) At a column length of 4 m, compute the maximum stress the wood column can theoretically develop and the maximum load the column can support without failure by buckling.

Solution (a) The stress (F/A) that a column with ends free to rotate can support without buckling is found by Euler's column formula:

$$\frac{F}{A} = \frac{\pi^2 E}{(l/r)^2} = \frac{\pi^2(11\ 000\ \text{MN/m}^2)}{(3\ \text{m}/0.0433\ \text{m})^2} = 22.6\ \text{MN/m}^2$$

The maximum load the column can support is found by rearranging terms in the $\sigma = F/A$ formula and substituting F/A for σ:

$$F = \sigma A = \frac{F}{A}A = 22.6\ \text{MN/m}^2(0.150\ \text{m} \cdot 0.150\ \text{m}) = 0.51\ \text{MN} = 510\ \text{kN}$$

(b)
$$\frac{F}{A} = \frac{\pi^2 E}{(l/r)^2} = \frac{\pi^2(11\ 000\ \text{MN/m}^2)}{(4\ \text{m}/0.0433\ \text{m})^2} = 12.7\ \text{MN/m}^2$$

$$F = \sigma A = \frac{F}{A}A = 12.7\ \text{MN/m}^2(0.150\ \text{m} \cdot 0.150\ \text{m}) = 0.29\ \text{MN} = 290\ \text{kN}$$

As demonstrated by application of Euler's column formula in Example 17.4, a longer column cannot support as great a load as a shorter column. A longer column fails at a lighter load.

17.4 ECCENTRIC LOADING AND BENDING OF A COLUMN

Gravity loads transferred to a column produce an end load on the column that imposes a downward-acting force through the column. When the line of action of the force coincides with the intersection of the x and y axes of the column section, the column is said to be axially or axially loaded. An *axial* or *concentric load* exists when the end load produces a force that acts through the longitudinal center of the column. Such loads are frequently referred to as *centroidally applied axial loads* because the

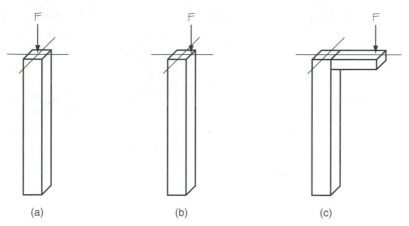

FIGURE 17.1 Loads on a column can be axial or concentric (a) or eccentric. Eccentric loading can be caused by an off balance load (b) or a bending moment induced on the column (c).

load produces a force that has a line of action that passes through the centroid of the member section and parallels the length of the member (Figure 17.1a).

On the other hand, an *eccentric load* is produced when the line of action of the force produced by the column end load does not pass through the centroid of the member section or does not parallel the longitudinal axis of the column (Figure 17.1b). Eccentric loading can be caused by an off-balance load or a bending moment induced on the column, as shown in Figure 17.1c. An eccentric load produces bending stresses in the column in addition to compressive stresses. Generally, bending stress increases the tendency to buckle; the column will buckle under a lighter end load than if the column was only axially loaded. In unique instances, eccentric loading can also counter the buckling effect.

A column or columnlike structural member (i.e., wall stud, wall, etc.) can also be subjected to a lateral force that causes it to deflect. Wind and soil produce a lateral force on a wall. A force acting laterally tends to increase the tendency of the column to buckle. Since a column generally fails once it has buckled, deflection from a lateral load has an influence on the maximum end load a column can carry.

17.5 COLUMN SLENDERNESS

At first glance it may appear that the primary tendency of a column to collapse by either buckling or crushing is dependent only upon the length of the column. However, length alone does not determine the governing mode of failure of a column. Instead, column slenderness is one of the determining factors in identifying the potential mode of failure. Column *slenderness* is the relationship between a column's

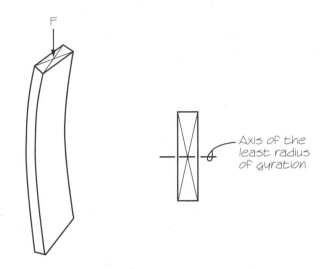

FIGURE 17.2 A rectangular column buckles in the direction of its least lateral dimension, unless laterally braced.

Axis of the least radius of gyration

unsupported length and its lateral or section properties. Columns are defined as being long or short based on their slenderness and not specifically on length alone. *Short columns* have a tendency to fail due predominately to crushing of the material. On the other hand, *long columns* have a tendency to fail primarily by buckling. Classification of a column as being long or short is determined by slenderness.

For example, a 2 × 12 member has about the same cross-sectional area as does a 4 × 6 member. However, under a gradually increasing end load, the 2 × 12 member will buckle and fail well before a 4 × 6 member of the same length. Furthermore, depending on column length, it is possible that the 2 × 12 member will buckle and fail while the 4 × 6 member may not buckle but ultimately fail by compression (crushing). Classification of column *slenderness* is the relationship between a column's unsupported length and its lateral or section properties.

17.6   COLUMN-SECTION PROFILES

All unbraced axially loaded long columns with rectangular section profiles have the greatest tendency to buckle in the direction of their least dimension, as shown in Figure 17.2. Consider a long 2 × 12 member supporting a gradually increasing load; the member will ultimately buckle such that the 2-in dimension gives way. It would fail in the direction of its least dimension.

Metal section profiles such as W and HP shapes are more intricate than a simple rectangular section. For these types of sections, a measure of stiffness is the radius of gyration. As presented in Chapter 14, the radius of gyration is a section property that serves as a measure of the section capacity to resist buckling independent of the material's elastic behavior. A column section with a high value of radius of gyration in all directions is more resistant to buckling than is a section with a low radius of gyration.

A yard or meter stick that stands vertically on the ground while a load is applied to the top end buckles in the direction of its least radius of gyration, the least dimension of the stick cross section. A long, slender compression member with an irregular section profile (i.e., I, H, C, and T shapes) has a tendency to buckle in the direction of the least radius of gyration (r) of the section. For example, a W24×62 is deep ($d = 23\frac{3}{4}$ in) and relatively narrow ($b = 7$ in) in section profile; its radii of gyration are 9.23 in about the x axis and 1.38 in about the y axis. This section profile will tend to buckle more easily about the y axis.

The perfect column section has a section profile where the radii of gyration are equal in all directions. Thus circular column sections such as round tubing or pipe serve as the most effective section. Circular column sections are, however, difficult to secure to the flat surfaces of other types of structural members and are used only when connections can be made easily. As a result, symmetrical or near-symmetrical column sections, such as square, nearly square rectangular, and W- and HP-shapes, are best suited for columns. Their ability to resist buckling is equal or nearly equal in both the x and y axes (Figure 17.3). A fabricated column with a built-up section of a W- or HP-shape and thick plates having a section with radii of gyration that are nearly equal work well in carrying the heavy loads of skyscrapers and bridges.

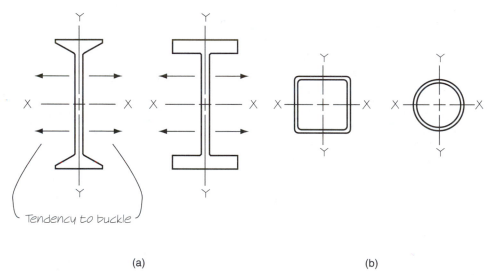

(a)   (b)

FIGURE 17.3 (a) Columns have the greatest tendency to buckle in the direction of the least radius of gyration of the column section. (b) Symmetrical sections, such as square and circular sections, are best suited for columns, because their ability to resist buckling is equal in all axes or directions.

■ EXAMPLE 17.5

A W14×26 steel section has radii of gyration of 5.54 in based on a neutral axis that is perpendicular to the section web (the x axis in Figure 17.3) and 1.04 in based on a neutral axis that is parallel to the web (the y axis in Figure 17.3). Determine in which direction the section will have the greatest tendency to fail by buckling.

Solution A ratio of the radii of gyration for the W14×26 section is found by

$$\frac{5.54 \text{ in}}{1.04 \text{ in}} = 5.33$$

The W14×26 section is 5.33 times stiffer against buckling about the x axis, as shown in Figure 17.3. Therefore, the section will have the greatest tendency to fail by buckling in a direction that is perpendicular to the section web (perpendicular to the y axis, as shown in Figure 17.3).

17.7   COLUMN END CONNECTIONS

The types of connections that secure the column ends in place affect its tendency to buckle. For example, a column with fully restrained ends, such as ends set substantially in concrete, will buckle less easily than a column that has ends that are free to rotate (Figure 17.4). Similarly, a column that is braced at its center tends to

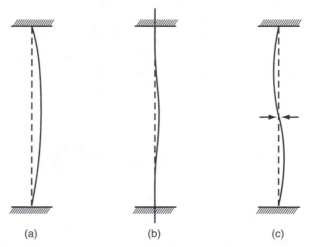

(a)   (b)   (c)

FIGURE 17.4   Column end connections affect the column's tendency to buckle. A column that has ends that are free to rotate (a) will buckle more easily than a column that has fixed ends (b). Column bracing tends to reduce the tendency to buckle (c).

buckle less easily than does a column that is left unbraced. The manner in which a column is secured greatly affects its capability to carry a load.

17.8 DYNAMICS OF COLUMN LOADING

Like beams, the impact loading of a column influences its tendency to fail by compression or buckling, especially with slender columns. A quick increase in load, such as a weight being dropped on the column end, tends to cause rapid failure. In comparison, a slow, gradual end-load buildup typically results in the column supporting a greater load before eventually failing.

17.9 COLUMN DESIGN

Columns that are loaded axially generally fail by a combination of buckling or crushing. Buckling and crushing are two very different actions. Crushing relates to strength against compressive forces. Buckling relates to the elastic resistance of the column material (i.e., modulus of elasticity) and column section (i.e., radius of gyration) against lateral deflection. Only excessively long columns or very short columns collapse by a single mode of failure.

Columns rarely carry only axial loads, particularly when they are rigidly connected to beams or girders supporting a roof or floor system. A bending moment induced on a column in addition to its axial end load induces shear stresses and bending stresses in the column material. A column supporting an axial load and a bending moment tends to buckle under a lighter load than does a column that is axially loaded. The combined effect of axial loading and a bending moment is treated with combined-load equations or load-moment interaction graphs that reduce the permissible load that an axially loaded column can safely support.

A column section and material are selected through careful investigation of the potential modes of failure. Investigation typically involves an examination of a combination of crushing, buckling, and bending from applied moments or lateral loads.

EXERCISES

17.1 A very short, 4 × 6 wood column is axially loaded. The column has a section with actual dimensions of $3\frac{1}{2}$ in × $5\frac{1}{2}$ in. The maximum allowable compressive stress the wood species can develop is 600 lb/in². Determine the maximum allowable load the 4 × 6 column can safely support.

17.2 A steel W14×131 section is used as a very short, axially-loaded column. The steel used in the column has an ultimate compressive strength of 76 ksi (kip/in²), a yield stress of

36 ksi, and an allowable stress of 21.6 ksi. The cross-sectional area of the W14×131 section is 38.8 in^2.

(a) Determine the ultimate load the column can support.

(b) Determine the maximum load the column can support with only elastic deformation.

(c) Determine the maximum allowable load the column can support based on allowable stress design.

17.3 A very short 12 in × 12 in concrete column is axially loaded. The concrete has an ultimate compressive strength of 4000 lb/in^2.

(a) Determine the ultimate load the unreinforced concrete column can theoretically support before it fails by crushing.

(b) Assume that the column is reinforced with eight No. 8 reinforcing bars (cross-sectional area = 0.79 in^2 per bar) placed vertically. Steel has an ultimate strength of 80 000 lb/in^2. Determine the ultimate load the reinforced concrete column can theoretically support before it fails by crushing.

17.4 An S24×121 steel section has radii of gyration of 9.43 in based on a neutral axis that is perpendicular to the section web (the x axis in Figure 17.3) and 1.53 in based on a neutral axis that is parallel to the web (the y axis in Figure 17.3). Identify the direction the section will have the greatest tendency to fail by buckling.

17.5 A 2 × 4 wood stud has radii of gyration of 1.01 in and 0.43 in. Identify the direction the section will have the greatest tendency to fail by buckling.

17.6 A 4 × 4 wood column has actual dimensions of $3\frac{1}{2}$ in × $3\frac{1}{2}$ in and a least radius of gyration (r) of 1.01 in. The wood species has a modulus of elasticity of 1 600 000 lb/in^2. The column has ends that are free to rotate.

(a) Using Euler's column formula, compute the stress (F/A) a 4 × 4 wood column can theoretically develop without failure by buckling at intervals of 10, 12, 14, and 16 ft of column length.

(b) Compute the maximum load a 4 × 4 wood column can theoretically develop without failure by buckling at intervals of 10, 12, 14, and 16 ft of column length.

Chapter 18
STRUCTURAL STEEL PRODUCTS

This chapter provides information on structural steel members. It serves as a prelude to the methods used in elementary design of simple structural steel beams and columns and steel joists introduced in Chapters 19 through 21.

18.1   TYPES OF STRUCTURAL STEEL

Steel is made chiefly from iron and between 0.12 and 1.7% carbon. Steel is frequently classified by its carbon content. High-carbon steels are strong, hard, and brittle, while low-carbon steels are softer, not as strong, and more ductile. Steel may also have small amounts of other elements, such as nickel, chromium, copper, and molybdenum. An alloy is a mix of two or more metals. *Alloy steels* contain one or more elements that give the steel special properties; nickel improves strength without increasing brittleness, copper enhances corrosion resistance, chromium improves abrasion and corrosion resistance, hardness, tensile strength, and elasticity, and so on.

Steels used in the construction industry are manufactured with varying chemical and physical properties that influence strength, corrosion resistance, and ability to be bolted, riveted, or welded. Specific grades of steel are designated by an American Society for Testing and Materials (ASTM) specification. Commonly specified structural steels are described in Table 18.1. A basic grade of structural steel that is widely used in building and bridge structures is ASTM A36 steel. A plain-carbon steel, ASTM A36 steel has a minimum yield point (F_y) of 36,000 lb/in^2 and an ultimate tensile strength (F_u) of between 58 000 and 80 000 lb/in^2.

High-strength low-alloy structural steels such as ASTM A572, A242, and A588 have minimum yield points of up to 50 000 lb/in^2. These high-strength steels permit use of steel members that are up to 40% lighter than structural members made from ASTM A36 steel. These grades of steel have been proven economical because of savings associated with the use of shallower members, reduction in dead loads, and shipping cost reduction. *Atmospheric corrosion-resistant steels,* such as ASTM A242 and A588, when used uncoated (bare), can be exposed to normal at-

Table 18.1 STEELS AVAILABLE FOR THE PRODUCTION OF ROLLED STRUCTURAL SHAPES, BARS, AND PLATES

ASTM specification	Steel type	Structural properties
A36	All-purpose, carbon structural steel	A carbon steel that has a minimum yield stress (F_y) of 36 000 lb/in^2 and an ultimate tensile strength (F_u) of between 58 000 and 80 000 lb/in^2. Properties for large sections have an F_y of 32 000 lb/in^2.
A529	High-strength structural steel	A carbon steel that has a minimum yield stress (F_y) of 42 000 lb/in^2 and a minimum ultimate tensile strength (F_u) of between 60 000 and 85 000 lb/in^2.
A441	High-strength low-alloy structural manganese vanadium steel	A low-alloy steel that is most commonly used with a minimum yield stress (F_y) of 50 000 lb/in^2 and a minimum ultimate tensile strength (F_u) of 70 000 lb/in^2 for smaller beam sections. Other properties are available for larger beam sections: F_y of 40 000 lb/in^2 and F_u of 60 000 lb/in^2; F_y of 42 000 lb/in^2 and F_u of 63 000 lb/in^2; and F_y of 46 000 lb/in^2 and F_u of 67 000 lb/in^2.
A572	High-strength low-alloy columbium-vanadium steel of structural quality	A high-strength low-alloy steel that is available in grade 42 with an F_y of 42 000 lb/in^2 and F_u of 60 000 lb/in^2; grade 50 with an F_y of 50 000 lb/in^2 and F_u of 65 000 lb/in^2; grade 60 with an F_y of 60 000 lb/in^2 and F_u of 75 000 lb/in^2; and grade 65 with an F_y of 65 000 lb/in^2 and F_u of 80 000 lb/in^2.
A242	High-strength low-alloy structural steel	A high-strength low-alloy steel that is most commonly used with a minimum yield stress (F_y) of 50 000 lb/in^2 and a minimum ultimate tensile strength (F_u) of 70 000 lb/in^2 for smaller beam sections. Other properties are available for larger beam sections: F_y of 42 000 lb/in^2 and F_u of 63 000 lb/in^2; and F_y of 46 000 lb/in^2 and F_u of 67 000 lb/in^2.
A588	High-strength low-alloy structural steel with a minimum yield point to 4 in thick	A high-strength low-alloy steel that is most commonly used with a minimum yield stress (F_y) of 50 000 lb/in^2 and a minimum ultimate tensile strength (F_u) of 70 000 lb/in^2 for smaller beam sections. Other properties are available for larger beam sections: F_y of 42 000 lb/in^2 and F_u of 63 000 lb/in^2; and F_y of 46 000 lb/in^2 and F_u of 67 000 lb/in^2.

mospheric conditions without excessive corrosion deterioration. Exposure to the atmosphere results in natural development of an oxide layer that forms on the surface of the steel. This oxide layer protects the steel from further corrosion deterioration. This property has proven economical because of a reduction in maintenance costs.

Designers rely on the success of existing and past structures when selecting a steel for a specific application. Selection of a particular type of steel is driven by certain desired characteristics, which include strength, elasticity, corrosion resistance, suitability for bolting or welding, and cost of manufacture and delivery.

18.2 STRUCTURAL STEEL SHAPES

An assortment of structural steel shapes are available, as shown in Figure 18.1. These shapes include W, S, M, and HP shapes, channels (C), miscellaneous channels (MC), structural tees cut from W, M, and S shapes (WT, MT, ST), single and double angles (L), structural tubing (TS), and pipe. Typically, wide-flange (W) and American Standard I beams (S) are used as beams. Structural steel columns are typically made of W, HP, and TS sections, due to the stout shapes available in these sections. Other shapes, such as channels (C), plates (P), angles (L), and tees (WT, ST, and MT), can be used to fabricate a structural member.

Structural steel shapes are produced from large stout steel slabs. In the secondary rolling mill, very hot slabs of steel pass back and forth through rollers that press the steel into long lengths of a specific shape. In addition to producing a homogeneous structural member, the rolling process forces the long molecular chains of steel to become oriented longitudinally in the member, thereby making the steel even better suited for use as a structural member. As a result of the rolling process, structural steel shapes are known as *hot-rolled steel* sections. Common hot-rolled steel shapes and their designations are described below.

- *Wide-flange (W) shapes.* The W shape has thick parallel flanges and a thinner web that form the shape of an H. The thick, wide flanges of the W shape enhance the moment of inertia of the section and make it more stable than other types of beam shapes. It is the most widely used structural section, typically used as beams and columns. All steel shapes are identified by their designation, such as W8×31, S12×35, or HP13×100. For

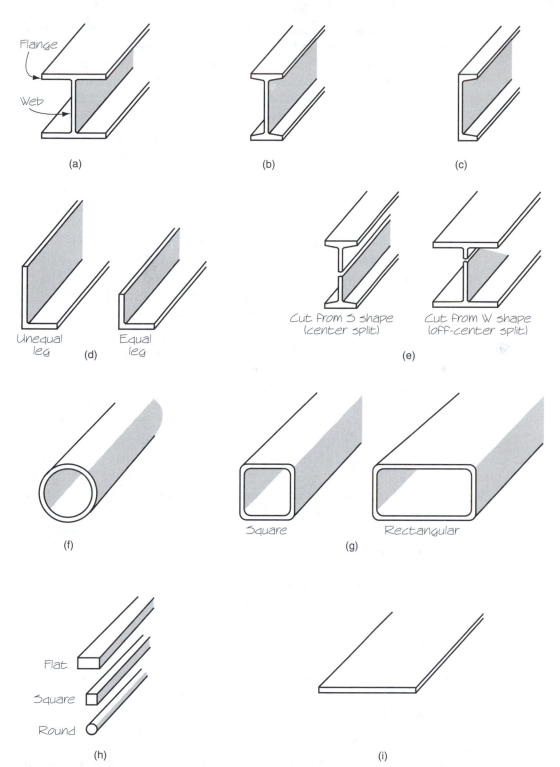

FIGURE 18.1 Structural steel shapes: (a) W shape; (b) S shape; (c) channel; (d) angles; (e) tees; (f) steel pipe; (g) structural tubing; (h) bars; (i) plate.

the W8×31 designation, the W indicates a wide-flange shape or W shape, the first digit, 8, refers to the nominal depth in inches, and the last digit, 31, refers to the approximate weight per foot of length. The digit associated with depth is nominal and may vary substantially from the specified (actual) beam depth, particularly with heavier sections; for example, the specified depth of a W14×211 is 15.72 in.

- *American Standard beam (S) shapes.* The S shape is the original I beam, the first steel section rolled in America. It has tapered flanges and a web that form the shape of an I. The flanges of the S shape are not as wide as those of a W shape. The inner faces of the flanges are sloped. An S12×35 designation is approximately 12 in deep and weighs about 35 lb/ft. The S shape is produced only in limited quantities because it has essentially been replaced by the W shape.

- *Bearing pile (HP) shapes.* The HP shape has flanges and a web of similar thickness that form the shape of an H. The heavy web and flanges of the HP shape make it suitable for foundation piles that are driven into the ground. Its stout shape also makes it suitable for columns. An HP13×100 has a nominal depth of 13 in and weighs about 100 lb per foot of length.

- *Miscellaneous (M) beam shapes.* An M beam shape is similar to a W shape but much lighter; it has parallel flanges and a web in the shape of an H. M shapes are placed in a miscellaneous grouping of shapes that cannot be designated as W, S, or HP shapes. An M14×18 has a depth of 14 in and weighs about 18 lb per foot of length. M shapes are used in light beam applications. Production of M shapes is infrequent, so the supply is limited.

- *American Standard channel (C) shapes.* The C shape has tapered flanges that extend out of only one side of the web in the form of a C. A C12×30 has a depth of exactly 12 in and weighs about 30 lb per foot of length. Channels are used as bracing, light beams, and stair stringers or are welded to other structural shapes to improve the strength of a column or beam.

- *Miscellaneous channel (MC) shapes.* The MC shape has tapered flanges that extend out of one side of the web like the C shape. It fits in a miscellaneous grouping of C-shaped members that cannot formally be designated as a C shape. A MC12×50 has a depth of exactly 12 in and weighs about 50 lb per foot of length. Manufacture of M shapes is intermittent, so availability is limited.

- *Angle (L) shapes.* Equal-leg angles have two legs of equal length in the shape of an L. Unequal-leg angles have two legs of unequal length. The legs of an angle are of equal thickness. An L8×4×1 angle has an 8-in leg and a 4-in leg, each of which is 1 in thick.

- *Structural tee (WT, MT, and ST) shapes.* Structural tees have a single flange and a weblike stem in the shape of a T. They are split (cut) from standard W, M, and S shapes. Although off-center splitting is possible, T shapes are generally cut into two like tees at the center of the web of another structural member. A WT8×25 is split from a W16×50. It has a depth of approximately 8 in and weighs about 25 lb per foot of length. These shapes are frequently used in large trusses or framing.

- *Structural tubing (TS) shapes.* TS shapes are hollow square or rectangular tubes. A TS6×6×$\frac{1}{2}$ is a square tube with outside dimensions 6 in × 6 in and a wall thickness of $\frac{1}{2}$ in. Square shapes work well as columns. Rectangular shapes work well as beams and bracing.

- *Steel pipe (P) shapes.* Steel pipe is a round tube. It is available in many sizes that relate to its outside diameter. The specified diameter will be less than the actual diameter. For a specific size, outside diameter does not vary; only wall thickness varies. Pipe strength is expressed by schedule number (a higher schedule number is a pipe with a thicker wall) or by weight (standard weight, extra strong, and double-extra strong). Pipe is used for columns, frames, and guardrails.

- *Steel bars and plates (PL).* Bars are available in round, square, and rectangular sections in various widths, lengths, thicknesses, and diameters. Plates have a width greater than 8 in. Plates are available in widths up to 60 in and in thicknesses in $\frac{1}{32}$-in increments up to $\frac{1}{2}$ in thick, $\frac{1}{16}$-in increments over $\frac{1}{2}$ to 1 in thick, $\frac{1}{8}$-in increments over 1 to 3 in thick, and $\frac{1}{4}$-in increments over 3 in thick. Bars and plates are specified by actual size.

Steel members are identified by their designation and length. A length of W8×31 that is exactly 20 ft long weighs approximately 620 lb (20 ft × 31 lb/ft = 620 lb). A W8×31 beam that is 20 ft 6 in long is written as W8×31×20′-6″ and is pronounced "W eight by thirty-one by twenty feet, six inches."

SI (metric) designations are also used in specifying steel shapes. A W8×31 steel section has the metric designation of W200×46.1. W200 is a soft conversion from the W8: 8 in · 25.4 mm/in = 203.2 mm, or approximately 200 mm. The 31 lb/ft is a soft conversion to 46.1 kg/m.

18.3   STRUCTURAL STEEL DESIGN STANDARDS

Founded in 1921, the American Institute of Steel Construction, Inc. (AISC), One East Wacker Drive, Suite 3100, Chicago, IL 60601-2001 (Web site address: www.aisc.org) is a nonprofit association representing and serving the fabricated structural steel industry in the United States. Its purpose is to expand the use of fabricated structural steel through research and development, education, technical assistance, standardization, and quality control.

AISC prescribes two techniques for design of structural steel members: the traditional *allowable stress design* (ASD) and the *load and resistance factor design* (LRFD). LRFD is based on strength and serviceability of steel combined with probability analysis as described in the provisions of the *LRFD Specification for Structural Steel Buildings* (AISC, 1993). ASD requires that the actual stress in a structural member not exceed a prescribed allowable stress. ASD is based on the provisions of the *ASD Specification for Structural Steel Buildings* (AISC, 1989).

The ASD method was once the principal design approach prescribed by AISC. It is now viewed as an alternative to the LRFD method. Although LRFD is the endorsed method, the traditional ASD method is still used in many segments of the industry because it is a less complicated method of design. ASD more closely represents the theory presented in earlier chapters of this book. It is presented in this book because its less complicated design approach serves as a good method of acquainting the technician with basic design principles of reinforced concrete design.

18.4 STEEL MANUALS

In the United States, AISC publishes the *Manual of Steel Construction,* sometimes referred to as the "Steel Construction Manual" or simply the "Steel Manual." The 9th Edition of the AISC ASD Manual is based on the allowable stress provisions of the 1989 ASD Specification for Structural Steel Buildings. The 2nd edition of the AISC LRFD Manual is based on the strength design provisions of the 1993 LRFD Specification for Structural Steel Buildings. In Canada, the Canadian Institute of Steel Construction (CISC) publishes the *Handbook of Steel Construction.* The 6th edition of the CISC Handbook contains detailed information required for designing and detailing of structural steel in metric (SI) units. These manuals contain sets of steel beam and column tables and computation techniques that assist the designer in selecting a beam or column section to carry a specific load and detailing and dimension information. These manuals serve as a reference for engineers, architects, detailers, drafters, contractors, building officials, and fabricators.

Abridged sets of beam and column tables for W shapes from the AISC 9th edition ASD Manual are provided in Appendix B. Table B.1 includes dimension and detailing information and structural properties of W-shaped sections. Table B.2 relates to steel beams carrying uniformly distributed loads. Finally, Table B.3 relates to steel columns carrying axial end loads.

EXERCISES

18.1 Refer to Table B.1 to identify the following dimensions of a W8×31 structural steel shape. Convert these dimensions to metric (SI) units of mm.
  (a) Overall depth   **(b)** Web thickness
  (c) Flange breadth   **(d)** Flange thickness

18.2 Using the *Manual of Steel Construction,* identify the following dimensions of an S12×35 structural steel shape. Convert these dimensions to metric (SI) units of mm.

    (a) Overall depth   **(b)** Web thickness
    (c) Flange breadth   **(d)** Flange thickness

18.3 Using the *Manual of Steel Construction,* identify the following dimensions of an HP13×100 structural steel shape. Convert these dimensions to metric (SI) units of mm.
    (a) Overall depth   **(b)** Web thickness
    (c) Flange breadth   **(d)** Flange thickness

18.4 Using the *Manual of Steel Construction,* identify the following dimensions of an M14×18 structural steel shape. Convert these dimensions to metric (SI) units of mm.
    (a) Overall depth   **(b)** Web thickness
    (c) Flange breadth   **(d)** Flange thickness

18.5 Using the *Manual of Steel Construction,* identify the following design properties of an S12×35 structural steel shape. Compare the difference between similar properties in the x axis and y axis by percentage.
    (a) Moment of inertia, I_x **(b)** Moment of inertia, I_y
    (c) Section modulus, S_x **(d)** Section modulus, S_y
    (e) Radius of gyration, r_x **(f)** Radius of gyration, r_y

18.6 Refer to Table B.1 to identify the following design properties of a W21×93 structural steel shape. Compare the difference between similar properties in the x axis and y axis by percentage.
    (a) Moment of inertia, I_x **(b)** Moment of inertia, I_y
    (c) Section modulus, S_x **(d)** Section modulus, S_y
    (e) Radius of gyration, r_x **(f)** Radius of gyration, r_y

18.7 Refer to Table B.1 to identify the *least* radius of gyration of a W21×93 structural steel shape.

18.8 A36 steel has a minimum yield stress (F_y) of 36 000 lb/in^2 and an ultimate tensile strength (F_u) of between 58 000 and 80 000 lb/in^2. Properties for large sections have an F_y value of 32 000 lb/in^2. Convert these properties of strength to metric (SI) units of MPa.

18.9 A588 steel is a high-strength low-alloy steel that is most commonly used with a minimum yield stress (F_y) of 50 000 lb/in^2 and a minimum ultimate tensile strength (F_u) of 70 000 lb/in^2 for smaller beam sections. Convert these properties of strength to metric (SI) units of MPa.

18.10 A designer must choose between A36 and A572 steel as a structural material in a large building structure. A36 steel is an all-purpose carbon structural steel that has a minimum yield stress (F_y) of 36 000 lb/in^2 and a minimum ultimate tensile strength (F_u) of 58 000 lb/in^2. A572 steel is a high-strength low-alloy steel that is available in grade 60 with an F_y value of 60 000 lb/in^2 and an F_u value of 75 000 lb/in^2. Compare the differences in yield stress and ultimate tensile strength between the two steels by percentage.

SPECIAL PROJECTS

Steel Project 1

    Objectives   The objectives of this project are to provide the student with experience in the methodology involved in determination of tributary loads of structural members, in development of free-body diagrams, and in selection of steel beams, columns, and joists based on gravity loading.

Design parameters   The framing plans of a structural steel building structure are shown. The two-story steel skeleton structure supports the second floor and low-slope (flat) roof. The first floor is a floating concrete slab that is isolated from the structure. The floor-to-floor and floor-to-roof distances are 14 ft. The columns rest on individual foundations. Exterior curtain walls rest on spandrel beams (SB) and spandrel girders (SG) around the building perimeter. The weight of the curtain walls is approximately 20 lb/ft^2 measured vertically. The specified compressive strength of the concrete foundation pedestals is 3000 lb/in^2. The loads of this structure are based on the following:

FLOOR:

Steel decking	1.5 lb/ft^2
Concrete slab (4 in):	48.0 lb/ft^2
HVAC, plumbing, electrical, lighting, ceiling	est. 15.0 lb/ft^2
Partition load	20.0 lb/ft^2
Live load (occupants, furniture, equipment, etc.)	75.0 lb/ft^2

ROOF:

Steel decking	1.5 lb/ft^2
Insulation (4 in)	0.75 lb/ft^2
Roof membrane with gravel	6.0 lb/ft^2
HVAC, plumbing, electrical, lighting, ceiling	est. 15.0 lb/ft^2
Roof load (live)	20.0 lb/ft^2
Snow load	30.0 lb/ft^2
Rain/ice load	30.0 lb/ft^2

Analysis   *Beams.* Beams and girders are to be wide-flange (W) sections of ASTM A242 steel (F_y = 50 000 lb/in^2, E = 29 000 000 lb/in^2). Section investigation includes constructing a free-body diagram of the beam, flexure analysis, web shear analysis, and deflection analysis where possible with the Δ_{max} equations provided in this book (otherwise, deflection should be neglected). The lightest acceptable structural member is to be specified.

Columns. Column-section investigation is to be based on tabular selection from AISC column tables with an effective length based on $K=1.0$, backed up with longhand analysis. The lightest acceptable W14 steel section of ASTM A242 steel (Fy = 50 000 lb/in^2, E = 29 000 000 lb/in^2) is to be specified. Analysis of column base plates is also required.

Joists. Joist-section investigation is to be based on tabular selection of K-series steel joists at 4-ft spacing. The lightest acceptable joist is to be specified. Analysis of bridging is also required.

The following load combinations of dead loads (D_{load}), live floor loads (L_{load}), roof loads ($L_{roof\ load}$), snow loads (S_{load}), rain/ice load (R_{load}), wind load

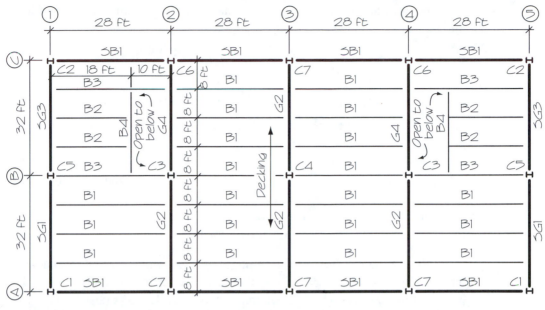

Floor framing plan

B — Beam
C — Column
G — Girder
SB — Spandrel beam
SG — Spandrel girder

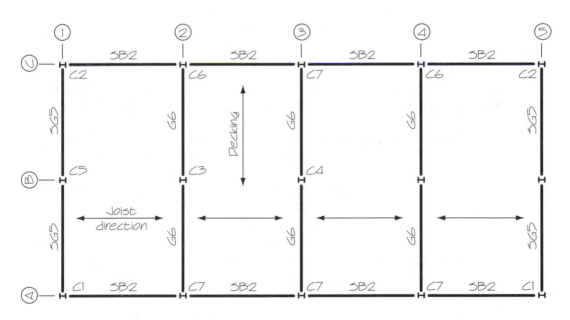

Roof framing plan

(W_{load}), and earthquake load (E_{load}) are to be used to determine the design load. Earthquake and wind loads are to be neglected.

$$L_{\text{design a}} = D_{load} + L_{load} + L_{\text{roof load}} \text{ or } S_{load}$$
$$L_{\text{design b}} = D_{load} + L_{load} + W_{load}$$
$$L_{\text{design c}} = D_{load} + L_{load} + W_{load} + 0.5S_{load}$$
$$L_{\text{design d}} = D_{load} + L_{load} + S_{load} + 0.5W_{load}$$
$$L_{\text{design e}} = D_{load} + L_{load} + S_{load} + E_{load}$$

Steel Project 2

Objectives   The objectives of this project are to provide the student with experience in the methodology involved in determination of tributary loads of structural members, in development of free-body diagrams, and in selection of steel beams, columns, and joists based on gravity loading.

Design parameters   The roof framing plan of a simple building structure is shown. The floor is a floating concrete slab that is isolated from the structure. The floor-to-roof distance is 16 ft. The columns rest on foundation pedestals. The specified compressive strength of the concrete foundation pedestals is 3000 lb/in^2. The roof loads of this structure are to be based on the following:

Steel decking	1.5 lb/ft^2
Insulation (4 in)	0.75 lb/ft^2
Single-ply membrane with gravel	6.0 lb/ft^2
HVAC, plumbing, electrical, lighting, ceiling	est. 20.0 lb/ft^2
Roof load (live)	20.0 lb/ft^2
Snow load	30.0 lb/ft^2
Rain/ice load	30.0 lb/ft^2

Rooftop HVAC units apply a 4-kip load at the center of the two girders. Assume that this load is carried by these two girders only.

Analysis   *Girders.* Girders are to be wide-flange sections. ASTM A36 steel $(F_y = 36\,000$ lb/in^2, $E = 29\,000\,000$ lb/in$^2)$ is specified for all structural steel. Section investigation is to include constructing a free-body diagram of the beam, flexure analysis, web shear analysis, and deflection analysis where possible with the Δ_{max} equations provided in this book (otherwise, deflection should be neglected). The lightest acceptable wide-flange structural member is to be specified.

Columns. Column-section investigation is to be based on tabular selection from AISC column tables with an effective length based on $K = 1.0$, backed up with longhand analysis. The lightest acceptable W14 steel section is to be specified. Analysis of column base plates is also required.

Joists. Joist-section investigation is to be based on tabular selection. Analysis is to be based on use of K-series steel joists at 4 ft spacing. Analysis of bridging is also required.

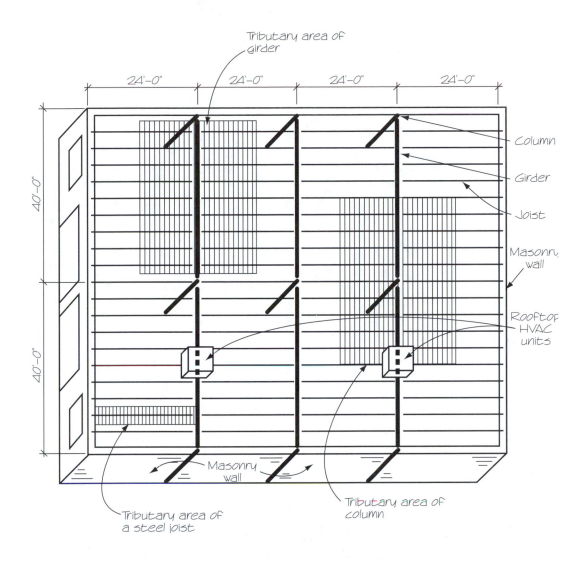

The following load combinations of dead loads (D_{load}), live floor loads (L_{load}), roof loads ($L_{roof\ load}$), snow loads (S_{load}), rain/ice load (R_{load}), wind load (W_{load}), and earthquake load (E_{load}) are to be used to determine the design load. Earthquake and wind loads are to be neglected.

$$L_{design\ a} = D_{load} + L_{load} + L_{roof\ load}\ or\ S_{load}$$
$$L_{design\ b} = D_{load} + L_{load} + W_{load}$$
$$L_{design\ c} = D_{load} + L_{load} + W_{load} + 0.5S_{load}$$
$$L_{design\ d} = D_{load} + L_{load} + S_{load} + 0.5W_{load}$$
$$L_{design\ e} = D_{load} + L_{load} + S_{load} + E_{load}$$

The finished envelope of a steel skeleton building.

The steel skeleton of a building under construction.

A corner of the steel skeleton structure during construction. Steel beams spaced repetitively below the floor deck make up the structure of the floor platforms. Open web steel joists will support a roof platform constructed of steel decking. Beams and joists supporting the roof and floor systems bear on spandrel girders. The girders span between vertically positioned columns and transfer their load to the columns.

Open web steel joists supporting steel roof decking.

Columns transfer building loads to individual foundations located at the base of each column.

A 4-inch-thick concrete topping is poured on top of the steel decking. The concrete bonds to the steel decking to improve structural integrity. Wire fabric reinforcement is cast in the concrete topping to reinforce the floor platform.

The girders transfer their load to the columns. Light-gauge steel framing is used to construct framing for exterior walls. Framed walls are non-load bearing.

Light-gauge steel framing is used to construct interior partitions that divide the interior space into rooms. Framed walls are non-load bearing.

Steel beams make up the structure of the floor platforms. Beams are spaced repetitively and support the floor deck.

A simple steel beam/girder connection. Both the beam and girder support the floor deck so the top side of the beam and girder are at the same elevation. Light-gauge steel framing is used to construct non-load bearing framing for exterior walls.

A simple steel beam/girder/column connection. Both the beam and girder support the floor deck and transfer their load to the column.

The steel skeleton of a second building under construction. A plastic film wraps the second story of the building to prevent heat loss while the concrete topping on the floor deck cures in cold weather. Repetitive framing surrounding much of the second story will support brick veneer.

Closely spaced steel joists support floor platforms constructed of steel decking. Joists supporting the floor system bear on girders. The girders span between vertically positioned columns and transfer their load to the columns.

A simple steel girder/column connection. Both girders support joists that support the floor deck. The top sides of the girders are at the same elevation.

Corrugated steel decking is spot welded to the topside of the steel joist. Wire fabric reinforcement is evident.

Concrete topping is poured on top of the steel decking to improve structural integrity. A column extends from the floor platform.

A column base plate is secured to the foundation with anchor bolts. Cement grout will be forced into the space between the plate and the concrete foundation. Once it sets, the grout provides a good bearing surface so the column load can be evenly transferred from the plate to the foundation.

Chapter 19
DESIGN OF SIMPLE STRUCTURAL STEEL BEAMS

An elementary design approach for structural steel beams using the allowable stress design method is introduced in this chapter. The reader is advised that the presentation does not provide full coverage of all design principles involved with design of steel structures. This book is intended to acquaint the technician with basic design principles used by an engineer. Additional coverage is beyond the intended breadth of the book.

19.1   SELECTION FROM BEAM TABLES

Table B.2 is a set of listings for use with beams carrying a uniformly distributed load across the entire span of the beam. These listings apply to W shapes and steels with a yield stress (F_y) of 36 000 lb/in^2 (36 ksi) only. In bold letters at the head of each listing is the steel section *designation* (i.e., W12 or W18), indicating the type of section. Each section has several section weights at the top of the table, in lb/ft. The main body of the listing consists of beam spans in feet (left column), the allowable *total* load in kip (center columns), and deflection in inches at the allowable total load (right column).

Following is the fundamental beam section selection procedure using the AISC table method for a beam carrying a uniformly distributed load:

1. Determine the beam span, in feet.
2. Find the *total* uniformly distributed load to be supported by the beam, in kip.
3. Refer to the appropriate beam span in each table and a select beam section from each designation (i.e., W8, W10, W12, etc.) that has an allowable total load equal to or greater than the *total* uniformly distributed load to be supported by the beam.

4. Select a beam section based on specified design criteria, such as weight or depth limitation.

5. Investigate deflection to ensure that maximum deflection (Δ_{max}) does not exceed maximum deflection permitted (D_{code}). If the section fails this investigation, a larger section is needed.

6. Investigate other factors related to steel beams.

■ EXAMPLE 19.1

Find a wide-flange (W) beam section to carry a uniformly distributed load of 1450 lb/ft, including beam weight, over a 20-ft span. Select the lightest section that supports the load safely without exceeding a deflection of $\ell/360$. ASTM A36 (F_y of 36 000 lb/in^2) steel is specified.

Solution

$$W = 1450 \text{ lb/ft} \times 20 \text{ ft} = 29 \text{ kip}$$

By referring to Table B.2, the following W-shape beam sections were selected based on the 20-ft span and an allowable load of 29.0 kip or more. The lightest beam section from each depth category that met the above-mentioned criteria was selected:

W8×48	W16×26 ← lightest section of the group
W10×39	W18×35
W12×30	W21×44
W14×30	W24×55 etc.

Although any section above is acceptable, the W16×26 section at 26 lb/ft is the lightest section of the group. It is investigated for deflection. Deflection is limited to $\ell/360$.

$$D_{code} = \frac{\ell}{360} = \frac{20 \text{ ft} \times 12 \text{ in/ft}}{360} = 0.667 \text{ in}$$

From the listing in Table B.2 for the W16×26 section and 20-ft span, it is found that maximum deflection is 0.61 in (from right column) at the allowable total load of 30 kip. Since deflection is proportional to uniformly distributed loading, maximum deflection for the W16×26 section supporting a 29-kip load uniformly distributed over a 20-ft span is found by

$$\Delta_{max} = \frac{29 \text{ kip}}{30 \text{ kip}} \times 0.61 \text{ in} = 0.589 \text{ in}$$

Δ_{max} does not exceed D_{code}, so the W16×26 section is acceptable.

When the beam loading condition does not involve only a uniformly distributed load across its entire span, an acceptable beam section must be selected using a calculation approach. This procedure involves sizing the beam section for flexure, deflection, and shear. This technique is the same approach used by AISC to assemble the tables in the Steel Manual.

The fundamental calculation design procedure involves the following sequence of steps:

1. Solve for maximum bending moment (M).
2. Use the flexure formula, $S = M/F_b$, to find the required section modulus (S). For ASTM A36 steel, the allowable bending stress (F_b) is 24 000 lb/in^2 or 24 ksi. For other structural steels, it is found by $F_b = \frac{2}{3} F_y$.
3. Select a beam section for each designation that has a section modulus (S) equal to or greater than required in step 2. Refer to Table B.1.
4. Investigate deflection to ensure that maximum deflection (Δ_{max}) does not exceed maximum deflection permitted (D_{code}). If the section fails this investigation, a larger section is needed.
5. Investigate other factors related to steel beams.

■ EXAMPLE 19.2

Find a wide-flange (W) beam section to support a 30-kip load at the center of a 20-ft span. Select the lightest section that supports the load safely without exceeding a deflection of $\ell/360$. ASTM A36 steel is specified.

Solution

1. Solve for maximum bending moment (M). From Table A.5:

$$M_{max} = \frac{PL}{4} = \frac{30 \text{ kip} \cdot 20 \text{ ft}}{4} = 150 \text{ f-kip}$$

2. Use the flexure formula to find the required section modulus (S):

$$S = \frac{M}{F_b} = \frac{150 \text{ ft–kip} \cdot 12 \text{ in/ft}}{24 \text{ ksi}} = 75.0 \text{ in}^3$$

3. Select a beam section that has a section modulus (S) equal to or greater than required. By referring to Table B.1, the following W-shape beam sections were selected based on a section modulus of 75.0 in^3 or more. The lightest beam section from each designation category that met the above-mentioned criteria was selected.

W10×68	W18×46
W12×58	W21×44 ← lightest beam section of group
W14×53	W24×55
W16×50	W27×84 etc.

The W21×44 has the lightest beam section of the group. It has a section modulus of 81.6 in^3, which is slightly greater than required. This section must be investigated for deflection.

4. Investigate deflection. Deflection is limited to $\ell/360$:

$$D_{code} = \frac{\ell}{360} = \frac{20 \text{ ft} \cdot 12 \text{ in/ft}}{360} = 0.667 \text{ in}$$

Maximum deflection (Δ_{max}) under this loading condition may be found by using one of the derived formulas in Table A.5. For the W21×44 section, the moment of inertia (I) is found in Table B.1 to be 843 in^4. The modulus of elasticity for steel is 29 000 ksi. 29 000 ksi is used rather than 29 000 000 lb/in^2 to ensure proper canceling of dimensional units. The 20-ft span must be converted to inches, to ensure proper canceling of units.

$$\Delta_{max} = \frac{P\ell^3}{48EI}$$

$$= \frac{30 \text{ kip} \cdot (20 \text{ ft} \cdot 12 \text{ in/ft})^3}{48 \cdot 29\,000 \text{ ksi} \cdot 843 \text{ in}^4}$$

$$= 0.353 \text{ in}$$

Δ_{max} does not exceed D_{code}, so the W21×44 section meets the deflection requirement.

19.3   OTHER FACTORS IN STEEL BEAM DESIGN

In addition to designing for flexure and deflection, the designer must also be concerned with the following additional factors:

Local buckling   Local buckling is buckling of a constituent area of a section profile such as a beam flange or web because it is under excessive compression stress. This type of buckling can result in failure of a beam at loads lower than the beam could theoretically carry if local buckling did not develop. To limit localized buckling, the following limitation applies:

$$\frac{b_f}{2t_f} \leq \frac{95}{\sqrt{F_y}}$$

For ASTM A36 steel with an F_y value of 36 kip/in^2, $bf/2t_f$ should not exceed 15.8: $95/\sqrt{36 \text{ kip/in}^2} = 15.8$. Values of $b_f/2t_f$ for standard steel sections are given in Table B.1.

Web shear An additional investigation is required that ensures that the beam section selected can resist shear forces acting on the beam web. Simply, the maximum vertical shear (V_{max}) experienced by the beam section should not exceed the allowable shear strength (V_s):

$$V_s = 0.55F_y dt_w$$

where V_s is the allowable shear strength, in lb or kip, V_{max} is the maximum vertical shear, in lb or kip, d is the beam section depth, in inches, and t_w is the beam section web thickness, in inches. The beam section must be increased if maximum vertical shear (V_{max}) experienced by the beam section exceeds the allowable shear strength (V_s).

■ EXAMPLE 19.3

Investigate the shear strength of the W21×44 section selected in Example 19.2. The beam will support a 30-kip load at the center of the 20-ft span. ASTM A36 ($F_y = 36$ ksi) steel is specified.

Solution From Table A.5:

$$V_{max} = \frac{P}{2} = \frac{30 \text{ kip}}{2} = 15 \text{ kip}$$

From Table B.1:

$$d = 20.66 \text{ in} \quad \text{and} \quad t_w = 0.350 \text{ in}$$
$$V_s = 0.55 \, F_y dt_w = 0.55(36 \text{ ksi} \cdot 20.66 \text{ in} \cdot 0.350 \text{ in}) = 143.2 \text{ kip}$$

The W21×44 section passes the shear investigation, because V_{max} does not exceed V_s.

Buckling of compression flange Under load, the beam flange experiencing compression stresses has a tendency to buckle, unless buckling is prevented (Figure 19.1b). Additional bracing of the compression flange may be required, even though the beam section adequately meets other design criteria (i.e., flexure, deflection, web shear, etc.). Usually, longer beam lengths require bracing, while very short lengths may not. Options include encasing the beam in concrete and adding stiffeners or lateral supports, such as floor framing, at intervals across the beam length (Figure 19.2).

In the AISC tables, maximum unbraced length of the compression flange (L_c) is provided for beam sections computed with a bending stress (F_b) of 24 000 lb/in^2. The W16×26 section selected in Example 19.1 has an L_c value of 5.6 ft. Since the beam will span 20 ft, lateral bracing is required.

Web crippling and yielding   Excessively large reactions or concentrated loads may cause crippling of the web (Figure 19.1c). Metal plates called *stiffeners* are welded to the section's web and inner flanges to resist web crippling. Associated calculations are too involved to cover adequately in a book of this breadth, but the reader should be aware of this concern.

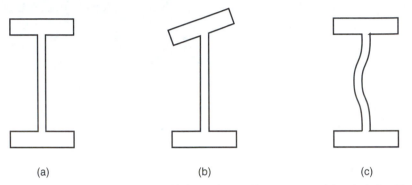

(a)   (b)   (c)

FIGURE 19.1   Modes of failure of a steel beam section (a) include buckling of the compression flange (b) and crippling of the beam web (c).

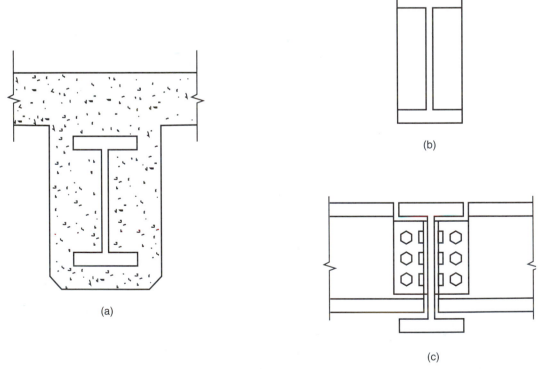

(a)

(b)

(c)

FIGURE 19.2   Buckling of the compression flange may be prevented by encasing the beam in concrete (a) or bracing the beam with stiffeners (b) or other members (c).

Bearing surfaces and plates   A steel beam transfers its load to the reaction supports. The reaction supports exert opposite forces called reactions that are equal in magnitude. A bearing stress develops where reactions contact the beam. Each reaction force must be adequately distributed at these bearing areas to prevent excessive bearing stress and crushing failure of the reaction support material.

A reaction support is usually a wall, pier, column, or girder that is made of steel, concrete, masonry, and in some instances, wood. Steel is much stronger in compression than concrete, masonry, and wood. To prevent crushing of these reaction support materials, bearing plates are frequently needed (Figure 19.3). It is larger in area than the beam flange so that the plate can distribute the reaction force over a larger bearing area.

Bearing plates are typically embedded in concrete or masonry at the appropriate elevation before beam placement. This greatly simplifies placement of beams, so bearing plates are even specified in instances when they are not required to limit bearing stress.

In the absence of more stringent requirements, the following bearing stresses apply: The allowable bearing stress (F_p) for brick in cement mortar is 250 lb/in^2. For sandstone and limestone, the allowable bearing stress (F_p) is 400 lb/in^2. In instances when a steel beam is permitted to bear on wood, the allowable bearing stress (F_p) of wood is equal to the compression stress perpendicular to the grain ($F_c \perp$), a value dependent on wood species and grade.

The allowable bearing stress of concrete is decreased from the specified compressive strength (f'_c) of concrete, the largest compressive stress that the concrete

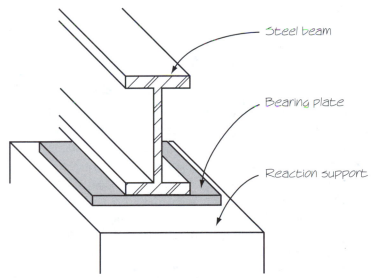

FIGURE 19.3   A bearing plate is a steel plate located at the reaction support.

must develop without failure in laboratory testing at 28 days after curing (see Chapter 26). The compressive strength specified for concrete used in walls and piers is typically 2500, 3000, 4000, or 5000 lb/in^2.

In cases when the bearing surface area of the concrete reaction support member (A_{bs}) is greater than the bearing plate bearing area (A_{bp}), the allowable bearing stress may be increased by the following expression up to $F_p = 0.7f'_c$:

$$F_p = 0.35f'_c \sqrt{\frac{A_{bs}}{A_{bp}}}, \text{ not to exceed } F_p = 0.7f'_c$$

In those rare cases when the bearing plate covers the entire bearing surface area of the reaction support ($A_{bp} = A_{bs}$), the allowable bearing stress (F_p) of concrete is simply

$$F_p = 0.35f'_c$$

Allowable bearing stress (F_p) governs the minimum size of a steel bearing plate area (A_{bp}) for a specific support reaction (R):

$$A_{bp} \geq \frac{R}{F_p}$$

The bearing plate area (A_{bp}) is based on its bearing breadth (B) and bearing length (N):

$$A_{bp} = B \cdot N$$

Plate thickness is computed with the assumption that the portion of the plate that extends past the beam flange bends upward beyond the edges of the beam flange. The required thickness of the bearing plate (t_{bp}) is determined by the following expression:

$$t_{bp} \geq \sqrt{\frac{3f_p n^2}{F_b}}$$

In bearing plate design (Figure 19.4), the designer typically rounds plate breadth (B) and length (N) in 1-in increments. Steel plate thicknesses are available and specified in $\frac{1}{32}$-in increments up to $\frac{1}{2}$ in thick, $\frac{1}{16}$-in increments over $\frac{1}{2}$ to 1 in thick, $\frac{1}{8}$-in increments over 1 to 3 in thick, and $\frac{1}{4}$-in increments over 3 in thick. Plates are available in widths up to 60 in and limited by lengths that can be transported by truck. Plate width and length are custom cut.

In addition to a satisfactory bearing plate area and thickness, the plate must be long enough to prevent web yielding and crippling at the reaction supports. Associated calculations are too involved to cover adequately in a book of this breadth, but the reader should be aware of this concern.

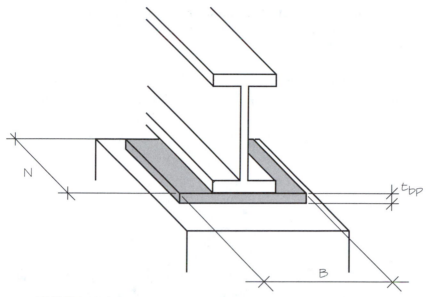

FIGURE 19.4   Layout of the dimensions for design of a steel bearing plate.

The following variables apply in basic design of steel bearing plates:

A_{bp} area of the bearing plate, in in^2

A_{bs} area of the bearing surface of concrete, in in^2

B bearing breadth of the bearing plate, in inches

F_b allowable bending stress of the bearing plate, in lb/in^2: $F_b = 0.75F_y$; for A36 steel F_b is 27 kip/in^2 or 27 000 lb/in^2

F_p allowable bearing stress for a specific reaction support material, in lb/in^2

f_p actual bending stress of the bearing plate, in lb/in^2: $f_p = R/(N \cdot B)$

F_y minimum yield strength of the steel, in kip/in^2 or lb/in^2

k distance from the bottom side of the beam flange to the top of the bottom web fillet, in inches (this value is specific to a beam section profile; it is given in Table B.1)

N bearing length of the bearing plate, in inches

R reaction, in lb

n $n = (B/2) - k$

t_{bp} required bearing plate thickness, in inches

■ EXAMPLE 19.4

Design a steel bearing plate for the reactions of the W21×44 section selected in Example 19.2. The beam reactions are 15.5 kip each, including the weight of the beam and beam fireproofing. The beam will rest on a bearing wall composed of brick with cement mortar. The plate will have an available bearing length of 6 in. ASTM A36 ($F_y = 36$ ksi) steel is specified for the bearing plate.

Solution With an allowable bearing stress for brick in cement mortar of 250 lb/in^2 and a reaction of 15 500 lb, the minimum area of the bearing plate is found by

$$A_{bp} \geq \frac{R}{F_p} = \frac{15\ 500\ \text{lb}}{250\ \text{lb/in}^2} = 62.0\ \text{in}^2$$

With an available bearing length (N) of 6 in, the expression $A_{bp} = B \cdot N$ is rearranged to find the minimum bearing breadth of the bearing plate (B):

$$B = \frac{A_{bp}}{N} = \frac{62.0\ \text{in}^2}{6\ \text{in}} = 10.33\ \text{in}$$

Assume that a 6 in $\times$ 11 in bearing plate is specified. The actual bending stress of the bearing plate is found by

$$f_p = \frac{R}{N \cdot B} = \frac{15\ 500\ \text{lb}}{6\ \text{in} \cdot 11\ \text{in}} = 235\ \text{lb/in}^2$$

In Table B.1, for a W21$\times$44, k is $1\frac{3}{16}$ in. The variable n is found by

$$n = \frac{B}{2} - k = \frac{11\ \text{in}}{2} - 1\tfrac{3}{16}\ \text{in} = 4.31\ \text{in}$$

The required bearing plate thickness (t_{bp}) is found by

$$t_{bp} \geq \sqrt{\frac{3f_p n^2}{F_b}}$$

$$= \sqrt{\frac{3(235\ \text{lb/in}^2)(4.31\ \text{in})^2}{27\ 000\ \text{lb/in}^2}} = 0.696\ \text{in}$$

With plate thicknesses around 0.696 available in increments of $\frac{1}{16}$ in, a $\frac{3}{4}$-in-thick plate is needed. A 6$\times$11$\times\frac{3}{4}$ bearing plate should be specified.

Cost Although the previous examples based their selection criteria on the lightest acceptable beam section, beam weight frequently plays only a small role in the overall cost of a building. It is true that the cost of structural steel members is based primarily on weight, so a lighter beam costs less than a heavier member. However, the depth of a structural member influences headroom and thus the cost of the building envelope.

In Example 19.2, a W12$\times$58 may have been selected over the lighter but deeper W21$\times$44 that was selected because of weight alone. The heavier W12$\times$58 is about 9 in smaller and would eliminate an additional course of masonry that is required for the W21×44. Cost savings in wall height may outweigh cost savings of a lighter, yet deeper beam section.

Beam weight   In design, the weight of the beam itself and any fireproofing must be considered in the analysis. Initially, the designer may estimate beam weight or simply neglect it. However, once a beam section is selected, additional calculations must be made to ensure that these additional weights are not neglected.

EXERCISES

19.1 A simple steel beam with a span of 16 ft will support a uniformly distributed load of 2375 lb/ft. The beam is fabricated of ASTM A36 ($F_y = 36\,000$ lb/in^2) steel. Using the AISC tables in Table B.2, find the following.
(a) The lightest W-shape section that will carry the load safely
(b) The lightest W-shape section, if depth is limited to a maximum of 12.00 in

19.2 A simple steel beam with a span of 24 ft will support a uniformly distributed load of 2375 lb/ft. The beam is fabricated of ASTM A36 ($F_y = 36\,000$ lb/in^2) steel. Using the AISC tables in Table B.2, find the lightest W-shape section that will carry the load safely.

19.3 A simple steel beam with a span of 32 ft will support a uniformly distributed load of 2375 lb/ft. The beam is fabricated of ASTM A36 ($F_y = 36\,000$ lb/in^2) steel. Using the AISC tables in Table B.2, find the lightest W-shape section that will carry the load safely.

19.4 A simple steel beam with a span of 48 ft will support a uniformly distributed load of 2375 lb/ft. The beam is fabricated of ASTM A36 ($F_y = 36\,000$ lb/in^2) steel. Using the AISC tables in Table B.2, find the lightest W-shape section that will carry the load safely.

19.5 A simple steel beam with a span of 60 ft will support a uniformly distributed load of 2375 lb/ft. The beam is fabricated of ASTM A36 ($F_y = 36\,000$ lb/in^2) steel. Using the AISC tables in Table B.2, find the lightest W-shape section that will carry the load safely.

19.6 In Exercises 19.1 through 19.5, the uniformly distributed load of 2375 lb/ft carried by the simple beam remained constant but the span varied. Compare the weight of the lightest W-shape sections found for each span to the beam section found in Exercise 19.1 by percentage.

19.7 A steel beam with a span of 20 ft will support a concentrated load of 20 kip at its center. The beam is fabricated of ASTM A36 ($F_y = 36\,000$ lb/in^2) steel. Use the calculation approach to find the lightest W-shape section that will carry the load safely. Include flexure (bending), deflection, and web shear investigations in your analysis.

19.8 Much like Exercise 19.7, a steel beam with a span of 20 ft will support a concentrated load of 20 kip at its center. However, in this case, the beam is fabricated of ASTM A572 ($F_y = 50\,000$ lb/in^2) steel. Use the calculation approach to find the lightest W-shape section that will carry the load safely. Include flexure (bending), deflection, and web shear investigations in your analysis.

19.9 A steel beam with a span of 20 ft will support *both* a concentrated load of 20 kip at the beam center *and* a uniformly distributed load of 2 kip/ft across the entire beam length. The beam is fabricated of ASTM A36 ($F_y = 36\,000$ lb/in^2) steel. Use the calculation approach to find the lightest W-shape section with a depth not to exceed 20.00 in deep that will carry the load safely. Include flexure (bending), deflection, and web shear investigations in your analysis.

19.10 Much like Exercise 19.9, a steel beam with a span of 20 ft will support *both* a concentrated load of 20 kip at the beam center *and* a uniformly distributed load of 2 kip/ft across the entire beam length. However, in this case, the beam is fabricated of ASTM A572 ($F_y = 50\,000$ lb/in^2) steel. Use the calculation approach to find the lightest W-shape section with a depth not to exceed 20.00 in deep that will carry the load safely. Include flexure (bending), deflection, and web shear investigations in your analysis.

19.11 Design a steel bearing plate for the reactions of the section selected in Exercise 19.7. The beam will rest on a bearing wall composed of brick with cement mortar. The plate will have an available bearing length of 7 in. ASTM A36 ($F_y = 36$ ksi) steel is specified for the bearing plate.

19.12 Design a steel bearing plate for the reactions of the section selected in Exercise 19.9. The beam will rest on a sandstone bearing wall. The plate will have an available bearing length of 7 in. ASTM A36 ($F_y = 36$ ksi) steel is specified for the bearing plate.

Chapter 20
DESIGN OF OPEN-WEB STEEL JOISTS

An elementary design approach for steel joists is introduced in this chapter. The reader is advised that the presentation does not provide full coverage of all design principles involved with design of steel structures. This book is intended to acquaint the technician with basic design principles used by an engineer. Additional coverage is beyond the intended breadth of the book.

20.1 STEEL JOISTS

Open-web steel joists, also called *bar joists* or simply *steel joists,* are prefabricated trusses. As a truss, the individual components of a steel joist experience either tensile or compressive stresses, but not bending stresses. Steel joists are generally used in a repetitive manner to support roof and floor systems structurally. These joists can span large distances very economically, which allows for design of very efficient structural systems.

Steel joists are generally composed of parallel upper and lower chords and an open web (Figure 20.1). The *chords* run longitudinally across the full length of the member. They are designed to resist tensile and compressive stresses that develop when the joist undergoes bending. The *web* simply maintains the distance between the chords. It keeps the chords from coming together when the joist is under load. Chords may be composed of steel structural angles, tees, or bars. The web is usually made from steel bars, although angles may be used on larger members. Except for the web itself, the web area remains open, and thus sometimes, steel joists are referred to as open-web steel joists.

The Steel Joist Institute (SJI), 3127 Tenth Avenue North Extension, Myrtle Beach, SC 29577-6760 (Web site address: www.steeljoist.org), is a nonprofit organization of active joist manufacturers that was founded in 1928, five years after the first open-web steel joist was manufactured. Over the years, SJI has expanded its focus to include research relating to steel joists and joist girders, providing technical counseling, promoting the use of steel joists through a national advertising program, and developing other training and research aids.

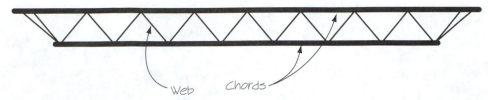

FIGURE 20.1 Steel joists are composed of parallel upper and lower chords and an open web. The *chords* run longitudinally across the full length of the member. They are designed to resist tensile and compressive stresses that develop when the joist undergoes bending. The *web* simply maintains the distance between the chords. To minimize lateral and torsional buckling, open web steel joists must be braced with bridging.

20.2 JOIST DESIGNATIONS

Specifications for steel joists are governed by SJI. Individual manufacturers may have additional specifications. SJI specifies that steel joists fit within different designation categories called a *series*. Each series is described below.

- *Open-web joists (K series).* These are the smallest and lightest joists. They are available in lengths of up to 48 ft and depths of 8 to 30 in. Designations in this series range from 8K1 to 30K12.
- *Long-span joists (LH series).* These are available in lengths from 25 to 96 ft and depths of 18 to 48 in. Designations in this series range from 18LH02 to 48LH17.
- *Deep long-span joists (DLH series).* These are available in lengths from 96 to 144 ft and depths of 52 to 72 in. Designations in this series range from 52DLH10 to 72DLH19.
- *Super long-span joists (SLH series).* These are available in lengths over 144 ft and depths of 80 to 120 in. Designations in this series range from 80SLH15 to 120SLH25.
- *Joist girders.* These are primary framing members designed for the direct support of steel joists. They are available in depths from 20 through 72 in in 4-in increments.

Steel joists are identified by a specific three-digit designation, such as 12K5, 48LH02, or 72DLH19. In a 12K5, the K indicates that the joist is from the K series, 12 refers to the nominal joist depth in inches, and 5 refers to the web section classification. A larger web section classification corresponds to a heavier web section and thus a stronger joist for the specific depth designation.

Table B.4 provides the maximum allowable uniformly distributed loads for K-series steel joists for various spans. Loads are expressed in pounds per foot (lb/ft) of joist span. Approximate joist weight is also provided in this table and is expressed in pounds per foot (lb/ft) of joist span.

The upper value for a specific span and joist designation represents the *total* safe uniformly distributed load that can be carried by the joist in pounds per foot of joist span. The lower value, expressed in *italic,* represents the *live* load that produces a deflection of $\ell/360$ of the span. Deflection from the live load of floor systems and roof systems with an underside plaster ceiling must not exceed $\ell/360$ of the span. In other cases, deflections no greater than $\ell/240$ of the span are acceptable. Live loads with a deflection not to exceed $\ell/240$ of the span may be obtained by multiplying the lower value, expressed in *italic,* by a factor of 1.5. Total loads must not be exceeded for either deflection limitation.

For the K-series joists outlined in this book, joist ends must, in most instances, bear at least 4 in over a masonry or concrete support and be anchored to the bearing plate. In the case of steel reaction supports, K-series joists must bear at least $2\frac{1}{2}$ in over the reaction support. The design span is computed from the centerline of a steel bearing support (i.e., the center of a steel beam or joist girder supporting a joist) and 4 in from the edge of a masonry or concrete bearing support (i.e., a wall surface).

Following is the design procedure used to determine joist size for joists carrying uniformly distributed loads:

1. Determine design span, spacing and deflection limitations (i.e., $\ell/360$ or $\ell/240$).
2. Calculate the uniformly distributed total and live roof or floor load, in lb/ft^2.
3. Calculate the total and live roof or floor load per foot of joist span, in lb/ft.
4. Find the joist designations (Table B.4) that meet or exceed the load after subtracting joist weight for the span and any deflection or depth criteria.
5. Select the lightest joist based on deflection or depth criteria.

■ EXAMPLE 20.1

A roof system will cover a masonry building with interior dimensions of 27'-4" × 43'-4". Joists will span the shorter dimension. The roof (membrane, insulation, and deck) and ceiling (suspended ceiling, lights, ductwork, etc.) systems have a combined dead load of 15 lb/ft^2. Assume a live load of 30 lb/ft^2. Deflection for the live load should not exceed $\ell/360$ of the span. Due to the structural properties of the metal roof deck, a joist spacing of 4 ft is under consideration.

(a) Select the lightest K-series joist designation that can carry the load safely.
(b) Due to headroom requirements and material costs associated with building envelope height, joist depth should not exceed 16 in.

Solution (a) The joist spacing is provided in the problem statement (4-ft spacing). The joist ends will bear on masonry walls, so the clear span must be increased by 4 in at each joist end:

$$27'\text{-}4'' + 4 \text{ in} + 4 \text{ in} = 28 \text{ ft design span}$$

The live uniformly distributed load computed is 30 lb/ft^2 and dead load is 15 lb/ft^2. The total uniformly distributed load is computed by

$$\text{total uniformly distributed load} = 15 \text{ lb/ft}^2 + 30 \text{ lb/ft}^2 = 45 \text{ lb/ft}^2$$

The live load per foot of joist span is computed by

$$4 \text{ ft} \times 30 \text{ lb/ft}^2 = 120 \text{ lb/ft of joist span}$$

The total load per foot of joist span is computed by

$$4 \text{ ft} \times 45 \text{ lb/ft}^2 = 180 \text{ lb/ft of joist span}$$

From Table B.4 for the 28-ft span, the following joist sections are selected. They are the lightest acceptable joists in their joist group (i.e., 18K, 20K, 22K, etc.). Their total allowable load per foot of joist span is not less than 180 lb/ft after subtracting the weight per foot of joist. Also, their live load per foot of joist span is not less than 120 lb/ft after subtracting the weight per foot of joist.

	16K4:	249 lb/ft of joist span − 7.0 lb/ft joist weight = 242 lb/ft (versus 180 lb/ft) 138 lb/ft of joist span − 7.0 lb/ft joist weight = 131 lb/ft (versus 120 lb/ft)
lightest →	18K3:	234 lb/ft of joist span − 6.6 lb/ft joist weight = 227.4 lb/ft (versus 180 lb/ft) 151 lb/ft of joist span − 6.6 lb/ft joist weight = 144.4 lb/ft (versus 120 lb/ft)
	20K3:	261 lb/ft of joist span − 6.7 lb/ft joist weight = 254.3 lb/ft (versus 180 lb/ft) 189 lb/ft of joist span − 6.7 lb/ft joist weight = 182.3 lb/ft (versus 120 lb/ft)
	22K4:	348 lb/ft of joist span − 8.0 lb/ft joist weight = 340 lb/ft (versus 180 lb/ft) 270 lb/ft of joist span − 8.0 lb/ft joist weight = 262 lb/ft (versus 120 lb/ft)

Chap. 20   Design of Open-Web Steel Joists

Joist weight increases beyond the 22K joist group, so further analysis is neglected. Any of these joists would be acceptable, but the 18K3 is the lightest.

(b) From previous analysis, the 16K4 is the lightest joist not exceeding 16 in deep. It is heavier than the 18K3 selected previously.

20.4 OTHER FACTORS IN STEEL JOIST DESIGN

The designer must also be concerned with the following additional factors in joist design.

Cost Overall joist weight affects cost, since steel is sold by weight. A designer may select the lightest joist designation that can carry the load. In selecting the most economical joist system, however, the designer must also consider the depth of the joist, the span, and the number of joists required. Depth of a joist affects overall building height and headroom, thereby influencing the cost of the building envelope. Additionally, the designer must consider joist spacing. A larger spacing interval usually means a smaller number of heavier joists. Savings in joist system weight (fewer joists means less weight) and joist system cost usually results from larger spacing. Loss of headroom due to a deeper joist system and the need for a taller structure and heavier decking reduces savings.

■ EXAMPLE 20.2

Much like in Example 20.1, a roof system will cover a masonry building with interior dimensions of 27'-4" × 43'-4". The roof (membrane, insulation, and deck) and ceiling (suspended ceiling, lights, ductwork, etc.) systems have a combined dead load of 15 lb/ft^2. Assume a live load of 30 lb/ft^2. Deflection for the live load should not exceed $\ell/360$ of the span. Again, a joist spacing of 4 ft is being considered. In this example, however, joists will span the longer dimension. Select the lightest K-series joist designation that can carry the load safely.

Solution The joist spacing is provided in the problem statement (4-ft spacing). Again, the joist ends will bear on masonry walls, so the span must be increased by 4 in at each joist end:

$$43'\text{-}4'' + 4 \text{ in} + 4 \text{ in} = 44 \text{ ft design span}$$

From Example 20.1, the total uniformly distributed load is 45 lb/ft^2, the uniformly distributed live load is 30 lb/ft^2, the total load per foot of joist span is 180 lb/ft of joist span, and the live load per foot of joist span is 120 lb/ft of joist span.

From Table B.4 for the 44-ft span, the following joist sections are se-lected. They are the lightest acceptable joists in their joist group (i.e., 18K, 20K, 22K, etc.). Their total load per foot of joist span is not less than 180 lb/ft after subtracting the weight per foot of joist. Also, their live load per foot of joist span is not less than 120 lb/ft after subtracting the weight per foot of joist.

Initially, it appears that the 22K10 is acceptable, but the live load per foot of joist span is less than 120 lb/ft after subtracting the weight per foot of joist. It is unacceptable because the deflection limitation is exceeded:

22K10:   272 lb/ft of joist span − 12.6 lb/ft joist weight
= 259.4 lb/ft (versus 180 lb/ft)
128 lb/ft of joist span − 12.6 lb/ft joist weight
= 115.4 lb/ft (versus 120 lb/ft)

The next largest joist in the 22K group is selected. Other groups require sim-ilar selections:

24K10:   298 lb/ft of joist span − 13.1 lb/ft joist weight
= 284.9 lb/ft (versus 180 lb/ft)
154 lb/ft of joist span − 13.1 lb/ft joist weight
= 140.9 lb/ft (versus 120 lb/ft)

lightest→   26K7:   251 lb/ft of joist span − 10.9 lb/ft joist weight
= 240.1 lb/ft (versus 180 lb/ft)
131 lb/ft of joist span − 10.9 lb/ft joist weight
= 120.1 lb/ft (versus 120 lb/ft)

28K6:   220 lb/ft of joist span − 11.4 lb/ft joist weight
= 208.6 lb/ft (versus 180 lb/ft)
137 lb/ft of joist span − 11.4 lb/ft joist weight
= 125.6 lb/ft (versus 120 lb/ft)

Either the 24K10, 26K7, or 28K6 joist section is acceptable; the 26K7 is the lightest and the 24K10 is the shallowest.

As shown by comparing the results in Examples 20.1 and 20.2, joist span af-fects the depth of the joist required for a specific load. Simply, a longer joist span may mean fewer joists, but for a specific load it results in a deeper joist section. A deeper joist system requires a taller structure to achieve headroom requirements. The need for a taller structure significantly reduces any savings from fewer joists.

In Example 20.2 it first appears that the 22K10 is acceptable; the table lists its live load for a deflection of $\ell/360$ of the span at 128 lb/ft of joist span. This is greater than the anticipated live load of 120 lb/ft. However, after subtracting the weight of the joist (12.6 lb/ft), the 22K10 can only support a live load of 115.4 lb/ft. It is unacceptable because the $\ell/360$ deflection limitation is exceeded.

End supports   Much like a steel beam, a steel joist transfers its end loads to reaction supports. A reaction support is usually a wall, pier, column, or girder that is made of steel, masonry, or concrete. A bearing stress develops where the joist ends contact the reaction support material. Steel is much stronger in compression than are concrete and masonry. To prevent crushing of these reaction support materials, bearing plates are needed on masonry and concrete reaction supports. A bearing plate is a thick steel plate located at the reaction support. The joist end rests on the plate! It is larger in area than the joist end, so the plate can distribute the end loads over a larger bearing area.

Design of bearing plates for steel joists follows the same procedure as that outlined for steel beams (see Chapter 19). For the K-series joists outlined in this book, joist ends must, in most instances, bear at least 4 in over a masonry or concrete support and be anchored to the bearing plate. In the case of steel reaction supports, K-series joists must bear at least $2\frac{1}{2}$ in over the reaction support.

20.5   BRIDGING

As a deep, narrow beamlike member, a steel joist is not stiff enough to resist lateral forces created when it bends under load. If it is not braced, it will buckle laterally. Additionally, eccentric or misaligned loading may cause the beam to twist across a portion of its length, thereby causing a torsional buckling effect. These modes of failure were discussed in Chapter 16.

To minimize concerns with lateral and torsional buckling, open-web steel joists must be braced with bridging. *Bridging* is composed of light structural members such as angles, rods, bars, or tubing that is located perpendicular to repetitive joists and spans between the joists.

Horizontal bridging consists of two steel members, one attached to the top chord and one attached to the bottom chord, that run horizontally and continuously through the joists. Attachment to each joist is made by a welded or mechanical connection. The connection must be capable of withstanding no less than a 700-lb force. A horizontal bridging member is designed such that the ratio of its unbraced length (ℓ) in inches to its least radius of gyration (r) does not exceed 300:

$$\frac{\ell}{r} \le 300 \quad \text{or} \quad r \ge \frac{\ell}{300}$$

If a round bar is used, its diameter shall be at least $\frac{1}{2}$ in. in diameter.

Diagonal bridging consists of two steel members in a cross-braced pattern. Each member is attached to the top chord of one joist and the bottom chord of the other joist. The cross-braced members can be attached at their center intersecting point. All attachments are made by a welded or mechanical connection. A diagonal bridging member is selected such that the ratio of its unbraced length (ℓ) in inches to its least radius of gyration (r) does not exceed 200:

$$\frac{\ell}{r} \leq 200 \quad \text{or} \quad r \geq \frac{\ell}{200}$$

Where cross-braced members are attached at their center intersecting point, unbraced length (ℓ) is the distance from the center point of intersection to the chord of the joist, usually half of the diagonal length.

■ **EXAMPLE 20.3**

26K7 steel joists are specified (from Example 20.2) for a design span of 44 ft and a joist spacing of 4 ft. Determine the bridging required.

Solution   The web section number of a 26K7 is No. 7. From Table B.5, for a No. 7 web section and a span of 44 ft:

> Three rows of horizontal bridging are required for spans over 33 through 46 ft.
>
> Diagonal bridging is not required, as it is only needed for four or more rows.

Bridging should be evenly spaced across the joist span. With three rows of bridging needed, four spaces across the 44-ft joist span are required. Bridging must be spaced at 11 ft apart:

$$\frac{44 \text{ ft}}{4} = 11 \text{ ft}$$

With joists spaced at 4 ft, the length of horizontal bridging is 4 ft:

$$4 \text{ ft} \cdot 12 \text{ in/ft} = 48 \text{ in}$$

A horizontal bridging member must have a ratio of its unbraced length (ℓ) in inches to its least radius of gyration (r) that does not exceed 300:

$$r \geq \frac{\ell}{300}$$

$$\frac{\ell}{300} = \frac{48 \text{ in}}{300} = 0.160 \text{ in}$$

$$r \geq 0.160 \text{ in}$$

A structural member, such as a bar, angle, or pipe, with a least (smallest) radius of gyration greater than or equal to 0.160 in is acceptable. From Table B.6, an L1 × 1 × $\frac{1}{8}$ steel angle ($r_{\text{axis } z\text{-}z} = 0.196$ in) or a $\frac{1}{2}$ in-diameter steel pipe ($r = 0.261$ in) would be acceptable.

20.1 A steel joist system will have a design span of 32 ft. The roof (membrane, insulation, and deck) and ceiling (suspended ceiling, lights, ductwork, etc.) systems have a combined dead load of 20 lb/ft^2. Assume a live load of 30 lb/ft^2. Deflection for the live load should not exceed $\ell/360$ of the span. Due to the structural properties of the metal roof deck, a joist spacing of 5 ft is under consideration.

    (a) Select the lightest K-series joist designation that can carry the load safely regardless of depth.

    (b) Due to headroom requirements and material costs associated with building envelope height, joist depth should not exceed 20 in. Select the lightest K-series joist designation that can carry the load safely and not exceed 20 in. in depth.

    (c) With the steel joists selected in parts a and b, determine the bridging requirement.

20.2 A steel joist system will have a design span of 24 ft. The roof (membrane, insulation, and deck) and ceiling (suspended ceiling, lights, ductwork, etc.) systems have a combined dead load of 25 lb/ft^2. Assume a snow load of 30 lb/ft^2. Deflection for the live load should not exceed $\ell/360$ of the span. Due to the structural properties of the metal roof deck, a joist spacing of 4 ft is under consideration.

    (a) Select the lightest K-series joist designation that can carry the load safely regardless of depth.

    (b) Due to headroom requirements and material costs associated with building envelope height, joist depth should not exceed 16 in. Select the lightest K-series joist designation that can carry the load safely and not exceed 16 in. in depth.

    (c) With the steel joists selected in parts a and b, determine the bridging requirement.

20.3 A steel joist system will have a design span of 24 ft. The floor (concrete topping on steel decking) and ceiling (suspended ceiling, lights, ductwork, etc.) systems have a combined dead load of 70 lb/ft^2. Assume a live load of 100 lb/ft^2. Deflection for the live load should not exceed $\ell/360$ of the span.

    (a) Select the lightest K-series joist designation that can safely carry the load at a joist spacing of 12 in.

    (b) Select the lightest K-series joist designation that can safely carry the load at a joist spacing of 16 in.

    (c) Select the lightest K-series joist designation that can safely carry the load at a joist spacing of 24 in.

    (d) Assuming that the joists are repetitively placed in a masonry building that measures 24 ft × 32 ft, determine the number of joists required for each spacing.

    (e) Determine the lightest joist system.

20.4 28K6 steel joists are specified for a design span of 32 ft and a joist spacing of 4 ft.

    (a) Determine the required number of rows of bridging.

    (b) Determine the unbraced length of the bridging.

    (c) Determine the required least radius of gyration for the bridging required.

    (d) Determine the lightest angle section that can be used for bridging.

20.5 28K6 steel joists are specified for a design span of 52 ft and a joist spacing of 4 ft.

    (a) Determine the required number of rows of bridging.

    (b) Determine the unbraced length of the types of bridging required.

    (c) Determine the required least radius of gyration for the required bridging.

    (d) Determine the lightest angle section that can be used for bridging.

The finished envelope of a masonry building with a steel joist roof system.

The masonry building under construction.

A part of the roof truss system during construction. Repetitively spaced steel joists will support a roof platform constructed of steel decking. Stacks of steel decking before placement are evident. Steel joists bear on joist girders that span between vertically positioned columns.

Steel joist ends bearing on top of a masonry wall. Steel decking creates a steel roof platform.

Steel joists are individually welded to steel bearing plates. Each plate is anchored to cement grout placed during assembly of the masonry wall.

Steel joists can also bear in pockets in a masonry wall. Bearing plates are placed and anchored in the pockets when the masonry is laid.

Steel joists bearing in pockets in a masonry wall.

Large trusses called joist girders span between columns and support the steel joists.

Joist girders bear on and are connected to columns.

Interior columns transfer their load to individual column foundations.

Due to the heavy end load of a joist girder, a steel column is sometimes needed to support the girder instead of allowing it to bear on the wall. The girder reaction would exceed the strength of the masonry wall.

Exterior columns transfer their load to a foundation pedestal that is an integral part of the foundation.

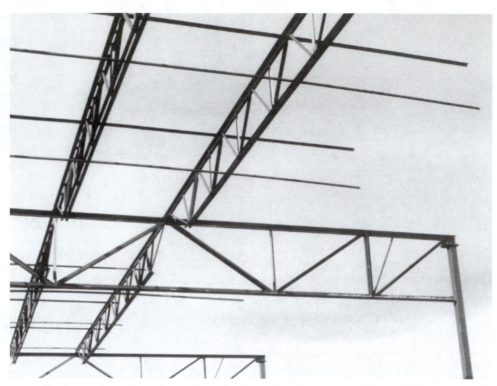

Slender steel members that make up horizontal bridging span steel joists. Bridging is spaced repetitively perpendicular to the joists, and runs continuously from joist to joist. Typically it is welded to the top and bottom chords of each joist.

Diagonal bridging is attached to the top chord of one joist and bottom chord of the next joist.

Joists with larger spans require diagonal bridging near the center of their span.

385

Chapter 21
DESIGN OF SIMPLE STRUCTURAL STEEL COLUMNS

An elementary design approach for structural steel columns using the allowable stress design method is introduced in this chapter. The reader is advised that the presentation does not provide full coverage of all design principles involved with design of steel structures. This book is intended to acquaint the technician with basic design principles used by an engineer. Additional coverage is beyond the intended breadth of the book.

21.1   COLUMN TABLES

As with structural steel beams, tabular data for column sections are available in the AISC *Manual of Steel Construction.* Tables for steel columns that were extracted from this resource are given in Table B.3.

The column section *designation* (i.e., W12, W14, W16, etc.) is found at the head of each AISC column table. On the left side of the main body of the table, effective length, in feet, is found. The main body of the table indicates the *maximum allowable axial load,* in kip, at a specific effective column length, designation, and weight for steels with a yield stress (F_y) of 36 and 50 ksi.

The *effective column length (KL)* is found by multiplying the effective column length factor by the column length *(L),* in feet. The *effective column length factor (K)* is a multiplier that accounts for constraints at the column ends and their effect on limiting buckling of the column. For example, a column with well-fixed ends, such as one with ends that extend substantially into concrete, is less likely to buckle than a column with ends that are free to rotate.

Theoretical values of effective column length factor *(K)* range from 0.5 to 2.0. Recommended design values are based on real-world connections, so they differ slightly from the theoretical values (Table 21.1). A column with fully restrained ends typically has a *K* value of less than 1.0, so its effective column length *(KL)* is

Table 21.1 EFFECTIVE COLUMN LENGTH FACTOR (*K* VALUE) FOR COLUMNS WITH DIFFERENT END CONNECTIONS

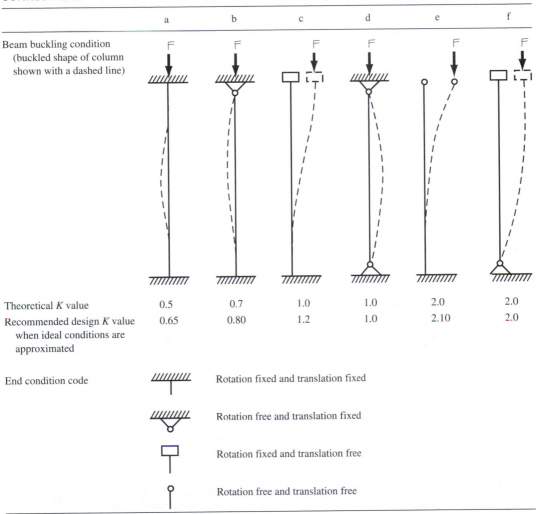

	a	b	c	d	e	f
Theoretical *K* value	0.5	0.7	1.0	1.0	2.0	2.0
Recommended design *K* value when ideal conditions are approximated	0.65	0.80	1.2	1.0	2.10	2.0

Source: AISC Manual of Steel Construction, 9th Edition

less than the actual column length (*L*). Since the end restraints limit buckling, the column can carry a greater axial load than a column without end restraints (Table 21.1). Conversely, a column that has ends that are free to rotate (i.e., pivot) and/or translate (i.e., move in one plane) has a greater potential for buckling. Such a column has a *K* value greater than 1.0 and will only support a much lighter axial load in comparison to a column with restrained ends (*K* < 1.0).

Steel column section selection using the AISC column tables is based on the following approach:

1. Determine the effective column length (KL), where column length (L) is in feet.
2. Determine the axial load to be carried by the column, in kip.
3. Select the lightest column sections for each column section designation (i.e., W8, W10, etc.) that will carry the load for the effective column length.

■ EXAMPLE 21.1

Select a W-shape column section that will support a load of 217 kip. Assume an actual column length of 8 ft, $K = 1.0$, and ASTM A36 ($F_y = 36$ ksi) steel.

Solution

$$KL = 1.0 \cdot 8 \text{ ft} = 8 \text{ ft}$$

From Table B.3, the following column sections for $KL = 8$ ft were selected as the lightest section in each designation:

W8×40	allowable axial load of 218 kip
W10×45	allowable axial load of 247 kip
W12×40	allowable axial load of 217 kip
W14×43	allowable axial load of 230 kip

Any of these column sections would be acceptable.

In this approach a column section is already selected and the allowable axial load is determined. Trial and error must be used to match a column section with a loading condition. Following is the calculation procedure:

1. Determine the unbraced length of the column (ℓ), in inches.
2. Find the slenderness ratio (SR):

$$\text{SR} = \frac{K\ell}{r}$$

where effective column length factor (K), unbraced column length (ℓ), in inches, and the least radius of gyration (r) of the column section selected are the variables. The radius of gyration is an index of stiffness related to buckling against an axis of a specific column section. The lowest of the $r_{y\text{-}y}$ or $r_{x\text{-}x}$ values should be used unless it is braced in that direction. Refer to Table B.1.

3. Determine the largest effective slenderness ratio (Cc):

$$C_c\sqrt{\frac{2\pi^2E}{F_y}}$$

where pi (π) is the mathematical constant of 3.1416, the modulus of elasticity of steel (E), in lb/in^2, and the minimum yield strength of the steel (F_y), in lb/in^2, are the variables. For ASTM A36 steel, $E = 29\,000\,000$ lb/in^2 and $F_y = 36\,000$ lb/in^2, so C_c is 126.1.

4. Determine the maximum allowable axial load (F), in pounds:

$$F = F_a \cdot A$$

where A is the section area, in in^2, and F_a is found by the appropriate formula:

(a) Short column formula ($K\ell/r$ is less than C_c):

$$F_a = \frac{\{1 - [(K\ell/r)^2/2C_c^2]\}F_y}{FS}$$

where:

$$FS = \frac{5}{3} + \frac{3(K\ell/r)}{8C_c} - \frac{(K\ell/r)^3}{8C_c^3}$$

(b) Long column formula ($K\ell/r$ is greater than C_c but not to exceed 200):

$$F_a = \frac{12\pi^2E}{23(K\ell/r)^2}$$

In the formulas above, we have:

A	column section area, in in^2
C_c	largest effective slenderness ratio (126.1 for ASTM A36 steel)
E	modulus of elasticity, in lb/in^2 (29 000 000 lb/in^2 for steel)
F	maximum allowable axial load, in lb
F_a	axial stress permitted, in lb/in^2
F_y	minimum yield stress of the steel, in lb/in^2 (36 000 lb/in^2 for ASTM A36 steel)
FS	factor of safety
K	effective column length factor

$K\ell/r$ slenderness ratio (SR)

ℓ unbraced column length, in inches

r least radius of gyration (i.e., lowest $r_{y\text{-}y}$ or $r_{x\text{-}x}$ value for section), in inches

π the mathematical constant pi (roughly the number 3.1416)

■ **EXAMPLE 21.2**

Determine the maximum allowable axial load for a W12×40 column that has an unbraced length of 8 ft. The column is made of ASTM A36 steel and is restrained at its ends such that $K = 1.0$.

Solution

1. Determine the unbraced length of the column, provided in the problem statement as 8 ft:

$$\ell = 8 \text{ ft} \cdot 12 \text{ in/ft} = 96 \text{ in}$$

2. Find the slenderness ratio (SR). The lowest of the $r_{y\text{-}y}$ or $r_{x\text{-}x}$ values is $r_{y\text{-}y} = 1.93$ in:

$$\text{SR} = \frac{K\ell}{r} = \frac{1.0 \cdot 96 \text{ in}}{1.93 \text{ in}} = 49.74$$

3. Determine the largest effective slenderness ratio (C_c). For ASTM A36 steel, $C_c = 126.1$.

4. Determine the maximum allowable axial load (F), in pounds. Since $K\ell/r$ is less than C_c, the short column formula must be used:

$$\text{FS} = \frac{5}{3} + \frac{3(K\ell/r)}{8C_c} - \frac{(K\ell/r)^3}{8C_c^3}$$

$$= \frac{5}{3} + \frac{3(49.74)}{8(126.1)} - \frac{(49.74)^3}{8(126.1)^3} = 1.807$$

$$= \frac{\{1 - [(49.74)^2 / 2(126.1)^2]\}\, 36\,000 \text{ lb /in}^2}{1.807} = 18\,373 \text{ lb /in}^2$$

The cross-sectional area of a W12×40 is 11.775 in²:

$$F = F_a \cdot A = 18\,373 \text{ lb/in}^2 \cdot 11.775 \text{ in}^2 = 216\,342 \text{ lb}$$

A W12×40 column with an unbraced length of 8 ft can support an axial load of about 217 kip.

■ EXAMPLE 21.3

Determine the maximum allowable axial load for a W12×40 column that has an unbraced length of 8 ft. The column is made of ASTM A36 steel. Unlike the similar column in Example 21.2, this column is restrained at its ends such that $K = 2.0$.

Solution

1. Determine the unbraced length of the column, provided in the problem statement as 8 ft:

$$\ell = 8 \text{ ft} \cdot 12 \text{ in/ft} = 96 \text{ in}$$

2. Find the slenderness ratio (SR). The lowest of the $r_{y\text{-}y}$ or $r_{x\text{-}x}$ values is $r_{y\text{-}y} = 1.93$ in:

$$\text{SR} = \frac{K\ell}{r} = \frac{2.0 \cdot 96 \text{ in}}{1.93 \text{ in}} = 99.48$$

3. Determine the largest effective slenderness ratio (C_c). For ASTM A36 steel, $C_c = 126.1$.

4. Determine the maximum allowable axial load (F), in pounds. Since $K\ell/r$ is less than C_c, the short column formula must be used:

$$\text{FS} = \frac{5}{3} + \frac{3(K\ell/r)}{8C_c} - \frac{(K\ell/r)^3}{8C_c^3}$$

$$= \frac{5}{3} + \frac{3(99.48)}{8(126.1)} - \frac{(99.48)^3}{8(126.1)^3} = 1.901$$

$$F_a = \frac{\{1 - [(K\ell/r)^2/2C_c^2]\}F_y}{\text{FS}}$$

$$= \frac{\{1 - [(99.48)^2/2(126.1)^2]\}36\,000 \text{ lb/in}^2}{1.901} = 13\,044 \text{ lb/in}^2$$

The cross-sectional area of a W12×40 is 11.775 in²:

$$F = F_a \cdot A = 13\,044 \text{ lb/in}^2 \times 11.775 \text{ in}^2 = 153\,593 \text{ lb}$$

A W12×40 column with an unbraced length of 8 ft can support an axial load of about 217 kip.

As demonstrated in Examples 21.2 and 21.3, the way in which a column is connected will affect the maximum load it will support. Even though these columns are the same length, the column in Example 21.3 ($K = 2.0$) has a greater tendency to buckle than the column in Example 21.2 ($K = 1.0$) because its KL value is greater.

■ EXAMPLE 21.4

Determine the maximum allowable axial load for a W12×40 column that has an unbraced length of 16 ft. The column is made of ASTM A36 steel and is restrained at its ends such that $K = 1.0$.

Solution

1. Determine the unbraced length of the column, provided in the problem statement as 16 ft:

$$\ell = 16 \text{ ft} \cdot 12 \text{ in/ft} = 192 \text{ in}$$

2. Find the slenderness ratio (SR). The lowest of the $r_{y\text{-}y}$ or $r_{x\text{-}x}$ values is $r_{y\text{-}y} = 1.93$ in:

$$\text{SR} = \frac{K\ell}{r} = \frac{1.0 \cdot 192 \text{ in}}{1.93 \text{ in}} = 99.48$$

3. Determine the largest effective slenderness ratio (C_c). For ASTM A36 steel, $C_c = 126.1$.

4. Determine the maximum allowable axial load (F), in pounds. Since $K\ell/r$ is less than C_c, the short column formula must be used:

$$\text{FS} = \frac{5}{3} + \frac{3(K\ell/r)}{8C_c} - \frac{(K\ell/r)^3}{8C_c^3}$$

$$= \frac{5}{3} + \frac{3(99.48)}{8(126.1)} - \frac{(99.48)^3}{8(126.1)^3} = 1.901$$

$$F_a = \frac{\{1 - [(K\ell/r)^2/2C_c^2]\}F_y}{\text{FS}}$$

$$= \frac{\{1 - [(99.48)^2/2(126.1)^2]\}36\,000 \text{ lb/in}^2}{1.901} = 13\,044 \text{ lb/in}^2$$

The cross-sectional area of a W12×40 is 11.775 in²:

$$F = F_a \cdot A = 13\,044 \text{ lb/in}^2 \times 11.775 \text{ in}^2 = 153\,593 \text{ lb}$$

A W12×40 column with an unbraced length of 16 ft can support an axial load of about 154 kip.

■ EXAMPLE 21.5

Determine the maximum allowable axial load for a W12×40 column that has an unbraced length of 24 ft. The column is made of ASTM A36 steel and is restrained at its ends such that $K = 1.0$.

Solution

1. Determine the unbraced length of the column, provided in the problem statement as 24 ft:

$$\ell = 24 \text{ ft} \cdot 12 \text{ in/ft} = 288 \text{ in}$$

2. Find the slenderness ratio (SR). The lowest of the $r_{y\text{-}y}$ or $r_{x\text{-}x}$ values is $r_{y\text{-}y} = 1.93$ in:

$$Sr = \frac{K\ell}{r} = \frac{1.0 \cdot 288 \text{ in}}{1.93 \text{ in}} = 149.2$$

3. Determine the largest effective slenderness ratio (C_c). For ASTM A36 steel, $C_c = 126.1$.

4. Determine the maximum allowable axial load (F), in pounds. Since $K\ell/r$ is greater than C_c but less than 200, the long column formula must be used:

$$
\begin{aligned}
F_a &= \frac{12\pi^2 E}{23(K\ell/r)^2} \\
&= \frac{12\pi^2(29\ 000\ 000 \text{ lb/in}^2)}{23(149.2)^2} \\
&= 6708 \text{ lb/in}^2
\end{aligned}
$$

The cross-sectional area of a W12×40 is 11.775 in²:

$$F = F_a \cdot A = 6708 \text{ lb/in}^2 \times 11.775 \text{ in}^2 = 78\ 986 \text{ lb}$$

A W12×40 column with an unbraced length of 24 ft can support an axial load of about 79 kip.

The columns in Examples 21.4 and 21.5 are 16 ft and 24 ft long, respectively. The maximum axial load of 154 kip for the longer column is considerably less than the 217 kip that can be carried by the same W12×40 section at half the effective length (see Example 21.3). Unbraced column length affects the load that a column will carry safely.

21.4   COLUMN BASE PLATES

A steel column typically transfers the load it supports to a concrete or masonry foundation. To prevent crushing of the foundation material, a steel base plate at the base of a column is needed to distribute the column load over a larger bearing area.

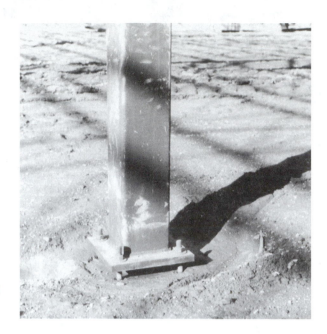

A column base plate is secured to the foundation with anchor bolts. Cement grout will be forced into the space between the plate and the concrete foundation. Once it sets, the grout provides a good bearing surface so the column load can be evenly transferred from the base plate to the foundation.

The plate is typically welded to the end of the column as shown in Figure 21.1. Once the structure has been erected, cement grout is forced between the column base plate and the bearing surface of the foundation. Once it sets, grout creates a good bearing surface, so the column load is evenly transferred from the base plate to the foundation. The plate is typically anchored to the foundation with anchor bolts.

In the absence of more stringent requirements, the following bearing stresses apply: The allowable bearing stress (F_p) for brick in cement mortar is 250 lb/in². For sandstone and limestone, the allowable bearing stress (F_p) is 400 lb/in². The allowable bearing stress of concrete is reduced from the compressive strength (f'_c) specified for concrete, the largest compressive stress that the concrete must develop without failure in laboratory testing at 28 days after curing (see Chapter 26). The compressive strength specified for concrete used in foundations is typically 2500, 3000, or 4000 lb/in². In those rare instances when the base plate covers the entire bearing surface area of the foundation ($A_{bp} = A_{bs}$), the allowable bearing stress (F_p) of concrete is based on

$$F_p = 0.35f'_c$$

In most cases, the bearing surface area of the concrete foundation member (A_{bs}) is greater than the base plate bearing area (A_{bp}), so the allowable bearing stress may be increased by the following expression up to $F_p = 0.7f'_c$:

$$F_p = 0.35f'_c\sqrt{\frac{A_{bs}}{A_{bp}}}$$

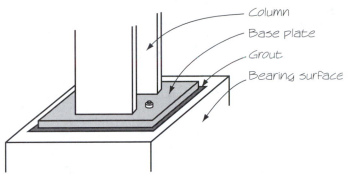

FIGURE 21.1 A steel column base plate is typically welded to the bottom end of the column. Once the structure has been erected, cement grout is forced between the column base plate and the bearing surface of the foundation. Once it sets, grout creates a good bearing surface so the column load is evenly transferred from the base plate to the foundation. The plate is typically anchored to the foundation with anchor bolts.

To limit the bearing stress of the foundation material to an acceptable level, the column base plate must be larger in area than the column section profile. The allowable bearing stress of concrete or masonry (F_p) governs the minimum size of a steel base plate bearing area (A_{bp}) for a specific axial column load including the weight of the column (P). The plae must extend beyond the breadth and depth of the section profile to accomodate welding the plate to the section.

$$A_{bp} \geq \frac{P}{F_p}$$

Plate thickness is computed with the assumption that the portion of the plate that extends past the column section profile bends upward beyond the edges of the column section. The required thickness of the plate (t_p) is determined by the largest thickness found by the following expressions:

$$t_p \geq 2m \sqrt{\frac{F_p}{F_y}} \quad \text{or} \quad t_p \geq 2n \sqrt{\frac{F_p}{F_y}}$$

Dimensions m and n are as shown in Figure 21.2. The minimum yield stress (F_y) is for the steel used in the column base plate.

Steel plate thicknesses are available in $\frac{1}{32}$-in increments up to $\frac{1}{2}$ in thick, $\frac{1}{16}$-in increments over $\frac{1}{2}$ to 1 in thick, $\frac{1}{8}$-in increments over 1 to 3 in thick, and $\frac{1}{4}$-in increments over 3 in thick. Plates are available in widths up to 60 in and limited by lengths that can be transported by truck. Plate width and length are custom cut. In base plate design, the designer typically rounds the plate width and length in inch increments.

FIGURE 21.2 Layout of the dimensions for design of a column base plate.

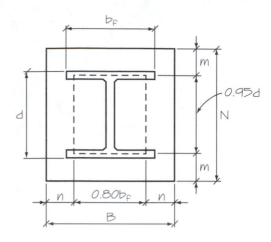

■ EXAMPLE 21.6

Design a column base plate of A36 steel ($F_y = 36\,000$ lb/in^2) for a W14×120 section profile supporting an axial load of 550 kip. The column will transfer the load to a 42 in × 42 in concrete foundation pedestal. The specified compressive strength of the concrete is 3000 lb/in^2.

Solution First assume that the bearing surface area of the concrete foundation member (A_{bs}) is significantly greater than the base plate bearing area (A_{bp}), so the allowable bearing stress may be increased by

$$F_p = 0.7f'_c = 0.7(3000\ \text{lb/in}^2) = 2100\ \text{lb/in}^2$$

The minimum base plate bearing area ($A_{bp\ min}$) and approximate dimensions are found by

$$A_{br\ min} = \frac{P}{F_p} = \frac{550\,000\ \text{lb}}{2100\ \text{lb/in}^2} = 261.9\ \text{in}^2$$
$$\sqrt{261.9\ \text{in}^2} = 16.2\ \text{in}$$

From Table B.1, the dimensions of a W14×120 section profile are a flange breadth (b_f) of 14.67 in and a depth (d) of 14.48 in. With the base plate required to have dimensions of at least 16.2 in × 16.2 in, a 16 in × 18 in plate is specified:

$$A_{bp} = B \cdot N = 16\ \text{in} \cdot 18\ \text{in} = 288\ \text{in}$$
$$m = \frac{N - 0.95d}{2} = \frac{18\ \text{in} - 0.95(14.48\ \text{in})}{2} = 2.12\ \text{in}$$
$$n = \frac{B - 0.8b_f}{2} = \frac{16\ \text{in} - 0.8(14.67\ \text{in})}{2} = 2.13\ \text{in}$$

The minimum plate thickness is found by

$$t_p = 2n \sqrt{\frac{F_p}{F_y}} = 2(2.12 \text{ in}) \sqrt{\frac{2100 \text{ lb/in}^2}{36\,000 \text{ lb/in}^2}} = 1.02 \text{ in}$$

$$t_p = 2m \sqrt{\frac{F_p}{F_y}} = 2(2.13 \text{ in}) \sqrt{\frac{2100 \text{ lb/in}^2}{36\,000 \text{ lb/in}^2}} = 1.03 \text{ in}$$

A $1\frac{1}{8}$-in-thick 16 in $\times$ 18 in plate is acceptable.

EXERCISES

21.1 Compute the maximum allowable axial load, in kips, of a W10×49 column with an unbraced length of 14 ft and $K = 1.0$. ASTM A36 ($F_y = 36\,000$ lb/in^2) steel is specified. Compare the results with those in the AISC tables.

21.2 Much as in Exercise 21.1, compute the maximum allowable axial load, in kips, of a W10×49 column with an unbraced length of 14 ft. ASTM A36 ($F_y = 36\,000$ lb/in^2) steel is specified. In this exercise, however, $K = 2.0$. Compare the results with those in the AISC tables.

21.3 Much as in Exercise 21.1, compute the maximum allowable axial load, in kips, of a W10×49 column with an unbraced length of 14 ft. ASTM A36 ($F_y = 36\,000$ lb/in^2) steel is specified. In this exercise, however, $K = 0.5$. Compare the results with those in the AISC tables.

21.4 Much as in Exercise 21.1, compute the maximum allowable axial load, in kips, of a W10×49 column with an unbraced length of 14 ft and $K = 1.0$. In this exercise, however, ASTM A572 ($F_y = 50\,000$ lb/in^2) steel is specified. Compare the results with those in the AISC tables.

21.5 In Exercises 21.1 through 21.4, the column section remained the same but the effective column length factor (K) or type of steel varied. Compare the maximum allowable axial load for each exercise to the load for the column in Exercise 21.1.
(a) How is the maximum allowable axial load of a column affected when K decreases?
(b) How is the maximum allowable axial load of a column affected when K increases?
(c) How is the maximum allowable axial load of a column affected when yield strength of steel increases?

21.6 Compute the maximum allowable axial load, in kips, of a W14×120 column with an unbraced length of 18 ft and $K = 1.0$. ASTM A36 ($F_y = 36\,000$ lb/in^2) steel is specified. Compare the results with those in the AISC tables.

21.7 Much as in Exercise 21.6, compute the maximum allowable axial load, in kips, of a W14×120 column with an unbraced length of 18 ft. ASTM A36 ($F_y = 36\,000$ lb/in^2) steel is specified. In this exercise, however, $K = 1.2$.
(a) Compare the results with those in the AISC tables.
(b) Compare the results with those in Exercise 21.6.

21.8 A 20-ft column will support an axial load of 500 kip, including the weight estimated for the column. ASTM A36 ($F_y = 36\,000$ lb/in^2) steel is specified. The effective column length factor (K) is 1.2. Determine the lightest W14 column section that will support the load.

21.9 A 16-ft column will support an axial load of 800 kip, including the weight estimated for the column. ASTM A36 (F_y = 36 000 lb/in^2) steel is specified. The effective column length factor (K) is 1.0. Determine the lightest W14 column section that will support the load.

21.10 A base plate is needed for a column with a W12×50 section profile supporting an axial load of 200 kip, including the weight estimated for the column. ASTM A36 (F_y = 36 000 lb/in^2) steel is specified. The base plate will transfer the load to a 24 in × 24 in concrete foundation. The compressive strength specified for the concrete is 3000 lb/in^2. Design a rectangular base plate for this column.

21.11 Much as in Exercise 21.10, a base plate is needed for a column with a W12×50 section profile supporting an axial load of 200 kip, including the weight estimated for the column. ASTM A36 (F_y = 36 000 lb/in^2) steel is specified. In this exercise, however, the base plate will transfer the load to a concrete foundation pedestal that will be the same size as the base plate. The compressive strength specified for the concrete is 3000 lb/in^2. Design a rectangular base plate for this column.

21.12 Much as in Exercise 21.10, a base plate is needed for a column with a W12×50 section profile supporting an axial load of 200 kip, including the weight estimated for the column. ASTM A36 (F_y = 36 000 lb/in^2) steel is specified. The base plate will transfer the load to a 24 in × 24 in concrete foundation. In this exercise, however, the compressive strength specified for the concrete is 2500 lb/in^2. Design a rectangular base plate for this column.

21.13 A 20-ft column will support an axial load of 350 kip, including the weight estimated for the column. ASTM A36 (F_y = 36 000 lb/in^2) steel is specified. The effective column length factor (K) is 1.0. The base plate will transfer the load to a 30 in × 30 in concrete foundation pedestal. The compressive strength specified for the concrete is 4000 lb/in^2.
(a) Determine the lightest W14 column section that will support the load.
(b) Design a square base plate for this column.

21.14 A column with an effective length (KL) of 20 ft will support an axial load of 1400 kip, including the weight estimated for the column. ASTM A572 (F_y = 50 000 lb/in^2) steel is specified. The base plate will transfer the load to a 5 ft × 5 ft concrete foundation pedestal. The compressive strength specified for the concrete is 3000 lb/in^2.
(a) Determine the lightest W14 column section that will support the load.
(b) Design a square base plate for this column.

Chapter 22
TIMBER AND ENGINEERED WOOD PRODUCTS

This chapter provides information on timber and engineered wood products. It serves as a prelude to the methods used in elementary design of simple timber beams, joists, rafters, and columns introduced in Chapters 23 to 25.

22.1 SAWN LUMBER

Sawn lumber is cut from logs, dried, and shaped into standard sections suitable for construction. Common wood species used for solid lumber include Douglas Fir, Western Hemlock, Southern Pine, Eastern Spruce, Red or White Oak, and Southern Cypress. Redwood, Cedar, or Cypress may be used as structural members in high-moisture exposure applications, due to the ability of these species to resist decay at a high moisture content.

Many factors influence the mechanical properties of lumber, including species, density, moisture content, slope of grain, and defects such as knots and checks (splits). Research and testing show that mechanical properties of wood, especially strength and stiffness, vary considerably by species. For example, Douglas Fir is stiffer and slightly stronger than Western Hemlock. Wood from old-growth forests tends to be more dense and thus stronger and stiffer than the new-growth timber that is cultivated on tree farms.

22.2 ENGINEERED WOOD PRODUCTS

Recent concerns with increased cost, decreased availability, and environmental issues related to lumber industry practices have caused a number of *engineered wood products* to become available and serve as good substitutes for solid sawn lumber. Fabricated at the factory and delivered to the construction site, manufacture and inspection of engineered wood members can be better controlled. In addition, engineered wood members can be fabricated into sections that are difficult to obtain with solid sawn lumber, making them more useful in unique structural applications. Common engineered wood members include:

- *Glued-laminated timber.* Usually referred to as *gluelam* members, these members are used as beams, columns, and arches. Glued laminated members are manufactured at a timber laminating plant. They are composed of laminations that do not exceed 2 inches in net thickness which are securely bonded together with adhesives to form the member. Laminations may be comprised of pieces end-joined to form any length, of pieces placed or glued edge to edge to make wider ones, or of pieces bent to a curved form during gluing. The grain of all laminations is approximately parallel longitudinally.

- *Laminated veneer lumber (LVL).* Sometimes referred to by the trade name *Microllam,* these members are generally used for headers, lintels, beams, and joists. Laminated veneer lumber beams are made from thin laminations (like plywood plies) that are placed in a grain-oriented position and glued together under pressure.

- *Parallel strand lumber (PSL).* These are frequently referred to by the trade name *Parallam.* These members are composed of slender wood strands that are oriented longitudinally and glued under pressure using microwave energy. They serve as an alternative to gluelams.

- *Laminated strand lumber (LSL).* A newer product that is sometimes referred to by the trade name *TimberStrand.* These members are composed of elongate flakelike wood strands that are oriented longitudinally and flat along the depth of the member and are glued under pressure. LSL members serve as an alternative to solid wood beams and gluelams.

- *I joists.* These are commonly known on the construction site by the trade names *TMI* or *TJI.* I joists are commonly used as an alternative to solid lumber members, such as joists on floor and roof decks. I joists are comprised of solid or laminated flanges that are secured to a plywood or an oriented strand board web. The flanges provide strength against bending stresses at the extreme fibers, and the web serves to hold the flanges apart.

- *Trusses and open-web joists.* These members have become a cost-effective alternative to roof and floor framed systems (rafters and joists). Use of small individual components to fabricate the member decreases weight and allows for large spans. Each element of these members is designed to withstand either tensile or compressive stresses, but not bending stresses.

Fabrication of engineered wood products is the responsibility of the manufacturer. Standards of manufacture and use are typically established by professional and trade organizations. APA–The Engineered Wood Association, Inc. (formerly the American Plywood Association), Tacoma, Washington (Web site address: www.apawood.org), is a source of information about engineered wood products such as plywood, oriented strand board, glued laminated timber, composite panels, and wood I joists. The American Institute of Timber Construction, Inc. (AITC), 7012 South Revere Parkway, Suite 140, Englewood, CO 80112 (Web site address: www.aitc-glulam.org) is the national technical trade association of the structural glued-laminated (glulam) timber industry. The building designer, architect, or engineer specifies and uses engineered wood products according to the manufacturer's specifications.

22.3 LUMBER SIZES AND GRADES

Dimension lumber is a surfaced sawn wood product. It is identified by its *nominal* section size. When dry, a 2×10 has an *actual* section dimension of $1\frac{1}{2}$ in $\times 9\frac{1}{4}$ in. The need to surface all sides of rough sawn lumber results in a member with an actual size that is smaller than the nominal size. The *National Grading Rule for Dimension Lumber* classifies and grades dimension lumber by several basic size categories related to intended end uses.

- *Dimension lumber structural grades.* These include surfaced softwood products of nominal thickness from 2 to 4 inches in thickness by 2 in and wider. These grades are intended for use as general framing members, including beams, joists, planks, rafters, and studs. Products are available in a variety of lengths, beginning at 6 ft and increasing in multiples of 2 ft. *Structural light framing grades* are used in light framing sizes for engineered systems, trusses, and multistory projects. *Light framing grades* are suited for general framing applications such as wall framing, plates, sills, cripples, blocking, and so on. *Stud grade* is used in interior and exterior wall framing.

- *Structural joists and planks.* This is a category of dimension lumber products (2×5 through 4×18) intended to fit structural applications for lumber 5 in and wider, such as floor joists, ceiling joists, roof rafters, headers, small beams, trusses, and general framing.

- *Beam and stringer grade.* These are large solid wood members that are 5 in and thicker (nominal) with a width more than 2 in greater than the thickness (i.e., 6×10, 8×12, etc.). Wood used for beams and stringers is graded to withstand flexure stresses caused by bending across the thicker portion of the beam section.

- *Post and timber grade.* These are large solid wood members that are 5 in $\times$ 5 in and larger (nominal), with a width not more than 2 in greater than the thickness (i.e., 6×6, 6×8, etc.). Wood used for these members is graded predominantly to resist compression stresses but not bending stresses.

In addition to classification by intended use, lumber is graded for characteristics of defects such as position and tightness of knots, splits, and checks. *Visual*

An inspector with a trained eye visually grades each piece of dimension lumber after it is rough cut, dried, and dressed at the lumber mill.

Dimension lumber is stacked by grade before it is bundled and shipped from the lumber mill.

grading involves observation and grading each piece of lumber by the trained eye of an inspector. *Machine stress-rated* (MSR) lumber involves nondestructive testing of each piece of lumber with mechanical stress-rating equipment to measure its stiffness and other physical working properties before it is subjected to visual inspection.

22.4 MOISTURE CONTENT

Sawn lumber and other wood products change dimensionally with change in moisture content; wood will swell with increase in moisture content and shrink with decrease in moisture content. Because of its cell structure, wood shrinks primarily in width and thickness and very little in length. Lumber attempts to reach equilibrium with the constantly changing relative humidity of the air in its immediate environment; usually, it stays between 8% in dry environments and 12% in humid environments. By definition, *green wood* has a moisture content above 19%.

Freshly cut lumber contains much moisture and is dried (seasoned) in the air or in an ovenlike kiln before surfacing or use. In structural grades, the term *dry* indicates a wood product that was either kiln- or air-dried to a moisture content of 19% or less before surfacing. The strength and stiffness of wood generally decreases with increase in moisture content.

Symbols Used in Structural Wood Design

In the chapters on timber and wood, the symbol F is used as a variable quantity for allowable stresses of wood known as base design properties (i.e., F_b, F_t, F_c, etc.). In the first 17 chapters, the Greek lowercase letter sigma (σ) was used to symbolize a variable quantity for stress because of its wide acceptance and universal use in general engineering. An exception to the use of σ is made in this and later chapters because different symbols are used in this branch of the structural engineering profession.

The symbol F was used in earlier chapters to identify a variable quantity for a concentrated force. Different symbols related to type of loads are also used in this chapter instead of F, such as P for a concentrated load, w for a uniformly distributed load, and W for a total uniformly distributed load. Although it can lead to some confusion, every effort is made to use trade-specific symbols in this book.

Thousands of tests of defect-free wood specimens were used to compile *clear wood strength* properties for most wood species. *Base design properties* for strength and stiffness in sawn lumber are established by decreasing the ideal clear wood strength values to values that include a factor of safety. *Design Values for Wood Construction* (1993) is a supplement to the NDS that contains design values for sawn lumber and glued laminated timber.

Base design values for sawn lumber are categorized into six basic properties as noted below and illustrated in Figure 22.1 to 22.6.

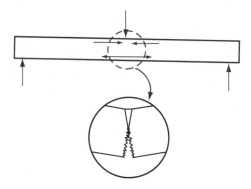

FIGURE 22.1 Exaggerated illustration of failure from extreme fiber stress in bending (F_b).

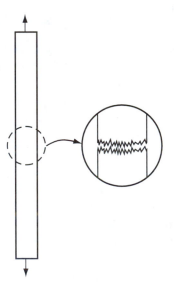

FIGURE 22.2 Exaggerated illustration of failure from tension stress parallel to the grain (F_t).

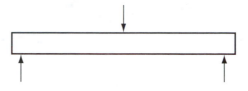

FIGURE 22.3 Exaggerated illustration of failure from horizontal shear stress (F_v).

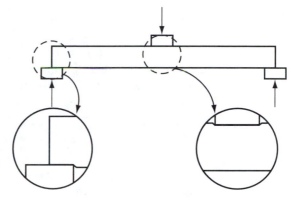

FIGURE 22.4 Exaggerated illustration of failure from compression stress perpendicular to the grain ($F_{c\perp}$).

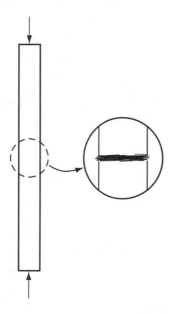

FIGURE 22.5 Exaggerated illustration of failure from compression stress parallel to the grain (F_c)

$E = 2,000,000$ psi / deflection = 1"

$E = 1,000,000$ psi / deflection = 2"

FIGURE 22.6 Exaggerated illustration of the effect of changing the modulus of elasticity (E).

- *Extreme fiber stress in bending* (F_b), in lb/in^2. As the beam undergoes loading, it bends. Although all fibers experience tensile or compressive stresses due to bending, the extreme fibers experience the greatest stresses. It is at the extreme fibers that a beam section first experiences flexure (bending) failure. Values of F_b are based on use in design as either repetitive or single members. *Repetitive* members are three or more 2- to 4-in thick members such as joists, rafters, truss chords, studs, planks, or decking that work together to support the design load. These repetitive elements, positioned side by side, share the load. The likelihood of defects occurring in the same location on all repetitive members is low, so the strength of the entire assembly is enhanced. *Single* members are solid beams and girders that carry the load alone. They are more likely to have defects in critical areas.

- *Tension stress parallel to the grain* (F_t), in lb/in^2. When a wood member is pulled with forces directed longitudinally with the grain, it experiences this type of stress. Tensile stresses act across the full section and over the entire length of the member. These stresses tend to pull the member apart. Wood offers its greatest tensile strength parallel to the grain.

- *Horizontal shear stress* (F_v), in lb/in^2. As a beam bends, theoretical fibers that run longitudinally through the beam tend to slide past one another, much like sheets in a ream of paper slip as the ream is curled. These shear stresses vary considerably across the length of the member.

- *Compression stress perpendicular to the grain* $(F_c \perp)$, in lb/in^2. As a joist or beam rests at its ends or a load rests on the member, a crushing force develops at the points of contact. Stress developed by this force is compressive in nature and acts perpendicular to the grain. Seasoned lumber is about 50% stronger than green lumber.

- *Compression stress parallel to the grain* (F_c), in lb/in^2. A column or post carrying a load experiences stresses parallel to the grain that are compres-

Farmed timber is a production crop grown and harvested from tree farms. Trees in photograph are 10 years old. Inferior trees will be cut and removed to make room for efficient growth of choice trees.

sive in nature. Compressive stresses act across the full section and over the entire length of the member.

- *Modulus of elasticity* (E), in lb/in^2. The modulus of elasticity serves as an index of material stiffness. Douglas Fir has a modulus of elasticity that is about 60% greater than the modulus of elasticity of Redwood. A Redwood beam experiencing the same loading condition as a beam of Douglas Fir will exhibit 60% more deflection than the Douglas Fir beam.

22.6   OLD-GROWTH VERSUS FARMED LUMBER

Timber from old-growth forests grows more slowly in a natural environment, while select new-growth timber is cultivated on tree farms. Timber from old-growth forests tends to be more dense, and thus stronger and stiffer, than new-growth timber. Due to environmental issues concerning the harvesting of old-growth timbers, new-growth lumber is widely available in the marketplace. New-growth lumber typically has base design values that are significantly lower than the old-growth properties cited in older-model building codes and other references. Caution should be used by the designer when designing to the limit of older design values.

Adjustment factors are used as multipliers to increase or decrease base design values (i.e., F_b, E, F_c, etc.) for special conditions. Common adjustment factors include:

- *Load duration factor* (C_D). Wood has the ability to carry substantially greater maximum loads for short durations (a few weeks) versus long durations (several years). A load duration factor (C_D) is used when additional intermittent loading is anticipated, such as under snow and construction loading conditions. Tabulated design values for F_b and F_v for *normal duration loading* ($C_D = 1.00$) apply to normal occupancy (live-load) conditions over a 10-year period. *Snow loading* ($C_D = 1.15$) is for a two-month snow loading condition. *Construction* or *seven-day loading* ($C_D = 1.25$) represents loading that may occur during construction.

- *Repetitive member factor* (C_r). The $C_r = 1.15$ factor is applied to the extreme fiber stress in bending (F_b) for repetitive 2- to 4-in-thick members such as joists, rafters, truss chords, studs, planks, decking, or similar members that are three or more in number and in contact or spaced no more than 24 in on center and joined by floor, roof, or other load-distributing elements adequate to support the design load.

- *Size factor* (C_F). This factor is applied to the extreme fiber stress in bending (F_b). For dimension lumber, it is available in tabular form based on grade of lumber. For timbers, beams, stringers, or posts greater than 12 in. in depth (d), it is based on:

$$C_F = \left(\frac{12}{d}\right)^{1/9} = \left(\frac{12}{d}\right)^{0.1111}$$

- *Wet service factor* (C_M). This factor is applied when the moisture content of the lumber is anticipated to be above 19% for an extended time period.

- *Flat-use factor* (C_{fu}). This factor is applied to the extreme fiber stress in bending (F_b) when dimension lumber is used with the load applied to the flat face such as decking.

When appropriate for the condition, relevant adjustment factors (C) are applied to the base design values (F) to determine the adjusted base design value (F_{adj}) as follows:

$$F_{\text{adj}} = (C_r)(C_F)(C_M)(C_{fu})(C_D)F$$

Adjustment factors specific to use and size of lumber or wood product are provided in Tables C.6 to C.8. The interested reader is directed to the building code for other adjustment factors.

■ EXAMPLE 22.1

Adjust the base design value for extreme fiber stress in bending (F_b) for a 6×16 beam. The wood species and grade under consideration is visually graded Douglas Fir-Larch, No. 2. The member will be used in a high-moisture application.

Solution From Table C.11, a 6×16 has a depth (d) of $15\frac{1}{2}$ in. The C_F is found by computation:

$$C_F = \left(\frac{12}{d}\right)^{1/9} = \left(\frac{12}{15.5}\right)^{1/9} = 0.972$$

From Table C.7 because the member is 5 in $\times$ 5 in or larger, the wet service factor (C_M) for F_b for a 6-in-thick member is 1.00, and visually graded Douglas Fir-Larch No. 2 has an $F_b = 875$ lb/in^2. $F_{b\text{-adj}}$ is found through computation:

$$F_{b\text{-adj}} = (C_F)(C_M)F_b = 0.972 \cdot 1.00 \cdot 875 \text{ lb/in}^2 = 851 \text{ lb/in}^2$$

In Example 22.1 the extreme fiber stress in bending (F_b) must be decreased to about 97% of the base design value due to corrections for size and moisture exposure.

■ EXAMPLE 22.2

Adjust the base design value for extreme fiber stress in bending (F_b) for a beam constructed of two 2×10 members. The wood species and grade under consideration is visually graded Douglas Fir-Larch, No. 2 grade. The member will be used in a high-moisture application.

Solution In Table C.6, $C_F = 1.1$ for F_b is found for a 2×10 member with a nominal depth (d) of 10 in and the wet service factor (C_M) for $F_b = 0.85$. $F_b = 825$ lb/in^2 for Douglas Fir-Larch No. 2 grade. $F_{b\text{-adj}}$ is found through computation:

$$F_{b\text{-adj}} = (C_F)(C_M)F_b = 1.1 \cdot 0.85 \cdot 825 \text{ lb/in}^2 = 771 \text{ lb/in}^2$$

In Example 22.2 the extreme fiber stress in bending (F_b) must be decreased to about 93% of the base design value, due to correction for moisture exposure. In fact, without a high-moisture application, the size factor (C_F) would have increased F_b.

22.8   STRUCTURAL WOOD DESIGN STANDARDS

The *National Design Specification (NDS) for Wood Construction* (1991) is the nationally recognized guide for wood structural design. It includes general requirements, design provisions and formulas, and data on sawn lumber, structural glued-laminated timber, round timber piles, and connections. It is available from the

American Wood Council (AWC), the wood products group of the American Forest and Paper Association, Inc. (AF&P) (Web site address: www.afandpa.org). The NDS is based on allowable stress design. It is currently adopted in all three major model building codes in the United States.

The *Load and Resistance Factor Design (LRFD) Manual for Engineered Wood Construction* (1996) provides a comprehensive guide for a more sophisticated approach to design of engineered wood structures. LRFD is based on strength and serviceability of wood combined with probability analysis as described in the AF&PA/ASCE 16-95 Standard for Load and Resistance Factor Design (LRFD) for Engineered Wood Construction.

While the LRFD approach is more sophisticated and is becoming more popular in engineering education, the NDS method is still used in industry. The NDS approach more closely represents the theory presented in early chapters of this book. Therefore, the ASD design approach is introduced in the chapters on wood.

EXERCISES

22.1 Compare the extreme fiber stress in bending (F_b) of No. 2 Douglas Fir-Larch to the extreme fiber stresses in bending of No. 2 Hemlock-Fir, Redwood, and Southern Pine by percentage. Rank these species by strength against bending.

22.2 Compare the tension stress parallel to the grain (F_t) of No. 2 Douglas Fir-Larch to the tension stresses parallel to the grain of No. 2 Hemlock-Fir, Redwood, and Southern Pine by percentage. Rank these species by tensile strength.

22.3 Compare the horizontal shear stress (F_v) of No. 2 Douglas Fir-Larch to the horizontal shear stresses of No. 2 Hemlock-Fir, Redwood, and Southern Pine by percentage. Rank these species by ability to resist horizontal shear.

22.4 Compare the compression stress perpendicular to the grain ($F_c^{\perp}$) of No. 2 Douglas Fir-Larch to the compression stresses perpendicular to the grain of No. 2 Hemlock-Fir, Redwood, and Southern Pine by percentage. Rank these species by compressive strength perpendicular to the grain.

22.5 Compare the compression stress parallel to the grain (F_c) of No. 2 Douglas Fir-Larch to the compression stresses perpendicular to the grain of No. 2 Hemlock-Fir, Redwood, and Southern Pine by percentage. Rank these species by compressive strength perpendicular to the grain.

22.6 Compare the modulus of elasticity (E) of No. 2 Douglas Fir-Larch to the moduli of elasticity of No. 2 Hemlock-Fir, Redwood, and Southern Pine by percentage. Rank these species by stiffness.

22.7 A 6 × 16 beam will be used in a high-moisture application. The wood species and grade is visually graded Douglas Fir-Larch, No. 2 grade. Adjust the base design value for extreme fiber stress in bending (F_b).

22.8 Adjust the base design value for extreme fiber stress in bending (F_b) for a 6 × 16 beam. The member will be used in a high-moisture application. The wood species and grade is visually graded Hemlock-Fir, No. 2 grade.

22.9 Adjust the base design value for extreme fiber stress in bending (F_b) for a beam constructed of three 2 × 8 members. The member will be used in a high-moisture application. The wood species and grade under consideration is visually graded Hemlock-Fir, No. 2 grade.

22.10 Adjust the base design value for extreme fiber stress in bending (F_b) for a 24-in-deep glued-laminated timber beam. The member will be used in a low-moisture application.

22.11 Adjust the base design value for compression stress parallel to the grain (F_c) for a 4 × 4 column. The member will be used in a high-moisture application. The wood species and grade under consideration is visually graded Douglas Fir-Larch, No. 2 grade.

22.12 Adjust the base design value for compression stress parallel to the grain (F_c) for a 4 × 4 column. The member will be used in a high-moisture application. The wood species and grade under consideration is visually graded Redwood, No. 2 grade.

SPECIAL PROJECTS

Wood Project 1

Objectives   The objectives of this project are to provide the student with experience in the methodology involved in determination of tributary loads of structural members, in development of free-body diagrams and in basic selection of wood (glued-laminated) beams and columns based on gravity loading.

Design Parameters   The framing plans of a glued-laminated building structure are shown. The two-story skeleton structure and exterior masonry walls support the second floor and low-slope (flat) roof. The first floor is a floating concrete slab that is isolated from the structure. The floor-to-floor and floor-to-roof distances are 14 ft. The columns rest on individual foundations. Loads of this structure are to be based on the following:

FLOOR:

Structural wood decking with finish-floor covering	11.5 lb/ft^2
HVAC, plumbing, electrical, lighting, ceiling	est. 15.0 lb/ft^2
Partition load (contingency for future)	20.0 lb/ft^2
Live load (occupants, furniture, equipment, etc.)	100.0 lb/ft^2

ROOF:

Structural wood roof decking	6.0 lb/ft^2
Insulation (4 in)	0.75 lb/ft^2
Roof membrane with gravel	6.0 lb/ft^2
HVAC, plumbing, electrical, lighting, ceiling	est. 15.0 lb/ft^2
Roof load (live)	20.0 lb/ft^2
Snow load	30.0 lb/ft^2
Rain/ice load	30.0 lb/ft^2

Analysis   *Beams.* All girders (i.e., G1, G2, and G3 members) shall be 8 $\frac{3}{4}$-in-wide glued-laminated members from stock sizes. All beams (i.e., FB1, FB2, RB1,

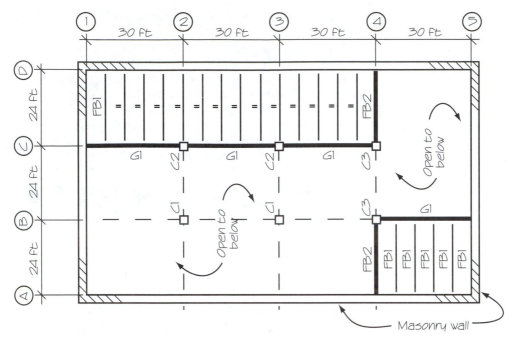

Floor framing plan

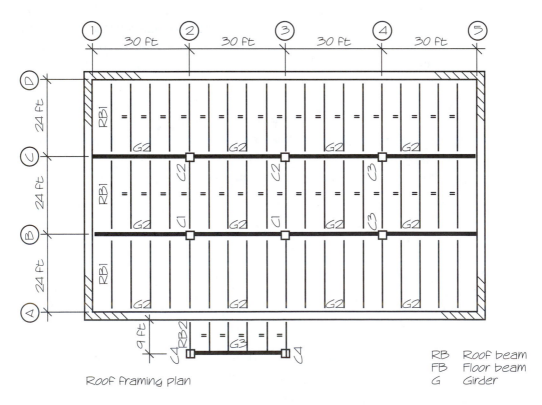

Roof framing plan

RB   Roof beam
FB   Floor beam
G   Girder

and RB2 members) are to be 6 $\frac{3}{4}$-in-wide glued-laminated members from stock sizes. In addition to floor loads, beams FB2 carry a 140-lb/ft load from interior walls bearing on the member. The shallowest acceptable structural member is to be specified. Section investigation is to include constructing a free-body diagram of the beam, flexure analysis, shear analysis, and deflection analysis where possible, with the Δ_{max} equations provided in this book (otherwise, deflection should be neglected). Girders and beams are also to be examined for lateral stability.

Columns. All columns (i.e., C1, C2, C3, and G4 members) are to be computed using the alternate column formula. Glued-laminated members are to be specified from near-square stock sizes (i.e., 6 $\frac{3}{4}$ in $\times$ 6.0 in, 6 $\frac{3}{4}$ in $\times$ 7.5 in, 8 $\frac{3}{4}$ in $\times$ 9.0 in, etc.); that is, dimensions of b and d should not vary by more than 1 $\frac{1}{2}$ in. The minimum column section dimensions are 6 $\frac{3}{4}$ in $\times$ 6.0 in. Unbraced column lengths are as follows: 28 ft for columns marked C1, 14 ft for columns marked C2 and C3, and 18 ft for columns marked C4. An effective length of $K_e = 1.0$ is to be used.

The following load combinations of dead loads (D_{load}), live floor loads (L_{load}), roof loads ($L_{roof\ load}$), snow loads (S_{load}), rain/ice load (R_{load}), wind load (W_{load}), and earthquake load (E_{load}) are to be used to determine the design load. Earthquake and wind loads are neglected.

$$L_{design\ a} = D_{load} + L_{load} + L_{roof\ load}\ or\ S_{load}$$
$$L_{design\ b} = D_{load} + L_{load} + W_{load}$$
$$L_{design\ c} = D_{load} + L_{load} + W_{load} + 0.5S_{load}$$
$$L_{design\ d} = D_{load} + L_{load} + S_{load} + 0.5W_{load}$$
$$L_{design\ e} = D_{load} + L_{load} + S_{load} + E_{load}$$

Wood Project 2

Objectives   The objectives of this project are to provide the student with experience in the methodology involved in determination of tributary loads and selection of simple wood framing members based on gravity loading.

Design parameters   The transverse section of a simple residence is shown. Following is a description of the design parameters for the framing systems.

Floor system.   The live load is 40 lb/ft^2 and the dead load is anticipated to be less than 10 lb/ft^2. Deflection should not exceed $\ell/360$. Framing material is specified as visually graded Hemlock-Fir, No. 2 grade and should be sized for normal-duration loading.

Ceiling system.   The live load is 10 lb/ft^2 and the dead load is anticipated to be less than 10 lb/ft^2. Deflection should not exceed $\ell/240$. Framing material is specified as visually graded Hemlock-Fir, No. 2 grade and should be sized for normal-duration loading.

Roof system. Live load for wind and snow is 30 lb/ft^2. The dead load is anticipated to be less than 15 lb/ft^2. Deflection should not exceed $\ell/240$. Framing material is specified as visually graded Hemlock-Fir, No. 2 grade and should be sized for normal-duration loading.

Analysis: Usint the appropriate table in the appendix, select the smallest section of the following:

    a. Floor joist at 16 in o.c. spacing.

    b. ceiling joist at 16 in o.c. spacing.

    c. rafter at 16 in o.c. spacing.

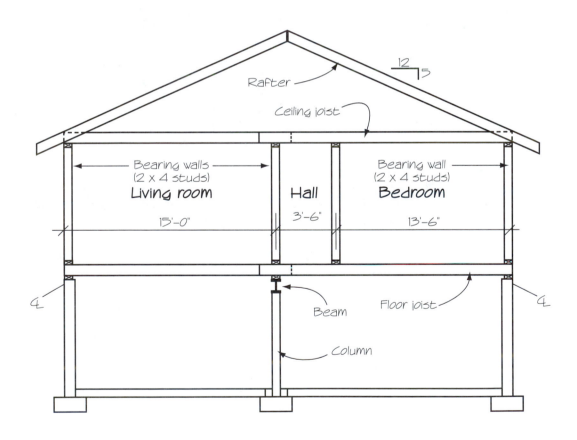

Chapter 23
DESIGN OF SIMPLE TIMBER BEAMS

An elementary design approach for timber beams is introduced in this chapter. The intent is to acquaint the technician with basic design principles and computation techniques used by an engineer. The reader is advised that the presentation does not provide full coverage of all design principles involved with timber design. Additional coverage is beyond the intended breadth of this book.

23.1   BASIC BEAM DESIGN PROCEDURE

Selection of a wood member for use as a beam, joist, or girder initially involves examining flexure, shear, and deflection. The approach is essentially the same for solid dimension lumber and engineered structural wood products such as glued-laminated timber beams, laminated veneer lumber beams, and parallel strand lumber and timber strand beams.

Following is the fundamental beam section selection procedure:

1. Solve for maximum bending moment (M_{max}) and maximum shear force (V_{max}).
2. Use the flexure formula, $S = M_{max}/F_b$, to find the required section modulus (S). Apply adjustment factors to F_b as appropriate.
3. Select a beam section that has a section modulus (S) equal to or greater than required, as calculated in step 2 above.
4. Investigate the horizontal shear stress (v) to ensure that v does not exceed F_v, the maximum permitted: $v = 3V_{max}/2bd$, where b is the beam breadth and d is the beam depth of a rectangular section. Apply adjustment factors to F_v as appropriate. If the section fails this investigation, a larger section is needed.
5. Investigate deflection to ensure that maximum deflection (Δ_{max}) does not exceed the maximum deflection permitted (D_{code}). Apply adjustment factors to E as appropriate. If the section fails this investigation, a larger section is needed.
6. Investigate other factors, such as bearing stresses and lateral stability.

■ EXAMPLE 23.1

Select a 12-in (nominal)-deep solid beam section that safely supports a uniformly distributed load of 1500 lb/ft on a simple beam spanning 5 ft. The wood species specified is visually graded Douglas Fir-Larch, No. 2 grade. The beam will be used in a dry application under normal-duration loading. Deflection should not exceed $\ell/360$.

Solution

1. Solve for maximum bending moment (M) and maximum shear force (V_{max}). With the appropriate derived formula in Table A.5:

$$M_{max} = \frac{WL}{8} = \frac{(1500 \text{ lb/ft} \cdot 5 \text{ ft})(5 \text{ ft} \cdot 12 \text{ in/ft})}{8} = 56\,250 \text{ in-lb}$$

With the appropriate derived formula in Table A.5:

$$V_{max} = \frac{W}{2} = \frac{1500 \text{ lb/ft} \cdot 5 \text{ ft}}{2} = 3750 \text{ lb}$$

2. Use the flexure formula, $S = M_{max}/F_b$, to find the required section modulus (S). A beam of dimension lumber (2 to 4 in wide) is investigated first: From Table C.6, for Douglas Fir-Larch, No. 2, $F_b = 825$ lb/in². Since the member is 12 in deep, $C_F = 1.00$. The beam will be used in a dry application under normal-duration loading, so the wet service and load duration factors do not apply. Therefore, F_b does not need to be adjusted.

$$S = \frac{M_{max}}{F_b} = \frac{56\,250 \text{ in-lb}}{825 \text{ lb/in}^2} = 68.18 \text{ in}^3$$

3. Select a beam section that has a section modulus (S) equal to or greater than required. From Table C.1, a 4 × 12 with a section modulus of 73.83 in³ is selected. It is the thinnest section that is 12 in deep and that has a section modulus that is equal to or greater than the required 64.29 in³.

4. Investigate the horizontal shear stress (v) to ensure that v does not exceed F_v. From Table C.6, for Douglas Fir-Larch, No. 2, $F_v = 95$ lb/in². For the 4 × 12 beam section:

$$v = \frac{3V_{max}}{2bd} = \frac{3 \cdot 3750 \text{ lb}}{2 \cdot 3.5 \text{ in} \cdot 11.25 \text{ in}} = 142.86 \text{ lb/in}^2$$

The 4 × 12 beam section fails the shear investigation because v exceeds F_v. The beam section must be increased. An 8 × 12 is selected and is investigated for flexure and shear. Different values for F_b and F_v are needed because the section under investigation is greater than

4 in wide. From Table C.7, for Douglas Fir-Larch, No. 2, $F_b = 875$ lb/in² and $F_v = 85$ lb/in².

$$S = \frac{M_{max}}{F_b} = \frac{56\,250 \text{ in-lb}}{875 \text{ lb/in}^2} = 64.29 \text{ in}^3$$

$$v = \frac{3V_{max}}{2bd} = \frac{3 \cdot 3750 \text{ lb}}{2 \cdot 7.5 \text{ in} \cdot 11.50 \text{ in}} = 65.2 \text{ lb/in}^2$$

The 8 × 12 section passes the flexure and shear investigation.

5. Investigate deflection to ensure that maximum deflection (Δ_{max}) does not exceed the maximum deflection permitted (D_{code}). Deflection should not exceed $\ell/360$:

$$D_{code} = \frac{\ell}{360} = \frac{5 \text{ ft} \cdot 12 \text{ in/ft}}{360} = 0.1667 \text{ in}$$

With the appropriate derived formula in Table A.5, $\Delta_{max} = 5Wl^3/384EI$. From Table C.1, $I_{8\times12} = 950.55$ in⁴. From Table C.7, $E = 1\,300\,000$ lb/in²:

$$\begin{aligned}
\Delta_{max} &= \frac{5Wl^3}{384EI} \\
&= \frac{5(1500 \text{ lb/ft} \cdot 5 \text{ ft})(5 \text{ ft} \cdot 12 \text{ in/ft})^3}{384 \cdot 1\,300\,000 \text{ lb/in}^2 \cdot 950.55 \text{ in}^4} \\
&= 0.0171 \text{ in}
\end{aligned}$$

The 8 × 12 passes the deflection investigation because Δ_{max} does not exceed D_{code}. The 8 × 12 is an acceptable beam section.

■ EXAMPLE 23.2

Select a 15-in (10 laminations)-deep glued-laminated beam that safely supports a uniformly distributed load of 350 lb/ft on a beam spanning 24 ft. The beam will be used in a dry application under normal-duration loading. Deflection should not exceed $\ell/360$.

1. Solve for maximum bending moment (M) and maximum shear force (V_{max}). With the appropriate derived formula in Table A.5:

$$M_{max} = \frac{WL}{8} = \frac{(350 \text{ lb/ft} \cdot 24 \text{ ft})(24 \text{ ft} \cdot 12 \text{ in/ft})}{8} = 302\,400 \text{ in-lb}$$

With the appropriate derived formula in Table A.5:

$$V_{max} = \frac{W}{2} = \frac{350 \text{ lb/ft} \cdot 24 \text{ ft}}{2} = 4200 \text{ lb}$$

2. Use the flexure formula, $S = M/F_b$, to find the required section modulus (S). From Table C.8, for glued-laminated timbers, $F_b = 2000$ lb/in^2. The beam will be used in a dry application under normal-duration loading, so the wet service and load-duration factors do not apply. Since the section is 15 in deep, the size factor, C_F, must be applied:

$$C_F = \left(\frac{12}{d}\right)^{1/9} = \left(\frac{12}{15}\right)^{1/9} = 0.976$$

$$F_{b\,adj} = (C_F)F_b = (0.976)2000 \text{ lb/in}^2 = 1952 \text{ lb/in}^2$$

$$S = \frac{M_{max}}{F_b} = \frac{302\,400 \text{ in-lb}}{1952 \text{ lb/in}^2} = 154.9 \text{ in}^3$$

3. Select a beam section that has a section modulus (S) equal to or greater than required. From Table C.2, a $5\frac{1}{8} \times 15$ glued-laminated timber is selected with a section modulus of 192.2 in^3. It is selected because it is the thinnest section that is 15 in deep with a section modulus that is equal to or greater than the required 154.9 in^3.

4. Investigate the horizontal shear stress (v) to ensure that v does not exceed F_v. From Table C.8 for glued-laminated timbers, $F_v = 165$ lb/in^2. For the $5\frac{1}{8} \times 15$ beam section:

$$v = \frac{3V_{max}}{2bd} = \frac{3 \cdot 4200 \text{ lb}}{2 \cdot 5.125 \text{ in} \cdot 15.00 \text{ in}} = 81.95 \text{ lb/in}^2$$

This section passes the shear investigation, because v is less than F_v.

5. Investigate deflection to ensure that maximum deflection (Δ_{max}) does not exceed deflection permitted (D_{code}). Deflection should not exceed $\ell/360$:

$$D_{code} = \frac{\ell}{360} = \frac{24 \text{ ft} \cdot 12 \text{ in/ft}}{360} = 0.800 \text{ in}$$

With the appropriate derived formula in Table A.5, $\Delta_{max} = 5\,W\ell^3/384EI$. From Table C.2, $I_{5\frac{1}{8} \times 15} = 1441.4$ in^4. From Table C.8, $E = 2\,000\,000$ lb/in^2.

$$\Delta_{max} = \frac{5W\ell^3}{384EI}$$

$$= \frac{5(350 \text{ lb/ft} \cdot 24 \text{ ft})(24 \text{ ft} \cdot 12 \text{ in/ft})^3}{384 \cdot 2\,000\,000 \text{ lb/in}^2 \cdot 1441.4 \text{ in}^4} = 0.906 \text{ in}$$

The $5\frac{1}{8} \times 15$ fails the deflection investigation because Δ_{max} exceeds D_{code}. The beam section must be increased to reduce Δ_{max}. A $6\frac{3}{4} \times 15$ with $I = 1898.4$ in^4 is investigated:

Chap. 23   Design of Simple Timber Beams

$$\Delta_{max} = \frac{5W\ell^3}{384EI}$$

$$= \frac{5(350 \text{ lb} \cdot 24 \text{ ft})(24 \text{ ft} \cdot 12 \text{ in/ft})^3}{348 \cdot 2\,000\,000 \text{ lb/in}^2 \cdot 1898.4 \text{ in}^4}$$

$$= 0.688 \text{ in}$$

The $6\frac{3}{4} \times 15$ passes the deflection investigation because Δ_{max} does not exceed D_{code}. The $6\frac{3}{4} \times 15$ is an acceptable beam section.

23.2   GANG-LAMINATED BEAM SECTIONS

Wood members are sometimes combined together to produce a single section known as a gang-laminated member as shown in Figure 23.1. A *gang-laminated* member is composed of two or more individual members, such as 2 × 12s or LVLs that are lapped and secured together to produce a stronger member. Examples of common gang-laminated beam sections might include a wood girder composed of three 2 × 12s that are lapped, glued and nailed, and positioned to rest on edge at the bearing points, or two LVL sections secured together to create a header above an overhead door or window opening.

If adequately laminated together, a gang-laminated member may be treated as a single section in structural analysis. For example, a structural member composed of two 2 × 12s has a section modulus and a moment of inertia that is twice the value of a single 2 × 12. Therefore, the gang-laminated section composed of two 2 × 12s is twice as stiff and twice as strong as a single 2 × 12.

To treat a gang-laminated member as a single unit in structural analysis, the designer must specify and the contractor must assure that the member is constructed adequately. Proper methods of nailing, screwing, or bolting gang-laminated members are generally provided with the manufacturer's technical literature for the specific wood product. In exterior applications where

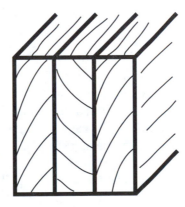

FIGURE 23.1   Wood members are sometimes combined together to produce a single section known as a gang-laminated member.

gang-laminated members are exposed to moisture from rainfall and snowmelt, provisions must be made to drain or cover joint surfaces adequately, as these surfaces tend to retain moisture and prematurely decay.

■ **EXAMPLE 23.3**

Determine the section modulus (S) and the moment inertia (I) of three 2×12 sections that are adequately secured together. Assume that the sections are joined at their flat surfaces such that they can act as a single girder resting on the edge at the girder ends.

Solution   From Table C.1, for single 2×12s:

$$S_{2 \times 12} = 31.64 \text{ in}^3 \qquad I_{2 \times 12} = 177.98 \text{ in}^4$$

Therefore,

$$S_{3\text{-}2 \times 12} = 3S = 3 \cdot 31.64 \text{ in}^3 = 94.92 \text{ in}^3$$
$$I_{3\text{-}2 \times 12} = 3I = 3 \cdot 177.98 \text{ in}^4 = 533.94 \text{ in}$$

■ **EXAMPLE 23.4**

Design a beam section of gang-laminated 2×12s that can safely support a uniformly distributed load of 540 lb/ft resting on a beam spanning 8 ft. The gang-laminated 2×12s will rest on edge, with the individual 2×12s sufficiently glued and screwed together. The wood species specified is visually graded Douglas Fir-Larch, No. 2 grade. The beam will be used in a dry application under normal-duration loading. Deflection should not exceed $\ell/360$.

1. Solve for maximum bending moment (M) and maximum shear force (V_{max}). With the appropriate derived formula in Table A.5:

$$M_{max} = \frac{WL}{8} = \frac{(540 \text{ lb/ft} \cdot 8\text{ft})(8 \text{ ft} \cdot 12 \text{ in/ft})}{8} = 51\,840 \text{ in-lb}$$

With the appropriate derived formula in Table A.5:

$$V_{max} = \frac{W}{2} = \frac{540 \text{ lb/ft} \cdot 8 \text{ ft}}{2} = 2160 \text{ lb}$$

2. Use the flexure formula, $S = M/F_b$, to find the required section modulus (S). From Table C.6, for Douglas Fir-Larch, No. 2, $F_b = 825$ lb/in². Since the sections are 12 in deep and 2 in thick, $C_F = 1.0$. The beam will be used in a dry application under normal-duration loading, so the wet service and load-duration factors do not apply. F_b does not need to be adjusted.

$$S = \frac{M_{max}}{F_b} = \frac{51\,840 \text{ in-lb}}{825 \text{ lb/in}^2} = 62.8 \text{ in}^3$$

3. Select a beam section composed of gang-laminated 2 × 12s that has a combined section modulus (S) equal to or greater than required. From Table C.1, a single 2 × 12 has a section modulus of 31.64 in^3.

$$\frac{62.8 \text{ in}^3}{31.64 \text{ in}^3} = 1.98 \text{ or two } 2 \times 12\text{s are required}$$

The combined section modulus of two 2 × 12s is

$$2 \cdot 31.64 \text{ in}^3 = 63.28 \text{ in}^3$$

4. Investigate the horizontal shear stress (v) to ensure that v does not exceed F_v. From Table C.6, for Douglas Fir-Larch, No. 2, $F_v = 95$ lb/in^2. For the beam section of two 2 × 12s:

$$v = \frac{3V_{\max}}{2bd} = \frac{3 \cdot 2160 \text{ lb}}{2 \cdot (2 \cdot 1.5 \text{ in}) \cdot 11.25 \text{ in}} = 96.0 \text{ lb/in}^2$$

This section fails the investigation, because v exceeds F_v. An additional 2 × 12 must be added to the section. A gang-laminated section composed of three 2 × 12s is investigated:

$$v = \frac{3V_{\max}}{2bd} = \frac{3 \cdot 2160 \text{ lb}}{2 \cdot (3 \cdot 1.5 \text{ in}) \cdot 11.25 \text{ in}} = 64.0 \text{ lb/in}^2$$

The section of three 2 × 12s passes the shear investigation, because v does not exceed F_v.

5. Investigate deflection to ensure that maximum deflection ($\Delta_{\max}$) does not exceed deflection permitted (D_{code}). Deflection should not exceed $\ell/360$.

$$D_{\text{code}} = \frac{\ell}{360} = \frac{8 \text{ ft} \cdot 12 \text{ in/ft}}{360} = 0.267 \text{ in}$$

With the appropriate derived formula in Table A.5, $\Delta_{\max} = 5W\ell^3/384EI$. From Table C.1, $I_{2\times12} = 177.98$ in^4. From Table C.6, $E = 1\,600\,000$ lb/in^2:

$$\Delta_{\max} = \frac{5W\ell^3}{384EI}$$
$$= \frac{5(540 \text{ lb/ft} \cdot 8 \text{ ft})(8 \text{ ft} \cdot 12 \text{ in/ft})^3}{384 \cdot 1\,600\,000 \text{ lb/in}^2 \cdot (3 \cdot 177.98 \text{ in}^4)}$$
$$= 0.0583 \text{ in}$$

The section of three 2 × 12s passes the deflection investigation, because $\Delta_{\max}$ does not exceed D_{code}. Three 2 × 12s properly secured is an acceptable gang-laminated beam section.

■ **EXAMPLE 23.5**

Design a beam section of gang-laminated 2 × 12s that can safely support a uniformly distributed load of 540 lb/ft resting on a beam spanning 8 ft. The gang-laminated 2 × 12s will rest on edge, with the individual 2 × 12s sufficiently glued and screwed together. The wood species specified is visually graded Redwood, No. 2 grade, open grain. The beam will be used in an outdoor (high-moisture) application under normal-duration loading. Deflection should not exceed $\ell/360$.

1. Solve for maximum bending moment (M) and maximum shear force (V_{max}). With the appropriate derived formula in Table A.5:

$$M_{max} = \frac{WL}{8} = \frac{(540 \text{ lb/ft} \cdot 8 \text{ ft})(8 \text{ ft} \cdot 12 \text{ in/ft})}{8} = 51\ 840 \text{ in-lb}$$

With the appropriate derived formula in Table A.5:

$$V_{max} = \frac{W}{2} = \frac{540 \text{ lb/ft} \cdot 8 \text{ ft}}{2} = 2160 \text{ lb}$$

2. Use the flexure formula, $S = M/F_b$, to find the required section modulus (S). From Table C.6, for Redwood, No. 2, open grain, $F_b = 725$ lb/in^2. Since the sections are 12 in deep and 2 in thick, $C_F = 1.0$. The beam will be used in a high-moisture application under normal-duration loading, so the wet service factor of $C_M = 0.85$ applies for F_b. F_b must be adjusted:

$$F_{b\ adj} = (C_M)F_b = (0.85)725 \text{ lb/in}^2 = 616.3 \text{ lb/in}^2$$
$$S = \frac{M_{max}}{F_b} = \frac{51\ 840 \text{ in-lb}}{616.3 \text{ lb/in}^2} = 84.1 \text{ in}^3$$

3. Select a beam section composed of gang-laminated 2 × 12s that has a combined section modulus (S) equal to or greater than required. From Table C.1, a single 2 × 12 has a section modulus of 31.64 in^3:

$$\frac{84.1 \text{ in}^3}{31.64 \text{ in}^3} = 2.7 \text{ or three } 2 \times 12 \text{s required}$$

The combined section modulus of three 2 × 12s is

$$3 \cdot 31.64 \text{ in}^3 = 94.92 \text{ in}^3$$

4. Investigate the horizontal shear stress (v) to ensure that v does not exceed F_v. From Table C.6, for Redwood, No. 2, open grain, $F_v = 80$ lb/in^2. The beam will be used in a high-moisture application under normal-duration loading, so the wet service factor of $C_M = 0.97$ applies for F_v. F_v must be adjusted:

$$F_{v\text{-adj}} = (C_M)F_v = (0.97)80 \text{ lb/in}^2 = 77.6 \text{ lb/in}^2$$

The shear investigation for the beam section of three 2 × 12s:

$$v = \frac{3V_{max}}{2bd} = \frac{3 \cdot 2160 \text{ lb}}{2 \cdot (3 \cdot 1.5 \text{ in}) \cdot 11.25 \text{ in}} = 64.00 \text{ lb/in}^2$$

The section of three 2 × 12s passes the shear investigation, because v does not exceed F_v.

5. Investigate deflection to ensure that maximum deflection (Δ_{max}) does not exceed deflection permitted (D_{code}). Deflection should not exceed $\ell/360$:

$$D_{code} = \frac{\ell}{360} = \frac{8 \text{ ft} \cdot 12 \text{ in/ft}}{360} = 0.267 \text{ in}$$

With the appropriate derived formula in Table A.5, $\Delta_{max} = 5W\ell^3/384EI$. From Table C.1, $I_{2\times12} = 177.98 \text{ in}^4$. From Table C.6, for Redwood, No. 2, open grain: $E = 1\,000\,000 \text{ lb/in}^2$. The beam will be used in a high-moisture application, so the wet service factor of $C_M = 0.90$ applies to E. E must be adjusted:

$$E_{adj} = (C_M)E = (0.90)1\,000\,000 \text{ lb/in}^2 = 900\,000 \text{ lb/in}^2$$

$$\begin{aligned}
\Delta_{max} &= \frac{5W\ell^3}{384EI} \\
&= \frac{5(540 \text{ lb/ft} \cdot 8 \text{ ft})(8 \text{ ft} \cdot 12 \text{ in/ft})^3}{384 \cdot 900\,000 \text{ lb/in}^2 \cdot (3 \cdot 177.98 \text{ in}^4)} \\
&= 0.1036 \text{ in}
\end{aligned}$$

The section of three 2 × 12s passes the deflection investigation, because Δ_{max} does not exceed D_{code}.

Three 2 × 12s properly secured is an acceptable gang-laminated beam section.

23.3   OTHER FACTORS IN TIMBER BEAM DESIGN

In addition to designing for flexure, shear, and deflection, the designer must investigate the following additional factors.

Reactions and load bearings   As discussed in Chapter 16, a bearing stress develops where loads and reactions contact the beam, causing the beam material to undergo a crushing effect. The designer must ascertain that the beam material at these bearing contact areas does not become crushed. Stress developed by loads and reactions is compressive in nature and acts perpendicular to the grain. Thus it is related to the design property of compression stress perpendicular to the grain ($F_c\perp$).

For a bearing area of any length at the end of the beam or for contact areas that are 6 in or greater in bearing length at any location other than the end of the beam, the allowable bearing stress ($F_c\perp^*$) is equal to the compression stress perpendicular to the grain ($F_c\perp$); that is, $F_c\perp^* = F_c\perp$. Any adjustment factors to $F_c\perp$ should be applied as appropriate.

In cases when the bearing length (l_b) is less than 6 in *and* at least 3 in from the beam end, the allowable bearing stress ($F_c\perp^*$) can be increased by the following expression, where any adjustment factors to $F_c\perp$ should be applied as appropriate:

$$F_c\perp^* = \frac{F_c\perp(\ell_b + \frac{3}{8})}{\ell_b}$$

For a bearing contact area (A_b), with allowable bearing stress ($F_c\perp^*$) and the reaction (R) or concentrated load (P) known, the following expressions apply:

$$F_c\perp^* \geq \frac{R}{A_b}$$

$$F_c\perp^* \geq \frac{P}{A_b}$$

By rearranging terms in these expressions, the minimum bearing area ($A_{b\,min}$) may be found by

$$A_{b\,min} = \frac{R}{F_c\perp^*} = \frac{P}{F_c\perp^*}$$

Since beam breadth (b) and the reaction (R) or concentrated load (P) are typically known, the minimum bearing length ($\ell_{b\text{-}min}$) may be found by the following expression:

$$\ell_{b\,min} = \frac{R}{bF_c\perp^*} = \frac{P}{bF_c\perp^*}$$

■ EXAMPLE 23.6

As shown in Example 23.1, an 8 × 12 solid beam section will safely support a uniformly distributed load of 1500 lb/ft on a simple beam spanning 5 ft. The specified wood species is visually graded Douglas Fir-Larch, No. 2 grade. The beam will be used in a dry application under normal-duration loading.
(a) Determine the minimum bearing area at the reactions.
(b) Determine the minimum bearing length at the reactions.
Solution (a) From Table C.7, the compression stress perpendicular to the grain ($F_c\perp$) for Douglas Fir-Larch, No. 2 grade, is 625 lb/in^2. No adjustment

factors apply for dry applications under normal-duration loading. Since the bearing area is at the reactions (end of the beam), $F_c\perp^* = F_c\perp$. The minimum bearing area ($A_{b\ min}$) is found by

$$R = \frac{W}{2} = \frac{1500\ \text{lb/ft} \cdot 5\ \text{ft}}{2} = 3750\ \text{lb}$$

$$A_{b\ min} = \frac{R}{F_c\perp^*} = \frac{3750\ \text{lb}}{625\ \text{lb/in}^2} = 6\ \text{in}^2$$

(b) From above, $F_c\perp^*$ is 625 lb/in² and R is 3750 lb. The actual breadth (b) of an 8 × 12 is 7.5 in. The minimum bearing length ($\ell_{b\ min}$) is found by

$$\ell_{b\ min} = \frac{R}{bF_c\perp^*} = \frac{3750\ \text{lb}}{7.5\ \text{in} \cdot 625\ \text{lb/in}^2} = 0.8\ \text{in}$$

In Example 23.6 it was found that a bearing length of 0.8 inch is needed to bear the reaction force without crushing the material. The designer should leave additional length due to the realities of construction tolerances. As a rule of thumb, beam ends should have no less than $1\frac{1}{2}$ inches bearing on wood or metal, and no less than 3 inches resting on masonry.

■ EXAMPLE 23.7

A 4 × 16 beam supports a concentrated load of 4000 lb at the center of the span of the simple beam. The specified wood species is visually graded, open grain, Redwood, No. 2 grade. The bearing length of the load is 2 in. The bearing length at the reaction supports is 3.5 in.
(a) Determine if the bearing length is adequate at the reactions.
(b) Determine if the bearing length is adequate at the load.

Solution (a) From Table C.6, the compression stress perpendicular to the grain ($F_c\perp$) for open-grain Redwood No. 2 is 425 lb/in². No adjustment factors apply. The actual breadth (b) of a 4 × 16 is 3.5 in. The minimum bearing length ($\ell_{b\ min}$) of the reaction is found by

$$R = \frac{P}{2} = \frac{4000\ \text{lb}}{2} = 2000\ \text{lb}$$

$$\ell_{b\ min} = \frac{R}{bF_c\perp^*} = \frac{2000\ \text{lb}}{3.5\ \text{in} \cdot 425\ \text{lb/in}^2} = 1.34\ \text{in}$$

The bearing length at the reactions is adequate.

(b) The bearing length (ℓ_b) is less than 6 in and at least 3 in from the beam end, so the allowable bearing stress ($F_c\perp^*$) must be increased:

$$F_c\perp^* = \frac{F_c\perp(\ell_b + \frac{3}{8})}{\ell_b} = \frac{425\ \text{lb/in}^2(2\ \text{in} + \frac{3}{8})}{2\ \text{in}} = 505\ \text{lb/in}^2$$

The minimum bearing length ($\ell_{b\text{-min}}$) of the concentrated load is found by

$$P = 4000 \text{ lb}$$

$$\ell_{b\,\text{min}} = \frac{P}{bF_c\!\perp^*} = \frac{4000 \text{ lb}}{3.5 \text{ in} \cdot 505 \text{ lb/in}^2} = 2.26 \text{ in}$$

The bearing length at the load is inadequate.

In Example 23.7, the 2-in bearing length at the load is inadequate; the wood will be crushed. The load must be transferred to a larger contact area; a larger bearing length is required. A steel plate located between the load and the beam can be used to create a larger bearing length and bearing contact area.

Lateral stability   As discussed in Chapter 16, deep rectangular beams will rotate or buckle laterally if left unsupported. Under load, the fibers of the beam that develop compressive stresses (the top fibers of a simple beam) tend to buckle laterally like a column. To minimize lateral buckling, deep, narrow beams are typically secured to a rigid deck and/or secured with bridging.

The following ratios serve as guidelines for providing adequate lateral support for common rectangular wood beams having *nominal* depth (d)-to-breadth (b) ratios:

$d/b = 2$ Lateral support is not required.

$d/b = 3$ Beam ends should be held in position (i.e., ends secured by full solid blocking, bridging, or anchoring ends into other structural members).

$d/b = 4$ Beam ends should be held in position.

$d/b = 5$ At least one beam edge should be held in line for its entire length.

$d/b = 6$ Diagonal bridging at intervals not exceeding 8 ft is required, or both beam edges should be held in line for the entire beam length.

$d/b = 7$ Both beam edges should be held in line.

■ **EXAMPLE 23.8**

In Example 23.1 an 8 × 12 was an acceptable beam section. Determine the lateral support requirements for this beam.

Solution   The 8 × 12 beam section has a nominal breadth (b) of 8 in by a nominal depth (d) of 12 in. The depth (d)-to-breadth (b) ratio is

$$\frac{d}{b} = \frac{12 \text{ in}}{8 \text{ in}} = 1.5$$

Lateral support of this beam is not required because the depth-to-breadth ratio is less than 2.

The 8 × 12 beam section in Example 23.8 is relatively stout; its depth-to-breadth ratio is less than 2. It can naturally resist lateral buckling and rotation, so lateral support is not required.

■ EXAMPLE 23.9

Determine the lateral support requirements for a 4 × 16 beam section.

Solution   The 4 × 16 beam section has a nominal breadth (b) of 4 in by a nominal depth (d) of 14 in. The depth-to-breadth ratio is

$$\frac{d}{b} = \frac{14 \text{ in}}{4 \text{ in}} = 3.5$$

Lateral support of the 4 × 16 beam is required because the depth-to-breadth ratio is 3.5. The beam ends should be held in position by solid blocking, bridging, or nailing or bolting into other structural members.

EXERCISES

23.1 Select a 4-in-nominal-wide solid beam section that safely supports a uniformly distributed load of 500 lb/ft across the entire length of a simple span of 10 ft. The wood species specified is visually graded Hemlock-Fir, No. 2 grade. The beam will be used in a dry application under normal-duration loading. Deflection should not exceed $\ell/360$.

23.2 Select a $3\frac{1}{8}$-in-wide glued-laminated (gluelam) beam that safely supports a uniformly distributed load of 500 lb/ft across the entire length of a simple span of 10 ft. The beam will be used in a dry application under normal-duration loading. Deflection should not exceed $\ell/360$.

23.3 Select a beam section consisting of two $1.8E$ laminated-veneered lumber (LVL) members that safely supports a uniformly distributed load of 500 lb/ft across the entire length of a simple span of 10 ft. The beam will be used in a dry application under normal-duration loading. Deflection should not exceed $\ell/360$.

23.4 Design a beam section of gang-laminated 2 × 12s that can safely support a uniformly distributed load of 500 lb/ft across the entire length of a simple span of 10 ft. The multiple 2 × 12s will rest on edge, with the individual 2 × 12s sufficiently glued and screwed together. The wood species specified is visually graded Hemlock-Fir, No. 2 grade. The beam will be used in a dry application under normal-duration loading. Deflection should not exceed $\ell/360$.

23.5 Select a $3\frac{1}{2}$-in-wide, $1.3E$ laminated-strand lumber (LSL) beam that safely supports a uniformly distributed load of 500 lb/ft across the entire length of a simple span of 10 ft. The beam will be used in a dry application under normal-duration loading. Deflection should not exceed $\ell/360$.

23.6 A beam section will be custom cut from $3\frac{1}{2}$-in-wide commercial-grade parallel-strand lumber (PSL) stock. The beam will support a uniformly distributed load of 500 lb/ft across the entire length of a simple span of 10 ft. It will be used in a dry application under normal-duration loading. Deflection should not exceed $\ell/360$. Determine the minimum required depth of the member for the breadth provided.

23.7 A 4 × 12 simple beam carries a 150 lb/ft uniformly distributed load over its entire span of 16'-6". The wood species specified is visually graded Hemlock-Fir, No. 2 grade. The bearing length at the reactions is 2 in.

(a) Determine if the bearing length at the reactions is adequate.

(b) Determine the lateral support requirements for the beam section.

23.8 A 4 × 16 simple beam carries a 500-lb/ft uniformly distributed load over its entire span of 16'-6". The wood species specified is visually graded Hemlock-Fir, No. 2 grade. The bearing length at the reactions is 2 in. Determine if the bearing length at the reactions is adequate.

23.9 A $5\frac{1}{8}$ × 18 glued-laminated beam carries three 1500-lb concentrated loads at fourth points of its span.

(a) Determine the bearing lengths required at the reactions.

(b) Determine the bearing lengths required at the 1500-lb concentrated loads.

(c) Determine the lateral support requirements for the beam section.

Chapter 24
DESIGN OF WOOD JOISTS AND RAFTERS

Selection of wood joists and rafters using the tabular method is introduced in this chapter. The reader is advised that the presentation does not provide full coverage of all design principles involved with timber design. This book is intended to acquaint the technician with basic design principles and computation techniques used by an engineer. Additional coverage is beyond the intended breadth of the book.

24.1   JOIST AND RAFTER TABLES

A *joist* is a repetitive member that supports loads like a beam (Figure 24.1). Joists are used to support floors, ceilings, and flat roofs. A *rafter* is a sloping repetitive member that is used to support the deck and covering of a sloping roof system.

Floor joists are horizontally positioned framing members that are repetitively spaced. Typically, a panel product such as plywood is glued and nailed to the top surface of the joists to create a structural floor deck.

Rafters are sloped framing members that are repetitively spaced. In the right side of the photograph, rafters support a roof system.

Solid wood joists and rafters are commonly used in light (residential) building construction because of ease of use and availability. As a result, *span tables* have been developed to eliminate the need for extensive calculations in sizing these members. Maximum span and spacing are found with the base design values of modulus of elasticity (E) and extreme fiber stress in bending (F_b).

For joists, span is measured as the distance from face to face of the reacting supports, such as a top plate, plus one-half the length of bearing length at each end. For example, two 2×4 ($3\frac{1}{2}$-in-wide) walls spaced exactly 12 ft apart measured from inside stud face to inside stud face create a span of $12'$-$3\frac{1}{2}''$: $12'$-$0'' + 1\frac{3}{4}'' + 1\frac{3}{4}'' = 12'$-$3\frac{1}{2}''$. As a rule of thumb, joist ends should have no less than $1\frac{1}{2}$ in bearing on wood or metal and no less than 3 in resting on masonry.

For rafters, the span is the horizontally projected distance from the center of the bearing wall to the centerline of the ridge board. The rafter span does not include the overhanging length of the rafter. The actual length of the rafter is much greater than the span.

Following is the joist or rafter section and spacing selection procedure:

1. Identify the *design* span of the member and the *design* modulus of elasticity (E) and *design* extreme fiber stress in bending (F_b) for the species and grade of wood specified. Design F_b and E values should be the repetitive member values, as joists and rafters are repetitive members. Values applicable only for repetitive members are given in Table C.9.

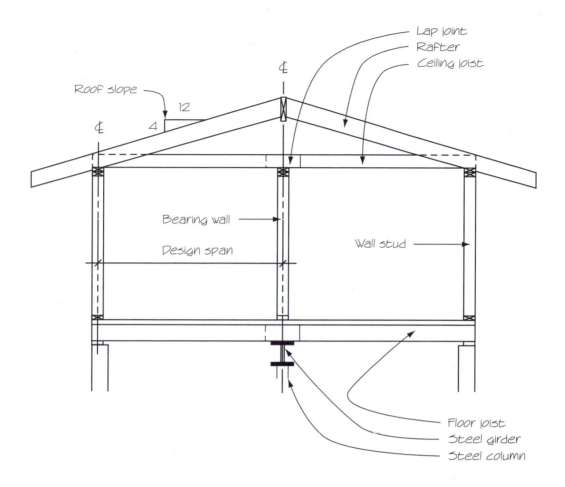

FIGURE 24.1 A *joist* is a repetitive member that supports loads like a beam. Joists are used to support floors, ceilings, and flat roofs. A *rafter* is a sloping repetitive member that is used to support the deck and covering of a sloping roof system.

2. Select the appropriate joist or rafter table. Verify that the table subtitle information meets the loading condition and other criteria precisely. Selected tables are provided in Tables C.10 to C.12. Additional tables for various loading conditions are found in the model building codes. The following loading conditions apply as a guide only:

  (a) For residences, minimum live loads are generally:

Floors spaces:	40 lb/ft^2
Attic spaces:	10 lb/ft^2
Roof areas (snow):	30 lb/ft^2

(b) For residences, minimum dead loads are generally:

Carpeted floors:	7 lb/ft²
Ceramic tile floors:	10 lb/ft²
Insulated attic floors:	10 lb/ft²
Composition shingles:	7 lb/ft²
Wood composite/shake:	7 lb/ft²
Tile/slate roof:	15 lb/ft²

3. Select a joist or rafter section at the desired on-center spacing that permits the design span and has required F_b and E values that do not exceed the design F_b and E values of the specified species and grade:

design span of structural member $\leq$ allowable span in span tables

design F_b and E values of wood $\geq$ required F_b and E values in span tables

■ EXAMPLE 24.1

A floor joist system for a residence will span between two bearing walls spaced at 12'-3½" on center. Joist material is specified as visually graded Douglas Fir-Larch, No. 2 grade. Live-load requirements for furniture and occupants is 40 lb/ft². The dead load is anticipated to be less than 10 lb/ft². Deflection should not exceed $\ell/360$. The joist will be used under normal-duration loading. Determine the smallest joist section at 16-in.-o.c. spacing.

Solution From Table C.9, for Douglas Fir-Larch, No. 2 grade, the design $E = 1\,600\,000$ lb/in². For normal-duration loading, the design $F_b = 1310$ lb/in² for a 2 × 6, 1210 lb/in² for a 2 × 8, 1105 lb/in² for a 2 × 10, and 1005 lb/in² for a 2 × 12.

Table C.10 applies to the design conditions described. From this table, select the shallowest member that meets the design span and the required F_b and E. The 2 × 6 is not acceptable. It does not meet the design span of 12'-3½" at 16-in-o.c. spacing. A 2 × 8 at 16-in-o.c. spacing allows a span of up to 12'-7" at $E = 1.5$ million lb/in². Since this is less than the design E of $1\,600\,000$ lb/in², E is acceptable. At the $E = 1.5$ million lb/in² column, the required $F_b = 1202$ lb/in² for 16-in-o.c. spacing. The design $F_b = 1210$ lb/in² (from Table C.10 for a 2 × 8) is greater than the required $F_b = 1202$ lb/in² (from Table C.9). F_b is acceptable.

A 2 × 8 joist section permits the design span of 12'-3½" at 16-in-o.c. spacing because it has required F_b and E values that do not exceed the design F_b and E values of the species and grade specified. A 2 × 8 section at 16-in-o.c. spacing is acceptable.

■ EXAMPLE 24.2

A ceiling joist system for a residence will span 12'-3½". Joist material is specified as visually graded Douglas Fir-Larch, No. 2 grade or better. The live load is 10 lb/ft². The dead load is anticipated to be less than 10 lb/ft². Deflection should not exceed $\ell/240$. The joist will be used in a dry application under normal-duration loading. Determine the smallest joist section at 16-in-o.c. spacing.

Solution From Table C.9, for Douglas Fir-Larch, No. 2 grade, the design $E = 1\ 600\ 000$ lb/in². For normal-duration loading, the design $F_b = 1510$ lb/in² for a 2 × 4, 1310 lb/in² for a 2 × 6, 1210 lb/in² for a 2 × 8, and 1105 lb/in² for a 2 × 10.

Table C.11 applies to the design conditions described. From this table select the shallowest member that meets the design span and the required F_b and E. The 2 × 4 is not acceptable. It does not meet the design span of 12'-$3\frac{1}{2}''$ at 16-in-o.c. spacing. A 2 × 6 at 16-in-o.c. spacing allows a span of up to 15'-2" at $E = 1.0$ million/in². At the $E = 1.0$ million lb/in² column, the required $F_b = 909$ lb/in² for 16-in-o.c. spacing. The design $F_b = 1310$ lb/in² (from Table C.11 for a 2 × 6) is greater than or equal to the required $F_b = 1310$ lb/in² (from Table C.9).

A 2 × 6 joist section at 16-in-o.c. spacing permits the design span and has the required F_b and E values that do not exceed the design F_b and E values of the specified species and grade. A 2 × 6 section is acceptable.

■ **EXAMPLE 24.3**

A roof rafter system for a residence will span 12'-$3\frac{1}{2}''$ horizontally. The rafter material is specified as visually graded Douglas Fir-Larch, No. 2 grade or better. The live load for wind and snow is 30 lb/ft². The dead load is anticipated to be less than 15 lb/ft². The deflection should not exceed $\ell/240$. The rafter will be used in a dry application under normal-duration loading. Determine the smallest rafter section at 16-in-o.c. spacing.

Solution From Table C.9, for Douglas Fir-Larch, No. 2 grade, the design $E = 1\ 600\ 000$ lb/in². For normal-duration loading, the design $F_b = 1310$ lb/in² for a 2 × 6, 1210 lb/in² for a 2 × 8, 1105 lb/in² for a 2 × 10, and 1005 lb/in² for a 2 × 12.

Table C.12 applies to the design conditions described. From this table select the shallowest member that meets the design span and the required F_b and E. The 2 × 6 is not acceptable. It only allows a span of 10'-5" at 16-in-o.c. spacing for the design $F_b = 1310$ lb/in². The 2 × 8 is acceptable. At the required $F_b = 1100$ lb/in², a 2 × 8 at 16-in-o.c. spacing allows a span of up to 12'-8". The design $F_b = 1210$ lb/in² (from Table C.12, for a 2 × 8) is greater than or equal to the required $F_b = 1100$ lb/in² (from Table C.9). The design $E = 1\ 600\ 000$ lb/in² is greater than or equal to the required $E = 0.77$ million lb/in² at 16-in-o.c. spacing.

A 2 × 8 rafter section at 16-in-o.c. spacing is acceptable for the specified 12′-3½″ design span. It has required F_b and E values that do not exceed the design F_b and E of the specified species and grade. The 2 × 8 section is acceptable.

24.2 LATERAL STABILITY

As discussed in Chapter 16, deep rectangular beams will buckle or rotate laterally if left unsupported. Under load, the fibers of the beam that develop compressive stresses (the top fibers of a simple beam) tend to buckle laterally like a column. To minimize lateral buckling, deep, narrow beams are typically secured to a rigid deck and/or secured with bridging.

The following ratios serve as guidelines for providing adequate lateral support for common rectangular wood joists and rafters having *nominal* depth (*d*)-to-breadth (*b*) ratios:

$d/b = 2$ Lateral support is not required.

$d/b = 3$ Ends should be held in position (i.e.,secured by solid blocking, bridging, or nailing into a header).

$d/b = 4$ Ends should be held in position.

$d/b = 5$ One edge should be held in line for its entire length (i.e., secured to plywood subfloor).

$d/b = 6$ Diagonal bridging at intervals not exceeding 8 ft is required.

$d/b = 7$ Both edges should be held in line.

■ EXAMPLE 24.4

Determine the lateral support requirements for the following joist sizes:

(a) 2 × 4
(b) 2 × 6
(c) 2 × 8
(d) 2 × 10
(e) 2 × 12
(f) 2 × 14

Solution   (a) The nominal size of a 2 × 4 section is breadth (*b*) of 2 in by depth (*d*) of 4 in:

$$\frac{d}{b} = \frac{4 \text{ in}}{2 \text{ in}} = 2$$

Lateral support of a 2 × 4 joist is not required.

(b) The nominal size of a 2 × 6 section is breadth (*b*) of 2 in by depth (*d*) of 6 in:

$$\frac{d}{b} = \frac{6 \text{ in}}{2 \text{ in}} = 3$$

Ends of a 2 × 6 joist should be secured.

(c) The nominal size of a 2 × 8 section is breadth (*b*) of 2 in by depth (*d*) of 8 in:

$$\frac{d}{b} = \frac{8 \text{ in}}{2 \text{ in}} = 4$$

The 2 × 8 joist should be held in line.

(d) The nominal size of a 2 × 10 section is breadth (*b*) of 2 in by depth (*d*) of 10 in:

$$\frac{d}{b} = \frac{10 \text{ in}}{2 \text{ in}} = 5$$

One edge of the 2 × 10 joist should be held in line (i.e., properly secured to plywood subfloor).

(e) The nominal size of a 2 × 12 section is breadth (*b*) of 2 in by depth (*d*) of 12 in:

$$\frac{d}{b} = \frac{12 \text{ in}}{2 \text{ in}} = 6$$

For a 2 × 12 joist, diagonal bridging at intervals not exceeding 8 ft is required.

(f) The nominal size of a 2 × 14 section is breadth (*b*) of 2 in by depth (*d*) of 14 in:

$$\frac{d}{b} = \frac{14 \text{ in}}{2 \text{ in}} = 7$$

Both edges of the 2 × 14 joist should be held in line (i.e., properly secured to top and bottom edge).

Properly secured (nailed and glued) plywood structural subflooring adequately keeps joists up to 2 × 10 in a floor deck system from buckling laterally. For a 2 × 12, diagonal bridging at intervals not exceeding 8 ft is required; a span of over 8 ft requires one row of bridging at the center point of the span, a span of over 16 ft requires two rows of bridging at the third points of the span, and so on.

EXERCISES

24.1 Floor joists will be used in a residential floor system. The live-load requirement for furniture and occupants is 40 lb/ft². The dead load is anticipated to be less than 10 lb/ft². Deflection should not exceed $\ell/360$. The joists will be used under normal-duration loading.

Joist material is specified as visually graded Hemlock-Fir, No. 2 grade. Determine the maximum span for the following joist spacings:

(a) 12 in o.c.

(b) 16 in o.c.

(c) 19.2 in o.c.

24.2 Much as in Exercise 24.1, floor joists will be used in a residential floor system. The live-load requirements for furniture and occupants is 40 lb/ft^2. The dead load is anticipated to be less than 10 lb/ft^2. Deflection should not exceed $\ell/360$. The joists will be used under normal-duration loading. In this case, however, joist material is specified as visually graded Douglas Fir, No. 2 grade. Determine the maximum span for the following joist spacings:

(a) 12 in o.c.

(b) 16 in o.c.

(c) 19.2 in o.c.

24.3 Much as in Exercise 24.1, floor joists will be used in a residential floor system. The live-load requirements for furniture and occupants is 40 lb/ft^2. The dead load is anticipated to be less than 10 lb/ft^2. Deflection should not exceed $\ell/360$. The joists will be used under normal-duration loading. In this case, however, joist material is specified as visually graded Southern Pine, No. 2 grade. Determine the maximum span for the following joist spacings:

(a) 12 in o.c.

(b) 16 in o.c.

(c) 19.2 in o.c.

24.4 Much as in Exercise 24.1, floor joists will be used in a residential floor system. The live-load requirements for furniture and occupants is 40 lb/ft^2. The dead load is anticipated to be less than 10 lb/ft^2. Deflection should not exceed $\ell/360$. The joists will be used under normal-duration loading. In this case, however, joist material is specified as visually graded Redwood, No. 2 grade. Determine the maximum span for the following joist spacings:

(a) 12 in o.c.

(b) 16 in o.c.

(c) 19.2 in o.c.

24.5 A floor joist system for a residence will span between two bearing walls spaced at 13′-8″ o.c. Joist material is specified as visually graded Hemlock-Fir, No. 2 grade. The live-load requirements for furniture and occupants is 40 lb/ft^2. The dead load is anticipated to be less than 10 lb/ft^2. Deflection should not exceed $\ell/360$. The joist will be used under normal-duration loading. Determine the smallest joist section at 16-in-o.c. spacing.

24.6 A ceiling joist system for a residence will span 13′-8″. Joist material is specified as visually graded Hemlock-Fir, No. 2 grade. The live load is 10 lb/ft^2. The dead load is anticipated to be less than 10 lb/ft^2. Deflection should not exceed $\ell/240$. The joist will be used in a dry application under normal-duration loading. Determine the smallest joist section at 16-in-o.c. spacing.

24.7 A roof rafter system for a residence will span 13′-8″. The rafter material is specified as visually graded Hemlock-Fir, No. 2 grade. The live load for wind and snow is 30 lb/ft^2. The dead load is anticipated to be less than 15 lb/ft^2. Deflection should not exceed $\ell/240$. The rafter will be used in a dry application under normal-duration loading. Determine the smallest rafter section at 16-in-o.c. spacing.

Chapter 25
DESIGN OF SIMPLE TIMBER COLUMNS

An elementary design approach for timber columns is introduced in this chapter. The intent is to review a typical design approach to acquaint the technician with design principles and to demonstrate common design concerns. The reader is advised that the presentation does not provide full coverage of all design principles involved with timber design. Additional coverage is beyond the intended breadth of this book.

25.1 BASIC COLUMN DESIGN PROCEDURE

The most commonly used wood column is cut and shaped from solid wood and is square or rectangular in section, such as a 4 × 4 or 4 × 6. Although round column sections are available, rectangular or square sections are typically preferred in building structures, as it is easier to secure other members to the flat surfaces offered by this shape. Posts, studs, and pillars fall into the category of columns and are designed using similar design principles.

Selection of a solid-wood section for use as a column or columnlike member is based on the slenderness of a member, its tendency to buckle, and the applied end load. Following is a typical procedure used to determine the maximum allowable axial end load of a solid-wood column:

1. Determine the slenderness ratio (SR) of the column, based on the unbraced column length (ℓ), in inches, and its least side dimension (d) of a rectangular column section, in inches. For round column sections, $d = 0.886\,D$, where D is the round column diameter, in inches. If the column's least dimension is laterally braced, such a stud secured to sheathing and gypsum board, the widest dimension should be used for d.

$$SR = \frac{\ell}{d}$$

2. Determine the K constant, based on the modulus of elasticity (E) and the base compressive stress parallel to the grain (F_c) for the grade of wood being specified, both expressed in lb/in^2.

$$K = 0.671\sqrt{\frac{E}{F_c}}$$

3. Determine the maximum allowable axial load (P), in pounds, using the appropriate formula:

(a) Short column formula (SR of 11 or less):

$$P = F_c \cdot A$$

(b) Intermediate column formula (SR greater than 11 but less than K):

$$P/A = F_c\left\{1 - \left[\frac{1}{3}\left(\frac{SR}{K}\right)^4\right]\right\}$$

$$P = P/A \cdot A$$

(c) Long column formula (SR of K or greater):

$$P/A = \frac{0.30E}{(SR)^2}$$

$$P = P/A \cdot A$$

In the formulas above, the variables are:

A cross-sectional area of the column section, in in^2

d least side dimension (d), in inches (The widest section dimension should be used for d if the column's least dimension is laterally braced, such as a stud secured to sheathing and gypsum board or a column that is laterally braced.)

E modulus of elasticity, in lb/in^2

F_c base compressive stress parallel to the grain, in lb/in^2

K a constant equal to $0.671\sqrt{E/F_c}$ (do not confuse this K with the design buckling factor K_e)

ℓ unsupported column length, in inches

P maximum allowable axial load, in lb

P/A maximum allowable stress, in lb/in^2

Column analysis involves calculation of an allowable unit compressive stress (P/A) based on the loading condition and properties of the beam material and section. The allowable unit stress is the maximum stress that the column section is permitted to support. Once calculated, the allowable unit stress (P/A) must be multiplied by the section area of the column (A).

■ EXAMPLE 25.1

Determine the maximum allowable axial load that may be carried by an unbraced 2 × 4 column that is 8 ft long. Assume the following properties of wood: $F_c = 600$ lb/in² and $E = 1\ 700\ 000$ lb/in².

Solution

1. Determine the slenderness ratio (SR) of the column. Since the column is unbraced, $d = 1.5$ in, the least dimension of a 2 × 4 section:

$$SR = \frac{\ell}{d} = \frac{8\ \text{ft} \cdot 12\ \text{in/ft}}{1.5\ \text{in}} = 64.00$$

2. Determine the K constant:

$$K = 0.671 \sqrt{\frac{E}{F_c}} = 0.671 \sqrt{\frac{1\ 700\ 000\ \text{lb/in}^2}{600\ \text{lb/in}^2}} = 35.7$$

3. Determine the maximum allowable axial load (P). Use the long column formula, because SR is greater than K:

$$P/A = \frac{0.30E}{(SR)^2} = \frac{0.30(1\ 700\ 000\ \text{lb/in}^2)}{(64.00)^2} = 124.5\ \text{lb/in}^2$$

$$P = P/A \cdot A = 124.5\ \text{lb/in}^2 \cdot (1.5\ \text{in} \cdot 3.5\ \text{in}) = 654\ \text{lb}$$

The column can carry an axial end load of up to 654 lb.

Column analysis involves calculation of an allowable unit compressive stress (F'_c) based on the loading condition and properties of the beam material and section. The allowable unit stress is the maximum stress that the column section is permitted to support. Once calculated, the allowable unit stress (F'_c) must be multiplied by the section area of the column.

■ EXAMPLE 25.2

Determine the maximum allowable axial load that may be carried by a 2 × 4 stud that is 8 ft long and is secured by wall sheathing and gypsum wallboard. Assume the following properties of wood: $F_c = 600$ lb/in² and $E = 1\ 700\ 000$ lb/in².

Solution

1. Determine the slenderness ratio (SR) of the column. Since the column is braced with wall material, $d = 3.5$ in, the widest dimension of a 2 × 4 section:

$$SR = \frac{\ell}{d} = \frac{8 \text{ ft} \cdot 12 \text{ in/ft}}{3.5 \text{ in}} = 27.4$$

2. Determine the K constant.

$$K = 0.671 \sqrt{\frac{E}{F_c}} = 0.671 \sqrt{\frac{1\,700\,000 \text{ lb/in}^2}{600 \text{ lb/in}^2}} = 35.7$$

3. Determine the maximum allowable axial load (P), in pounds, using the appropriate formula. Use the intermediate column formula, because SR is greater than 11 but less than K.

$$P/A = F_c \left\{ 1 - \left[\frac{1}{3} \left(\frac{SR}{K} \right)^4 \right] \right\} = 600 \text{ lb/in}^2 \left\{ 1 - \left[\frac{1}{3} \left(\frac{27.4}{35.7} \right)^4 \right] \right\} = 530.6 \text{ lb/in}^2$$

$$P = P/A \times A = 530.6 \text{ lb/in}^2 \times (1.5 \text{ in} \times 3.5 \text{ in}) = 2786 \text{ lb}$$

The column can carry an axial end load of up to 2786 lb.

As illustrated in the examples above, columns with like sections and lengths can have a wide disparity in allowable axial loads depending on whether the column is unbraced or laterally supported. Lateral bracing always strengthens a column.

25.2   ALTERNATIVE COLUMN DESIGN PROCEDURE

Following is an alternative procedure prescribed by the Uniform Building Code (1997). It uses a single formula to determine the maximum allowable axial end load of a solid-wood column. In the alternative procedure, the allowable unit stress (F'_c) in lb/in^2 of the cross-sectional area of a square or rectangular column is found by a single formula:

$$F'_c = F_c^* \left[\frac{1 + (F_{cE}/F_c^*)}{2c'} \right] - \sqrt{\left[\frac{1 + (F_{cE}/F_c^*)}{2c'} \right]^2 - \frac{(F_{cE}/F_c^*)}{c'}}$$

This complex formula simplifies to the following streamlined equation:

$$F'_c = F_c^* \left[(Q) - \sqrt{(Q)^2 - \frac{F_{cE}/F_c^*}{c'}} \right] \quad \text{where } Q = \frac{1 + (F_{cE}/F_c^*)}{2c'}$$

In the formulas above, the variables are:

c' 0.8 for sawn lumber

        0.9 for glued-laminated timbers

d least side dimension (d), in inches (the widest section dimension should be used for d if the column's least dimension is laterally braced, such as a stud secured to sheathing and gypsum board or a column that is laterally braced)

E modulus of elasticity, in lb/in^2

F_c' allowable unit stress, lb/in^2

F_c^* tabulated base design value for compression stress parallel to grain (F_c) multiplied by all applicable adjustment factors (see Tables C.6 to C.8)

F_{cE} $K_{cE}(E)/(\ell_e/d)^2$

K_{cE} 0.3 for visually graded lumber

    0.418 for products such as machine stressed lumber or glued-laminated timbers

K_e design buckling factor (see Table 25.1)

ℓ_e effective length of the compression member, in inches, $\ell_e = K_e\ell$

ℓ unbraced column length, in inches

Q $[1 + (F_{cE}/F_c^*)]/2c'$

The maximum allowable axial load (P) a column can safely support is found by the formula below, where A is the section area, in inches, and F_c' is the allowable unit stress, in lb/in^2:

$$P = F_c' \cdot A$$

where P is the maximum allowable axial load, in lb; $P = F_c' \times A$; F_c' is the allowable unit stress, lb/in^2; and A is the area of the column section, in in^2.

■ EXAMPLE 25.3

Determine the maximum allowable axial load that may be carried by an unbraced 4 × 4 column that is 8 ft long. Column material is specified as visually graded Douglas Fir-Larch, No. 2 grade. The column will be used in a dry application under normal-duration loading. Column connections are free to rotate and fixed in translation.

Solution   From Table C.6, for Douglas Fir-Larch, No. 2, $F_c = 1350$ lb/in^2 and the size factor $C_F = 1.15$ for a 4-in-wide member. The column will be used in a dry application, so the wet service factor does not apply. F_c is adjusted to

$$F_c^* = (C_F)F_c = (1.15)1350 \text{ lb/in}^2 = 1553 \text{ lb/in}^2$$

For visually graded lumber, $K_{cE} = 0.3$. From Table C.6, for Douglas Fir-Larch, No. 2, $E = 1\,600\,000$ lb/in^2. From Table C.11, a 4 × 4 is actually 3.5 in × 3.5 in, so $d = 3.5$ in. For sawn lumber, $c' = 0.8$. The effective length,

Table 25.1 DESIGN BUCKLING FACTOR (K_e) FOR WOOD COLUMNS WITH DIFFERENT END CONNECTIONS

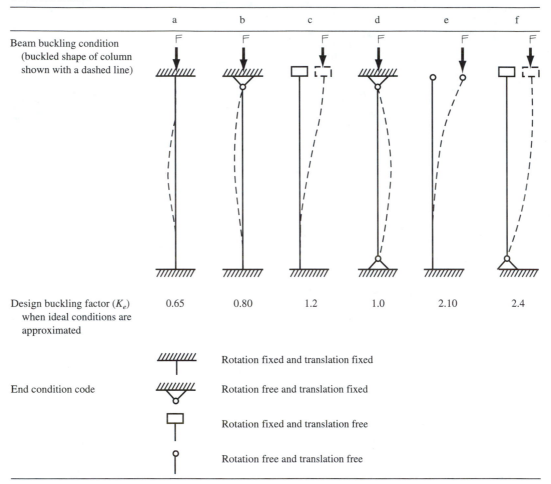

	a	b	c	d	e	f
Beam buckling condition (buckled shape of column shown with a dashed line)						
Design buckling factor (K_e) when ideal conditions are approximated	0.65	0.80	1.2	1.0	2.10	2.4

End condition code

Rotation fixed and translation fixed

Rotation free and translation fixed

Rotation fixed and translation free

Rotation free and translation free

Source: Reproduced from the 1997 edition of the Uniform Building Code™, copyright © 1997, with permission of the publisher, the International Conference of Building Officials. The International Conference of Building Officials assumes no responsibility for the accuracy or summaries provided in this book.

$\ell_e = K_e \ell = 1.0 \cdot 8 \text{ ft} \cdot 12 \text{ in/ft} = 96 \text{ in}$, assuming that the top and bottom connections are free to rotate and fixed in translation ($K_e = 1.0$):

$$F_{cE} = \frac{K_{cE}(E)}{(\ell_e/d)^2} = \frac{0.3(1\,600\,000)}{(96/3.5)^2} = 638 \text{ lb/in}^2$$

$$F_c' = F_c^* \left\{ \left[\frac{1 + (F_{cE}/F_c^*)}{2c'} \right] - \sqrt{\left[\frac{1 + (F_{cE}/F_c^*)}{2c'} \right]^2 - \frac{(F_{cE}/F_c^*)}{c'}} \right\}$$

$$= 1553 \text{ lb/in}^2 \left\{ \left[\frac{1 + (638/1553)}{2 \cdot 0.8} \right] - \sqrt{\left[\frac{1 + (638/1553)}{2 \cdot 0.8} \right]^2 - \frac{(638/1553)}{0.8}} \right\}$$

$$= 571.5 \text{ lb/in}^2$$

$$P = F_c' \cdot A = 571.5 \text{ lb/in}^2 \cdot (3.5 \text{ in} \cdot 3.5 \text{ in}) = 7001 \text{ lb}$$

The maximum allowable axial load of an 8-ft-long 4×4 column is 7001 lb.

■ EXAMPLE 25.4

Determine the maximum allowable axial load that may be carried by an unbraced 4×4 column that is 12 ft long (50% longer than the column in Example 25.3). Column material is specified as visually graded Douglas Fir-Larch, No. 2 grade. The column will be used in a dry application under normal-duration loading. Column connections are free to rotate and fixed in translation.

Solution As shown in Example 25.3, F_c is adjusted: $F_c^* = (C_F)F_c = (1.15)$ 1350 lb/in^2 = 1553 lb/in^2. For visually graded lumber, $K_{cE} = 0.3$. From Table C.6, for Douglas Fir-Larch, No. 2, $E = 1\,600\,000$ lb/in^2. From Table C.11, a 4×4 is actually 3.5 in $\times$ 3.5 in, so $d = 3.5$ in. For sawn lumber, $c' = 0.8$. The effective length, $\ell_e = K_e\ell = 1.0 \cdot 12$ ft $\cdot$ 12 in/ft = 144 in, assuming that the top and bottom connections are free to rotate and fixed in translation ($K_e = 1.0$).

$$F_{cE} = \frac{K_{cE}(E)}{(\ell_e/d)^2} = \frac{0.3(1\,600\,000)}{(144/3.5)^2} = 284 \text{ lb/in}^2$$

$$F_c' = F_c^* \left\{ \left[\frac{1 + (F_{cE}/F_c^*)}{2c'} \right] - \sqrt{ \left[\left(\frac{1 + (F_{cE}/F_c^*)}{2c'} \right) \right]^2 - \frac{(F_{cE}/F_c^*)}{c'} } \right\}$$

$$= 1553 \text{ lb/in}^2 \left\{ \left[\frac{1 + (284/1553)}{2 \cdot 0.8} \right] - \sqrt{ \left[\frac{1 + (284/1553)}{2 \cdot 0.8} \right]^2 - \frac{(284/1553)}{0.8} } \right\}$$

$$= 272.4 \text{ lb/in}^2$$

$$P = P = F_c' \cdot A = 272.4 \text{ lb/in}^2 \cdot (3.5 \text{ in} \cdot 3.5 \text{ in}) = 3337 \text{ lb}$$

The maximum allowable axial load of a 12-ft-long 4×4 column is 3337 lb.

■ EXAMPLE 25.5

Determine the maximum allowable axial load that may be carried by an unbraced 4×4 column that is 4 ft long (half the size of the column in Example 25.3). Column material is specified as visually graded Douglas Fir-Larch, No. 2 grade. The column will be used in a dry application under normal-duration loading. Column connections are free to rotate and fixed in translation.

Solution As shown in Example 25.3, F_c is adjusted: $F_c^* = (C_F)F_c = (1.15)$ 1350 lb/in^2 = 1553 lb/in^2. For visually graded lumber, $K_{cE} = 0.3$. From Table

C.6, for Douglas Fir-Larch, No. 2, $E = 1\,600\,000$ lb/in². From Table C.1, a 4×4 is actually 3.5 in $\times$ 3.5 in, so $d = 3.5$ in. For sawn lumber, $c' = 0.8$. The effective length, $\ell_e = K_e\ell = 1.0 \cdot 4$ ft $\cdot$ 12 in/ft $= 48$ in, assuming that the top and bottom connections are free to rotate and fixed in translation ($K_e = 1.0$).

$$F_{cE} = \frac{K_{cE}(E)}{(\ell_e/d)^2} = \frac{0.3(1\,600\,000)}{(48/3.5)^2} = 2552 \text{ lb/in}^2$$

$$F_c' = F_c^* \left\{ \left[\frac{(1 + (F_{cE}/F_c^*))}{2c'} \right] - \sqrt{\left[\frac{(1 + (F_{cE}/F_c^*))}{2c'} \right]^2 - \frac{(F_{cE}/F_c^*)}{c'}} \right\}$$

$$= 1553 \text{ lb/in}^2 \left\{ \left[\frac{(1 + 2552/1553)}{2 \cdot 0.8} \right] - \sqrt{\left[\frac{1 + (2552/1553)}{2 \cdot 0.8} \right]^2 - \frac{(2552/1553)}{0.8}} \right\}$$

$$= 1290 \text{ lb/in}^2$$

$$P = P = F_c' \cdot A = 1290 \text{ lb/in}^2 \cdot (3.5 \text{ in} \cdot 3.5 \text{ in}) = 15\,803 \text{ lb}$$

The maximum allowable axial load of a 4-ft-long 4×4 column is 15 803 lb.

A long column cannot carry the same load as a short column of the same section because a long column has a greater tendency to fail by buckling. This is evident when comparing analysis results in Examples 25.3, 25.4, and 25.5. Because it is shorter, the 4′-0″ column in Example 25.5 can carry more than twice the load of the 8′-0″ column in Example 25.3 and almost five times the load as the 12′-0″ column in Example 25.4. The longer 12′-0″ column in Example 25.5 has the greatest tendency to fail by buckling and thus can safely carry a smaller load.

EXERCISES

25.1 An unbraced 4×4 column is 10 ft long. Column material is specified as visually graded Douglas Fir, No. 2 grade. Column connections are free to rotate and fixed in translation ($K_e = 1.0$). The column will be used in a dry application and under normal-duration loading.
   (a) Determine the maximum allowable axial load that can be carried by the column.
   (b) Determine the maximum allowable axial load that can be carried by the column using the alternate design procedure.

25.2 Much as in Exercise 25.1, an unbraced 4×4 column is 10 ft long. The column will be used in a dry application and under normal-duration loading. Column connections are free to rotate and fixed in translation ($K_e = 1.0$). In this case, however, the column material is specified as visually graded Hemlock-Fir, No. 2 grade. Determine the maximum allowable axial load that can be carried by the column using the alternative design procedure.

25.3 Much as in Exercise 25.1, an unbraced 4×4 column is 10 ft long. The column will be used in a dry application and under normal-duration loading. The column material is specified as visually graded Douglas Fir, No. 2 grade. In this case, however, column connections are fixed in rotation and translation ($K_e = 0.65$). Determine the maximum allowable axial load that can be carried by the column using the alternative design procedure.

25.4 Much as in Exercise 25.1, an unbraced 4 × 4 column is 10 ft long. The column connections are free to rotate and fixed in translation ($K_e = 1.0$). The column will be used in a dry application and under normal-duration loading. In this case, however, column material is specified as visually graded Redwood, No 2. Determine the maximum allowable axial load that can be carried by the column using the alternative design procedure.

25.5 Much as in Exercise 25.4, an unbraced 4 × 4 column is 10 ft long. The column material is specified as visually graded Redwood, No. 2 grade. Column connections are free to rotate and fixed in translation ($K_e = 1.0$). The column will be used under normal-duration loading. In this case, however, the column will be used in a wet (outdoor) application. Determine the maximum allowable axial load that can be carried by the column using the alternative design procedure.

25.6 Compare the maximum allowable axial loads that can be carried by the columns of Exercises 25.1 to 25.5.
   (a) How does a change in material influence the axial load a column can support?
   (b) How does the type of column connections influence the axial load a column can support?
   (c) How does a wet (outdoor) application influence the axial load a column can support?

25.7 An unbraced $5\frac{1}{8}$ × 9.0 glued-laminated timber column is 16 ft long. The column will be used in a dry application and under normal-duration loading. Column connections are free to rotate and fixed in translation ($K_e = 1.0$). The glued-laminated timber has a compression stress parallel to grain (F_c^*) of 2100 lb/in^2. Determine the maximum allowable axial load that can be carried by the column using the alternative design procedure.

25.8 A $5\frac{1}{8}$-in-thick glued-laminated timber will be used as a 24-ft-long column. The column will support a roof and floor load. A roof load of 16 kip will be applied to the top end of the column. A floor load of 38 kip will be applied to the column at a distance of 13′-4″ ft from the column base. All loads are assumed to be axial. The column will be used in a dry application and under normal-duration loading. The column connections are free to rotate and fixed in translation ($K_e = 1.0$). The specified glued-laminated timber has a compression stress parallel to grain (F_c^*) of 2100 lb/in^2. From stock glued-laminated timber sizes in Table C.2, determine the minimum column section dimensions at the base of the column using the alternative design procedure.

25.9 Develop a table of and compare safe loads that can be supported by unbraced 4 × 4 columns of various lengths. Base computations on visually graded Douglas Fir, No. 2, a dry application, normal-duration loading, and column connections that are free to rotate and fixed in translation ($K_e = 1.0$).
   (a) Compute the maximum allowable axial loads from 8 to 24 ft long at intervals of 2 ft and place the results in tabular form.
   (b) Draw a graph of maximum allowable axial load (y axis) versus unbraced column length (x axis) in feet.

25.10 Develop a table of and compare safe loads that can be supported by unbraced 4 × 4 columns of various lengths. Base computations on visually graded Hemlock-Fir, No. 2, a dry application, normal-duration loading, and column connections that are free to rotate and fixed in translation ($K_e = 1.0$).
   (a) Compute maximum allowable axial loads from 8 to 24 ft long at intervals of 2 ft and place the results in tabular form.

(b) Draw a graph of maximum allowable axial load (y axis) versus unbraced column length (x axis) in feet.

25.11 Develop a table of and compare safe loads that can be supported by unbraced 4×4 columns of various lengths. Base computations on visually graded Redwood, No. 2, a dry application, normal-duration loading, and column connections that are free to rotate and fixed in translation ($K_e = 1.0$).

(a) Compute maximum allowable axial loads from 8 to 24 ft long at intervals of 2 ft and place the results in tabular form.

(b) Draw a graph of maximum allowable axial load (y axis) versus unbraced column length (x axis) in feet.

25.12 Develop a table of and compare safe loads that can be supported by unbraced 4×4 columns of various lengths. Base computations on visually graded Spruce-Pine-Fir, No. 2, a dry application, normal-duration loading, and column connections that are free to rotate and fixed in translation ($K_e = 1.0$).

(a) Compute maximum allowable axial loads from 8 to 24 ft long at intervals of 2 ft and place the results in tabular form.

(b) Draw a graph of maximum allowable axial load (y axis) versus unbraced column length (x axis) in feet.

Chapter 26
REINFORCED CONCRETE PRINCIPLES

This chapter provides information on steel reinforcement and concrete used in reinforced concrete members. It serves as a prelude to the methods used in design of simple nonprestressed reinforced concrete beams and columns that are introduced in Chapters 27 and 28.

26.1   CONCRETE

Concrete used in the building construction industry is made from a mix of hydraulic cement, aggregate, water, and other inert materials. Hydraulic cement, the key ingredient in concrete, is made by heating a mixture of limestone and clay until it almost fuses and then grinding it to a fine powder. Portland cement, the most common hydraulic cement, was patented by English bricklayer Joseph Aspdin in 1824. It gets its name because its light gray appearance is much like the color of the cliffs of the Isle of Portland in England.

Portland cement is made by mixing and then heating substances containing lime, silica, alumina, and iron oxide. Gypsum is added during the grinding process. When mixed with water, hydraulic cement begins to cure. The cementitious paste of Portland cement and water, as it cures, bonds the fine (sand) and coarse (gravel) aggregate to produce a strong, hard conglomerate.

Concrete sets up and cures through a chemical reaction called *hydration.* Although the hydration reaction begins immediately when water and cement are mixed, the concrete does not begin to set up noticeably for about an hour. Actual strength development of concrete occurs over a period of days and weeks and continues for as long as moisture is present. Strength development of a concrete batch will vary with water content, type of concrete, temperature of the concrete, mix and sizes of aggregate, how well the concrete was worked (vibrated), number of days after mixing, and any admixtures included in the mix.

The *ultimate compressive strength* of concrete is the laboratory-determined compressive stress at which a concrete specimen fails after curing for 28 days. The *specified compressive strength* (f'_c) of concrete is the largest stress that the concrete

Concrete test specimens are cast in forms when the concrete is delivered to the construction site. Concrete specimens will be taken to a laboratory, where they will be tested to determine the compressive strength of the concrete.

must develop without compressive failure in laboratory testing at 28 days after curing. The concrete supplier must deliver concrete that meets or exceeds the compressive strength specified. Design values for the specified compressive strength of cast-in-place concrete generally range from 2500 to 4000 lb/in^2. With precast concrete members, specified compressive strengths are typically 4000 to 6000 lb/in^2. The 28-day compressive strength of concrete that is in a solid state is generally no less than 2000 lb/in^2, while the tensile strength is about 200 lb/in^2.

In the United States, concrete specimens in the shape of a 6-in-diameter $\times$ 12-in-tall cylinder are tested to determine compressive strength of concrete. At the construction site at the time the concrete is delivered, cylindrical forms are each filled in three layers with fresh concrete. Each layer is rodded with a metal bar to produce a consistent specimen. After initial curing at the construction site, the concrete specimens are removed to the laboratory, where they continue to cure. Each specimen is eventually compression tested to failure.

For example, 28 days after a specimen is cast, it is removed from the form and compression tested to certify the strength of the concrete. Some specimens are tested before day 28 to confirm that concrete strength is developing at the appropriate rate. The analysis results of a typical compressive stress test used to determine the ultimate compressive strength of a concrete specimen are shown in Example 26.1. The compression test procedure was discussed in Chapter 10.

■ EXAMPLE 26.1

A 6.0-in-diameter concrete specimen fails at a force of 125 500 lb in a compression test exactly four weeks (28 days) after it was mixed and cast into a specimen. Calculate the ultimate compressive strength for this concrete mix.

Solution

$$A = \frac{\pi D^2}{4} = \frac{\pi (6.0 \text{ in})^2}{4} = 28.3 \text{ in}^2$$

$$\sigma_{\text{ultimate strength}} = \frac{F}{A} = \frac{125\ 500 \text{ lb}}{28.3 \text{ in}^2} = 4435 \text{ lb/in}^2$$

Assume that the engineer designed a concrete member on the basis of a specified compressive strength (f'_c) of 4000 lb/in^2 and that the concrete delivered to the construction site and used to construct this member had a laboratory-determined ultimate compressive strength of 4435 lb/in^2 (as in Example 26.1). The concrete delivered would have been acceptable because the ultimate compressive strength attained the specified compressive strength. Had the concrete not attained the specified compressive strength—if it had an ultimate compressive strength of, say, 3435 lb/in^2—the concrete would have been unacceptable. Several potentially costly scenarios result from this undesirable situation, such as reevaluating the required strength of the existing member to ascertain whether the weaker concrete will be adequate, renovating to strengthen the member, or removal of the unsatisfactory concrete.

Concrete members can be cast-in-place or precast. A *cast-in-place* concrete member is cast in its final position in the building structure. A *precast* concrete member is cast and then moved to its final position. A precast member is generally cast at a manufacturing plant and trucked to the construction site, where it is lifted by crane into its final position. *Site-cast* concrete members are cast on the construction site and lifted into their final position.

26.2 REINFORCED CONCRETE

Concrete resists compressive forces well but fails easily in tension: Concrete is about 10 times stronger in resisting compression stress than tensile stress. In reinforced concrete, a reinforcement such as steel is added to create a composite *reinforced concrete* member.

Typically, reinforcement is designed and placed to resist tensile and shearing stresses while concrete cast around and bonded to the reinforcement resists compressive stresses. Reinforcement can also be used to withstand excessive compressive stresses. Vertical reinforcement in a reinforced column permits use of smaller column sections because the steel reinforcement can safely carry greater compression stresses than concrete. Compression reinforcement is used to resist excessive compressive stresses that develop at the point of application of a heavy concentrated load. Additionally, reinforcement is used to control cracks from shrinking of concrete during curing and from temperature change, or to transfer the load from one member to another.

Reinforced concrete members are generally designed in terms of the theory that steel reinforcement resists tensile stresses and concrete withstands compressive stresses; therefore, main reinforcement is generally called *tensile reinforcement. Temperature reinforcement* is a supplementary type of reinforcement that is used to minimize the damaging effects of tensile stresses from temperature variations that develop when the concrete cures. Other types of reinforcement, such as stirrups and ties, are used to counter other types of stresses that develop in a reinforced concrete member.

Tensile reinforcement can be nonstressed or prestressed. *Nonstressed concrete* is a method of reinforcing concrete in which the reinforcement is placed without any induced stresses in the reinforcement except those stresses caused by the applied load. *Prestressed concrete* is a method of reinforcing concrete in which tensile stresses are induced in the steel reinforcement. Steel *tendons* composed of cable, wire, or bars are embedded in the concrete, and a tensile force is applied to the tendons by stretching the reinforcement and placing clamps at the ends.

Tensioning the reinforcement introduces compressive stresses in the concrete. Under load, the compressive stresses already developed in the "tensile" fibers of the concrete beam must first be relieved before any tensile stresses begin to develop in these fibers; that is, fibers in a beam that are normally under tension actually begin in compression and are not in tension until the applied load first relieves these compression stresses. Thus a prestressed member can support a greater load. *Pretensioning* is a method of prestressing the reinforcement in the concrete before the concrete is placed. *Post-tensioning* is a method of prestressing the reinforcement in the concrete after it has hardened.

The designer bases analysis of reinforced concrete structural members on specified strength of the concrete, the strength of the reinforcement, and any prestressing of tensile reinforcement. In this book only nonstressed members are examined.

26.3   REINFORCEMENT

Steel is frequently used as a concrete reinforcement because it offers high tensile strength and has thermal expansion properties that are very comparable to those of concrete. *Reinforcing bars,* known in the reinforced concrete industry as *rebars* or simply as *bars,* are round steel bars that are deformed on their surface. Surface deformations on the bars allow the concrete to grip the steel. Bar diameter (D_b) is based on the bar gauge number. As a rule of thumb, a single gauge in U.S. Customary System units relates to roughly $\frac{1}{8}$ inch in diameter:

$$D_b = \frac{\text{Bar No.}}{8} \quad \text{or} \quad \text{Bar No.} = 8(D_b)$$

A No. 11 bar is eleven $\frac{1}{8}$ths, or about $1\frac{3}{8}$ in diameter, a No. 5 bar is $\frac{5}{8}$ in diameter, and so on. Information on bar sizes is given in Table B.7.

The reinforcing steel industry has adopted a soft conversion to classify metric bars. Metric bar gauges are expressed in SI units, but the bar does not change from the original diameter. For example, a No. 3 steel bar having a diameter of 0.375 in (9.5 mm) has a metric bar designation of No. 10 because 9.5 mm is approximately 10 mm.

The strength of bars used as longitudinal tensile and web reinforcement is based on the grade specified. Grades of steel commonly specified for nonprestressed concrete reinforcement includes grades 40, 60, and 75 in U.S. customary system units and grades 300, 420, and 520 in SI units. The yield strengths (f_y) and allowable tensile stresses (f_s) specified for these grades are given in Table 26.1.

Table 26.1 TENSILE PROPERTIES OF COMMONLY SPECIFIED GRADES OF STEEL USED FOR NONPRESTRESSED CONCRETE REINFORCEMENT

| | | Steel grade | | | | | |
| | | U.S. Customary units (lb/in^2) | | | Metric (SI) units (MPa) | | |
Property	Symbol	40	60	75	300	420	520
Specified yield strength	f_y	40 000	60 000	75 000	300	420	520
Specified allowable tensile strength	f_s	20 000	24 000	24 000	150	168	208

Structural welded wire reinforcement (WWR), formerly called welded wire mesh (WWM) or welded wire fabric (WWF), consists of a steel wire fabric configured in the form of a screenlike grid with square or rectangular openings. It is available in sheets or rolls. WWR is typically used to increase the tensile strain capacity of the concrete slab. Wire is either smooth (W) or deformed (D). Wire size is expressed by a number that indicates its cross-sectional area in 0.01 in^2. A W1.4 wire is a smooth wire having a cross-sectional area of 1.4/100 in^2 or 0.014 in^2. This system has replaced the old steel wire gauge system where wire was specified by gauge number (i.e., No. 10 gauge); a smaller gauge number indicated a heavier wire. A typical designation for welded wire reinforcement is WWR W1.4 $\times$ W1.4–4 $\times$ 12, which indicates that the W1.4 wire is spaced 4 in longitudinally and 12 in transversely.

Fiber concrete reinforcement is a modern alternative to welded wire reinforcement and serves as a better method of reducing the formation of settlement and shrinkage cracks. Bundled multifilament polypropylene fibers are added to the concrete batch during mixing in proportions of about 1.5 lb of fibers per yard of concrete (0.9 kg/m^3). Agitation of mixing causes the bundled fibers to separate into single fiber strands that are uniformly distributed in all directions throughout the concrete. As concrete hardens and shrinks or expands and contracts differentially, microscopic cracks develop. When a growing microcrack intersects a fiber strand, it is obstructed and prevented from developing into a larger crack. Fiber reinforcement in concrete serves to minimize cracking due to drying shrinkage and thermal expansion/contraction, provides reduction of permeability, and increases impact capacity, shatter resistance, abrasion resistance, and toughness.

26.4 CONCRETE COVER

To take advantage of its tensile strength and to prevent surface cracking, steel reinforcement is typically placed as close to the surface of the member as possible. Steel corrodes easily in the presence of moisture, which can lead to corrosion failure of the steel. Steel also loses strength rapidly if exposed to high temperatures, so an adequate concrete cover limits the temperature increase of steel reinforcement in a fire. Encasing steel reinforcement with concrete is necessary to ensure that the reinforcement is adequately bonded to the concrete and protected from moisture and fire. The distance between the outside concrete surface and steel reinforcement is called *concrete cover.*

Requirements for concrete cover vary by exposure to soil and weather. Reinforced concrete that is not exposed to weather or soil generally requires a minimum concrete cover of $\frac{3}{4}$ in (20 mm) for slabs and walls and $1\frac{1}{2}$ in (40 mm) for large structural members such as beams and columns. For form-cast reinforced concrete members that are exposed to weather or soil, the general requirement is a minimum concrete cover of $1\frac{1}{2}$ in (40 mm) for No. 5 bars and smaller and 2 in (50 mm) for bars larger than No. 5. Due to the effect that soil has on surface curing and moisture penetration, a concrete cover of 3 in (70 mm) or more is frequently required when the concrete is cast in direct contact with soil. Fiber concrete reinforcement does not require concrete cover.

Proper spacing of bars is necessary to ensure that fresh concrete can be placed adequately without being obstructed by the reinforcement and that the reinforcement will bond properly to the concrete. With longitudinal tensile reinforcement in beams and columns, *concrete cover spacing* generally refers to the distance measured surface to surface. A spacing of $1\frac{1}{2}$ times the bar diameter and not less than $1\frac{1}{2}$ in (40 mm) is required in columns. In beams, a spacing of one bar diameter and not less than 1 in (25 mm) is generally required. Other spacings are required for main and temperature reinforcement in slabs and walls.

26.5 BONDING STRESS AND DEVELOPMENT LENGTH

Reinforcing bars are deformed to help the concrete bond to the reinforcement surface. When reinforcement is under tension or compression, a *bonding stress* develops at the interfacing surfaces of the reinforcement and concrete due to the pulling or pushing action of the reinforcement. Concrete must adequately grip the reinforcement when this stress at the surface of the reinforcement develops. Excessive bonding stress can cause tear-out of the reinforcement. Concrete encasing the reinforcement must be capable of resisting the bonding stress that develops.

Reinforcement must be large enough in section and of sufficient length to keep bonding stresses at acceptable levels. The *development length* of reinforcement is the minimum length the reinforcement must be embedded in the concrete to develop sufficient strength at the section experiencing the greatest bonding stress. The development length of reinforcement must be computed by the designer to ensure that the bonding stress is not exceeded.

In reinforced concrete beams under load, bonding stress occurs at the surfaces of the tensile reinforcement. Usually, the development length of the tensile reinforcement in long beams and slabs is adequate. In short beams under heavy load or where the required development length cannot be achieved with tensile reinforcement length, use of hooks is required. *Hooks* are 90°, 135°, or 180° bends at the end of the reinforcement.

In columns where tensile stresses are minimal, bonding stresses occur at the surface of the reinforcement but tend to be less than in beams. However, since reinforced concrete columns are generally poured in separate foundation-to-floor or floor-to-floor pours, it is necessary to ensure that there is adequate development length of reinforcement beyond the construction joint. In addition to the longitudinal bars that extend vertically beyond a construction joint created by the separate pours, short reinforcement bars called dowels can also be used. *Dowels* are additional reinforcing bars used to attain the required development length of reinforcement when primary reinforcement cannot.

26.6 REINFORCED CONCRETE DESIGN STANDARDS

The American Concrete Institute, Inc. (ACI), P.O. Box 9094, Farmington Hills, MI 48333 (Web site address: www.aci-int.org), was founded in 1905 as a nonprofit association that serves the concrete industry. The concrete industry is a segmented industry that includes cement manufacturers, aggregate quarries and producers, ready-mixed concrete suppliers, reinforcing steel manufacturers, equipment suppliers, additive manufacturers, form material suppliers, design architects and engineers, contractors, subcontractors, skilled labor, construction managers, inspectors, and material testing laboratories.

ACI gathers and distributes information on the improvement of design, construction, and maintenance of concrete products and structures. Much of the work of ACI is done through volunteer committees. ACI Committee 318 is one of over 100 technical committees that develops building code requirements for reinforced concrete. *ACI 318, Building Code Requirements for Reinforced Concrete,* covers the proper design and construction of buildings of reinforced concrete. The code is fully updated every six years.

The ACI 318 Code specifies two approaches for design of reinforced concrete members: the *strength design* method and the alternative *working stress design* method. The working stress design method limits the stress to which the steel and concrete is subjected. It ensures that the allowable stress of the steel and concrete are not exceeded. Working stress design is based on a factor of safety and thus is limited to only a fraction of the ultimate strength of the material.

The strength design method better approximates the load a member can support without failure and frequently results in a more slender member. The working stress method was once the only method prescribed by ACI. It is now only used as an alternative design method for nonprestressed beams. The working stress design method for design of a tied column is no longer permitted.

The working stress design method for a simple beam is presented in this book because of its less complicated design approach. A theoretical approach founded on the strength design method of designing a tied column will be introduced. This introduction serves as a method of acquainting the technician with basic principles of reinforced concrete design.

EXERCISES

26.1 A 6.0-in-diameter concrete specimen fails at a force of 140 500 lb in a compression test exactly four weeks (28 days) after it was mixed and formed into a specimen. Calculate the ultimate compressive strength for this concrete mix.

26.2 An engineer designs a reinforced concrete member on the basis of concrete with a specified compressive strength (f'_c) of 3000 lb/in^2. A 6.0-in-diameter specimen of the concrete fails at a force of 102 500 lb in a compression test exactly four weeks (28 days) after it was mixed and cast into a specimen.
   (a) Calculate the ultimate compressive strength for this concrete mix.
   (b) Does the concrete meet minimum design requirements for compressive strength?

26.3 An engineer designs a reinforced concrete member on the basis of concrete with a specified compressive strength (f'_c) of 4000 lb/in^2. A 6.0-in-diameter specimen of the concrete fails at a force of 108 500 lb in a compression test exactly four weeks (28 days) after it was mixed and cast into a specimen.
   (a) Calculate the ultimate compressive strength for this concrete mix.
   (b) Does the concrete meet minimum design requirements for compressive strength?
   (c) Since the test was conducted four weeks after the concrete was cast as a structural building member, what options are available to resolve any problems that may result?

26.4 Approximate the diameter of No. 3 through No. 11, No. 14, and No. 18 steel reinforcing bars. Compare your results to the nominal diameters provided in Table B.7.

26.5 Convert the diameter of No. 3 through No. 11, No. 14, and No. 18 steel reinforcing bars to SI units, in mm. Compare your results to the SI bar sizes provided in Table B.7.

SPECIAL PROJECTS

Reinforced Concrete Project I

Objectives   The objectives of this project are to provide the student with experience in the methodology involved in determination of tributary loads of structural members, in development of free-body diagrams, and in basic selection of reinforced concrete beams and columns based on gravity loading.

Design parameters   Partial floor and roof framing plans of a precast concrete building structure are shown in Figure SP26.1. The three bays shown in the framing

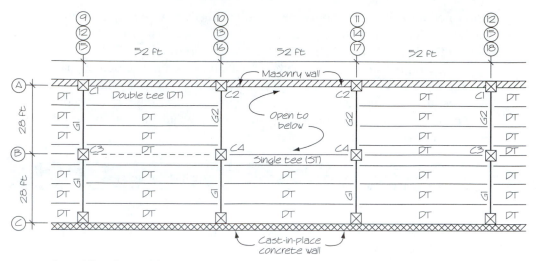

Second-floor framing plan

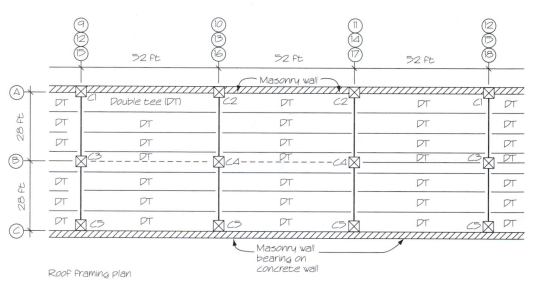

Roof framing plan

FIGURE SP26.1 Reinforced Concrete Project 1

plans repeat in the grid pattern shown. The two-story reinforced concrete structure has a second floor and low-slope (flat) roof. The first floor is a floating concrete slab that is isolated from the structure. Second-floor and roof systems are constructed of 24-in-deep double tees covered with a 4-in-thick concrete topping that support the roof and floor loads. The double tees rest on and transfer the roof and floor loads to the girders. Loads are uniformly distributed to the girders. The girder ends bear on column haunches that hold the top of the girders at an elevation that is 28 in below the floor or roof elevation. Columns C1 through C4 are two stories tall and rest on individual foundations. Columns C5 are one story tall and rest on a cast-in-place reinforced concrete wall. The floor-to-floor and floor-to-roof distances are 16 ft. Loads of this structure are to be based on the following:

FLOOR:

Structural double or single tees (DT or ST) with concrete topping	80.0 lb/ft^2
HVAC, plumbing, electrical, lighting, ceiling	est. 25.0 lb/ft^2
Partition load	20.0 lb/ft^2
Live load (occupants, furniture, equipment, etc.)	100.0 lb/ft^2

ROOF:

Structural double tees (DT) with lightweight (insulating) concrete topping	60.0 lb/ft^2
Roof membrane with gravel	6.0 lb/ft^2
HVAC, plumbing, electrical, lighting, ceiling	est. 15.0 lb/ft^2
Roof load (live)	30.0 lb/ft^2
Snow load	30.0 lb/ft^2
Rain/ice load	30.0 lb/ft^2

Analysis *Girders.* To simplify design and ensure production quality, all girders are to have dimensions and reinforcement based on the largest girder load. The breadth of all girders is to be 16 in. The shallowest acceptable structural member is to be specified. Concrete with a specified compressive strength (f'_c) of 4000 lb/in^2 and grade 60 steel is specified. Four longitudinal tensile bars are to be used. If required, No. 3 stirrups will be specified. In addition to floor loads, girders carry an 800 lb/ft load from interior masonry walls bearing on the member and a 600-lb/ft estimated load for their own weight. Concrete with a specified compressive strength (f'_c) of 4000 lb/in^2 and grade 60 steel is specified. Four longitudinal tensile bars are to be used. If required, No. 3 stirrups will be specified.

Columns. To simplify design and ensure quality control, all columns are to have dimensions of 16 in × 16 in and the same reinforcement based on the largest axial load. Analysis should be based on a 16-ft column length. Concrete with a specified compressive strength (f'_c) of 4000 lb/in^2 and grade 60 steel is specified.

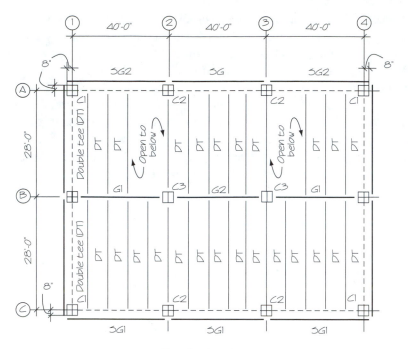

Floor framing plan

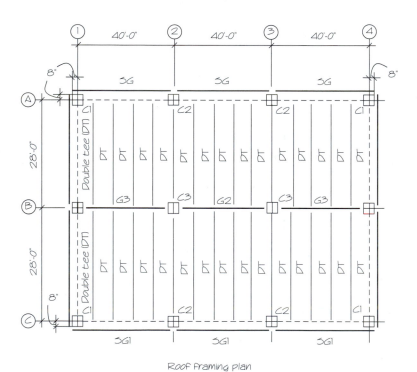

Roof framing plan

FIGURE SP26.2 Reinforced Concrete Project 2

Eight longitudinal bars and No. 3 ties are to be used. In addition to axial loads, columns have a 250-lb/ft estimated load for their own weight.

Reinforced Concrete Project 2

Objectives   The objectives of this project are to provide the student with experience in the methodology involved in determination of tributary loads of structural members, in development of free-body diagrams, and in basic selection of reinforced concrete beams and columns based on gravity loading.

Design Parameters   Framing plans of a simple precast concrete building structure are shown in Figure SP26.2. The three-story reinforced concrete structure has three floors and a low-slope (flat) roof. It is similar to the building shown on pages 461-471. The first floor is a floating concrete slab that is isolated from the structure. The second- and third-floor systems and the roof system are constructed of 20-in-deep double tees

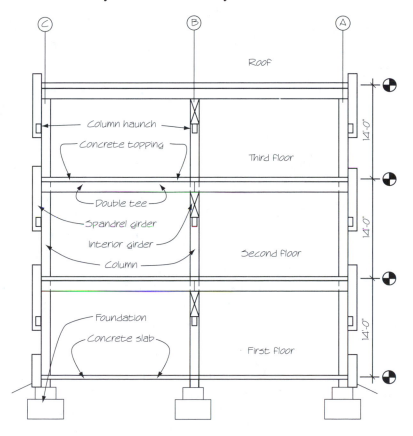

Transverse building section

FIGURE SP26.2   Reinforced Concrete Project 2 continued

covered with a 4-in-thick concrete topping that support the roof and floor loads. The interior ends of the double tees rest on and transfer the roof and floor loads to the interior girders (G1). The exterior ends of the double tees rest in pockets in the spandrel (exterior) girders. Floor and roof loads from the double tees are distributed uniformly to the girders. The ends of the girders bear on column haunches. The floor-to-floor and floor-to-roof distances are 14 ft. Loads of this structure are to be based on the following:

FLOOR:

Structural double or single tees (DT or ST) with concrete topping	80.0 lb/ft^2
HVAC, plumbing, electrical, lighting, ceiling	est. 25.0 lb/ft^2
Partition load	20.0 lb/ft^2
Live load (occupants, furniture, equipment, etc.)	100.0 lb/ft^2

ROOF:

Structural double tees (DT) with lightweight (insulating) concrete topping	60.0 lb/ft^2
Roof membrane with gravel	6.0 lb/ft^2
HVAC, plumbing, electrical, lighting, ceiling	est. 15.0 lb/ft^2
Roof load (live)	30.0 lb/ft^2
Snow load	30.0 lb/ft^2
Rain/ice load	30.0 lb/ft^2

Analysis *Girders.* To simplify design and ensure production quality, all interior girders are to have dimensions and reinforcement based on the largest interior girder load, and all spandrel girders are to have dimensions and reinforcement based on the largest spandrel girder load. Assume a design span of 40 ft for all girders and a span of 28 ft for double tees. The breadth of interior girders is to be 16 in. The depth of spandrel girders is to be 8 ft with a minimum breadth of 8 in. Concrete with a specified compressive strength (f'_c) of 4000 lb/in^2 and grade 60 steel are specified. Three longitudinal tensile bars are to be used. If required, No. 3 stirrups will be specified. In addition to floor loads, spandrel girders carry a 200-lb/ft load from windows and a 750-lb/ft estimated load for their own weight. Interior girders have an additional 750-lb/ft estimated load for their own weight. Concrete with a specified compressive strength (f'_c) of 4000 lb/in^2 and grade 60 steel is specified. Four longitudinal tensile bars are to be used. If required, No. 3 stirrups will be specified.

Columns. To simplify design and ensure production quality, all columns are to have dimensions of 16 in × 16 in and the same reinforcement based on the largest axial load. Analysis should be based on a 14-ft column length. Concrete with a specified compressive strength (f'_c) of 4000 lb/in^2 and grade 60 steel is specified. Eight longitudinal bars and No. 3 ties are to be used. In addition to axial loads, columns have a 250-lb/ft estimated load for their own weight.

The finished envelope of a precast concrete building.

A transverse view of the structural system of a precast concrete structure during construction four months before the photograph above was taken. Concrete double tees make up the structure of the floor and roof platforms and span transversely between girders. Inverted tee girders in the center of the structure and exterior spandrel girders span longitudinally between vertically positioned columns.

Columns are vertically positioned structural members that support girders. Columns transfer building loads to individual foundations located at the base of each column.

A foundation pedestal. Anchor bolts extend from the top of the foundation pedestal to accept and secure the column base.

Steel base plates are cast into the bottom of the concrete column. Holes in the plates accept anchor bolts when the column is secured to the foundation pedestal.

Columns are vertically positioned structural members that support horizontally positioned girders. The girders support double tees that make up the floor and roof structural platform. Temporary column bracing and cabling support the columns during erection.

Square-shaped column haunches are cast as an integral part of the column. A column haunch serves as a ledger to provide support for the girders. U-shaped hooks called picks are cast in concrete structural members to make erection easier. These picks are cut and removed after the member is placed.

Girders span longitudinally between columns and bear on column haunches.

A concrete double tee that makes up the floor and roof structure. U-shaped picks are cast in the double tee to make erection easier.

Concrete double tees bear on a haunch that runs continuously along the bottom of both sides of the inverted tee girder. The haunch is an integral part of the girder.

The extended webs of double tees bear in pockets on the inside of spandrel girders. See the left side of the upper spandrel girder. The pockets are hollow rectangular recesses in the inside of spandrel girders that are formed when the girder is cast.

Spandrel girders span between columns and carry the load of the double tees (not shown) of the floor or roof system. The decorative finish on the spandrel girder is cast as an integral part of the member.

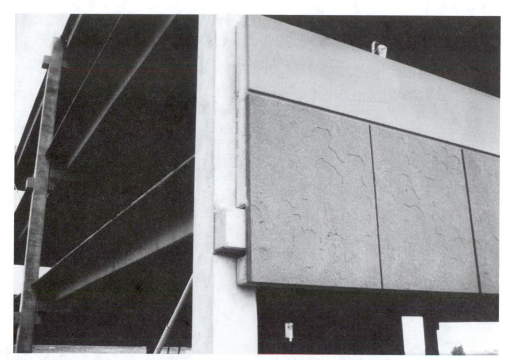

Spandrel girders bear on column haunches.

The back side of the spandrel girder is secured to the column by welding a small angle to steel plates cast in the panel and column.

Reinforced concrete wall panels are structural members that can be used to enclose stair systems or elevator hoist ways. As shown, double tees can bear in pockets or haunches in vertical wall panels.

The top side of the double tees and girder during erection. The rough surface of the topside of the members develops a better bond with concrete topping that is poured on top of the members. Stirrups (reinforcement) extend from the inverted tee girder and bond to the concrete topping.

The top side of concrete double tees before placement of wire fabric reinforcement and concrete topping.

A 2- to 4-inch-thick concrete topping is poured on top of the double tees that make up the floor structure. The concrete bonds to the floor members to improve structural integrity.

A temporary construction joint separating pours of concrete topping. Wire fabric reinforcement is cast in the concrete topping to reinforce the roof and floor platforms. Wood forms will be removed.

The finished concrete topping on the floor.

Window framing and glass are placed between the spandrel girders to completely enclose the structure.

An interior view of the spandrel girders and windows. Exterior walls will be framed and insulated, and a suspended acoustical ceiling will be hung to finish the interior space.

Light-gauge steel framing is used to construct interior partitions that divide the interior space into rooms. Framed walls are nonload bearing.

Steel stairs are enclosed by concrete wall panels.

A view of the roof. Concrete panels create an open penthouse that encloses rooftop mechanical equipment. Note how the spandrel girders extend above the roofline to create a parapet wall above the roof surface.

A multistory cast-in-place concrete structure during construction.

A corner of the cast-in-place concrete structure during construction. Forms are being placed on the upper story.

Column reinforcement.

Column placement of a cast-in-place concrete structure during construction.

Reinforcement of a beam and column connection. Square pans that form the waffle appearance on the underside of a two-way joist floor system (see next photo) are evident.

The bottom side of the second-story floor platform of a two-way joist cast-in-place concrete structure during construction. A concrete beam spans between the columns.

Chapter 27
DESIGN OF SIMPLE REINFORCED CONCRETE BEAMS

An elementary method of designing simple nonprestressed reinforced concrete beams is introduced in this chapter. The working stress design method for a simple beam is presented because of its less complicated design approach. The intent is to review a typical design approach to acquaint the technician with design principles and demonstrate common design concerns. The reader is advised that this presentation does not provide full coverage of design principles involved with reinforced concrete design. Full coverage is beyond the intended breadth of this book.

27.1   TENSILE REINFORCEMENT

A concrete beam *without* reinforcement will fail by flexural stress of the tension fibers because the capacity of concrete to resist tensile stress is much less than its capacity to resist compressive stress. So, like other reinforced concrete members, a reinforced concrete beam is designed with the theory that steel reinforcement resists tensile stresses and concrete withstands compressive stresses.

In reinforced concrete beams, *tensile bars* are placed so that they run longitudinally through the entire length of the beam, as shown in Figure 27.1. Tensile bars are positioned in the extreme fiber on the tension side of the neutral axis so that the reinforcement can resist the greatest tensile stresses. Adequate concrete cover beyond the longitudinal tensile reinforcement is required for proper bonding of the concrete to the steel and for fire and moisture protection of the steel. For simple beams, longitudinal tensile reinforcement is located at the bottom side of the member. Overhanging or cantilever beams develop tensile stresses on the top side of the member. In such beams, longitudinal tensile reinforcement shifts to the top side of the member in areas where tensile stresses occur above the neutral axis.

The effective depth and breadth of the beam is defined as the working zone of the beam section, the section that resists flexural stresses. The *breadth (b)* is the actual width of cross section of the beam. The *effective depth (d)* of the beam sec-

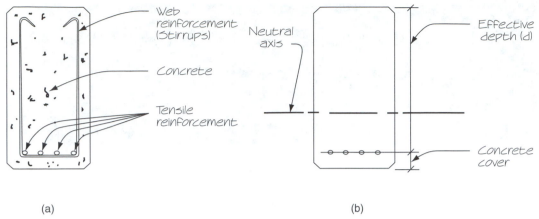

(a)   (b)

FIGURE 27.1   (a) Cross section of a simple rectangular reinforced concrete beam, illustrating the position of tensile bars and web reinforcement (stirrups); (b) effective depth of a concrete beam and the approximate location of the neutral axis.

tion is measured from the centroid of the longitudinal tensile reinforcement to the extreme fiber in compression on the opposite side of the section. Concrete cover that protects the reinforcement is located outside the effective depth and is excluded from the working zone. This placement allows more concrete area in the working zone to resist compression stresses and the steel to resist tensile stresses. Placement of longitudinal tensile reinforcement is near the extreme fiber in tension. This location moves the neutral axis of bending closer to that fiber.

27.2   WEB REINFORCEMENT

Differences in the mechanical properties of steel and concrete exist. Steel is much better than concrete at resisting tensile and shear stresses. Due to these dissimilar properties, a simple reinforced concrete beam may develop diagonal cracks near the bearing points (reactions). These cracks are due to the combined effect of vertical shear and tensile stresses; this type of stress is commonly called *diagonal tension* or simply *shear.* To counter this diagonal tension cracking effect, web reinforcement is typically needed in the beam.

   Web reinforcement typically consists of individual U-shaped bars called *stirrups* that are positioned perpendicular to the longitudinal tensile reinforcement and vertically wrap the tensile bars as shown in Figure 27.1a. Other methods of providing web reinforcement include use of welded wire fabric (WWF), spiral reinforcement, and longitudinal reinforcement that parallels longitudinal tensile reinforcement but is bent at an angle of 30° or more measured from the longitudinal

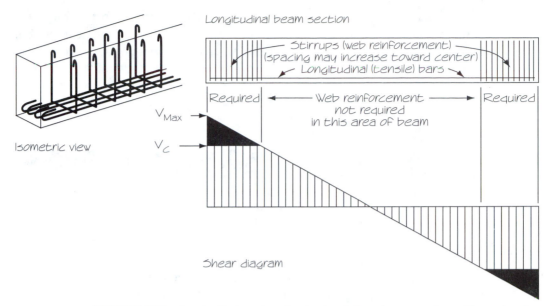

FIGURE 27.2 Longitudinal section of a simple reinforced concrete beam, illustrating the approximate location of longitudinal tensile reinforcement and web reinforcement (stirrups) extending from beam ends toward beam center. Web reinforcement is needed to resist excessive diagonal tension stresses that develop near the beam ends.

tensile reinforcement. A combination of stirrups and bent longitudinal reinforcement is frequently used.

Stirrup reinforcement grade is no greater than grade 60 steel; grade 75 steel does not easily bend in the form of a stirrup without development of excessive bending stresses at the bend. Standard stirrups are limited to No. 8 bars and smaller; they are generally composed of No. 3 bars in small to medium-sized members and No. 4 bars in large members. U-shaped stirrups have hooks at the end to enhance bonding with the concrete. A 135° hook plus an extension of $6d_b$ typically exists at the free ends of the ties. A 90° hook with the $6d_b$ extension is limited to No. 5 bars and smaller. Larger bars through No. 8 require a $12d_b$ extension.

For a member carrying a uniformly distributed load, the shear force is greatest at the beam reactions and decreases toward the center of beam length. The need for web reinforcement is greatest at the beam ends. This need decreases as the section under consideration is moved toward the longitudinal center of beam. Required spacing of web reinforcement increases toward the center of the beam. As shown in Figure 27.2, when moving the section toward the center of the member, shear stresses fall below an excessive level ($v_x < v_c$, as discussed below) and web reinforcement to counter this diagonal tension cracking effect is no longer needed.

Web reinforcement (stirrups) is placed at specified spacings (s). *Web reinforcement spacing* is the maximum permissible spacing measured perpendicular to the longitudinal reinforcement on either side of the stirrup. Web reinforcement should extend as close to the extreme fibers as concrete cover requirements permit. To toughen the beam by ensuring that diagonal cracking does not occur, spacing of web reinforcement is limited to $d/2$, where d is the effective depth of the beam section. In instances when shear stress is excessive ($v > v_c$, as discussed below), spacing of web reinforcement is limited to $d/4$.

27.3 BASIC BEAM DESIGN PROCEDURE

A sample calculation procedure based on the working stress method and used to size a simple nonstressed rectangular reinforced concrete beam is introduced below. Selection of p, j, and R properties for use in computations of rectangular concrete beam sections with tension reinforcement are found in Table 27.1. The properties found in this table are based on a designer specified strength of concrete (f'_c) and steel reinforcement grade (f_s) used in the beam.

1. Identify or specify the properties of the specified compressive strength (f'_c) of the concrete and of the grade of steel reinforcement specified. Note the associated allowable tensile stress and yield strength (f_s and f_y) of the steel reinforcement and p, j, and R properties.

2. Compute the minimum effective depth (d) or breadth (b) of the beam section, where $M = Rbd^2$. At the center of a simple beam carrying a uniformly distributed and/or concentrated load, this equation becomes $M_{max} = Rbd^2$. To determine the minimum effective depth (d) for a specified breadth (b), the equation $M_{max} = Rbd^2$ is rearranged algebraically and used as

$$d = \sqrt{\frac{M_{max}}{Rb}}$$

To determine the minimum breadth (b) for a specified effective depth (d), the equation $M_{max} = Rbd^2$ is rearranged algebraically and used as

$$b = \frac{M_{max}}{Rd^2}$$

Once minimum breadth (b) or effective depth (d) of the beam section are computed, it is necessary to add the weight of the concrete and revise the bending moment computation.

3. Compute the required total (all bars) cross-sectional area of the nonprestressed longitudinal tensile reinforcement (A_s):

$$A_s = pdb$$

Table 27.1   SECTION PROPERTIES FOR RECTANGULAR CONCRETE BEAMS

Grade of Reinforcement	Allowable Tensile Stress by Grade of Reinforcement, f_s (lb/in²)	Specified Compressive Strength of Concrete f'_c (lb/in²)	Properties		
			p	j	R
40	20 000	2500	0.010 19	0.8792	179.2
60	24 000		0.007 53	0.8929	161.4
40	20 000	3000	0.012 93	0.8723	225.6
60	24 000		0.009 59	0.8863	204.0
40	20 000	4000	0.018 84	0.8605	324.2
60	24 000		0.014 06	0.8750	295.3

or

$$A_s = \frac{M_{\text{max}}}{f_s j d}$$

Refer to Table B.7 and select bars that have a total cross-sectional area for all bars that is equal to or greater than A_s.

4. Investigate critical shear stress to determine whether web reinforcement is required. Critical shear stress (v_d) is the vertical shear stress at a distance d (the effective depth) away from the beam support:

$$v_d = \frac{V_d}{bd}$$

If the critical shear stress (v_d) is greater than the permissible shearing stress (v_c), web reinforcement is required. The critical shear stress (v_d) must not exceed the maximum permissible shearing stress carried by concrete with web reinforcement (v_c cannot exceed $v_{c\,\text{max}}$).

The permissible shearing stress carried by concrete without web reinforcement (v_c) for beams of normal-weight concrete subjected to flexure and shear is computed by: $v_c = 1.1\sqrt{f'_c}$. The maximum permissible shearing stress carried by concrete with web reinforcement ($v_{c\,\text{max}}$) for beams of normal-weight concrete subjected to flexure and shear is computed by $v_{c\,\text{max}} = v_c + 4.4\sqrt{f'_c}$. Permissible shearing stresses carried by concrete with and without web reinforcement for common specified compressive strengths of concrete are provided in Table 27.2.

5. Design web reinforcement, if required.

   (a) Compute the required distance that web reinforcement must extend into beam (ℓ_s) measured from surface of the beam support (the reac-

Table 27.2 PERMISSIBLE SHEARING STRESS CARRIED BY CONCRETE WITHOUT AND WITH WEB REINFORCEMENT FOR COMMON SPECIFIED COMPRESSIVE STRENGTHS ($\ell f'_c$) OF CONCRETE

Permissible Shearing Stress Carried by Concrete	Symbol	Specified Compressive (lb/in²) Strength (f'_c) of Concrete		
		2500	3000	4000
Without web reinforcement	v_c	55	60	70
With web reinforcement	$v_{c\ max}$	275	300	350

tion) toward the longitudinal center of the beam where $v'_{max} = v_{max} - v_c$ and where $v_{max} = V_{max}/bd$:

$$\ell_s = \left(\frac{\ell}{2}\right)\left(\frac{v'_{max}}{v_{max}}\right)$$

(b) Compute the web reinforcement spacing (s) required at distance d (the effective depth) measured from the beam support by the following equation:

$$s = \frac{A_v f_s}{b(v_d - v_c)}$$

Spacing of web reinforcement should not exceed s. Additionally, spacing of web reinforcement should not exceed $d/2$ or 24 in, where d is the effective depth of the beam section. In instances where shear stress is excessive (when $v_d > v_c$), spacing of web reinforcement is limited to $d/4$. When $v_d - v_c$ exceeds $2\sqrt{f'_c}$, spacing of web reinforcement should not exceed $d/4$ or 12 in.

Spacing of web reinforcement in areas of the beam beyond d is based on shear stress at the section under consideration (v_x):

$$s = \frac{A_v f_s}{b(v_x - v_c)}$$

(c) The cross-sectional area of two legs of a stirrup must be investigated for adequacy against shear stress by two equations. Refer to Table B.7 and select bars for web reinforcement (stirrups) that have a cross section area that is equal to or greater than A_v. The minimum area of web reinforcement (A_v) should be at least:

$$A_v = \frac{50bs}{f_y}$$

The minimum area of web reinforcement (A_v) for web reinforcement perpendicular to the longitudinal reinforcement is found by

$$A_v = \frac{(v_d - v_c)s}{f_s}$$

6. Investigate bond stress by computing the *development length* (ℓ_d) of the longitudinal reinforcement; the minimum length the tensile reinforcement must be embedded in the concrete:

$$\ell_d = \frac{0.04A_b f_y}{\sqrt{f'_c}}$$

In the formulas above, we have:

A_b cross-sectional area of a single nonprestressed longitudinal tensile reinforcement bar, in in^2

A_s total (all bars) cross-sectional area of the nonprestressed longitudinal tensile reinforcement, in in^2

A_v cross-sectional area of both legs of web reinforcement (stirrups), in in^2

b breadth (width) of the actual beam cross section, in inches

d effective depth of the beam cross section measured from the centroid of the longitudinal tensile reinforcement to the extreme fiber in compression on the opposite side of the section, in inches

f'_c compressive strength specified for the concrete, in lb/in^2

f_s allowable tensile stress of reinforcement based on steel grade, in lb/in^2 (see Table 26.1)

f_y yield strength specified for nonprestressed reinforcement based on steel grade, in lb/in^2 (see Table 26.1)

j property found in Table 27.1 based on f_s and f'_c, specified in lb/in^2

k property found in Table 27.1 based on f_s and f'_c, specified in lb/in^2

ℓ total beam length, in inches

ℓ_d development length of the longitudinal tensile reinforcement, in inches

ℓ_s minimum distance that web reinforcement must extend into beam, measured from beam end, in inches

$M_{\max}$ maximum bending moment, in in-lb

p property given in Table 27.1 based on f_s and f'_c specified

R property given in Table 27.1 based on f_s and f'_c specified

s required web reinforcement (stirrup) spacing; the maximum permissible spacing on either side of the stirrup measured perpendicular to the longitudinal reinforcement, in inches

V_d shear force at distance d from the surface of the reaction support, in lb

V_{max} maximum design shear force, in lb

V_x shear force at the section under consideration (section x), in lb

v_c allowable shearing stress carried by concrete *without* web reinforcement for beams of normal-weight concrete subjected to flexure and shear is computed by $v_c = 1.1\sqrt{f'_c}$ (values for v_c based on commonly specified values of f'_c are listed in Table 27.2, in lb/in^2)

$v_{c\,max}$ maximum permissible shearing stress carried by concrete *with* web reinforcement for beams of normal-weight concrete subjected to flexure and shear; $v_{c\,max} = v_c + 4.4\sqrt{f'_c}$

v_d shear stress at distance d from the surface of the reaction support, in lb/in^2

v_{max} maximum shear stress (usually alongside the reaction support), in lb/in^2; $v_{max} = V_{max}/bd$

v'_{max} maximum shear stress to be carried by web reinforcement (stirrups) at beam ends, in lb/in^2; $v'_{max} = v_{max} - v_c$

v_x shearing stress at the section under consideration (section x), in lb/in^2; $v_x = V_x/bd$

■ EXAMPLE 27.1

A rectangular steel-reinforced concrete beam supports a uniformly distributed load of 3.5 kip/ft (including the beam weight) for a span of 16′-3″ and a concentrated load of 20 kip at the center of the span. The breadth of the beam is to be 10 in. Four longitudinal tensile bars are to be used. Concrete with a specified compressive strength (f'_c) of 3000 lb/in^2 and grade 40 steel for tensile and web reinforcement are to be used. Determine the minimum effective depth (d) of the beam section and specify the gauge number of the longitudinal tensile reinforcement.

Solution   From Table A.5, the appropriate derived formulas are used to determine maximum bending moment (M_{max}). For the uniformly distributed load, $M_{max} = WL/8$, and for the concentrated load, $M_{max} = PL/4$.

$$M_{max} = \frac{WL}{8} + \frac{PL}{4}$$

$$= \frac{(3\,500 \text{ lb/ft} \cdot 16.25 \text{ ft})(16.25 \text{ ft})}{8} + \frac{20\,000 \text{ lb} \cdot 16.25 \text{ ft}}{4} = 196\,777 \text{ lb-ft}$$

Converting M_{max} from lb-ft to lb-in:

$$M_{max} = 196\,777 \text{ ft-lb} \cdot 12 \text{ in/ft} = 2\,361\,324 \text{ lb-in}$$

The minimum effective depth (d) of the beam section is found by

$$d = \sqrt{\frac{M_{max}}{Rb}} = \sqrt{\frac{2\ 361\ 324\ \text{lb-in}}{225.6 \cdot 10\ \text{in}}} = 32.35\ \text{in}$$

The required cross-section area of the longitudinal tensile reinforcement (A_s) is found by

$$A_s = pdb = 0.01293 \cdot 32.35\ \text{in} \cdot 10\ \text{in} = 4.18\ \text{in}^2$$

Four tensile bars are specified in the statement of the problem. From Table B.7, four bars having a combined section area of at least 4.18 in² are selected. Thus four No. 10 bars are selected. They have a total cross-sectional area for all bars of 5.08 in² (5.08 in² ≥ 4.18 in²).

The beam in Example 27.1 is composed of a 10-in-wide concrete beam with a minimum effective depth of 32.35 in. The effective depth is measured from the centroid of the longitudinal tensile reinforcement to the extreme compression surface. Additional concrete cover below the longitudinal tensile reinforcement is required. The designer may choose to specify the beam as a 12 in × 36 in section with tensile reinforcement placed at a depth of 33 in. This will change the calculation for the cross-sectional area required for longitudinal tensile reinforcement (A_s):

$$A_s = pdb = 0.01293 \cdot 33\ \text{in} \cdot 10\ \text{in} = 4.27\ \text{in}^2$$

Four No. 10 bars for longitudinal tensile reinforcement are still acceptable (5.08 in² ≥ 4.27 in²).

■ EXAMPLE 27.2

A rectangular steel-reinforced concrete beam supports a uniformly distributed load of 3.5 kip/ft (including the beam weight) for a span of 16'-3" and a concentrated load of 20 kip at the center of the span. The effective depth of the beam is to be 24 in. Four longitudinal tensile bars are to be used. Concrete with a specified compressive strength (f'_c) of 3000 lb/in² and grade 40 steel for tensile and web reinforcement are to be used. Determine the minimum breadth (b) of the beam section and specify the gauge number of the longitudinal tensile reinforcement.

Solution This is the same loading condition as Example 27.1, so M_{max} = 2 361 324 lb-in. The minimum breadth (b) of the beam section is found by

$$b = \frac{M_{max}}{Rd^2}$$
$$= \frac{2\ 361\ 324\ \text{lb-in}}{(225.6)(24\ \text{in})^2} = 18.2\ \text{in}$$

The cross-sectional area required for longitudinal tensile reinforcement (A_s) is found by

$$A_s = pdb = 0.01293 \cdot 24 \text{ in} \cdot 18.2 \text{ in} = 5.65 \text{ in}^2$$

Four tensile bars are specified in the statement of the problem. From Table B.7, four bars having a combined section area of at least 5.65 in² are selected: Four No. 11 bars are selected. They have a total cross-sectional area for all bars of 6.24 in² (6.24 in² ≥ 5.65 in²).

The beam in Example 27.2 is similar to the beam in Example 27.1; the properties of concrete and reinforcement remain the same. Unlike in Example 27.1, the effective depth has been specified and beam breadth is not known and must be computed in Example 27.2. A very wide beam section results.

■ EXAMPLE 27.3

A rectangular steel-reinforced concrete beam supports a uniformly distributed load of 3.5 kip/ft (including the beam weight) for a span of 16′-3″ and a concentrated load of 20 kip at the center of the span. The breadth of the beam is to be 10 in. Four longitudinal tensile bars are to be used. Concrete with a specified compressive strength (f'_c) of 4000 lb/in² and grade 60 steel for tensile and web reinforcement are to be used. Determine the minimum effective depth (d) of the beam section and specify the gauge number of the longitudinal tensile reinforcement.

Solution This is the same loading condition as Example 27.1, so $M_{max} = $ 2 361 324 lb-in. The minimum effective depth (d) of the beam section is found by

$$d = \sqrt{\frac{M_{max}}{Rb}} = \sqrt{\frac{2\ 361\ 324 \text{ lb-in}}{295.3 \cdot 10 \text{ in}}} = 28.28 \text{ in}$$

The cross-sectional area required for longitudinal tensile reinforcement (A_s) is found by

$$A_s = pdb = 0.01406 \cdot 28.28 \text{ in} \cdot 10 \text{ in} = 3.98 \text{ in}^2$$

Four tensile bars are specified in the statement of the problem. From Table B.7, four bars having a combined section area of at least 3.98 in² are selected: Four No. 9 bars are selected. They have a total cross-sectional area for all bars of 4.00 in² (4.00 in² ≥ 3.98 in²).

The beam in Example 27.3 is similar to the beam in Example 27.1 except that the properties of concrete and reinforcement have been increased to higher strengths. As demonstrated when comparing the results of these examples, an increase in the strength properties of concrete and reinforcement appears to be more economical; a reduction in concrete volume and reinforcing bar gauge tends to result. However, a stronger concrete and higher grade of steel typically costs more to produce. The designer must be conscious of the cost in materials and weigh costs differences in achieving the most economical design.

■ EXAMPLE 27.4

As in Example 27.1, a rectangular steel-reinforced concrete beam supports a uniformly distributed load of 3.5 kip/ft (including the beam weight) for a span of 16'-3" and a concentrated load of 20 kip at the center of the span. Concrete with a specified compressive strength (f'_c) of 3000 lb/in^2 and grade 40 steel for tensile and web reinforcement are to be used. The breadth of the beam is 10 in and the effective depth was rounded to 33 in. Four No. 10 longitudinal tensile bars are specified.

(a) Determine whether web reinforcement is required.

(b) Determine the spacing, size, and length of web reinforcement.

Solution   (a) Before computing the critical shear stress (v_d), the shear force (V_d) at a distance of 33 in (d = 33 in) or 2.75 ft away from the beam support must be found. The left reaction R_1 is found by

$$R = \frac{W}{2} + \frac{P}{2} = \frac{3500 \text{ lb/ft} \cdot 16.25 \text{ ft}}{2} + \frac{20\ 000 \text{ lb}}{2} = 38\ 438 \text{ lb}$$

The shear force (V_d) at a distance of 2.75 ft away from the beam support is found by the method presented in Chapter 11:

$$V @ x = \Sigma F_{up} - \Sigma F_{down} \text{ to the left of section } x$$
$$V @ 2.75 \text{ ft} = 38\ 438 \text{ lb} - (3500 \text{ lb/ft} \cdot 2.75 \text{ ft}) = 28\ 813 \text{ lb}$$

The critical shear stress (v_d) is found by

$$v_d = \frac{V_d}{bd} = \frac{28\ 813 \text{ lb}}{10 \text{ in} \cdot 33 \text{ in}} = 87.3 \text{ lb/in}^2$$

In Table 27.2 for 3000-lb/in^2 concrete, v_c is 60 lb/in^2. Since v_d is greater than v_c, web reinforcement (stirrups) is required.

(b) The maximum shear force is equal to the reaction found above, 38 438 lb. The maximum shear stress (v_{max}) is found by

$$v_{max} = \frac{V_{max}}{bd} = \frac{38\ 438 \text{ lb}}{10 \text{ in} \cdot 33 \text{ in}} = 116.5 \text{ lb/in}^2$$

The maximum shear stress to be carried by web reinforcement (v'_{max}) at beam ends is found by

$$v'_{max} = v_{max} - v_c = 116.5 \text{ lb/in}^2 - 60 \text{ lb/in}^2 = 56.5 \text{ lb/in}^2$$

The total beam length (ℓ) in inches is 195 in: 16.25 ft · 12 in/ft = 195 in. The required distance that web reinforcement must extend into the beam (ℓ_s) measured from the surface of the beam support (the reaction) toward the longitudinal center of the beam is found by

$$\ell_s = \left(\frac{\ell}{2}\right)\left(\frac{v'_{max}}{v_{max}}\right) = \left(\frac{195\ \text{in}}{2}\right)\left(\frac{56.5\ \text{lb/in}^2}{116.5\ \text{lb/in}^2}\right) = 47.3\ \text{in} = 3'\text{-}11\tfrac{1}{4}''$$

Web reinforcement must extend $3'\text{-}11\tfrac{1}{4}''$ into the beam from the beam ends.

Assume the use of No. 3 bars for stirrups. In Table B.7 the cross-sectional area of two legs of a No. 3 stirrup is found to be 0.22 in². The web reinforcement spacing (s) required at distance d is found by

$$\begin{aligned}
s &= \frac{A_v f_s}{b(v_d - v_c)} \\
&= \frac{0.22\ \text{in}^2 \cdot 20\,000\ \text{lb/in}^2}{10\ \text{in}(87.3\ \text{lb/in}^2 - 60\ \text{lb/in}^2)} \\
&= 16.1\ \text{in}
\end{aligned}$$

This spacing is extreme and unacceptable. In instances where shear stress is excessive (when $v_d > v_c$), spacing of web reinforcement is limited to $d/4$:

$$s = \frac{d}{4} = \frac{33\ \text{in}}{4} = 8.25\ \text{in}$$

Round stirrup spacing to 8 in.

Stirrups are required at 8.25-in-O.C. spacing for at least $3'\text{-}11\tfrac{1}{4}''$ (47.3 in) into the beam from both beam ends:

$$\frac{47.3\ \text{in}}{8\ \text{in}} = 5.91\ \text{spaces}$$

The first stirrup begins 4 in from the bearing point, so six stirrups at 8-in-O.C. spacing are required:

$$(6\ \text{stirrups} \cdot 8\ \text{in}) + 4\ \text{in} = 52\ \text{in} = 4'\text{-}4''$$

$4'\text{-}4''$ is greater than the required $4'\text{-}9\tfrac{1}{2}''$.

The minimum area (A_v) of web reinforcement (two legs of a stirrup) is found by

$$A_v = \frac{50bs}{f_y} = \frac{50(10\ \text{in})(8\ \text{in})}{40\,000\ \text{lb/in}^2}\, 0.10\ \text{in}^2$$

The minimum area of web reinforcement (A_v) for web reinforcement (stirrups) perpendicular to the longitudinal reinforcement is found by

$$A_v = \frac{(v_d - v_c)bs}{f_s} = \frac{(87.3 \text{ lb/in}^2 - 60 \text{ lb/in}^2)(10 \text{ in})(8 \text{ in})}{20\,000 \text{ lb/in}^2} = 0.11 \text{ in}^2$$

Use of No. 3 bars for web reinforcement (stirrups) that have a cross-sectional area of 0.22 in^2 is acceptable because it is equal to or greater than A_v.

■ EXAMPLE 27.5

As in Example 27.1, a rectangular steel-reinforced concrete beam supports a uniformly distributed load of 3.5 kip/ft (including the beam weight) for a span of 16′-3″ and a concentrated load of 20 kip at the center of the span. Concrete with a specified compressive strength (f'_c) of 3000 lb/in^2 and grade 40 steel for tensile and web reinforcement are to be used. The breadth of the beam is 10 in and the effective depth was rounded to 33 in. Four No. 10 longitudinal tensile bars are specified. Investigate the development length (ℓ_d) of the longitudinal tensile reinforcement for adequacy.

Solution From Table B.7, the cross-sectional area of a No. 10 bar is 1.27 in^2. The development length (ℓ_d) of the tensile reinforcement (No. 10 bars) is found by

$$\ell_d = \frac{0.04 A_b f_y}{\sqrt{f'_c}} = \frac{0.04(1.27 \text{ in}^2)(40\,000 \text{ lb/in}^2)}{\sqrt{3000 \text{ lb/in}^2}} = 37.1 \text{ in}$$

The minimum length the longitudinal tensile reinforcement must be embedded in the concrete is 37.1 in (just over 3 ft). Reinforcement runs the entire beam length (16′-3″), so sufficient development length is provided.

■ EXAMPLE 27.6

Design a rectangular steel-reinforced concrete beam carrying a uniformly distributed load of 3 kip/ft (including the beam weight), for a span of 27′-6″. The breadth of the beam is 12 in. Four longitudinal tensile bars are to be used. Concrete with a specified compressive strength (f'_c) of 4000 lb/in^2 and grade 60 steel for tensile and web reinforcement are to be used.

Solution From Table A.5, the appropriate derived formula is used to determine maximum bending moment (M_{max}). The minimum effective depth (d) of the beam section is found by

$$M_{max} = \frac{WL}{8} = \frac{(3000 \text{ lb/ft} \cdot 27.5 \text{ ft})(27.5 \text{ ft} \cdot 12 \text{ in/ft})}{8} = 3\,403\,125 \text{ lb-in}$$

$$d = \sqrt{\frac{M_{max}}{Rb}} = \sqrt{\frac{3\,403\,125 \text{ lb-in}}{295.3 \cdot 12 \text{ in}}} = 30.99 \text{ in}$$

Assume that the effective depth (d) will be specified at 32 in. The required cross-sectional area of the longitudinal tensile reinforcement (A_s) is found by

$$A_s = pdb = 0.01406 \cdot 32 \text{ in} \cdot 12 \text{ in} = 5.40 \text{ in}^2$$

Four tensile bars are specified in the statement of the problem. From Table B.7, four bars having a combined section area of at least 5.40 in² are selected: Four No. 11 bars are selected. They have a total cross-sectional area for all bars of 6.24 in² (6.24 in² ≥ 5.40 in²).

Before computing the critical shear stress (v_d), the shear force (V_d) at a distance of 32 in $(d = 32$ in) or 2.67 ft away from the beam support must be found. The left reaction R_1 is found by

$$R_1 = \frac{W}{2} = \frac{3000 \text{ lb/ft} \cdot 27.5 \text{ ft}}{2} = 41\ 250 \text{ lb}$$

The shear force (V_d) at a distance of 2.67 ft away from the beam support is found by the method presented in Chapter 11:

$$V @ x = \Sigma F_{up} - \Sigma F_{down} \text{ to the left of section } x$$
$$V @ 2.67 \text{ ft} = 41\ 250 \text{ lb} - (3000 \text{ lb/ft} \cdot 2.67 \text{ ft}) = 33\ 240 \text{ lb}$$

The critical shear stress (v_d) is found by

$$v_d = \frac{V_d}{bd} = \frac{33\ 240 \text{ lb}}{(12 \text{ in} \cdot 32 \text{ in})} = 86.6 \text{ lb/in}^2$$

In Table 27.2 for 4000 lb/in² concrete, v_c is 70 lb/in². Since v_d is greater than v_c, web reinforcement (stirrups) is required.

The maximum shear force is equal to the reaction found above, 41 250 lb. The maximum shear stress (v_{max}) is found by

$$v_{max} = \frac{V_{max}}{bd} = \frac{41\ 250 \text{ lb}}{(12 \text{ in} \cdot 32 \text{ in})} = 107.4 \text{ lb/in}^2$$

The maximum shear stress to be carried by web reinforcement (v'_{max}) at beam ends is found by

$$v'_{max} = v_{max} - v_c = 107.4 \text{ lb/in}^2 - 70 \text{ lb/in}^2 = 37.4 \text{ lb/in}^2$$

The total beam length (ℓ) in inches is 330 in: 27.5 ft $\cdot$ 12 in/ft = 330 in.

The required distance that web reinforcement must extend into the beam (ℓ_s) measured from the surface of the beam support (the reaction) toward the longitudinal center of the beam is found by

$$\ell_s = \left(\frac{\ell}{2}\right)\left(\frac{v'_{max}}{v_{max}}\right) = \left(\frac{330 \text{ in}}{2}\right)\left(\frac{37.4 \text{ lb/in}^2}{107.4 \text{ lb/in}^2}\right) = 57.5 \text{ in} = 4'\text{-}9\frac{1}{2}''$$

Web reinforcement must extend 4'-10½″ into the beam from the beam ends.

Assume the use of No. 3 bars for stirrups. In Table B.7, the cross-sectional area of two legs of a No. 3 stirrup is found to be 0.22 in². The required web reinforcement spacing (s) at distance d is found by

$$s = \frac{A_v f_s}{b(v_d - v_c)}$$
$$= \frac{0.22 \text{ in}^2 \cdot 24\,000 \text{ lb/in}^2}{12 \text{ in}(86.6 \text{ lb/in}^2 - 70 \text{ lb/in}^2)}$$
$$= 26.5 \text{ in}$$

This spacing is extreme and unacceptable. In instances where shear stress is excessive (when $v_d > v_c$), the spacing of web reinforcement is limited to $d/4$:

$$s = \frac{d}{4} = \frac{32 \text{ in}}{4} = 8.0 \text{ in}$$

Stirrups are required at 8-in-O.C. spacing for at least 4′-9 ½″ (57.5 in) into the beam from both beam ends:

$$\frac{5.75 \text{ in}}{8 \text{ in}} = 7.19 \text{ spaces}$$

The first stirrup begins 4 in from the bearing point, so seven stirrups at 8-in-O.C. spacing are required:

$$(7 \text{ stirrups} \cdot 8 \text{ in}) + 4 \text{ in} = 60 \text{ in} = 5 \text{ ft}$$

5 ft is greater than the required 4′-9½″.

The minimum area (A_v) of web reinforcement (two legs of a stirrup) is found by

$$A_v = \frac{50bs}{f_y} = \frac{50(12 \text{ in})(8 \text{ in})}{60\,000 \text{ lb/in}^2} = 0.08 \text{ in}^2$$

The minimum area of web reinforcement (A_v) for web reinforcement (stirrups) perpendicular to the longitudinal reinforcement is found by

$$A_v = \frac{(v_d - v_c)bs}{f_s} = \frac{(86.6 \text{ lb/in}^2 - 70 \text{ lb/in}^2)(12 \text{ in})(8 \text{ in})}{24\,000 \text{ lb/in}^2} = 0.07 \text{ in}^2$$

Use for web reinforcement (stirrups) of No. 3 bars that have a cross-sectional area of 0.22 in² is acceptable because the area is equal to or greater than A_v.

The development length (ℓ_d) of the tensile reinforcement (No. 11 bars) is found by

$$\ell_d = \frac{0.04A_b f_y}{\sqrt{f'_c}} = \frac{0.04(1.56 \text{ in}^2)(60\,000 \text{ lb/in}^2)}{\sqrt{4000 \text{ lb/in}^2}} = 59.2 \text{ in}$$

The minimum length the longitudinal tensile reinforcement must be embedded in the concrete is 59.2 in (nearly 5 ft). Reinforcement runs the entire beam length (27′-6″), so sufficient development length is provided.

The beam in Example 27.6 is composed of a 12-in-wide concrete beam with a minimum effective depth of 32 in, measured from the longitudinal tensile reinforcement to the extreme compression surface (additional concrete protection will be required beyond the longitudinal tensile and web reinforcement); four No. 11 bars for longitudinal tensile reinforcement that run longitudinally at the tensile surface of the beam; and stirrups, serving as web reinforcement, spaced a maximum of 8 in at the beam ends with stirrups extending into the beam for 60 in from each end.

EXERCISES

27.1 A rectangular steel-reinforced concrete beam supports a uniformly distributed load of 4.5 kip/ft (including the beam weight) for a span of 28 ft. The breadth of the beam is 16 in. Concrete with a specified compressive strength (f'_c) of 4000 lb/in² and grade 60 steel for tensile and web reinforcement are specified. Determine the minimum effective depth (d) of the beam section.

27.2 Much as in Exercise 27.1, a rectangular steel-reinforced concrete beam supports a uniformly distributed load of 4.5 kip/ft (including the beam weight) for a span of 28 ft. The breadth (b) of the beam is 16 in. The effective depth (d) of the beam section was rounded to 34 in. Concrete with a specified compressive strength (f'_c) of 4000 lb/in² and grade 60 steel for tensile and web reinforcement are specified. Six longitudinal tensile bars are to be used. Specify the gauge number of the longitudinal tensile reinforcement.

27.3 Much as in Exercises 27.1 and 27.2, a rectangular steel-reinforced concrete beam supports a uniformly distributed load of 4.5 kip/ft (including the beam weight) for a span of 28 ft. Concrete with a specified compressive strength (f'_c) of 4000 lb/in² and grade 60 steel for tensile and web reinforcement are specified. The breadth (b) of the beam is 16 in and the effective depth (d) of the beam section is 34 in. Six No. 11 longitudinal tensile bars are to be used. If required, No. 4 stirrups will be specified. Determine if web reinforcement is required. If web reinforcement is required, determine:

(a) The minimum distance that web reinforcement must extend into the beam.

(b) The maximum web reinforcement (stirrup) spacing at the beam end.

27.4 Much as in Exercises 27.1 and 27.3, a rectangular steel-reinforced concrete beam supports a uniformly distributed load of 4.5 kip/ft (including the beam weight) for a span of 28 ft. Concrete with a specified compressive strength (f'_c) of 4000 lb/in² and grade 60 steel for tensile and web reinforcement are specified. The breadth (b) of the beam is

16 in and the effective depth (d) of the beam section is 34 in. Six No. 11 longitudinal tensile bars and No. 4 stirrups are specified. Determine if the cross section area (A_v) of both legs of the web reinforcement (stirrups) is acceptable.

27.5 Much as in Exercises 27.1 to 27.4, a rectangular steel-reinforced concrete beam supports a uniformly distributed load of 4.5 kip/ft (including the beam weight) for a span of 28 ft. Concrete with a specified compressive strength (f'_c) of 4000 lb/in^2 and grade 60 steel for tensile and web reinforcement are specified. The breadth (b) of the beam is 16 in and the effective depth (d) of the beam section is 34 in. Six No. 11 longitudinal tensile bars and No. 4 stirrups are specified. Determine the required development length (ℓ_d) of the longitudinal tensile reinforcement.

27.6 A rectangular steel-reinforced concrete beam supports a uniformly distributed load of 2875 lb/ft (including the beam weight) for a span of 24 ft. The breadth of the beam is 12 in. Concrete with a specified compressive strength (f'_c) of 3000 lb/in^2 and grade 60 steel for tensile and web reinforcement are specified. Four longitudinal tensile bars are to be used. If required, No. 3 stirrups will be used.

(a) Determine the minimum effective depth (d) of the beam section (round to the nearest inch increment).

(b) Specify the gauge number of the longitudinal tensile reinforcement.

(c) Determine whether web reinforcement is required.

(d) Determine the maximum spacing, size, and length of web reinforcement.

(e) Investigate the development length (ℓ_d) of the longitudinal tensile reinforcement for adequacy.

(f) Sketch a longitudinal section and a transverse section of the beam. Show beam size, effective depth, tensile (main) reinforcement size and location, and stirrup location and spacing (if needed).

Chapter 28
DESIGN OF SIMPLE REINFORCED CONCRETE COLUMNS

An elementary method of designing simple nonprestressed reinforced concrete columns is introduced in this chapter. The intent is to acquaint the technician with basic design principles used by an engineer so that the technician can gain an awareness of structural engineering and design. The reader is advised that this presentation does not provide full coverage of the design principles involved with reinforced concrete design; full coverage is beyond the intended breadth of this book.

28.1 TIED CONCRETE COLUMNS

A frequently used reinforced concrete column is the tied column. A *tied column* is made of steel reinforcement called longitudinal bars that run vertically and that are enclosed by ties or spirals before being encased in concrete. A tied column may be fabricated at a plant and delivered to the construction site or may be site cast as a monolithic member of the building structure. Tied columns may be square, rectangular, triangular, or circular in cross section.

28.2 LONGITUDINAL BAR REINFORCEMENT

Longitudinal bars are placed vertically and extend the full length of the column, as shown in Figure 28.1. These bars support part of the axial load. According to the ACI standard, the total cross-sectional area of the longitudinal bars (A_{st}) must not be less than 0.01 or greater than 0.08 of the gross column section area (A_g):

$$0.01 < P_g < 0.08 \qquad \text{where} \quad P_g = \frac{A_{st}}{A_g}$$

The minimum number of longitudinal bars within rectangular or circular ties is four bars. Within triangular ties at least three bars are required.

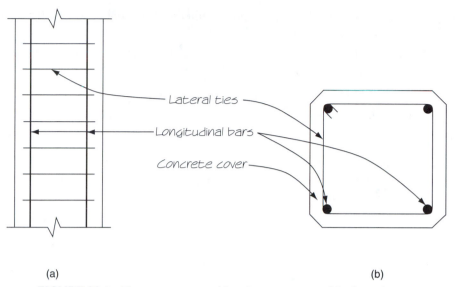

(a)　　　　　　　　　　　　　　　　　　　　　　　　　(b)

FIGURE 28.1　Transverse section (a) and a cross section (b) of a tied concrete column.

28.3　TIE AND SPIRAL REINFORCEMENT

Ties are individual loops of reinforcing bars or wire that encase the longitudinal bars as shown in Figure 28.1. *Spirals* are hoops of continuously wound reinforcement in the form of a springlike cylindrical helix that enclose the vertical longitudinal bars. Ties or spirals structurally wrap the vertical longitudinal bars and hold them in place during pouring of the concrete. Ties are horizontally positioned and wired to the longitudinal reinforcement at specific vertical spacings to hold them in place during the concrete pour. Some of the requirements for ties include:

- Grade of reinforcement for ties should not exceed grade 60 steel; grade 75 steel does not bend easily without development of excessive bending stresses.
- Standard ties are limited to No. 8 bars and smaller; they are generally composed of No. 3 bars in small to medium-sized members and No. 4 bars in large members.
- Longitudinal bars No. 10 and smaller must be enclosed with at least No. 3 ties. Longitudinal bars No. 11 and larger must be enclosed with at least No. 4 ties.
- Hooks at the ends of the tie enhance bonding with the concrete. A 135° hook plus an extension of $6d_b$ typically exists at the free ends of the ties. A 90°

hook with the $6d_b$ extension is limited to No. 5 bars and smaller. Larger bars through No. 8 require a $12d_b$ extension.

- Ties should be spaced not to exceed 16 longitudinal bar diameters, 48 tie bar diameters, or the least column dimension of the column.

Spirals are wired to the longitudinal reinforcement at the appropriate spacing and pitch to hold them in place during the concrete pour. Some of the specifications for spirals include:

- Grade of reinforcement for ties and spirals should not exceed grade 60 steel; grade 75 steel does not bend easily without development of excessive bending stresses.
- Spirals should not be less than $\frac{3}{8}$ inch in diameter in cast-in-place construction. They are usually $\frac{3}{8}$, $\frac{1}{2}$, or $\frac{5}{8}$ inch in diameter.
- Clear spacing between each spiral hoop should not exceed 3 in or be less than 1 in.

28.4   BASIC COLUMN DESIGN PROCEDURE

A concrete column is a long, slender member with a ratio of height to least lateral dimension of 3 or greater ($\ell/d \geq 3$) that is designed to support axial compressive loads. However, concrete columns rarely carry only axial loads, particularly when they are part of a monolithic floor/column or roof/column frame system. The eccentricity of the load on a concrete column induces shear stresses and bending stresses caused by moments in the column material. As a result of a myriad of variables related to compressive, shear, and bending stresses in columns, design methods for reinforced concrete columns are complex, usually involving use of design aids such as complex nomographs and tables or computer software. Therefore, an approach founded on the strength design method of designing a steel reinforced tied column is introduced below.

1. Compute the factored axial load (P_u) that can be supported by a tied concrete column:

$$P_u = \phi 0.80[(0.85f'_c(A_g - A_{st})) + f_y A_{st}]$$

2. The strength of the column calculated as the factored axial load (P_u) must be at least equal to the factored design load (L_{design}):

$$P_u \geq L_{design}$$

where the factored design load (L_{design}) is computed on the basis of the service dead load (D_{load}) and service live load (L_{load}) by the following equation:

$$L_{design} = 1.4D_{load} + 1.7L_{load}$$

3. Determine the tie spacing required. Tie spacing should not exceed the spacings determined by the following equations, whichever is the lowest:

> 16 longitudinal bar diameters
> least column dimension
> 48 tie diameters

4. Investigate bond stress by computing the development length of the longitudinal reinforcement, the minimum length the reinforcement must be embedded in the concrete. The length developed should be determined by the following equations or 8 in, whichever is greatest:

$$\ell_d = \frac{0.02 d_b f_y}{\sqrt{f'_c}}$$

$$\ell_d = 0.0003 f_y d_b$$

In the formulas above, we have

A_g	gross column section area, in in^2
A_{st}	total area of longitudinal reinforcement (all bars), in in^2
d_b	diameter of a single longitudinal reinforcement bar, in inches
D_{load}	service dead load, in lb
f'_c	specified compressive strength of the concrete, in lb/in^2
f_y	yield strength specified for nonprestressed reinforcement based on steel grade, in in^2 (see Table 26.1)
L_{design}	factored design load, in lb
L_{load}	service live load, in lb
ℓ_d	development length of the longitudinal reinforcement, in inches
P_g	ratio of the total cross-sectional area of the longitudinal bars (A_{st}) to the gross column section area (A_g): $P_g = A_{st}/A_g$ (as described above, must not be less than 0.01 or greater than 0.08)
P_u	factored axial load
ϕ	strength reduction factor
	$\phi = 0.70$ members in axial compression with flexure, such as tied columns
	$\phi = 0.75$ members in axial compression with flexure, such as columns with spiral reinforcement

The strength reduction factor (ϕ) is used in strength design to reduce the permissible design strength of a member. It is a type of factor of safety that accounts for variations in construction and flaws in material.

28.4 Basic Column Design Procedure

■ EXAMPLE 28.1

A 12 in × 12 in tied concrete column is 9 ft high. It is composed of four No. 7 longitudinal bars and No. 3 ties. Assume that the specified compressive strength of concrete is 4000 lb/in² and that grade 60 bars will be used.

(a) Compute the factored axial load that can be supported by this column.
(b) It is anticipated that the column will carry an 80-kip dead load and a 120-kip live load. Can the column carry this service load safely?
(c) Determine the tie spacing required for this column.
(d) Investigate the bond stress.

Solution (a) The strength reduction factor (ϕ) is 0.70 for members in axial compression with flexure. The gross column section area (A_g) is 144 in²: 12 in · 12 in = 144 in². From Table B.7 the total area of longitudinal reinforcement for four No. 7 bars is 2.41 in². The factored axial load is found by

$$P_u = \phi 0.80\{[0.85f_c'(A_g - A_{st})] + f_y A_{st}\}$$
$$= (0.70)(0.80)[(0.85 \cdot 4000 \text{ lb/in}^2)(144 \text{ in}^2 - 2.41 \text{ in}^2)$$
$$+ (60\,000 \text{ lb/in}^2 \cdot 2.41 \text{ in}^2)]$$
$$= 350\,563 \text{ lb} = 351 \text{ kip}$$

(b)

$$L_{design} = 1.4D_{load} + 1.7L_{load}$$
$$= 1.4(80 \text{ kip}) + 1.7(120 \text{ kip}) = 316 \text{ kip}$$

The factored design load (L_{design}) of 316 kip does not exceed the factored axial load that can be supported by this column (P_u) of 351 kip. The column can carry the load safely.

(c) The tie spacing required is limited by the least dimension of the following:

16 longitudinal bar diameters for No. 7 bars:	$16 \cdot \frac{7}{8}$ in =	14 in
least column dimension for a 12 in × 12 in column:		12 in
48 tie diameters for No. 3 ties:	$48 \cdot \frac{3}{8}$ in =	18 in

Tie spacing required is 12 in.

(d) The development length (ℓ_d) of the longitudinal reinforcement (No. 7 bars) is found by

$$\ell_d = \frac{0.02d_b f_y}{\sqrt{f_c'}} = \frac{0.02\left(\frac{7}{8}\right) \text{ in})(60\,000 \text{ lb/in}^2)}{\sqrt{4000 \text{ lb/in}^2}} = 16.6 \text{ in}$$

$$\ell_d = 0.0003 f_y d_b = 0.0003(60\,000 \text{ lb/in}^2)\left(\frac{7}{8} \text{ in}\right) = 15.78 \text{ in}$$

The minimum length the longitudinal reinforcement must be embedded in the concrete is 16.6 in. Longitudinal reinforcement must extend at least 16.6 in into the column above or the foundation below, assuming that these members have the same structural properties.

The 12 in × 12 in column in Example 28.1 requires concrete protection of about 2 in on all sides; $1\frac{1}{2}$ in plus one-half the bar diameter. Four No. 7 bars and No. 3 ties spaced at 12 in O.C. will be required.

■ EXAMPLE 28.2

A 12 in × 12 in tied concrete column is 9 ft high. It is composed of four No. 11 longitudinal bars and No. 3 ties. Assume that the specified compressive strength of concrete is 4000 lb/in^2 and that grade 60 bars will be used.
(a) Compute the factored axial load that can be supported by this column.
(b) Determine the tie spacing required for this column.
(c) Investigate the bond stress.

Solution (a) The strength reduction factor (ϕ) is 0.70 for members in axial compression with flexure. The gross column section area (A_g) is 144 in^2: 12 in · 12 in = 144 in^2. From Table B.7 the total area of longitudinal reinforcement for four No. 11 bars is 6.24 in^2. The factored axial load is found by

$$P_u = \phi 0.80\{[0.85f'_c(A_g - A_{st})] + f_y A_{st}\}$$
$$= (0.70)(0.80)[(0.85 \cdot 4000 \text{ lb/in}^2)(144 \text{ in}^2 - 6.24 \text{ in}^2)$$
$$+ (60\,000 \text{ lb/in}^2 \cdot 6.24 \text{ in}^2)]$$
$$= 471\,960 \text{ lb} = 472 \text{ kip}$$

(b) The tie spacing required is limited by the least dimension of the following:

16 longitudinal bar diameters for No. 7 bars:	$16 \cdot \frac{7}{8}$ in $=$	14 in
least column dimension for a 12 in × 12 in column:		12 in
48 tie diameters for No. 3 ties:	$48 \cdot \frac{3}{8}$ in $=$	18 in

The tie spacing required is 12 in.

(c) The development length (ℓ_d) of the longitudinal reinforcement (No. 7 bars) is found by

$$\ell_d = \frac{0.02d_b f_y}{\sqrt{f'_c}} = \frac{0.02\left(\frac{11}{8} \text{ in}\right)(60\,000 \text{ lb/in}^2)}{\sqrt{4000 \text{ lb/in}^2}} = 26.1 \text{ in}$$

$$\ell_d = 0.0003 f_y d_b = 0.0003(60\,000 \text{ lb/in}^2)\left(\frac{11}{8} \text{ in}\right) = 24.8 \text{ in}$$

The minimum length the longitudinal reinforcement must be embedded in the concrete is 26.1 in. Longitudinal reinforcement must extend at least 26.1 in into the column above or the foundation below, assuming that these members have the same structural properties.

The column in Example 28.2 is the same as the column in Example 28.1 except that the longitudinal bars are increased in gauge from No. 7 to No. 11 in Example 20.2. In comparing the factored axial loads that can be supported by each column it becomes evident that a column with more steel will support a greater load.

In Example 28.2, observe that the longitudinal steel reinforcement (A_{st}) comprises only 4.3% of the gross section area of the column (A_g), yet the steel reinforcement carries 44.4% of the axial load:

$$P_g = \frac{A_{st}}{A_g} = \frac{6.24 \text{ in}^2}{12 \text{ in} \cdot 12 \text{ in}} = 0.0433 = 4.3\%$$

$$P_{u \text{ steel}} = \phi 0.80(f_y A_{st})$$
$$= (0.70)(0.80)(60\,000 \text{ lb/in}^2 \cdot 6.24 \text{ in}^2)$$
$$= 209\,664 \text{ lb} = 209.7 \text{ kip}$$

$$\frac{209.7 \text{ kip}}{472 \text{ kip}} = 0.444 = 44.4\%$$

This analysis demonstrates that longitudinal reinforcement makes a significant contribution to the load that a column can support safely. It makes a case for the need for corrosion and fire protection of steel reinforcement in reinforced concrete members. Damage by corrosion and temperature increase of steel in a fire both result in a decrease in strength of reinforcement. A decrease in strength of reinforcement reduces the load that the reinforced concrete member can safely support. Failure from these modes can be catastrophic so corrosion and fire protection of steel reinforcement is important.

Also note that unapproved substitution of a lower-grade reinforcement on the construction site can prove catastrophic. For example, substitution of grade 40 steel for grade 60 steel reinforcement for the longitudinal bars in Example 28.2 results in a column that can carry only 85% of the load.

$$P_u = \phi 0.80\{[0.85 f_c'(A_g - A_{st})] + f_y A_{st}\}$$
$$= (0.70)(0.80)[(0.85 \cdot 4000 \text{ lb/in}^2)(144 \text{ in}^2 - 6.24 \text{ in}^2) + (40\,000 \text{lb/in}^2 \cdot 6.24 \text{ in}^2)]$$
$$= 402\,071 \text{ lb} = 402 \text{ kip}$$

$$\frac{402 \text{ kip}}{472 \text{ kip}} = 0.85 = 85\%$$

Proper inspection of reinforcement during construction is important.

EXERCISES

28.1 A 10 in × 10 in tied concrete column with four No. 14 longitudinal bars and No. 3 ties is under consideration. Is the total cross-sectional area of the longitudinal bars acceptable under the procedures outlined?

28.2 A 12 in $\times$ 12 in tied concrete column is composed of four No. 9 longitudinal bars and No. 3 ties. Assume that the compressive strength specified for the concrete is 4000 lb/in^2 and that grade 60 bars will be used.

(a) Compute the maximum factored axial load that can be supported by this column.

(b) Determine the tie spacing required for this column.

(c) Investigate the bond stress.

28.3 A 12 in $\times$ 12 in tied concrete column is composed of four No. 9 longitudinal bars and No. 3 ties. Assume that the compressive strength specified for the concrete is 3000 lb/in^2 and that grade 40 bars will be used.

(a) Compute the maximum factored axial load that can be supported by this column.

(b) Determine the tie spacing required for this column.

(c) Investigate the bond stress.

28.4 A 12 in $\times$ 16 in tied concrete column is composed of four No. 9 longitudinal bars and No. 3 ties. Assume that the compressive strength specified for the concrete is 3000 lb/in^2 and that grade 40 bars will be used.

(a) Compute the maximum factored axial load that can be supported by this column.

(b) Determine the tie spacing required for this column.

(c) Investigate the bond stress.

28.5 A 12 in $\times$ 12 in tied concrete column is composed of concrete with a specified compressive strength of 4000 lb/in^2 and grade 40 bars.

(a) Compute the maximum factored axial load that can be supported by this column with four No. 10 bars.

(b) Compute the maximum factored axial load that can be supported by this column with eight No. 10 bars.

(c) The column will carry a 100-kip dead load and a 180-kip live load. According to ACI standards, which of the two columns above can carry the combined load safely?

Chapter 29
MASONRY PRODUCTS AND CONSTRUCTION TECHNIQUES

Structural masonry units include brick, concrete block, stone, and clay tile. In this chapter we provide an overview of masonry products and construction techniques that can serve as a prelude to the methods used in elementary design of simple masonry walls and columns, introduced in Chapter 30.

29.1   Masonry Units

Masonry units are the oldest form of manufactured building products, dating to about 5000 B.C. In modern history, early building codes required that masonry walls be progressively thicker at their base. Masonry walls ranging from 1 ft thick at the top floor to 5 to 6 ft thick at their base were common in buildings 10 to 20 stories tall. Today, better quality assurance techniques in the manufacture of masonry products and more efficient design regulations allow for better use of masonry as a structural element in buildings.

By definition, a *solid* masonry unit has a net cross-sectional area of not less than 75% solid. A *hollow* unit has a net solid cross-sectional area less than 75% solid. The percentage of solid area is computed in all planes parallel to the bearing force (i.e., usually a plane parallel with the horizontal face of the unit). Since most masonry units are symmetrical along the horizontal plane, the net solid cross-sectional area is the gross area of the horizontal face of the unit minus the area of any voids.

Masonry units commonly specified for use in buildings include:

- *Clay masonry.* This type of masonry includes common (building) brick, face brick, structural and facing tile, and adobe units. Clay masonry is composed of clay, shale, or other minerals that are formed into the desired shape. *Unburned* clay masonry units such as adobe brick are rough formed to size and sun-dried before use. Unburned clay masonry is limited to applications where bearing wall thickness is at least 16 in and walls are limited to a height of one story. It is reserved for use in aesthetic and historical applications. *Burned* clay masonry units are formed by extrusion, molding, or dry-pressing and are fired in a kiln at temperatures between 1800 and 2100°F (980 and 1150°C) for a period of 2 to 5 days so the clay becomes vitrified (heat fused). Burned clay masonry products that are used in structural applications include brick and structural clay tiles:

- *Brick. Common brick* is the most widely used solid brick. It is available in three grades: SW, MW, and NW. Grade SW (severe weathering) has the highest resistance to moisture permeation and frost action. It can be used in all below- and above-grade applications. Grade MW (moderate weathering) brick can be exposed to temperatures below freezing where it is not in direct contact with moisture, so it is used in abovegrade applications. Grade NW (negligible weathering) brick should be used away from moisture and frost, such as in interior or backup masonry applications. *Face* or *facing brick* is manufactured under controlled conditions that regulate texture, color range, and size. It is available in two grades: SW and MW. *Hollow brick* is generally larger than solid brick with a net solid cross-sectional area that is less than 75% solid. It is available in SW and MW grades. Hollow units are classified as H40V for units with a total void area greater than 25% and less than 40% of the gross cross-sectional area, or H60V for units with a total void area greater than 40% and less than 60% of the gross cross-sectional area. The void area of the cells in hollow structural brick provides space for grout and reinforcement and allows for an increase in unit size without increasing the weight of the unit significantly.

- *Structural clay tile.* This is a hollow-celled structural masonry unit much like hollow brick but larger. Structural clay tile has thinner web and shell dimensions and is not as strong as hollow brick. *Load-bearing wall tile* is available in two grades: LBX for weather exposure and LB for interior or backup masonry applications. *Non-load-bearing wall tile* (NB) is used in non-load-bearing wall applications such as furring, backing, or fireproofing tile. *Facing tile* is manufactured under controlled conditions that maintain appearance: color uniformity and size. It is used to face walls and in interior partitions.

While brick and structural clay tile are both durable and aesthetically appealing, they are also well suited for many structural applications.

- *Concrete masonry.* Large masonry units made of portland cement, aggregate, and water are called *concrete masonry units* (CMUs). These units are sometimes referred to as concrete block. CMUs can be hollow or solid. The void area of the cells in hollow concrete masonry reduces the weight of the unit and provides space for grout and reinforcement. Load-bearing CMUs are available in two grades, N and S, and two types, I and II. Grades are classified as N-I,

N-II, S-I, and S-II. Grade N concrete masonry has the highest resistance to moisture permeation and frost action and can be used in above- and below-grade applications. Grade S concrete masonry should be used away from moisture and frost and only in abovegrade walls that are protected from weather. The moisture content of type I concrete masonry is controlled during manufacture, so these are higher-quality units. Smaller solid concrete masonry units are called *concrete brick*. Concrete brick is available in three grades: U, P, and G. Grade U is used in exterior architectural veneer and facing applications where texture, color uniformity, and size are important. It offers the highest resistance against frost action and moisture permeation. Grade P is for use in abovegrade applications, where it is protected from weather. Grade G is for use away from moisture, such as in interior partitions and backup masonry applications. Additionally, concrete masonry is classified by weight of concrete used in manufacturing the unit: lightweight (less than 105 lb/ft^3), medium weight (105 to 125 lb/ft^3), and normal weight (above 125 lb/ft^3).

- *Stone masonry.* Stone is a natural building material that can be used as structural masonry and in architectural facing and veneer applications. Granite, marble, sandstone, and limestone are popular types of structural building stone. Each type of stone has a grading system that is based on color uniformity, grain structure, working qualities, and strength characteristics. Stone masonry is categorized by its shape and how it is arranged in a masonry assembly. *Rubble masonry* is rounded fieldstone or angular quarry stone arranged in random or rough course configurations. *Ashlar masonry* is composed of stones that have been squared and set in random or uniform course configurations. The mason laying rubble and ashlar stone masonry arranges stone placement. *Cut* or *dimension stone masonry* is typically a veneer or facing stone that is fabricated and finished at the mill and constructed on site in a specific configuration according to shop drawings.

- *Gypsum masonry.* Gypsum masonry units are 2- to 6-in-thick panel-like hollow or solid units made of gypsum and fibrous materials. They are not commonly used in structural applications. Gypsum masonry is used in non-load-bearing applications such as interior partitions and as fireproofing of beams and columns, where it is not subjected to moisture exposure.

- *Glass block.* Glass blocks are made by fusing two square bowl-like sections of molded glass together. They are available in 3- and 4-in thicknesses with vertical face dimensions ranging from 3 in × 6 in to 12 in × 12 in. The translucent nature of glass block makes it well suited for natural illumination and solar glazing applications. It is not commonly used in structural applications.

Structural properties of masonry units vary by type and grade. The strength of a batch of masonry is determined through compression testing groups of masonry units (three or five units from a batch). Usually this is based on the minimum average compressive strength of the test group and the minimum compressive strength of an individual unit in the test group. Strength and water absorption of a specific masonry unit determine its grade. Selected structural properties of common clay and concrete masonry units are provided in Table 29.1.

The *actual dimensions* of a masonry unit relate to the measured size of the unit as manufactured. *Specified dimensions* are the stipulated dimensions for manufacture or construction of the masonry unit. Actual dimensions of a specific masonry unit will vary slightly from the specified dimensions from batch to batch and by manufacturer. Any variations from specified dimensions are limited by industry standards. The *nominal dimensions* of a masonry unit or assembly include the dimensions of the mortar joints; for example, a nominal 4-in-thick face brick has an actual thickness of $3\frac{1}{2}$ in plus $\frac{1}{2}$ in for the mortar joint thickness.

Masonry unit dimensions are expressed in order of thickness (t) by height (h) by length (l) (Figure 29.1). A standard 6-in CMU has specified dimensions of $5\frac{5}{8}$ in

Table 29.1 STRUCTURAL PROPERTIES OF SELECTED COMMON CLAY AND CONCRETE MASONRY UNITS

Type of masonry unit	Grade	Compressive strength (lb/in² or MPa)			
		Minimum average of tests		Minimum in individual test	
		U.S.	Metric	U.S.	Metric
Clay masonry products (five tests)					
Common (building) solid brick	SW (severe weathering)	3000	20.7	2500	17.2
	MW (moderate weathering)	2500	17.2	2200	15.2
	NW (negligible weathering)	1500	10.3	1250	8.6
Hollow brick	SW (severe weathering)	3000	20.7	2500	17.2
	MW (moderate weathering)	2500	17.2	2200	15.2
Structural clay load-bearing wall tile	LBX, end construction	1400	9.6	1000	6.9
	LBX, side construction	700	4.8	500	3.4
	LB, end construction	1000	6.9	700	4.8
	LB, side construction	700	4.8	500	3.4
Unburned clay masonry	—	300	2.1	250	1.7
Concrete masonry products (three tests)					
Concrete building brick	N	2500	17.2	2000	13.8
	S	1500	10.3	1250	8.6
Solid load-bearing CMU	N	1800	12.4	1600	11.0
	S	1200	8.3	1000	6.9
Hollow load-bearing CMU	N	1000	6.9	800	5.5
	S	700	4.8	600	4.1
Hollow non-load-bearing CMU	—	350	2.4	300	2.1

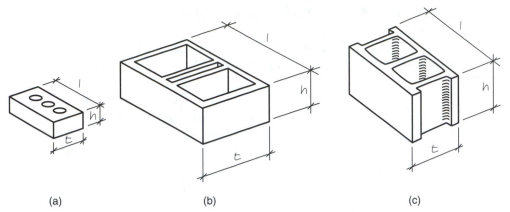

FIGURE 29.1 Dimensions of masonry unit dimensions are expressed in order of thickness (t) by height (h) by length (l): (a) solid brick; (b) hollow clay brick; (c) concrete masonry unit.

Table 29.2 SIZES OF SELECTED CLAY BRICK UNITS[a]

Type	Nominal dimensions			Joint thickness	Specified dimensions			Modular coursing, C
	w	h	l		w	h	l	
Modular	4 in	$2\frac{2}{3}$ in	8 in	$\frac{1}{2}$ in	$3\frac{1}{2}$ in	$2\frac{1}{4}$ in	$7\frac{1}{2}$ in	3C = 8 in
	100 mm	67 mm	200 mm	10 mm	90 mm	57 mm	190 mm	3C = 200 mm
Engineer Modular	4 in	$3\frac{1}{5}$ in	8 in	$\frac{1}{2}$ in	$3\frac{1}{2}$ in	$2\frac{3}{4}$ in	$7\frac{1}{2}$ in	5C = 16 in
	100 mm	80 mm	200 mm	10 mm	90 mm	70 mm	190 mm	5C = 400 mm
Roman	4 in	2 in	12 in	$\frac{1}{2}$ in	$3\frac{1}{2}$ in	$1\frac{5}{8}$ in	$11\frac{1}{2}$ in	2C = 4 in
	100 mm	50 mm	300 mm	10 mm	90 mm	40 mm	290 mm	2C = 100 mm
Norman	4 in	$2\frac{2}{3}$ in	12 in	$\frac{1}{2}$ in	$3\frac{1}{2}$ in	$2\frac{1}{4}$ in	$11\frac{1}{2}$ in	3C = 8 in
	100 mm	67 mm	300 mm	10 mm	90 mm	57 mm	290 mm	3C = 200 mm
Engineer Norman	4 in	$3\frac{1}{5}$ in	12 in	$\frac{1}{2}$ in	$3\frac{1}{2}$ in	$2\frac{3}{4}$ in	$11\frac{1}{2}$ in	5C = 16 in
	100 mm	80 mm	300 mm	10 mm	90 mm	70 mm	290 mm	5C = 400 mm
Utility	4 in	4 in	12 in	$\frac{1}{2}$ in	$3\frac{1}{2}$ in	$3\frac{1}{2}$ in	$11\frac{1}{2}$ in	1C = 4 in
	100 mm	100 mm	300 mm	10 mm	90 mm	90 mm	290 mm	2C = 200 mm
Standard[b]	100 mm	67 mm	200 mm	$\frac{1}{2}$ in	$3\frac{5}{8}$ in	$2\frac{1}{4}$ in	8 in	3C = 8 in
				10 mm	90 mm	57 mm	190 mm	3C = 200 mm
Engineer standard[b]	100 mm	80 mm	200 mm	$\frac{1}{2}$ in	$3\frac{5}{8}$ in	$2\frac{13}{16}$ in	8 in	5C = 16 in
				10 mm	90 mm	70 mm	190 mm	5C = 400 mm
King[b]	86 mm[c]	80 mm	255 mm[c]	$\frac{1}{2}$ in	3 in	$2\frac{3}{4}$ in	$9\frac{5}{8}$ in	5C = 16 in
				10 mm	76 mm[c]	70 mm	245 mm[c]	5C = 400 mm
Queen[b]	86 mm[c]	80 mm	213 mm[c]	$\frac{1}{2}$ in	3 in	$2\frac{3}{4}$ in	8 in	5C = 16 in
				10 mm	76 mm[c]	70 mm	203 mm[c]	5C = 400 mm

[a]Sizes are expressed in order of width (w) by height (h) by length (l).

[b]Metric nonmodular common brick sizes.

[c]Based on actual manufactured unit size.

Table 29.3 SIZES OF SELECTED HOLLOW STRUCTURAL BRICK UNITS (in)[a]

Type	Nominal dimensions			Joint thickness	Specified dimensions			Modular coursing, C	Minimum face shell thickness[b]
	w	h	l		w	h	l		
12-in-long	4	4	12	$\frac{1}{2}$	$3\frac{1}{2}$	$3\frac{1}{2}$	$11\frac{1}{2}$	$2C = 8$	$\frac{3}{4}$
standard	6	4	12	$\frac{1}{2}$	$5\frac{1}{2}$	$3\frac{1}{2}$	$11\frac{1}{2}$	$2C = 8$	1
	8	4	12	$\frac{1}{2}$	$7\frac{1}{2}$	$3\frac{1}{2}$	$11\frac{1}{2}$	$2C = 8$	$1\frac{1}{4}$
12-in-long	4	8	12	$\frac{1}{2}$	$3\frac{1}{2}$	$7\frac{1}{2}$	$11\frac{1}{2}$	$1C = 8$	$\frac{3}{4}$
super	6	8	12	$\frac{1}{2}$	$5\frac{1}{2}$	$7\frac{1}{2}$	$11\frac{1}{2}$	$1C = 8$	1
	8	8	12	$\frac{1}{2}$	$7\frac{1}{2}$	$7\frac{1}{2}$	$11\frac{1}{2}$	$1C = 8$	$1\frac{1}{4}$
16-in-long	4	4	16	$\frac{1}{2}$	$3\frac{1}{2}$	$3\frac{1}{2}$	$15\frac{1}{2}$	$2C = 8$	$\frac{3}{4}$
standard	6	4	16	$\frac{1}{2}$	$5\frac{1}{2}$	$3\frac{1}{2}$	$15\frac{1}{2}$	$2C = 8$	1
	8	4	16	$\frac{1}{2}$	$7\frac{1}{2}$	$3\frac{1}{2}$	$15\frac{1}{2}$	$2C = 8$	$1\frac{1}{4}$
	10	4	16	$\frac{1}{2}$	$9\frac{1}{2}$	$3\frac{1}{2}$	$15\frac{1}{2}$	$2C = 8$	$1\frac{3}{8}$
	12	4	16	$\frac{1}{2}$	$11\frac{1}{2}$	$3\frac{1}{2}$	$15\frac{1}{2}$	$2C = 8$	$1\frac{1}{2}$
16-in-long	4	8	16	$\frac{1}{2}$	$3\frac{1}{2}$	$7\frac{1}{2}$	$15\frac{1}{2}$	$1C = 8$	$\frac{3}{4}$
super	6	8	16	$\frac{1}{2}$	$5\frac{1}{2}$	$7\frac{1}{2}$	$15\frac{1}{2}$	$1C = 8$	1
	8	8	16	$\frac{1}{2}$	$7\frac{1}{2}$	$7\frac{1}{2}$	$15\frac{1}{2}$	$1C = 8$	$1\frac{1}{4}$
	10	8	16	$\frac{1}{2}$	$9\frac{1}{2}$	$7\frac{1}{2}$	$15\frac{1}{2}$	$1C = 8$	$1\frac{3}{8}$
	12	8	16	$\frac{1}{2}$	$11\frac{1}{2}$	$7\frac{1}{2}$	$15\frac{1}{2}$	$1C = 8$	$1\frac{1}{2}$

[a]Sizes are expressed in order of width (w) by height (h) by length (l).
[b]Minimum face shell thicknesses are for solid shell hollow brick units.

Table 29.4 SIZES OF SELECTED CONCRETE MASONRY UNITS (in)[a]

CMU size	Nominal dimensions			Joint thickness	Specified dimensions			Modular coursing, C	Minimum face shell thickness
	w	h	l		w	h	l		
4	4	8	16	$\frac{3}{8}$	$3\frac{5}{8}$	$7\frac{5}{8}$	$15\frac{5}{8}$	$1C = 8$	$\frac{3}{4}$
6	6	8	16	$\frac{3}{8}$	$5\frac{5}{8}$	$7\frac{5}{8}$	$15\frac{5}{8}$	$1C = 8$	1
8	8	8	16	$\frac{3}{8}$	$7\frac{5}{8}$	$7\frac{5}{8}$	$15\frac{5}{8}$	$1C = 8$	$1\frac{1}{4}$
10	10	8	16	$\frac{3}{8}$	$9\frac{5}{8}$	$7\frac{5}{8}$	$15\frac{5}{8}$	$1C = 8$	$1\frac{3}{8}$
12	12	8	16	$\frac{3}{8}$	$11\frac{5}{8}$	$7\frac{5}{8}$	$15\frac{5}{8}$	$1C = 8$	$1\frac{1}{2}$

[a]Sizes are expressed in order of width (w) by height (h) by length (l).

$\times$ $7\frac{5}{8}$ in $\times$ $15\frac{5}{8}$ in and nominal dimensions of 6 in $\times$ 8 in $\times$ 16 in, the difference being $\frac{3}{8}$-in-thick mortar joints on three sides. Specified and nominal sizes of selected clay brick and concrete masonry units are provided in Tables 29.2 to 29.4.

Masonry units in walls and columns are arranged in an assembly of continuous horizontal rows of masonry units called *courses* (Figure 29.2). Each course of masonry is bonded with a horizontal mortar joint called a *bed joint*. The individual masonry units in a course are bonded with vertical mortar joints called *head joints*.

29.1 Masonry Units

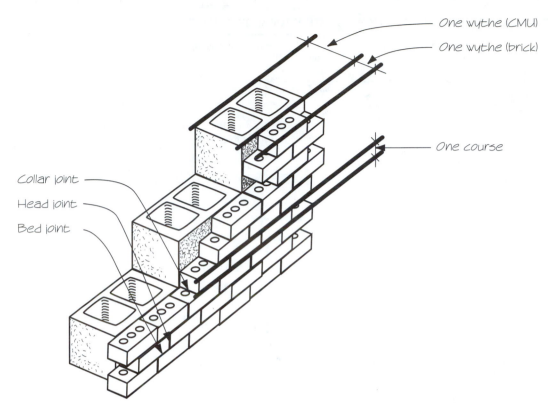

FIGURE 29.2 Masonry units in walls and columns are arranged in an assembly of continuous horizontal rows of masonry units called *courses*. A *wythe* of masonry is a layer of masonry that is one masonry unit thick. Masonry units are bonded with a horizontal mortar *joint*.

A *wythe* of masonry is a layer of masonry that is one masonry unit thick. A wall composed of a layer of face brick backed by a layer of CMU is two wythes thick; it has one wythe of face brick and one wythe of CMU. Multiple wythes of a masonry assembly are frequently bonded with a mortar joint between wythes called a *collar joint* or are separated with a cavity.

The method of bonding masonry units into a single masonry assembly will depend on the application, assembly type, and other factors. By definition, here are three types of bonds:

1. *Structural bond:* the method by which individual masonry units are interlocked or tied together to cause the entire masonry assembly to act as a single homogeneous structural unit. Structural bonding of masonry walls may be accomplished by overlapping or interlocking masonry units, by the use of metal ties embedded in connecting joints, and by the adhesion of grout to adjacent wythes of masonry.

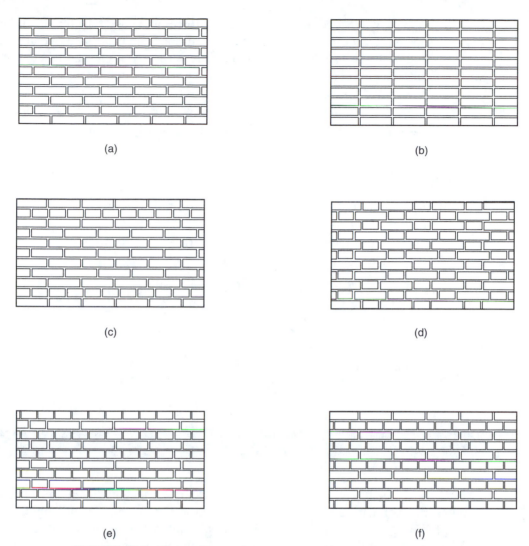

FIGURE 29.3 The stacking configuration created between a masonry unit in one course and another in the next course is called a *masonry pattern bond:* (a) running bond; (b) stack bond; (c) common or American bond; (d) Flemish bond; (e) English bond; (f) English cross or Dutch bond.

2. *Pattern bond:* the stacking configuration that is created by the masonry units and the mortar joints on the face of a wall. The pattern may be the result of the structural bond or may be decorative and unrelated to the structural bonding. There are five basic pattern bonds: running bond, common or American bond, Flemish bond, English bond, and stack bond (Figure 29.3). A *running bond* is an assembly of interlocking masonry units with each course staggered by a half unit. It is the strongest pattern bond

and provides good resistance to lateral forces and good transfer of bearing loads. A *stack bond* is an assembly of masonry units stacked one unit on top of the next so that the head joints are vertically aligned and there is no interlocking of units between courses. It provides little resistance to lateral forces and poor transfer of bearing loads. A structural masonry assembly constructed using a stack pattern bond requires horizontal reinforcement.

3. *Mortar bond:* the adhesion of mortar to the masonry units or to reinforcing steel.

■ EXAMPLE 29.1

Determine the actual height of a masonry assembly constructed of 51 courses of modular clay brick.

Solution From Table 29.2, modular clay brick has a coursing of $3C = 8$ in. The height of one course (C) is found by

$$3C = 8 \text{ in}$$
$$C = \frac{8 \text{ in}}{3}$$
$$= 2\tfrac{2}{3} \text{ in}$$

The height of the wall is found by

$$51 \text{ courses} \cdot 2\tfrac{2}{3} \text{ in} = 136 \text{ in} = 11'\text{-}4''$$

29.2 Mortar

Mortar is a cementitious mix of portland or masonry cement, hydrated lime or lime putty, fine aggregate, and water that bonds masonry units into a wall, column, or other building element. It is applied in a plastic state when the masonry units are laid. Properties of mortar that influence the structural performance of masonry are compressive strength, bond strength, and elasticity. Mortar cures to form a bond that joins the masonry units together as a homogeneous assembly. It also compensates for minor size variations in the individual masonry units, which allows for a uniform transfer of loads through the masonry assembly. A good mortar bond is very important to a well-functioning masonry assembly.

Mortar is available in five types: M, S, N, O, and K, although types M, S, and N are used in structural masonry applications. Average compressive strengths and mixing proportions by mortar type are provided in Table 29.5. Use of mortar type will vary by the structural and application requirements. Following is a general description of the applications of each type of mortar permitted by the building code and industry standards:

Table 29.5 AVERAGE COMPRESSIVE STRENGTHS, MIXING PROPORTIONS, AND COMMON APPLICATIONS BY MORTAR TYPE

Mortar type	Average compressive strength at 28 days		Mixing proportions (parts per volume)				Common applications
	lb/in^2	MPa	Portland cement	Hydrated or lime putty	Masonry cement	Aggregate	
M	2500	17.2	1	$\frac{1}{4}$	—	3	All exposures (i.e., belowgrade and weather exposed)
			1	—	1	6	
S	1800	12.4	1	$\frac{1}{2}$	—	$4\frac{1}{2}$	All exposures; resistance to lateral forces
			$\frac{1}{2}$	—	1	$4\frac{1}{2}$	
N	750	5.2	1	1	—	6	Abovegrade exposures
			—	—	1	3	
O	350	2.4	1	2	—	9	Interior, non-load-bearing applications
K	75	0.52	1	4	—	15	Non-load-bearing fill

- *Type M mortar.* This mortar type is specifically recommended for masonry belowgrade and in contact with earth, such as foundation walls, retaining walls, sewers, and manholes. It has high compressive strength and high resistance to moisture permeation and frost action.

- *Type S mortar.* This mortar type is suggested for use in reinforced masonry and unreinforced masonry assemblies, such as exterior walls, where maximum flexural strength is required. It has a high compressive strength and a high tensile (flexural) bond strength.

- *Type N mortar.* This mortar type is recommended for chimneys, parapet walls, exterior walls, and as masonry veneer over wood framing that are subject to severe exposure. It has medium bond and compressive strength characteristics suitable for general use in exposed masonry abovegrade. Type N mortar is generally not permitted in locations with a likelihood of significant seismic activity.

Mortar joint thickness varies with type of masonry unit and minor irregularities in individual masonry units. Joint thicknesses are generally $\frac{1}{4}$, $\frac{3}{8}$, and $\frac{1}{2}$ in for clay, concrete, and stone ashlar masonry. Because rubble masonry consists of rounded fieldstone or angular quarry stone, larger mortar joints are needed to compensate for size variations in the stone.

In structural masonry assemblies, on solid masonry units the bed (horizontal) joints and head (side) joints should be covered fully with mortar. This is referred to as *solid bedding.* On hollow masonry units, all head and bed joints should be filled solidly with mortar for a distance in from the face of the masonry unit not less than the thickness of the face shell. This is referred to as *face-shell bedding.* The webs of hollow masonry units generally do not receive mortar.

After the masonry units are laid and with the mortar still somewhat plastic, the face joints are tooled or troweled to a surface finish. *Concave* and *vee joints* are formed by the use of a steel round or V-shaped jointing tool. They are very effective in resisting rain penetration and are recommended for use in areas subjected to heavy rains and high winds. The *raked joint* is made by removing the surface of the mortar before it hardens. It is difficult to make this joint weathertight, so it is not recommended where heavy rain, high wind, or freezing is likely to occur. Control joints are also needed in masonry: *expansion joints* limit damage from expansion and contraction from temperature change, and *relief joints* attempt to control where cracking from movement occurs.

29.3 Grout

Grout is a cementitious mix of portland cement or blended cement, hydrated lime or lime putty, coarse or fine aggregate, and water. It is used in brick masonry to fill cells of hollow unit masonry or spaces between wythes of solid unit masonry. Grout increases the compressive, shear, and flexural strength of the masonry element and bonds steel reinforcement and masonry together.

Grout must have a compressive strength of at least 2000 lb/in^2 (14.0 MPa). Unlike mortar and concrete, grout is mixed with the highest acceptable water/cement ratio that does not allow segregation of mix ingredients when the grout is poured. The soupy consistency allows the grout to flow sufficiently to fully fill masonry cells and cavities. Once it is placed, the excess water in the grout is absorbed rapidly by the porous masonry. Absorption of excess water reduces the water/cement ratio of the cementitious mix to an acceptable level.

In grouted masonry assemblies, grout is poured as the masonry assembly is being constructed to prevent segregation of mix ingredients caused by ricochet off surfaces or the inertial drop of a deep pour. Generally, grout must be poured in horizontal intervals not exceeding 6 ft (1.83 m), unless approved by the building official. After the grout is poured and while it is still in a plastic state, it must be vibrated mechanically to remove voids.

29.4 Reinforcement

Masonry assemblies can be strengthened by fully or partially increasing the thickness of masonry units in the wall or by increasing the number of masonry unit wythes in the assembly. A thicker masonry assembly better resists lateral and buckling forces. Common types of masonry assemblies that reinforce walls include:

- *Pilaster:* a columnlike assembly that is an integral part of the wall. Pilasters are added to walls to counter lateral and buckling forces and to support concentrated loads applied to the slender wall.
- *Buttress:* a vertical mass that extends from the wall and reinforces the wall against thrusting forces and buckling.

Placed at intervals in a masonry wall, pilasters and buttresses strengthen the wall, thereby allowing the wall to be longer and taller for a specific thickness. The extended mass of a pilaster or buttress limits the practicality of their use in many applications.

The capability of a masonry assembly to resist lateral forces improves significantly by adding steel reinforcement much like reinforcement is added to strengthen concrete. Steel reinforcement for masonry construction consists of bars and wires. *Reinforcing bars* are used in masonry elements such as walls, columns, pilasters, and beams. Wires are used in masonry bed joints to reinforce individual masonry wythes or to tie multiple wythes together. *Joint reinforcement* is a wire assembly bent in a trusslike pattern. It is placed horizontally in the bed (horizontal) joint of masonry walls to provide lateral reinforcement. It is installed when the masonry units are laid. The vertical spacing of joint reinforcement must not exceed 16 in measured vertically and must engage all wythes of masonry in a wall.

Designated cells of hollow masonry wall assemblies or the cavity between masonry wythes can be filled with grout to strengthen the assembly. This method of strengthening a wall assembly is referred to as *grouted masonry.* Pouring grout into designated cells of a hollow masonry assembly improves the net area of the section and further bonds the masonry units. A *reinforced masonry* assembly is constructed by placing reinforcing steel in the cells of hollow masonry units or the wall cavity between masonry wythes before grout is added. The grout forms a bond between the steel bars and the masonry which strengthens the assembly. Several methods are used to reinforce a masonry assembly:

- *Vertical reinforcement* (Figure 29.4): steel rebars that extend vertically through cells of hollow masonry units or through the cavity between masonry wythes of a wall and are grouted in place. It provides lateral and thrust reinforcement.
- *Horizontal reinforcement* (Figure 29.4): steel rebars that extend horizontally through the cavity of a masonry wall and are grouted in place. It provides lateral and flexural reinforcement and reduces thermal expansion and stress.
- *Bond beams* (Figure 29.5): reinforced assembly that consists of a course of special U-shaped masonry units called lintel block. Steel rebars extend horizontally through the cavity of the lintel block and are grouted in place.

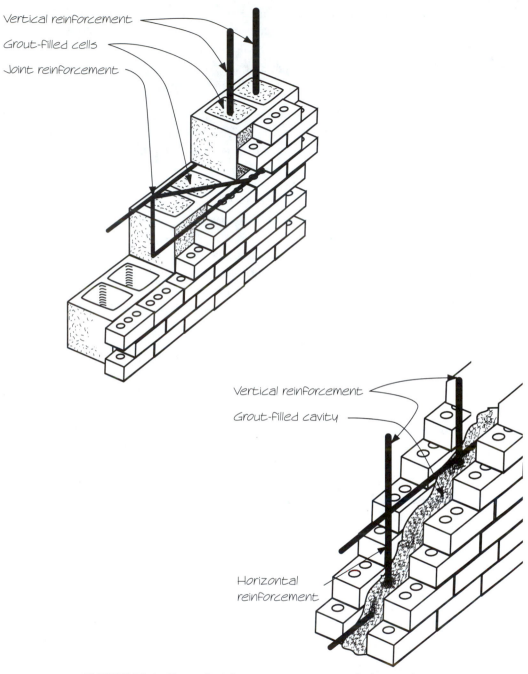

FIGURE 29.4 Types of reinforcement in masonry walls. *Joint reinforcement* is a wire assembly bent in a trusslike pattern that is placed horizontally in the bed (horizontal) joint of masonry walls. *Horizontal* and *vertical reinforcement* consist of steel rebars that extend horizontally through the cavity of a masonry wall that are grouted in place.

Labels in figure:
- Vertical reinforcement
- Grout-filled cells
- Joint reinforcement
- Vertical reinforcement
- Grout-filled cavity
- Horizontal reinforcement

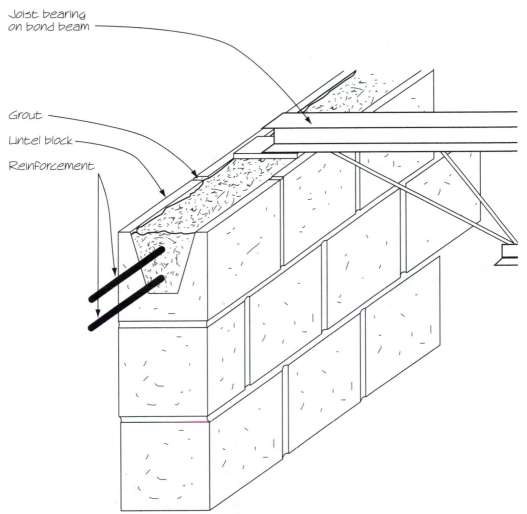

Joist bearing
on bond beam

Grout

Lintel block

Reinforcement

FIGURE 29.5 *Bond beams* are a continuous course of horizontally reinforced masonry that structurally wraps the masonry assembly.

Bond beams are generally located on the top course of masonry and as a bearing course upon which roof and floor structural members such as steel joists and concrete core slabs bear. A bond beam provides lateral and bearing reinforcement.

Vertical and horizontal reinforcement in masonry walls is required in zones with moderate to high seismic activity. It is also specified to resist wind and other lateral loads.

29.4 Reinforcement   **515**

A starter course of hollow brick in a single-wythe masonry wall. Vertical reinforcing bars extend from the concrete foundation walls through the cells of the brick. After more courses of masonry are laid, the cells with reinforcement will be grouted.

A hollow-unit masonry wall constructed of hollow brick. A vertical reinforcing bar extends through the cells of the hollow brick. Cells with vertical reinforcement will be grouted. Wire joint reinforcement is located between the fourth and fifth course.

Cells with vertical reinforcement are grouted.

A hollow-unit masonry wall constructed of concrete masonry units (CMU). Note how the mortar joints cover only the face shell thickness. Joint reinforcement is placed in the mortar joint. Cells with vertical reinforcement will be grouted.

A veneered masonry wall. Dark brick is placed in a running pattern bond and capped with a soldier course of light brick (left side of photo). Mortar joints cover the entire wythe of the brick. Brick is anchored to the framing with ties. An air space separates the brick and sheathing.

Brick veneer is non-load bearing and is backed by sheathing anchored to framing.

An unfinished control joint in a hollow-unit masonry wall constructed of concrete masonry units (CMU). The joint will limit damage from expansion and contraction due to temperature change or movement from settlement or heaving. Joint faces will be finished with caulk to match the appearance of mortar.

A finished control joint.

By definition, a *wall* is a vertical member of a building structure with a length that exceeds three times its thickness. A *column* is an isolated masonry assembly that has a length that does not exceed three times its thickness. Column and wall length is defined as the horizontal dimension measured at right angles to the thickness of the member or assembly. Many types of masonry wall and column assemblies have been used successfully (Figure 29.6).

Solid masonry walls   A solid masonry assembly is constructed of one or more wythes of hollow or solid masonry. A wall composed of hollow CMU, a wall composed of two wythes of closely placed brick, and a wall composed of a wythe of face brick backed by and bonded to a wythe of CMU are three examples of solid masonry construction. A *single-wythe* masonry wall is constructed of one wythe of hollow or solid masonry. *Multiwythe* masonry wall is an assembly comprised of at least two wythes of masonry that are bonded together with interlocking masonry wythes, a collar joint, or with ties or joint reinforcement. A *composite* masonry wall is a multiwythe wall in which at least one of the wythes is dissimilar to the other wythe or wythes with respect to type or grade of masonry unit or mortar. When two or more wythes of masonry are bonded structurally, the assembly is known as a *bonded* masonry wall. In an assembly known as a *grouted hollow-unit* masonry wall, grout is poured into designated cells of an assembly constructed of hollow masonry units. When reinforcing bars are placed in and extend through the cells of masonry and are embedded in grout, the assembly is known as a *reinforced hollow-unit* masonry wall.

Cavity masonry walls   A *cavity* masonry wall is a multiwythe assembly constructed of two wythes of hollow or solid masonry that are separated by a continuous 2- to $4\frac{1}{2}$-in airspace called a *cavity*. The masonry wythes must be structurally bonded across the cavity with metal ties or joint reinforcement. An example of this type of assembly is a wall composed of a wythe of face brick and a wythe of CMU that are separated by a cavity and structurally bonded with joint reinforcement. An assembly with an empty cavity is referred to as a *hollow multiwythe* masonry wall. The hollow cavity improves the insulating performance of the masonry assembly significantly. If the cavity is filled with grout, the assembly is known as a *grouted multiwythe* masonry wall. When reinforcing bars extend through the cavity and are embedded with grout, the masonry assembly is known as a *reinforced multiwythe* masonry wall.

Veneered walls   This type of assembly consists of a single wythe of non-load-bearing masonry called a *veneer* that is attached but not structurally bonded to framing or backup masonry and isolated from it with an airspace. In wood

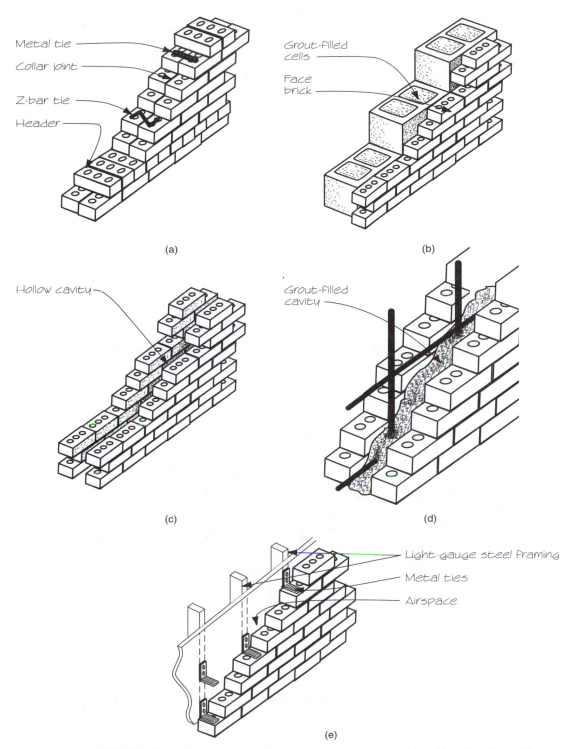

FIGURE 29.6 Selected types of masonry construction assemblies: (a) multiwythe solid masonry; (b) grouted hollow unit, multiwythe masonry; (c) hollow multiwythe masonry; (d) grouted multiwythe masonry; and (e) veneered masonry.

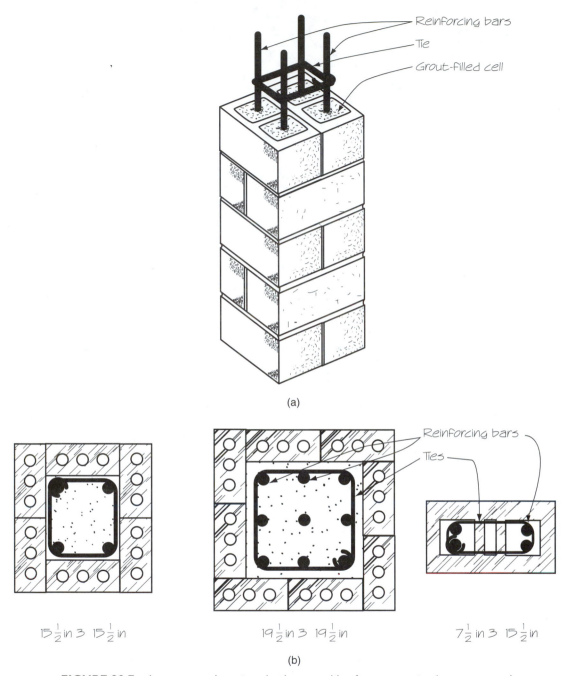

(a)

(b)

$15\frac{1}{2}$ in 3 $15\frac{1}{2}$ in $19\frac{1}{2}$ in 3 $19\frac{1}{2}$ in $7\frac{1}{2}$ in 3 $15\frac{1}{2}$ in

FIGURE 29.7 A masonry column is a slender assembly of masonry units that supports a large axial loads: Masonry units in a column are structurally bonded together with interlocking masonry wythes, a collar joint, ties or reinforcement. Masonry columns may also be constructed with a grout-filled cavity. Shown are: (a) grouted hollow masonry column; and (b) grouted brick columns.

frame and masonry construction, corrugated metal ties extend from the framing or backup masonry and are embedded in the mortar bed joints of the veneer. In light-gauge steel frame construction, anchors are used to span the airspace and attach the veneer to the framing.

Masonry columns   A masonry column (Figure 29.7) is a slender assembly of masonry units that supports large axial loads. Masonry units in a solid column are structurally bonded together with interlocking masonry wythes, a collar joint, ties, or reinforcement. Masonry columns may also be constructed with a grout-filled cavity.

29.6   Masonry Design Standards

A myriad of professional and trade organizations promote use of masonry and develop design standards. The Brick Institute of America (BIA), Reston, Virginia, (Web site address: www.buckley.com/bia), is a nonprofit trade association representing the manufacturers of brick in the United States for more than 60 years. BIA has pursued the goal of developing and improving markets for the brick industry. BIA provides considerable technical information on their web site. The National Concrete Masonry Association (NCMA; Web site: www.ncma.org), established in 1918, is the national trade association representing the concrete masonry industry. It is involved in a broad range of technical, research, marketing, government relations, and communications activities. NCMA offers a variety of technical services and design aids through publications, computer programs, slide presentations and technical training.

Three techniques are used in the design of masonry construction: *working stress design, strength design,* and *empirical design.* The empirical design technique is limited to use in zones with little likelihood of seismic activity. The strength design method is fairly complicated but results in more efficient members. The working stress design method more closely represents the concepts presented earlier in this book. Therefore, the working stress design method for unreinforced concrete walls and columns is introduced in Chapter 30.

The building code specifies many design parameters for construction of masonry walls and columns. In most cases the minimum thickness of masonry bearing walls should not be less than a nominal 6 in (153 mm). In some cases, nominal-4-in structural clay brick is allowed. Columns are limited to a minimum nominal dimension of 8 in (200 mm) for the least cross-sectional dimension. Also, the column slenderness ratio of height to least lateral dimension must not exceed 25. Reinforced columns must be reinforced with a minimum of four reinforcing bars. The area of reinforcement (A_s) must be at least 0.0025 but not more than 0.04 times the net cross-sectional area of the column.

In the chapters on structural masonry, the symbols f and F are used as a variable quantity for stresses such as specified and allowable compressive strengths (i.e., f'_m, F_a, etc.). In the first 17 chapters, the Greek lowercase letter sigma (σ) was used to symbolize a variable quantity for stress because of its wide acceptance and universal use in general engineering. An exception to the use of σ is made in this and later chapters because this branch of the structural engineering profession uses the symbol F for quantities of stress.

The symbol F *was* used in earlier chapters to identify a variable quantity for a concentrated force. Different symbols related to type of loads are also used in this chapter instead of F, such as P for a concentrated load, w for a uniformly distributed load, and W for a total uniformly distributed load. Although it can lead to some confusion, every effort is made to use trade-specific symbols in this book.

The strength of a masonry assembly is determined by laboratory testing. *Masonry prism testing* is a formal testing procedure that is used to determine the ultimate compressive strength of a specific masonry assembly. A *prism* is a masonry assembly that is constructed of the same materials with the same workmanship and under the same conditions that will be found at the job site. It must be constructed under the observation of an engineer or special inspector. A set of prisms is tested much like concrete specimens, to ascertain the strength properties of a specific masonry assembly.

The *compressive strength specified for masonry at the age of 28 days (f'_m)* is a type of ultimate compressive strength determined through masonry prism testing. It varies with type of mortar and strength and type of masonry unit. In the absence of a prism testing for a specific project, conservative design values for f'_m are specified in the building code. If special inspections of the masonry assembly by an engineer or special inspector are not conducted during construction, design values for f'_m must be reduced by one-half.

In design of masonry columns and walls, the *effective cross-sectional area (A_e)* is used in computing the load that the masonry assembly can safely support. It is based on the bedded area of the hollow masonry unit or the gross area of solid masonry *plus* any grouted area. The area is computed using specified (actual) dimensions of the masonry wall or column.

For solid masonry units, the effective cross-sectional area (A_e) is based on the gross area of the horizontal face of the unit *plus* any grouted area. The effective cross-sectional area (A_e) of hollow masonry units is based on the bedded area *plus* any grouted area. Bed joints of hollow masonry units are filled with mortar for a distance in from the face of the masonry unit not less than the thickness of the face shell. The webs of hollow masonry units do not receive mortar. Therefore, the effective cross-sectional area (A_e) of ungrouted hollow masonry units is based only on the face shell thickness and length of the faces. Minimum face shell thicknesses for hollow CMU are provided in Table 29.2. For cavity walls, the effective cross-sectional area (A_e) is based on the loaded wythe(s) only *plus* any grouted area of the cavity and hollow units.

Masonry joints are sometimes raked when the mortar is finished. A raked joint is a decorative joint where mortar is scraped from the joint so that it is recessed within the masonry column or wall face. Where bed joints are raked, the effective cross-sectional area must be reduced accordingly.

■ EXAMPLE 29.2

A wall is constructed of two wythes of solid modular (metric) clay brick. The unreinforced solid masonry wall has a 10-mm collar joint between the brick wythes.
(a) Determine the effective area per linear meter of wall.
(b) Determine the effective area per linear meter of wall with 6-mm-deep raked joints on the front and back faces of the wall.

Solution (a) Since the wall and masonry units are solid, the effective area of the wall is based on the gross area of the horizontal face of the masonry unit. From Table 29.2, the modular clay brick has a specified width of 90 mm. The specified width of two wythes of brick and the 10-mm collar joint is found by

$$\text{specified width} = 90 \text{ mm} + 90 \text{ mm} + 10 \text{ mm} = 190 \text{ mm} = 0.19 \text{ m}$$

The effective area (A_e) per linear meter of wall is found by

$$A_e = 0.19 \text{ m} \cdot 1 \text{ m} = 0.19 \text{ m}^2 \text{ per linear meter of wall}$$

(b) The effective area of the wall is based on the gross area of masonry less the area of the raked joints. From Table 29.2, the modular clay brick has a specified width of 90 mm. The specified width of two wythes of brick and the 10-mm collar joint less the two 6-mm raked joints is found by

$$\text{specified width} = 90 \text{ mm} + 90 \text{ mm} + 10 \text{ mm} - 6 \text{ mm} - 6 \text{ mm}$$
$$= 178 \text{ mm} = 0.178 \text{ m}$$

The effective area (A_e) per linear meter of wall is found by

$$A_e = 0.178 \text{ m} \cdot 1 \text{ m} = 0.178 \text{ m}^2 \text{ per linear meter of wall}$$

■ EXAMPLE 29.3

A masonry wall is constructed of one wythe of 8 in hollow concrete masonry units (CMUs).

(a) Determine the effective area per linear foot of wall if the wall is not grouted.

(b) Determine the effective area per linear foot of wall if the wall is fully grouted.

Solution (a) The effective cross-sectional area (A_e) of hollow masonry units is based on the bedded area: the face shell thickness and length of the faces. From Table 29.4, the minimum face shell thickness of an 8-in hollow CMU is $1\frac{1}{4}$ in. There are two faces (front and back) on which the mortar is applied. The effective area of the wall (A_e) per linear foot of wall is found by

$$A_e = 2 \text{ shell faces} \cdot 1\tfrac{1}{4} \text{ in face shell thickness} \cdot 12 \text{ in/ft} = 30.0 \text{ in}^2/\text{ft}$$

(b) Although the wall is constructed of hollow CMU, it is fully grouted. Its effective area is based on the gross area of masonry. In Table 29.4, the width of an 8-in CMU is $7\frac{5}{8}$ in. The effective area (A_e) per linear foot of wall is found by

$$7\tfrac{5}{8} \text{ in} \cdot 12 \text{ in/ft} = 91.5 \text{ in}^2/\text{ft}$$

Effective area dictates the load that a specific masonry assembly can support; a specific masonry assembly with a greater effective area can support a larger load. As shown in Example 29.3, a raked joint will reduce the effective area of a masonry assembly. Ultimately, it reduces the load that a wall or column can support. On the other hand, grouting a wall constructed of hollow masonry will increase effective area, as shown in Example 29.3, and will increase the load that the wall or column can support.

EXERCISES

29.1 Determine the actual height of a masonry assembly constructed of 48 courses of 6 in × 4 in × 16 in hollow structural clay brick.

29.2 Determine the actual height of a masonry assembly constructed of 26 courses of 6 in × 8 in × 12 in hollow structural clay brick.

29.3 Determine the actual height of a masonry assembly constructed of 26 courses of 10-in CMUs.

29.4 A masonry wall is constructed of one wythe of modular clay brick and one wythe of 8-in CMUs bonded with a $\frac{1}{2}$-in collar joint. Determine the actual thickness of the wall.

29.5 A hollow multiwythe masonry wall is constructed of one wythe of modular clay brick and one wythe of 6-in CMUs that are separated with a hollow cavity. Determine the minimum actual thickness of the wall.

29.6 A masonry wall is constructed of one wythe of king (metric) brick.
    (a) Determine the actual wall thickness.
    (b) Determine the effective area per linear meter of wall.
    (c) Determine the effective area per linear meter of wall with 5-mm-deep raked joints on the front and back faces of the wall.

29.7 A masonry wall is constructed of one wythe of 6-in CMUs.

    (a) Determine the actual thickness of the wall.

    (b) Determine the effective area per linear foot of wall if the CMUs are not grouted.

    (c) Determine the effective area per linear foot of wall if the CMUs are fully grouted.

29.8 A masonry wall is constructed of one wythe of 12-in CMUs.

    (a) Determine the actual thickness of the wall.

    (b) Determine the effective area of the masonry assembly per linear foot of wall if the CMUs are not grouted.

    (c) Determine the effective area of the masonry assembly per linear foot of wall if the CMUs are fully grouted.

Chapter 30
DESIGN OF SIMPLE MASONRY WALLS AND COLUMNS

Structural masonry walls and columns generally support gravity loads and thus must be designed to resist compressive forces. Additionally, they must be designed to withstand bending stresses from lateral and gravity loads. Principles of working stress design related to stresses from simple bearing and wind loads in columns and walls are introduced in this chapter. The intent is to acquaint the technician with basic masonry design principles so that the technician can gain an awareness of structural engineering and design. The reader is advised that this presentation does not provide full coverage of design principles involved with masonry design.

30.1 Basic Masonry Wall and Column Design Procedure

In working stress design of simple axially loaded masonry walls and columns, an *allowable centroidal axial load* (P or P_a) is determined using formulas prescribed in the building code. For masonry walls and columns, this allowable load is computed based on the height, section profile characteristics, and test strength of the masonry assembly. In the design of reinforced masonry columns, vertical reinforcement is also considered.

The allowable load that can be supported by a masonry wall or column is based on the *specified compressive strength of masonry at the age of 28 days* (f'_m), a conservative ultimate strength that varies with type of mortar and strength and type of masonry unit. As discussed in Chapter 29, f'_m is determined through masonry prism testing. Conservative values for f'_m are specified in the building code and are provided in Table 30.1. If special inspections of the masonry construction are not conducted by an engineer or special inspector, design values for f'_m must be reduced by one-half.

Columns and walls are relatively slender in comparison to height, so they have a tendency to fail by buckling. In Chapter 14, the radius of gyration (r) was defined

Table 30.1 SPECIFIED COMPRESSIVE STRENGTH OF MASONRY (f'_m) BASED ON SPECIFYING THE COMPRESSIVE STRENGTH OF MASONRY UNITS

Compressive strength of clay masonry units		Specified compressive strength of clay masonry, f'_m			
		Type M or S mortar[a]		Type N mortar[a]	
lb/in^2	MPa	lb/in^2	MPa	lb/in^2	MPa
Clay masonry[b,c]					
14 000 or more	96.46 or more	5300	36.52	4400	30.32
12 000	82.68	4700	32.38	3800	26.18
10 000	68.90	4000	27.56	3300	22.74
8 000	55.12	3360	23.08	2700	18.60
6 000	41.34	2700	18.60	2200	15.16
4 000	27.56	2000	13.78	1600	11.02
Concrete masonry[c,d]					
4800 or more	33.07	3000	20.67	2800	19.29
3750	25.84	2500	17.23	2350	16.19
2800	19.29	2000	13.78	1850	12.75
1900	13.09	1500	10.34	1350	9.30
1250	8.61	1000	6.89	950	6.55

Source: 1997 Uniform Building Code, Table 21-D.

[a]Mortar for unit masonry, production specification, as specified in Table 21-B of the 1994 Uniform Building Code. These values apply to portland cement-lime mortars without air-entraining materials.

[b]Compressive strength of solid clay masonry units is based on gross area, compressive strength of hollow clay masonry units is based on minimum net area; values may be interpolated.

[c]Compressive strength is for an assumed assemblage. The compressive strength of masonry, f'_m, is based on gross area strength when using solid units or solid grouted masonry and net area strength when using ungrouted hollow units.

[d]Values for compressive strength may be interpolated. In grouted concrete masonry, the compressive strength of grout should be equal to or greater than the compressive strength of the concrete masonry units.

as a section profile property that is a measure of the section capacity to resist buckling independent of the material's elastic behavior. Computations for masonry columns and walls are based on the radius of gyration of the specified masonry assembly. For hollow and grouted hollow masonry assemblies, the radius of gyration is given in Tables 30.2 to 30.4. For solid masonry assemblies that have a rectangular horizontal face, the radius of gyration can be approximated based on the specified wall width (w): $r = 0.289w$. This formula was introduced in Section 14.9.

Following is the working stress design method that is used to determine the allowable average axial compressive stress for centroidally applied axial loads and the design (maximum allowable) axial load that can be supported at the top of the column or wall.

30.1   Basic Masonry Wall and Column Design Procedure   **529**

Table 30.2 RADIUS OF GYRATION OF HOLLOW CONCRETE MASONRY UNITS

	Nominal width of wall (in)				
Grout spacing (in)	4	6	8	10	12
Solid grouted	1.04	1.62	2.19	2.77	3.34
16	1.16	1.79	2.43	3.04	3.67
24	1.21	1.87	2.53	3.17	3.82
32	1.24	1.91	2.59	3.25	3.91
40	1.26	1.94	2.63	3.30	3.97
48	1.27	1.96	2.66	3.33	4.02
56	1.28	1.98	2.68	3.36	4.05
64	1.29	1.99	2.70	3.38	4.08
72	1.30	2.00	2.71	3.40	4.10
No grout	1.35	2.08	2.84	3.55	4.29

Source: 1997 Uniform Building Code, Table 21-H-1.

Table 30.3 RADIUS OF GYRATION OF HOLLOW CLAY MASONRY UNITS
WITH A LENGTH OF 16 IN

	Nominal width of wall (in)				
Grout spacing (in)	4	6	8	10	12
Solid grouted	1.04	1.64	2.23	2.81	3.39
16	1.16	1.78	2.42	3.03	3.65
24	1.20	1.85	2.51	3.13	3.77
32	1.23	1.88	2.56	3.19	3.85
40	1.25	1.91	2.59	3.23	3.90
48	1.26	1.93	2.61	3.26	3.93
56	1.27	1.94	2.63	3.28	3.95
64	1.27	1.95	2.64	3.30	3.97
72	1.28	1.95	2.65	3.31	3.99
No grout	1.32	2.02	2.75	3.42	4.13

Source: 1997 Uniform Building Code, Table 21-H-2.

1. Determine the slenderness ratio (h'/r) of the column or wall.
2. Determine the maximum allowable centroidal axial load (P' or P_a) that can be carried by the wall or column by the appropriate approach.
   (a) For an unreinforced wall or column or a reinforced wall with a slenderness ratio (h'/r) less than or equal to 99 ($h'/r \le 99$); first determine the allowable average axial compressive stress for centroidally applied axial loads (F_a) by using the following formula:

$$F_a = 0.25f'_m \left[1 - \left(\frac{h'}{140r} \right)^2 \right]$$

Table 30.4 RADIUS OF GYRATION OF HOLLOW CLAY MASONRY UNITS
WITH A LENGTH OF 12 IN

Grout spacing (in)	Nominal width of wall (in)				
	4	6	8	10	12
Solid grouted	1.06	1.65	2.24	2.82	3.41
12	1.15	1.77	2.40	3.00	3.61
18	1.19	1.82	2.47	3.08	3.71
24	1.21	1.85	2.51	3.12	3.76
30	1.23	1.87	2.53	3.15	3.80
36	1.24	1.88	2.55	3.17	3.82
42	1.24	1.89	2.56	3.19	3.84
48	1.25	1.90	2.57	3.20	3.85
54	1.25	1.90	2.58	3.21	3.86
60	1.26	1.91	2.59	3.21	3.87
66	1.26	1.91	2.59	3.22	3.88
72	1.26	1.91	2.59	3.22	3.88
No grout	1.29	1.95	2.65	3.28	3.95

Source: 1997 Uniform Building Code, Table 21-H-3.

Determine the allowable centroidal axial load (P) that can be carried by the wall or column by using the following formula:

$$P' = F_a A_e$$

(b) For an unreinforced wall or column or a reinforced wall with a slenderness ratio (h'/r) greater than 99 ($h'/r > 99$); first determine the allowable average axial compressive stress for centroidally applied axial loads (F_a) by using the following formula:

$$F_a = 0.25f'_m\left[1 - \left(\frac{70r}{h'}\right)^2\right]$$

Determine the allowable centroidal axial load (P) that can be carried by the wall or column by using the following formula:

$$P' = F_a A_e$$

(c) For a reinforced column with a slenderness ratio (h'/r) less than or equal to 99 ($h'/r \leq 99$); determine the allowable centroidal axial load (P_a) by using the following formula:

$$P_a = (0.25f'_m A_e + 0.65A_s F_{sc})\left[1 - \left(\frac{h'}{140r}\right)^2\right]$$

(d) For a reinforced column with a slenderness ratio (h'/r) greater than 99 ($h'/r > 99$); determine the allowable centroidal axial load (P_a) by using the following formula:

30.1 Basic Masonry Wall and Column Design Procedure

531

$$P_a = (0.25f'_m A_e + 0.65A_s F_{sc})\left(\frac{70r}{h'}\right)^2$$

3. Determine the design (maximum allowable) axial load that can be supported at the top of the column or wall.

$$P = P' - \text{column or wall weight}$$

or

$$P = P_a - \text{column weight}$$

4. Investigate other factors related to the design of masonry walls and columns.

In the formulas above, we have:

A_e effective (net) cross-sectional area of the column or wall, in in^2 [The effective cross-sectional area is based on the bedded area of the hollow masonry unit (usually the face shell thickness and length) *plus* any grouted area or the gross area of solid units (Figure 30.1). The area is computed using specified dimensions of the masonry wall or column. Where bed joints are raked, the effective cross-sectional area must be reduced accordingly. For cavity walls, the effective area is based on the loaded wythe(s) only.]

A_s effective cross-sectional area of steel reinforcement in the column, in in^2; dependent on number and gauge of rebars specified (see Table B.7).

f'_m compressive strength specified for masonry at 28 days as determined through prism testing or as provided by the building code (see Table 30.1) (If special inspections of masonry construction are not conducted by an engineer or special inspector, design values for f'_m must be reduced by one-half.)

F_a allowable average axial compressive stress for centroidally applied axial loads (F_a), in lb/in^2

F_{sc} allowable compressive strength of column reinforcement, in lb/in^2 [$F_{sc} = 0.4f_y$ but cannot exceed 24 000 lb/in^2, where f_y is the yield stress of the deformed rebars. Therefore, F_{sc} is 16 000 lb/in^2 for grade 40 steel bars and 24 000 lb/in^2 for steel bars of grade 60 or greater.]

h' effective (unbraced) height of the wall or column, in inches, if the wall or column is supported at the top and bottom normal to the axis considered (If the top of the member is not supported normal to the axis being considered, the effective height is twice the actual height of the column or wall.)

P design (maximum allowable) axial load that can be carried at the top of a wall or column; includes the weight of the wall or column

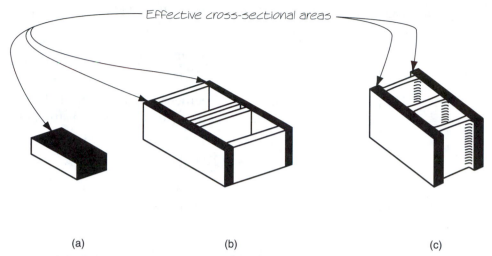

Effective cross-sectional areas

(a)   (b)   (c)

FIGURE 30.1   The effective cross-sectional area (A_e) is based on the bedded area of the hollow masonry unit, usually the shell thickness and length, *plus* any grouted area or the gross area of solid units: (a) solid brick; (b) hollow clay brick; (c) concrete masonry unit. The effective cross-sectional areas of the horizontal and vertical faces of common masonry units are highlighted in black.

P' allowable centroidal axial load that can be carried by an unreinforced wall or column; neglects the weight of the wall or column

P_a allowable centroidal axial load that can be carried by a reinforced masonry column; neglects the weight of the column

r radius of gyration of the specified masonry width (w) (For hollow masonry assemblies, see Tables 30.2 and 30.3; for solid masonry assemblies, it can be approximated by $r = 0.289w$, as introduced in Chapter 14.)

■ EXAMPLE 30.1

A 12′-0″-high wall is constructed of two wythes of solid modular clay brick and type N mortar ($f'_m = 2200$ lb/in^2). The unreinforced solid masonry wall is $7\frac{1}{2}$ in thick, including the $\frac{1}{2}$ in collar joint between the brick wythes. The wall carries a roof system at its top which supports the wall normal to its longitudinal axis. Special inspections of masonry will not be conducted during construction.

(a) Determine the allowable average axial compressive stress for centroidally applied axial loads.

(b) Determine the design axial load (P) that can be carried per linear foot of wall.

Solution   Because the masonry units and wall are solid, the radius of gyration (r) can be approximated by the following formula:

$$r = 0.289w$$
$$= 0.289(7\tfrac{1}{2} \text{ in}) = 2.17 \text{ in}$$

The effective (unbraced) height of the wall is 144 in. The slenderness ratio is found by

$$\frac{h'}{r} = \frac{144 \text{ in}}{2.17 \text{ in}} = 66.4$$

The compressive strength (f'_m) specified must be reduced to half of the tabular value because special inspections of masonry construction will not conducted.

$$\text{revised } f'_m = \frac{(2200 \text{ lb/in}^2)}{2} = 1100 \text{ lb/in}^2$$

With a slenderness ratio (h'/r) less than or equal to 99, the following formula is used:

$$F_a = 0.25 f'_m \left[1 - \left(\frac{h'}{140r} \right)^2 \right]$$
$$= 0.25(1100 \text{ lb/in}^2) \left[1 - \left(\frac{144 \text{ in}}{140 \cdot 2.17 \text{ in}} \right)^2 \right]$$
$$= 213 \text{ lb/in}^2$$

(b) The effective area (A_e) per linear foot of wall is $7\frac{1}{2}$ in $\cdot$ 12 in/ft $= 90.0$ in^2/ft. The allowable centroidal axial load is found by

$$P' = F_a A_e$$
$$= 213 \text{ lb/in}^2 \cdot 90.0 \text{ in}^2/\text{ft} = 19\,170 \text{ lb/ft of wall}$$

The weight of a solid brick wall is 80 lb/ft^2 (from Table 15.1). The height of the wall is 12 ft. The weight of the wall per foot is found by

$$80 \text{ lb/ft}^2 \cdot 12 \text{ ft/ft of wall} = 960 \text{ lb/ft of wall}$$

The design (maximum allowable) axial load that can be carried per foot of wall at the top of the wall is found by

$$P = P' - \text{wall weight}$$
$$= 19\,170 \text{ lb/ft of wall} - 960 \text{ lb/ft of wall}$$
$$= 18\,210 \text{ lb/ft of wall}$$

■ **EXAMPLE 30.2**

Much as in Example 30.1, a 12′-0″-high wall is constructed of two wythes of solid modular clay brick and type N mortar ($f'_m = 2200$ lb/in^2). The unreinforced solid masonry wall is $7\frac{1}{2}$ in thick, including the $\frac{1}{2}$ in collar joint between the brick wythes. The wall carries a load, but unlike the wall in Example 30.1, this wall is not supported laterally at the top. Special inspections of masonry will not be conducted during construction.

(a) Determine the allowable average axial compressive stress for centroidally applied axial loads.

(b) Determine the design axial load (P) that can be carried per linear foot of wall.

Solution (a) Because the masonry units and wall are solid, the radius of gyration (r) can be approximated by the following formula:

$$r = 0.289w$$
$$= 0.289(7\tfrac{1}{2} \text{ in}) = 2.17 \text{ in}$$

Since the wall is not supported at the top normal to the longitudinal axis, the effective height is twice the actual wall height. The effective height (h') of the wall is $2 \cdot 144$ in $= 288$ in. The slenderness ratio is found by

$$\frac{h'}{r} = \frac{288 \text{ in}}{2.17 \text{ in}} = 133$$

The compressive strength (f'_m) specified must be reduced to half of the tabular value because special inspections of masonry construction will not be conducted.

$$\text{revised } f'_m = \frac{(2200 \text{ lb/in}^2)}{2} = 1100 \text{ lb/in}^2$$

With a slenderness ratio (h'/r) greater than 99, the following formula is used:

$$F_a = 0.25f'_m \left[1 - \left(\frac{70r}{h'} \right)^2 \right]$$
$$= 0.25(1100 \text{ lb/in}^2) \left[1 - \left(\frac{70 \cdot 2.17 \text{ in}}{288 \text{ in}} \right)^2 \right]$$
$$= 199 \text{ lb/in}^2$$

(b) The effective area (A_e) per linear foot of wall is $7\tfrac{1}{2}$ in $\cdot$ 12 in/ft $= 90.0$ in²/ft. The allowable centroidal axial load is found by

$$P' = F_a A_e$$
$$= 199 \text{ lb/in}^2 \cdot 90.0 \text{ in}^2/\text{ft} = 17\,910 \text{ lb/ft of wall}$$

The weight of a solid brick wall is 80 lb/ft² (from Table 15.1). The height of the wall is 12 ft. The weight of the wall per foot is found by

$$80 \text{ lb/ft}^2 \cdot 12 \text{ ft/ft of wall} = 960 \text{ lb/ft of wall}$$

The design (maximum allowable) axial load that can be carried per foot of wall at the top of the wall is found by

$$P = P' - \text{wall weight}$$
$$= 17\,910 \text{ lb/ft of wall} - 960 \text{ lb/ft of wall}$$
$$= 16\,950 \text{ lb/ft of wall}$$

30.1 Basic Masonry Wall and Column Design Procedure

With the exception of how they are supported at their tops, the solid brick walls in Examples 30.1 and 30.2 are alike. Lack of lateral support at the top of the wall in Example 30.2 limits the allowable axial load that it can carry to only about 93% of the wall in Example 30.1, which is supported laterally at the top of the wall. An unsupported wall cannot carry as great a load as a similar wall that is supported at the top normal to its longitudinal axis.

■ **EXAMPLE 30.3**

A 12'-0"-high unreinforced hollow-unit masonry wall is constructed of one wythe of 8-in hollow concrete masonry units (CMUs) and type S mortar (f'_m = 2000 lb/in²). The wall is $7\frac{5}{8}$ in thick and is solid grouted. The wall supports a roof system, so it is supported laterally at the top. Special inspections of masonry will be conducted during construction.

(a) Determine the allowable average axial compressive stress per foot of masonry.

(b) Determine the design axial load (P) that can be carried per linear foot of wall.

Solution From Table 30.2, the radius of gyration (r) of an 8-in solid-grouted CMU wall is found to be 2.19 in. The effective (unbraced) height of the wall is 144 in. The slenderness ratio is found by

$$\frac{h'}{r} = \frac{144 \text{ in}}{2.19} = 65.8$$

With a slenderness ratio (h'/r) less than or equal to 99, the following formula is used:

$$F_a = 0.25f'_m\left[1 - \left(\frac{h'}{140r}\right)^2\right]$$
$$= 0.25(2000 \text{ lb/in}^2)\left[1 - \left(\frac{144 \text{ in}}{140 \cdot 2.19 \text{ in}}\right)^2\right]$$
$$= 390 \text{ lb/in}^2$$

(b) Although the wall is constructed of hollow CMUs, it is fully grouted; the effective area (A_e) per linear foot of wall is $7\frac{5}{8}$ in $\cdot$ 12 in/ft = 91.5 in²/ft. The allowable centroidal axial load is found by

$$P' = F_a A_e$$
$$= 390 \text{ lb/in}^2 \cdot 91.5 \text{ in}^2/\text{ft} = 35\ 685 \text{ lb/ft of wall}$$

The weight of an 8-in-thick solid-grouted concrete masonry wall is 80 lb/ft² (from Table 15.1). The height of the wall is 12.0 ft. The weight of the wall per foot is found by

$$80 \text{ lb/ft}^2 \cdot 12 \text{ ft/ft of wall} = 960 \text{ lb/ft of wall}$$

The design (maximum allowable) axial load that can be carried per foot of wall at the top of the wall is found by

$$P = P' - \text{wall weight}$$
$$= 35\ 685 \text{ lb/ft of wall} - 960 \text{ lb/ft of wall}$$
$$= 34\ 725 \text{ lb/ft of wall}$$

■ EXAMPLE 30.4

Much as in Example 30.3, a 12'-0"-high unreinforced hollow-unit masonry wall is constructed of one wythe of 8-in hollow concrete masonry units (CMUs) and type S mortar ($f'_m = 2000$ lb/in^2). The wall is $7\frac{5}{8}$ in thick. Unlike the wall in Example 30.3, the wall is not grouted. The wall supports a roof system, so it is supported laterally at the top. Special inspections of masonry will be conducted during construction.

(a) Determine the allowable average axial compressive stress per foot of masonry wall with no grout.

(b) Determine the design axial load (P) that can be carried per linear foot of wall.

Solution From Table 30.2, the radius of gyration (r) of an 8-in hollow CMU wall with no grout is found to be 2.84 in. The effective (unbraced) height of the wall is 144 in. The slenderness ratio is found by

$$\frac{h'}{r} = \frac{144 \text{ in}}{2.84} = 50.7$$

With a slenderness ratio (h'/r) less than or equal to 99, the following formula is used:

$$F_a = 0.25 f'_m \left[1 - \left(\frac{h'}{140r} \right)^2 \right]$$
$$= 0.25(2000 \text{ lb/in}^2)\left[1 - \left(\frac{144 \text{ in}}{140 \cdot 2.84 \text{ in}} \right)^2 \right]$$
$$= 434 \text{ lb/in}^2$$

(b) As discussed in Chapter 29, bed joints of hollow masonry units are filled with mortar for a distance in from the face of the masonry unit not less than the thickness of the face shell. The webs of hollow masonry units do not receive mortar. Therefore, the effective cross-sectional area (A_e) of hollow masonry units is based only on the face shell thickness and length of the faces. From Table 29.4, the minimum face shell thickness of an 8-in hollow CMUs is $1\frac{1}{4}$ in. There are two faces (front and back) on which the mortar is applied. The effective area (A_e) per linear foot of wall is found by

$$A_e = 2 \text{ shell faces} \cdot 1\frac{1}{4} \text{ in face shell thickness} \cdot 12 \text{ in/ft} = 30.0 \text{ in}^2/\text{ft}$$

The allowable centroidal axial load is found by

$$P' = F_a A_e$$
$$= 434 \text{ lb/in}^2 \cdot 30.0 \text{ in}^2/\text{ft} = 13\,020 \text{ lb/ft of wall}$$

The weight of an 8-in-thick hollow concrete masonry wall is 40 lb/ft^2 (from Table 15.1). The height of the wall is 12 ft. The weight of the wall per foot is found by

$$40 \text{ lb/ft}^2 \cdot 12 \text{ ft/ft of wall} = 480 \text{ lb/ft of wall}$$

The design (maximum allowable) axial load that can be carried per foot of wall at the top of the wall is found by

$$P = P' - \text{wall weight}$$
$$= 13\,020 \text{ lb/ft of wall} - 480 \text{ lb/ft of wall}$$
$$= 12\,540 \text{ lb/ft of wall}$$

When comparing the results of Examples 30.3 and 30.4, it is evident that a grouted wall can carry a greater axial load than a similar wall that is not grouted. In addition to improving the masonry bond, grout also improves wall strength.

■ EXAMPLE 30.5

A 12'-0"-high reinforced masonry column is constructed of two interlocking wythes of 6 in × 4 in × 12 in hollow structural brick units and type N mortar ($f'_m = 2200$ lb/in^2). The nominal 12 in × 18 in column has actual dimensions of $11\frac{1}{2}$ in × $17\frac{1}{2}$ in and is solid grouted. Eight No. 8 grade 60 steel rebars are used for vertical reinforcement. The column supports a floor system, so it is supported laterally at the top. Special inspections of masonry will be conducted during construction.
(a) Is the column height specified acceptable?
(b) Is the column reinforcement specified acceptable?
(c) Determine the allowable average axial compressive stress for a centroidally applied axial load.
(d) Determine the design axial load (P) that can be carried by the column.

Solution (a) From Chapter 29, the column slenderness ratio of height to least lateral dimension must not exceed 25.

$$\text{slenderness ratio} = \frac{12 \text{ ft} \cdot 12 \text{ in/ft}}{11\frac{1}{2} \text{ in}} = 12.5$$

The column height specified is acceptable.
(b) From Chapter 29, columns must be reinforced with a minimum of four reinforcing bars. The area of reinforcement (A_s) must be at least 0.0025 but not more than 0.04 times the net cross-sectional area of the column.

$$0.0025 \text{ in } (11\tfrac{1}{2} \text{ in} \cdot 17\tfrac{1}{2} \text{ in}) = 0.503 \text{ in}^2$$
$$0.04 \ (11\tfrac{1}{2} \text{ in} \cdot 17\tfrac{1}{2} \text{ in}) = 8.05 \text{ in}^2$$

From Table B.7, the cross-sectional area of a No. 8 rebar is 0.79 in². The effective cross-sectional area of steel reinforcement (A_s) of eight rebars is found by

$$A_s = 0.79 \text{ in}^2 \cdot 8 \text{ bars} = 6.08 \text{ in}^2$$

The column reinforcement specified is acceptable.

(c) From Table 30.3 the radius of gyration (r) of a 12-in hollow structural brick is found to be 3.41 in; the least column section dimension is used. The effective (unbraced) height of the column is 144 in. The slenderness ratio is found by

$$\frac{h'}{r} = \frac{144 \text{ in}}{3.41 \text{ in}} = 42.2$$

With a slenderness ratio (h'/r) less than or equal to 99 and the column reinforced, formula c is used. From above, the effective cross-sectional area of steel reinforcement (A_s) of eight rebars is 6.08 in². Column dimensions are $11\tfrac{1}{2}$ in by $17\tfrac{1}{2}$ in. The effective cross-sectional area of the column (A_e) is found by

$$A_s = 8 \text{ rebars} \cdot 0.79 \text{ in}^2/\text{rebar} = 6.32 \text{ in}^2$$
$$A_e = A - A_s = (11\tfrac{1}{2} \text{ in} \cdot 17\tfrac{1}{2} \text{ in}) - 6.32 \text{ in}^2 = 194.9 \text{ in}^2$$

The allowable centroidal axial load (P_a) is found using formula c:

$$P_a = (0.25 f'_m A_e + 0.65 A_s F_{sc}) \left[1 - \left(\frac{h'}{140r} \right)^2 \right]$$
$$= [0.25(2200 \text{ lb/in}^2)(194.9 \text{ in}^2) + 0.65(6.32 \text{ in}^2)(24\ 000 \text{ lb/in}^2)]$$
$$\left[1 - \left(\frac{144 \text{ in}}{140 \cdot 3.41 \text{ in}} \right)^2 \right]$$
$$= 187\ 060 \text{ lb}$$

(d) The weight of a 12-in-thick solid grouted hollow structural brick assembly is 120 lb/ft² (from Table 15.1). The column is 18 in (1.5 ft) long with a height of 12.0 ft. The overall weight of the column is found by

$$120 \text{ lb/ft}^2 \cdot 1.5 \text{ ft} \cdot 12 \text{ ft/ft of wall} = 2160 \text{ lb}$$

The design (maximum allowable) axial load that can be carried per foot of wall at the top of the wall is found by

$$P = P' - \text{column weight}$$
$$= 35\ 685 \text{ lb} - 2160 \text{ lb}$$
$$= 32\ 565 \text{ lb}$$

In theory, masonry wall and column assemblies develop only compression stresses and are designed to support axial compressive loads. However, in addition to compressive stresses, masonry walls and columns must be designed to withstand other stresses, such as those associated with flexure (bending) and bearing loads:

Flexure (bending) of unreinforced walls and columns   However, columns that are loaded by an eccentric axial load or by a large lateral load may fail by flexure. Wind and soil loads produce lateral flexure stresses in masonry walls and columns. In working stress design, lateral flexure (tension) stresses should not exceed allowable stresses provided in the building code that are dependent on mortar and unit type. For example, for masonry assemblies constructed of type M or S cement-lime and mortar cement, *allowable flexure stresses* (f_b) are 40 lb/in^2 for solid and 25 lb/in^2 for hollow masonry units where flexure is normal to the bed (horizontal) mortar joints such as wind and soil loads, and 80 lb/in^2 for solid and 50 lb/in^2 for hollow masonry units where flexure is normal to the head (vertical) joints. These stresses do not apply to masonry beams.

In the case of walls and columns, the flexure formula introduced in Chapter 16 applies:

$$F_b = \frac{M}{S}$$

where F_b is the flexure (tension) stress, lb/in^2, M is the bending moment, lb-in, and S is the section modulus, in^3.

■ EXAMPLE 30.6

A 12'-0"-high wall is constructed of two wythes of modular clay brick and type S mortar ($f'_m = 4000$ lb/in^2). The unreinforced solid masonry wall is $7\frac{1}{2}$ in thick, including the $\frac{1}{2}$-in collar joint between the brick wythes. The wall carries a roof system at its top that applies an axial gravity load of 2000 lb per linear foot of wall. The roof system laterally supports the top of the wall and a foundation system at grade supports the wall at the base. The design wind load is 25 lb/ft^2 (approximately 100 mile/hr wind). Special inspections of masonry will be conducted during construction. Investigate the lateral flexure stresses in the wall.

Solution   The wall will be treated as a beam spanning vertically from the roof to the foundation system as shown in Figure 30.2. Under the wind load, the wall will deflect laterally. The wind load is uniformly distributed over the entire wall, but to simplify the computations, the wall will be treated as having a length of 1 linear foot. Thus the wall is considered as a simple beam that is

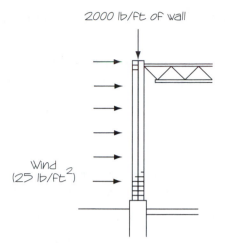

FIGURE 30.2 Wall in Example 30.6. The wall is treated as a beam spanning vertically from the roof to the foundation system. Under the wind load, the wall will deflect laterally.

12 ft long with a breadth of 1 ft and a depth of $7\frac{1}{2}$ in. For this 1-ft length of wall, the wind load exerts a force of 25 lb per foot of wall height. Using the appropriate formula from Table A.5, maximum bending moment (M_{max}) is found by

$$M_{max} = \frac{WL}{8} = \frac{(25 \text{ lb/ft} \cdot 12 \text{ ft})(12 \text{ ft} \cdot 12 \text{ in/ft})}{8} = 5400 \text{ lb-in}$$

From Table A.6 and as described in Chapter 14, the section modulus (S) of 1 linear foot (12 in) of the $7\frac{1}{2}$-in wall is found by

$$S = \frac{bd^2}{6} = \frac{12 \text{ in}(7\frac{1}{2} \text{ in})^2}{6}$$
$$= 112.5 \text{ in}^3$$

Bending (flexure) of the wall from the wind load is normal to the bed (horizontal) mortar joints, so the allowable flexure stress (F_b) is 40 lb/in^2. Maximum flexure (tension) stress is found by

$$F_b = \frac{M}{S} = \frac{5400 \text{ lb-in}}{112.5 \text{ in}^3}$$
$$= 48.0 \text{ lb/in}^2$$

Maximum bending moment and thus maximum flexure (tension) stress occurs at half the wall height. The weight of the wall and the roof load apply a compressive force that creates a compressive stress which counters the flexure (tension) stress. The weight of a nominal 8-in solid brick wall is 80 lb/ft^2 (from Table 15.1). The compressive load at half the wall height is found by

$$80 \text{ lb/ft}^2 \cdot 6 \text{ ft/ft of wall} = 480 \text{ lb/ft of wall}$$
$$2000 \text{ lb} + 480 \text{ lb} = 2480 \text{ lb}$$

The compressive stress (F_c) that counters the maximum flexure (tension) stress at half the wall height is found by

$$F_c = \frac{P}{A} = \frac{2480 \text{ lb}}{12 \text{ in} \cdot 7\frac{1}{2} \text{ in}}$$
$$= 27.6 \text{ lb/in}^2$$

Since compressive stress (F_c) counters the flexure (tension) stress, the actual maximum flexure (tension) stress is found by

$$F_b - F_c = 48.0 \text{ lb/in}^2 - 27.6 \text{ lb/in}^2 = 20.4 \text{ lb/in}^2$$

The allowable flexure stress (f_b) of 40 lb/in^2 is not exceeded.

In Example 30.6 it appears initially that the flexure stress (F_b) exceeds the allowable flexure stress (f_b). However, compressive stress (F_c) counters flexure (tension) stress, thereby reducing it to acceptable levels. The 2000-lb/ft roof load played a major role in resisting flexure. Without the roof load, the 480-lb weight of the wall alone would have been incapable of countering flexure:

$$F_c = \frac{P}{A} = \frac{480 \text{ lb}}{12 \text{ in} \cdot 7\frac{1}{2} \text{ in}}$$
$$= 5.3 \text{ lb/in}^2$$
$$F_b - F_c = 48.0 \text{ lb/in}^2 - 5.3 \text{ lb/in}^2 = 42.7 \text{ lb/in}^2$$

The allowable flexure stress (f_b) of 40 lb/in^2 would have been exceeded. Axial loads applied to walls and columns work to counter flexure (tension) stress.

Flexure (bending) of reinforced walls, columns, and beams   The working stress design of reinforced flexure members is complex. It follows a design method similar to the approach that was introduced in Chapter 27 for rectangular reinforced concrete beams. Design of rectangular masonry beams and other flexure members is not included because it is beyond the intended scope of this book.

Bearing   The allowable bearing stress (F_{br}) on a masonry element is related to a fraction of f'_m. When a member or bearing plate rests on the full area of masonry, the bearing stress will be

$$F_{br} = 0.26f'_m$$

In some cases when a member or bearing plate bears on one-third or less of the area of masonry, the allowable bearing stress may be increased to

$$F_{br} = 0.38f'_m$$

■ EXAMPLE 30.7

In Example 19.4 a 6 in × 11 in steel bearing plate was specified to transfer a 15 500-lb beam end load to a brick wall ($f'_m = 2700$ lb/in^2). Assuming that the bearing plate rests on more than one-third of the area of masonry and that special inspections of masonry will be conducted during construction, is the bearing stress acceptable?

Solution Both the allowable bearing stress (F_{br}) and the actual bearing stress ($\sigma_{bearing}$) are found by

$$F_{br} = 0.38f'_m = 0.38(2700 \text{ lb/in}^2) = 1026 \text{ lb/in}^2$$

$$\sigma_{bearing} = \frac{F}{A} = \frac{15\,500 \text{ lb}}{6 \text{ in} \cdot 11 \text{ in}} = 234 \text{ lb/in}^2$$

The actual bearing stress does not exceed the allowable bearing stress. The plate is acceptable.

EXERCISES

30.1 A 10'-0"-high wall is constructed of two wythes of solid Norman clay brick and type S mortar ($f'_m = 2700$ lb/in^2). The unreinforced solid masonry wall has a $\frac{1}{2}$-in collar joint bonding the brick wythes. The wall carries a roof system at its top which supports the wall normal to its longitudinal axis. Special inspections of masonry will not be conducted during construction.

Determine the allowable average axial compressive stress for centroidally applied axial loads.

30.2 Much as in Exercise 30.1, a 10'-0" high wall is constructed of two wythes of solid Norman clay brick and type S mortar ($f'_m = 2700$ lb/in^2). The unreinforced solid masonry wall has a $\frac{1}{2}$-in collar joint bonding the brick wythes. The wall carries a roof system at its top which supports the wall normal to its longitudinal axis. Unlike Example 30.1, special inspections of masonry will be conducted during construction.

Determine the allowable average axial compressive stress for centroidally applied axial loads.

30.3 A wall is constructed of two wythes of solid Norman clay brick and type S mortar ($f'_m = 27\,560$ kPa). The 6.0-m-high unreinforced solid masonry wall has a 10-mm collar joint bonding the brick wythes. The wall carries a roof system at its top which supports the wall normal to its longitudinal axis. Special inspections of masonry will not be conducted during construction.

Determine the allowable average axial compressive stress for centroidally applied axial loads.

30.4 A 16'-0"-high unreinforced hollow-unit masonry wall is constructed of one wythe of 10-in hollow concrete masonry units (CMU) and type M mortar ($f'_m = 2500$ lb/in^2). The wall is solid grouted. The wall supports a roof system, so it is laterally supported at the top. Special inspections of masonry will not be conducted during construction.

(a) Determine the allowable average axial compressive stress for centroidally applied axial loads.

(b) Determine the design axial load that can be carried per linear foot of wall.

30.5 Much as in Exercise 30.4, a 16'-0"-high unreinforced hollow-unit masonry wall is constructed of one wythe of 10 in hollow concrete masonry units (CMU) and type M mortar ($f'_m = 2500$ lb/in^2). The wall supports a roof system, so it is laterally supported at the top. Special inspections of masonry will not be conducted during construction. Unlike the wall in Exercise 30.4, the wall is not grouted.

    (a) Determine the allowable average axial compressive stress for centroidally applied axial loads.

    (b) Determine the design axial load that can be carried per linear foot of wall.

30.6 A 13'-8"-high reinforced masonry column is constructed of two interlocking wythes of 6 in hollow CMU and type S mortar ($f'_m = 2500$ lb/in^2). The 12 in × 24 in column has actual dimensions of $11\frac{5}{8}$ in by $23\frac{5}{8}$ in and is solid grouted. Eight No. 7 grade 40 steel rebars are used for vertical reinforcement. A roof system laterally supports the top of the wall and a foundation system at grade supports the wall at the base. Special inspections of masonry will not be conducted during construction.

    (a) Determine the allowable average axial compressive stress for a centroidally applied axial load.

    (b) It is anticipated that the column will carry an axial load of 22 500 lb. Can the column safely carry this load?

30.7 A 16'-0"-high wall is constructed of 8-in hollow CMU and type S mortar ($f'_m = 2500$ lb/in^2). The ungrouted single-wythe wall carries a roof system at its top that applies an axial gravity load to the wall. The design wind load is 23 lb/ft^2. Special inspections of masonry will be conducted during construction.

    (a) It is anticipated that the wall will carry an axial load of 3500 lb per linear foot of wall. Can the wall safely carry this load?

    (b) Investigate lateral flexure stresses in the wall. Can the wall safely handle the wind load?

30.8 A 6 in × 6 in steel bearing plate is specified to transfer a 20 000-lb load to a brick wall ($f'_m = 4000$ lb/in^2). Special inspections of masonry will be conducted during construction. Assuming that the bearing plate rests on more than one-third of the area of masonry, is the bearing stress acceptable?

Appendix A
USEFUL TABLES AND FORMULAS

Table A.1   CONVERSIONS FROM U.S. CUSTOMARY SYSTEM UNITS TO SI UNITS

Quantity	U.S. Customary units	Conversion factor: U.S. Customary system to SI	SI unit Units	SI unit Name (symbol)
Acceleration	ft/s^2	0.3048	m/s^2	—
Acceleration, angular	deg/s^2	0.017 453	rad/s^2	—
Amount of a substance[a]	—	—	mole	mole (mol)
Angle, plane	degree	0.017 453	rad	
Area	in^2	0.000 645 16	m^2	—
	ft^2	0.092 90		
	yd^2	0.836 13		
	acre	4046.87		
	square mile	2 950 000		
Bending moment	ft-lb	1.355 82	N-m or m-N	—
	in-lb	0.112 985		
	ft-kip	1355.82		
	in-kip	112.985		
Calorific value	btu/lb	2326	J/kg	
Density	lb/in^3	0.009 27	kg/m^3	—
	lb/ft^3	16.018 5		
Electric current[a]	coulomb/s	1.000	coulomb/s	ampere (A)
Energy	Btu	1055.06	N-m	joule (J)
Force	lb	4.448 22	$(kg\text{-}m)/s^2$	newton (N)
	kip (1000 lb)	0.004 448 22		
Heat	Btu	1055.06	N-m	joule (J)
Heat, specific	Btu/(lb-°F)	4186.8	J/(kg-°C)	—
Heat capacity, volumetric	$Btu/(ft^3\text{-}°F)$	67 066.1	$J/(m^3\text{-}°C)$	—
Heat transfer coefficient	$Btu/(hr\text{-}ft^2\text{-}°F)$	5.678 26	$W/(m^2\text{-}°C)$	—
Illuminance	$lumens/ft^2$ (foot-candle)	10.764	lux (lx)	
Length[a]	foot (ft)	0.304 80	meter (m)	—
	inch (in)	0.025 40		
	mile	1 609.34		
Light (luminous) intensity[a]	candela (cd)	1.000	candela (cd)	—

Table A.1 *(cont'd.)*

Quantity	U.S. Customary units	Conversion factor: U.S. Customary System to SI	SI unit Units	Name (symbol)
Load, uniformly distributed	lb/ft^2	47.9	N/m^2	—
	lb/ft^2	0.0479	kN/m^2	
Luminance	$cd\text{-}ft^2$	10.764	cd/m^2	—
Mass[a]	pound (lb)	0.453 592	kilogram (kg)	—
	ton (2000 lb)	907.185		
Modulus of elasticity	lb/in^2	6894.760	N/m^2	pascal (Pa)
	kip/in^2	6 894 760		
Moment of force	lb-ft	1.355 82	N-m or m-N	—
	lb-in	0.112 985		
	kip-ft	1355.82		
	kip-in	112.985		
Moment of inertia	in^4	0.000 000 416	m^4	—
		416 000	mm^4	
Power	Btu/h	0.293 071	J/s	watt (W)
	hp	745.7		
Pressure	lb/in^2	6894.760	N/m^2	pascal (Pa)
	kip/in^2	6 894 760		
Section modulus	in^3	0.000 016	m^3	—
Stress	lb/in^2	6894.760	N/m^2	pascal (Pa)
	kip/in^2	6 894 760		
Temperature[a]	R (rankine)	$T_C = 5/9(T_F - 32)$	K[a]	kelvin (K)
	°F		°C	Celsius (°C)
	$T_F = T_R - 459.7$	$T_K = T_C + 273.2$		
Temperature difference	R (rankine)	0.555 56	K	kelvin (ΔK)
	°F	or 5/9	°C	Celsius (Δ°C)
Thermal conductivity	$Btu/(hr\text{-}ft^2\text{-}°F)$	1.730 74	W/(m-°C)	—
Time[a]	hr (hour)	0.000 277 78	s (second)[a]	—
	m (minute)	0.016 666 67		
Torque	ft-lb	1.355 82	N-m	
	in-lb	0.112 985		
	ft-kip	1355.82		
	in-kip	112.985		
Velocity (speed)	ft/s	0.304 8	m/s	—
	ft/m	0.005 080		
Velocity, angular	deg/s	0.017 453	rad/s	—
Volume	in^3	0.000 016	m^3	—
	ft^3	0.028 317		
	yd^3	0.764 555		
	gal	0.003 786		
Volumetric flow rate	gal/m	0.063 090	m^3/s	—
	ft^3/s	0.028 317		
	ft^3/m	1.699 02		
Work	ft-lb	1.355 82	N-m	joule (J)

[a]Base unit in the SI system.

Table A.2 CONVERSIONS FROM SI UNITS TO U.S. CUSTOMARY SYSTEM UNITS

Quantity	SI unit Units	Name (symbol)	Conversion factor: SI to U.S. Customary system	U.S. Customary units
Acceleration	m/s^2	—	3.2808	ft/s^2
Acceleration (angular)	rad/s^2	—	57.2958	deg/s^2
Amount of a substance[a]	mole	mole (mol)	—	—
Angle (plane)	rad		57.2958	degree
Area	m^2	—	1550.0	in^2
			10.764	ft^2
			1.1960	yd^2
			0.000 247	acre
			0.000 000 339	square mile
Bending moment	N-m	—	0.737 561	ft-lb
			8.850 732	in/lb
			0.000 750	ft-kip
			0.008 851	in-kip
Calorific value	J/kg		0.000 430	Btu/lb
Density	kg/m^3	—	107.875	lb/in^3
			0.624 278	lb/ft^3
Electric current[a]	coulomb/s	ampere (A)	1.000	coulomb/s
Energy	N-m	joule (J)	0.000 947 813	Btu
Force	$(kg\text{-}m)/s^2$	newton (N)	0.224 809	lb
			224.809	kip (1000 lb)
Heat	N-m	joule (J)	0.000 947 813	Btu
Heat, specific	J/(kg-°C)	—	0.000 238 85	Btu/(lb-°F)
Heat capacity, volumetric	$J/(m^3\text{-}°C)$	—	0.000 0149	$Btu/(ft^3\text{-}°F)$
Heat transfer coefficient	$W/(m^2\text{-}°C)$	—	0.176 110	$Btu/(hr\text{-}ft^2\text{-}°F)$
Illuminance	lux (lx)	—	0.092 903	$lumens/ft^2$ (foot-candle)
Length[a]	meter (m)	—	3.280 840	feet (ft)
			39.370 08	inch (in)
			0.000 621 37	mile
Light (luminous) intensity[a]	candela (cd)	—	1.000	candela (cd)
Load, uniformly distributed	N/m^2		0.0209	lb/ft^2
	kN/m^2		20.88	lb/ft^2
Luminance	cd/m^2	—	0.092 903	cd/ft^2
Mass[a]	kilogram (kg)	—	2.2046	pound (lb)
			0.001 102	ton (2000 lb)
Modulus of elasticity	N/m^2	pascal (Pa)	0.000 145	lb/in^2
			0.000 000 145	kip/in^2
Moment of force	N-m	—	0.737 561	lb-ft
			8.850 73	lb-in
			0.000 738	kip-ft
			0.008 850	kip-in
Moment of inertia	m^4	—	2 403 846.154	in^4
	mm^4		0.000 002 40	

Table A.2 (*cont'd.*)

Quantity	SI unit Units	Name (symbol)	Conversion factor: SI to U.S. Customary system	U.S. Customary units
Power	J/s	watt (W)	3.412 142	Btu/h
			0.001 341	hp
Pressure	N/m^2	pascal (Pa)	0.000 145	lb/in^2
			0.000 000 145	kip/in^2
Section modulus	m^3	—	60 975.61	in^3
Stress	N/m^2	pascal (Pa)	0.000 145	lb/in^2
			0.000 000 145	kip/in^2
Temperature[a]	K[a]	kelvin (K)	$T_F = 9/5\, T_C + 32$	R (rankine)
	°C	Celsius (°C)		°F
	$T_K = T_C + 273.2$			$T_F = T_R - 459.7$
Temperature difference	K	kelvin (ΔK)		R (rankine)
	°C	Celsius (Δ°C)		°F
Thermal conductivity	W/(m-°C)	—	0.577 789	Btu/(hr-ft^2-°F)
Time[a]	s (second)[a]	—	3600	hr (hour)
			60	m (minute)
Torque	N-m		0.737 561	ft-lb
			8.131 073	in-lb
			0.000 738	ft-kip
			0.000 885	in-kip
Velocity (speed)	m/s	—	3.280 840	ft/s
				ft/m
Velocity, angular	rad/s	—	57.295 755	deg/s
Volume	m^3	—	60 975.61	in^3
			35.314 475	ft^3
			1.307 950	yd^3
			264.131	gal
Volumetric flow rate	m^3/s	—	15.850 372	gal/m
			35.314 475	ft^3/s
				ft^3/m
Work	N-m	joule (J)	0.737 561	ft-lb

[a]Base unit in the SI system.

Table A.3 AVERAGE PHYSICAL PROPERTIES OF COMMON ENGINEERING MATERIALS (U.S. CUSTOMARY SYSTEM)

Material	Allowable stress (lb/in²) Tension	Compression	Shear	Ultimate strength (lb/in²) Tension	Compression	Shear	Modulus of elasticity, E (lb/in²)	Density lb/ft³	lb/in³	Poisson's ratio, μ	Coefficient of linear thermal expansion, α (in/in-°F)
Aluminum	16 000	16 000	10 000	62 000	—	38 000	10 000 000	165	0.095 5	0.340	0.000 013
Brass	4 000	4 000	4 000	21 000	30 000	30 000	9 000 000	534	0.309 0	0.333	0.000 006
Cast iron	5 000–8 000	20 000–30 000	7 500–12 500	20 000	75 000	—	20 000 000	442	0.255 8	0.270	0.000 006
Clay brick (fired)	—	300–1 000	25	—	—	—	2 000 000	120	0.069 4	—	0.000 005
Concrete	25–100	250–1 000	—	250–1 500	2 500–10 000	—	2 000 000	144	0.083 3	0.10–0.20	0.000 006
Copper	8 000	8 000	—	32 000	32 000	—	16 000 000	556	0.321 8	0.333	0.000 010
Glass	—	—	—	30 000	—	—	10 000 000	162	0.093 8	0.244	0.000 005
Polyethylene											
Low density (LDPE)	—	—	—	—	—	—	14 000–50 000	57	0.033 0	—	0.000 100
High density (HDPE)	—	—	—	—	—	—	50 000–180 000	60	0.034 7	—	0.000 67
Rubber, synthetic	—	—	—	—	—	—	600–11 000	94	0.054 4	0.45–0.50	0.000 400
Steel, low carbon	24 000	24 000	14 500	58 000–80 000	—	—	29 000 000	489	0.283 0	0.303	0.000 007
Stone masonry											
Granite	—	—	—	—	—	—	—	165	0.095 5	—	0.000 004
Sandstone	—	—	—	—	—	—	—	140	0.081 0	—	0.000 005
Wood											
Douglas Fir, No. 2 grade	500[a] 825[b]	625[c] 1 350[d]	95	—	—	—	1 600 000	32	0.018 5	—	0.000 010–0.000 030
Redwood, No. 2 grade	425[a] 725[b]	425[c] 700[d]	80	—	—	—	1 000 000	26	0.015 0	—	0.000 010–0.000 030
Wrought iron	16 000	16 000	10 000	48 000	48 000	40 000	28 000 000	480	0.277 8	0.278	0.000 007

[a]Tension parallel with grain.

[b]Extreme fiber stress in bending.

[c]Compression perpendicular to grain.

[d]Compression parallel with grain.

Table A.4 AVERAGE PHYSICAL PROPERTIES OF COMMON ENGINEERING MATERIALS (SI UNITS)

Material	Allowable stress (MPa or MN/m²)			Ultimate strength (MPa or MN/m²)			Modulus of elasticity, E (MPa or MN/m²)	Density		Poisson's ratio, μ	Coefficient of linear thermal expansion, α m/m-°C or m/m-K
	Tension	Com-pression	Shear	Tension	Com-pression	Shear		kg/m³	Mg/m³		
Aluminum	110.3	110.3	69.0	427.5	—	262.0	69 000	2 640	2.64	0.340	0.000 023
Brass	27.6	27.6	27.6	144.8	207.0	207.0	108 000	8 560	8.56	0.333	0.000 011
Cast iron	34.5–55.2	138.0–207.0	51.1–103.4	137.9	517.0	—	138 000	7 080	7.08	0.270	0.000 011
Clay brick (fired)	—	2.07–6.90	0.172	—	—	—	14 000	1 920	1.92	—	0.000 009
Concrete	0.172–0.689	1.72–6.89	—	1.72–6.89	17.2–68.9	—	14 000	2 310	2.31	0.10–0.20	0.000 011
Copper	55.2	55.2	—	220.6	220.6	—	110 000	8 910	8.91	0.333	0.000 018
Glass	—	—	—	20.7	—	—	69 000	2 600	2.60	0.244	0.000 009
Polyethylene											
Low density (LDPE)	—	—	—	—	—	—	100–350	920	0.92	—	0.000 180
High density (HDPE)	—	—	—	—	—	—	350–1 250	960	0.96	—	0.000 120
Rubber, synthetic	—	—	—	—	—	—	4–75	1 500	1.50	0.45–0.50	0.000 222
Steel, low carbon	165.5	165.5	100.0	400.0–551.6	—	—	200 000	7 840	7.84	0.303	0.000 013
Stone masonry											
Granite	—	—	—	—	—	—	—	2 650	2.65	—	0.000 008
Sandstone	—	—	—	—	—	—	—	2 250	2.25	—	0.000 010
Wood											
Douglas Fir, No. 2 grade	3.45–5.69	4.31–9.31	0.66	—	—	—	11 000	510	0.51	—	0.000 017–0.000 054
Redwood, No. 2 grade	2.93–5.00	2.93–4.80	0.55	—	—	—	6 900	420	0.42	—	0.000 017–0.000 054
Wrought iron	110.3	110.3	68.9	331.0	331.0	275.8	193 000	7 690	7.69	0.278	0.000 013

Table A.5 DERIVED FORMULAS FOR COMMON SYMMETRICALLY LOADED BEAMS[a]

Beam type and derived formulas	Graphic description

Simple beam 1: concentrated force (*P*) at center

$$R = V_{max} = \frac{P}{2}$$

$$M_{max} = \frac{PL}{4}$$

$$\Delta_{max} = \frac{Pl^3}{48EI}$$

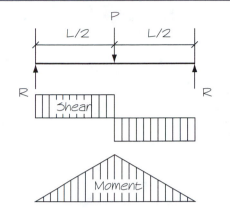

Simple beam 2: uniformly distributed force (*w*) across entire length

$$R = V_{max} = \frac{W}{2}$$

$$M_{max} = \frac{WL}{8} = \frac{wL^2}{8}$$

$$\Delta_{max} = \frac{5Wl^3}{384EI}$$

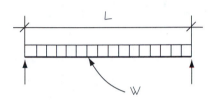

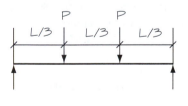

Simple beam 3: equal concentrated forces (*P*) at third points

$$R = V_{max} = P$$

$$M_{max} = \frac{PL}{3}$$

$$\Delta_{max} = \frac{23Pl^3}{648EI}$$

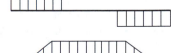

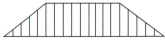

Table A.5 (*cont'd.*)

Beam type and derived formulas	Graphic description

Cantilever beam 1: concentrated force (*P*) at the free end

$$R = V_{max} = P$$

$$M_{max} = PL$$

$$\Delta_{max} = \frac{Pl^3}{3EI}$$

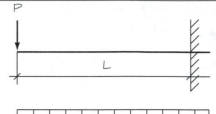

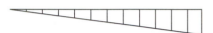

Cantilever beam 2: uniformly distributed force (*w*) across entire length

$$R = V_{max} = W = wL$$

$$M_{max} = \frac{WL}{2}$$

$$\Delta_{max} = \frac{Wl^3}{8EI}$$

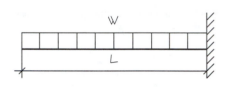

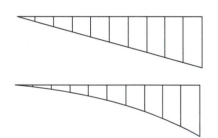

[a]Δ_{max} maximum allowable deflection, in

E modulus of elasticity for material, lb/in^2 or kip/in^2

I moment of inertia for beam section, in^4

l beam span, in ($l = L \times 12$, where L is in feet)

L beam span, ft

M_{max} maximum bending moment, lb-ft or kip-ft

P simple concentrated force, lb or kip

R reaction, lb or kip

V_{max} maximum vertical shear force, lb or kip

w uniformly distributed force, lb/ft or kip/ft

W total uniform force, lb or kip ($W = wl$)

Table A.6 FORMULAS FOR COMMON SECTIONS

Section	Property	Formula	Graphic description
Rectangle	Cross sectional area	$A = bd$	
	Most extreme fiber	$c = \dfrac{d}{2}$	
	Moment of inertia	$I = \dfrac{bd^3}{12}$	
	Section modulus	$S = \dfrac{bd^2}{6}$	
	Radius of gyration	$r = \dfrac{d}{\sqrt{12}}$	
Triangle	Cross sectional area	$A = \dfrac{bd}{2}$	
	Most extreme fiber	$c = \dfrac{2d}{3}$	
	Moment of inertia	$I = \dfrac{bd^3}{36}$	
	Section modulus	$S = \dfrac{bd^2}{24}$	
	Radius of gyration	$r = \dfrac{d}{\sqrt{18}}$	
Circle	Cross sectional area	$A = \pi R^2 = \dfrac{\pi D^2}{4}$	
	Most extreme fiber	$c = R = \dfrac{D}{2}$	
	Moment of inertia	$I = \dfrac{\pi R^4}{4} = \dfrac{\pi D^4}{64}$	
	Section modulus	$S = \dfrac{\pi R^3}{4} = \dfrac{\pi D^3}{32}$	
	Radius of gyration	$r = \dfrac{R}{2} = \dfrac{D}{4}$	
Trapezoid	Cross sectional area	$A = \dfrac{(b_1 + b_2)d}{2}$	
	Most extreme fiber	$c = \dfrac{(b_1 + 2b_2)d}{3(b_1 + b_2)}$	
	Moment of inertia	$I = \dfrac{(b_1^2 + 4b_1b_2 + b_2^2)d^3}{36(b_1 + b_2)}$	
	Section modulus	$S = \dfrac{(b_1^2 + 4b_1b_2 + b_2^2)d^3}{12(b_1 + b_2)}$	
	Radius of gyration	$r = \dfrac{d\sqrt{2(b_1^2 + 4b_1b_2 + b_2^2)}}{6(b_1 + b_2)}$	

Appendix B
PROPERTIES OF STEEL MEMBERS AND LOAD TABLES FOR BEAMS, COLUMNS AND JOISTS

Table B.1 DIMENSIONS AND DESIGN PROPERTIES OF SELECTED STEEL SECTION PROFILES

A	Cross sectional area (in.^{2})
E	Modulus of elasticity of steel (29 000 000 lb/in.^{2})
E_o	Distance from the center of the web to the shear center of a channel section (in.)
F_y	Specified minimum yield stress (lb/in.^{2})
F'_y	Theoretical yield stress at which the shape becomes noncompact as defined by flange criteria (1b/in.^{2})
I_x, I_y	Moment of inertia of a section (in.^{4})
R	Radius of fillet (in.)
S_x, S_y	Elastic section modulus (in.^{3}) (based on the exact theoretical value of 1)
T	Tangent distance on the web between fillets (in.)
$Z_x,$	Plastic section modulus (in.^{3})
a	Distance from web face to edge of flange (in.)
b_f	Width of flange (in.)
d	Depth of section (in.)
g	Usual gage in flange (in.)
k	Distance from outside of flange face to intersection of fillet with web (in.)
k_1	Distance from center line of web to intersection of fillet with flange (in.)
r_x, r_y	Radius of gyration (in.)
r_T	Radius of gyration about the Y-Y axis of a "T-section" consisting of the compression flange and fillets plus one-sixth of the web's clear distance between the flanges (in.). For beams with sloping flanges, the web's clear depth is the distance between the points at which the planes of the insides of the flanges intersect the plane of the face of the web.
t_t	Flange thickness for shapes with no flange slope, also, average thickness for shapes with sloped flanges (in.)
t_w	Web thickness (in.)
x, y	Distances from outside face of section to neutral axes Y-Y and X-X respectively (in.)

Taken from Bethlehem Steel *Structural Shapes* handbook.

WIDE FLANGE SHAPES

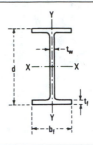

Theoretical Dimensions and Properties for **Designing**

Section Number	Weight per Foot	Area of Section	Depth of Section	Flange		Web Thickness	Axis X-X			Axis Y-Y			r_T
				Width	Thickness		I_x	S_x	r_x	I_y	S_y	r_y	
		A	d	b_f	t_f	t_w							
	lb	in.²	in.	in.	in.	in.	in.⁴	in.³	in.	in.⁴	in.³	in.	in.
W36 x 300		88.3	36.74	16.655	1.680	0.945	20300	1110	15.2	1300	156	3.83	4.39
280		82.4	36.52	16.595	1.570	0.885	18900	1030	15.1	1200	144	3.81	4.37
260		76.5	36.26	16.550	1.440	0.840	17300	953	15.0	1090	132	3.78	4.34
245		72.1	36.08	16.510	1.350	0.800	16100	895	15.0	1010	123	3.75	4.32
230		67.6	35.90	16.470	1.260	0.760	15000	837	14.9	940	114	3.73	4.30
W36 x 210		61.8	36.69	12.180	1.360	0.830	13200	719	14.6	411	67.5	2.58	3.09
194		57.0	36.49	12.115	1.260	0.765	12100	664	14.6	375	61.9	2.56	3.07
182		53.6	36.33	12.075	1.180	0.725	11300	623	14.5	347	57.6	2.55	3.05
170		50.0	36.17	12.030	1.100	0.680	10500	580	14.5	320	53.2	2.53	3.04
160		47.0	36.01	12.000	1.020	0.650	9750	542	14.4	295	49.1	2.50	3.02
150		44.2	35.85	11.975	0.940	0.625	9040	504	14.3	270	45.1	2.47	2.99
135		39.7	35.55	11.950	0.790	0.600	7800	439	14.0	225	37.7	2.38	2.93
W33 x 241		70.9	34.18	15.860	1.400	0.830	14200	829	14.1	932	118	3.63	4.17
221		65.0	33.93	15.805	1.275	0.775	12800	757	14.1	840	106	3.59	4.15
201		59.1	33.68	15.745	1.150	0.715	11500	684	14.0	749	95.2	3.56	4.12
W33 x 152		44.7	33.49	11.565	1.055	0.635	8160	487	13.5	273	47.2	2.47	2.94
141		41.6	33.30	11.535	0.960	0.605	7450	448	13.4	246	42.7	2.43	2.92
130		38.3	33.09	11.510	0.855	0.580	6710	406	13.2	218	37.9	2.39	2.88
118		34.7	32.86	11.480	0.740	0.550	5900	359	13.0	187	32.6	2.32	2.84
W30 x 211		62.0	30.94	15.105	1.315	0.775	10300	663	12.9	757	100	3.49	3.99
191		56.1	30.68	15.040	1.185	0.710	9170	598	12.8	673	89.5	3.46	3.97
173		50.8	30.44	14.985	1.065	0.655	8200	539	12.7	598	79.8	3.43	3.94
W30 x 132		38.9	30.31	10.545	1.000	0.615	5770	380	12.2	196	37.2	2.25	2.68
124		36.5	30.17	10.515	0.930	0.585	5360	355	12.1	181	34.4	2.23	2.66
116		34.2	30.01	10.495	0.850	0.565	4930	329	12.0	164	31.3	2.19	2.64
108		31.7	29.83	10.475	0.760	0.545	4470	299	11.9	146	27.9	2.15	2.61
99		29.1	29.65	10.450	0.670	0.520	3990	269	11.7	128	24.5	2.10	2.57

All shapes on these pages have parallel-faced flanges.

Extracted from Bethlehem Steel *Structural Shapes Handbook.* Reprinted with permission of Bethlehem Steel Corporation.

WIDE FLANGE SHAPES

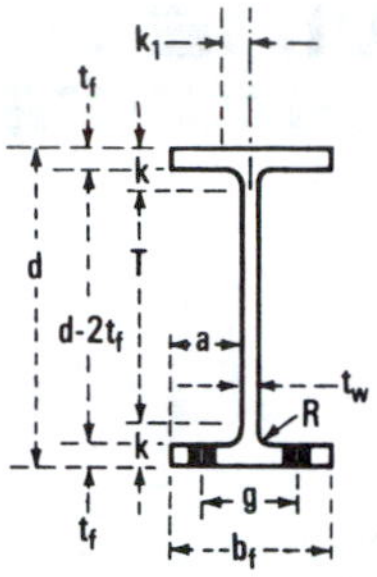

Approximate Dimensions for **Detailing**

Section Number	Weight per Foot	Depth of Section d	Flange Width b_f	Flange Thickness t_f	Web Thickness t_w	Half Web Thickness $\frac{t_w}{2}$	$d-2t_f$	a	T	k	k_1	R	Usual Flange Gage g
	lb	in.	in.	in.	in.	in.	in.	in.	in.	in.	in.	in.	in.
W36 x	300	36 3/4	16 5/8	1 11/16	15/16	1/2	33 3/8	7 7/8	31 1/8	2 13/16	1 1/2	0.95	5 1/2
	280	36 1/2	16 5/8	1 9/16	7/8	7/16	33 3/8	7 7/8	31 1/8	2 11/16	1 1/2	0.95	5 1/2
	260	36 1/4	16 1/2	1 7/16	13/16	7/16	33 3/8	7 7/8	31 1/8	2 9/16	1 1/2	0.95	5 1/2
	245	36 1/8	16 1/2	1 3/8	13/16	7/16	33 3/8	7 7/8	31 1/8	2 1/2	1 7/16	0.95	5 1/2
	230	35 7/8	16 1/2	1 1/4	3/4	3/8	33 3/8	7 7/8	31 1/8	2 3/8	1 7/16	0.95	5 1/2
W36 x	210	36 3/4	12 1/8	1 3/8	13/16	7/16	34	5 5/8	32 1/8	2 5/16	1 1/4	0.75	5 1/2
	194	36 1/2	12 1/8	1 1/4	3/4	3/8	34	5 5/8	32 1/8	2 3/16	1 3/16	0.75	5 1/2
	182	36 3/8	12 1/8	1 3/16	3/4	3/8	34	5 5/8	32 1/8	2 1/8	1 3/16	0.75	5 1/2
	170	36 1/8	12	1 1/8	11/16	3/8	34	5 5/8	32 1/8	2	1 3/16	0.75	5 1/2
	160	36	12	1	5/8	5/16	34	5 5/8	32 1/8	1 15/16	1 1/8	0.75	5 1/2
	150	35 7/8	12	15/16	5/8	5/16	34	5 5/8	32 1/8	1 7/8	1 1/8	0.75	5 1/2
	135	35 1/2	12	13/16	5/8	5/16	34	5 5/8	32 1/8	1 11/16	1 1/8	0.75	5 1/2
W33 x	241	34 1/8	15 7/8	1 3/8	13/16	7/16	31 3/8	7 1/2	29 3/4	2 3/16	1 3/16	0.70	5 1/2
	221	33 7/8	15 3/4	1 1/4	3/4	3/8	31 3/8	7 1/2	29 3/4	2 1/16	1 3/16	0.70	5 1/2
	201	33 5/8	15 3/4	1 1/8	11/16	3/8	31 3/8	7 1/2	29 3/4	1 15/16	1 1/8	0.70	5 1/2
W33 x	152	33 1/2	11 5/8	1 1/16	5/8	5/16	31 3/8	5 1/2	29 3/4	1 7/8	1 1/8	0.70	5 1/2
	141	33 1/4	11 1/2	15/16	5/8	5/16	31 3/8	5 1/2	29 3/4	1 3/4	1 1/16	0.70	5 1/2
	130	33 1/8	11 1/2	7/8	9/16	5/16	31 3/8	5 1/2	29 3/4	1 11/16	1 1/16	0.70	5 1/2
	118	32 7/8	11 1/2	3/4	9/16	5/16	31 3/8	5 1/2	29 3/4	1 9/16	1 1/16	0.70	5 1/2
W30 x	211	31	15 1/8	1 5/16	3/4	3/8	28 5/16	7 1/8	26 3/4	2 1/8	1 1/8	0.65	5 1/2
	191	30 5/8	15	1 3/16	11/16	3/8	28 5/16	7 1/8	26 3/4	1 15/16	1 1/16	0.65	5 1/2
	173	30 1/2	15	1 1/16	5/8	5/16	28 5/16	7 1/8	26 3/4	1 7/8	1 1/16	0.65	5 1/2
W30 x	132	30 1/4	10 1/2	1	5/8	5/16	28 5/16	5	26 3/4	1 3/4	1 1/16	0.65	5 1/2
	124	30 1/8	10 1/2	15/16	9/16	5/16	28 5/16	5	26 3/4	1 11/16	1	0.65	5 1/2
	116	30	10 1/2	7/8	9/16	5/16	28 5/16	5	26 3/4	1 5/8	1	0.65	5 1/2
	108	29 7/8	10 1/2	3/4	9/16	5/16	28 5/16	5	26 3/4	1 9/16	1	0.65	5 1/2
	99	29 5/8	10 1/2	11/16	1/2	1/4	28 5/16	5	26 3/4	1 7/16	1	0.65	5 1/2

WIDE FLANGE SHAPES

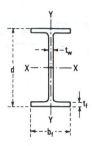

Theoretical Dimensions and Properties for **Designing**

Section Number	Weight per Foot	Area of Section	Depth of Section	Flange		Web Thick-ness	Axis X-X			Axis Y-Y			r_T
		A	d	Width b_f	Thick-ness t_f	t_w	I_x	S_x	r_x	I_y	S_y	r_y	
	lb	in.²	in.	in.	in.	in.	in.⁴	in.³	in.	in.⁴	in.³	in.	in.
W27 x **178**	52.3	27.81	14.085	1.190	0.725	6990	502	11.6	555	78.8	3.26	3.72
161	47.4	27.59	14.020	1.080	0.660	6280	455	11.5	497	70.9	3.24	3.70
146	42.9	27.38	13.965	0.975	0.605	5630	411	11.4	443	63.5	3.21	3.68
W27 x **114**	33.5	27.29	10.070	0.930	0.570	4090	299	11.0	159	31.5	2.18	2.58
102	30.0	27.09	10.015	0.830	0.515	3620	267	11.0	139	27.8	2.15	2.56
94	27.7	26.92	9.990	0.745	0.490	3270	243	10.9	124	24.8	2.12	2.53
84	24.8	26.71	9.960	0.640	0.460	2850	213	10.7	106	21.2	2.07	2.49
W24 x **162**	47.7	25.00	12.955	1.220	0.705	5170	414	10.4	443	68.4	3.05	3.45
146	43.0	24.74	12.900	1.090	0.650	4580	371	10.3	391	60.5	3.01	3.43
131	38.5	24.48	12.855	0.960	0.605	4020	329	10.2	340	53.0	2.97	3.40
117	34.4	24.26	12.800	0.850	0.550	3540	291	10.1	297	46.5	2.94	3.37
104	30 6	24.06	12.750	0.750	0.500	3100	258	10.1	259	40.7	2.91	3.35
W24 x **94**	27.7	24.31	9.065	0.875	0.515	2700	222	9.87	109	24.0	1.98	2.33
84	24.7	24.10	9.020	0.770	0.470	2370	196	9.79	94.4	20.9	1.95	2.31
76	22.4	23.92	8.990	0.680	0.440	2100	176	9.69	82.5	18.4	1.92	2.29
68	20.1	23.73	8.965	0.585	0.415	1830	154	9.55	70.4	15.7	1.87	2.26
W24 x **62**	18.2	23.74	7.040	0.590	0.430	1550	131	9.23	34.5	9.80	1.38	1.71
55	16.2	23.57	7.005	0.505	0.395	1350	114	9.11	29.1	8.30	1.34	1.68
W21 x **147**	43.2	22.06	12.510	1.150	0.720	3630	329	9.17	376	60.1	2.95	3.34
132	38.8	21.83	12.440	1.035	0.650	3220	295	9.12	333	53.5	2.93	3.31
122	35.9	21.68	12.390	0.960	0.600	2960	273	9.09	305	49.2	2.92	3.30
111	32.7	21.51	12.340	0.875	0.550	2670	249	9.05	274	44.5	2.90	3.28
101	29.8	21.36	12.290	0.800	0.500	2420	227	9.02	248	40.3	2.89	3.27
W21 x **93**	27.3	21.62	8.420	0.930	0.580	2070	192	8.70	92.9	22.1	1.84	2.17
83	24.3	21.43	8.355	0.835	0.515	1830	171	8.67	81.4	19.5	1.83	2.15
73	21.5	21.24	8.295	0.740	0.455	1600	151	8.64	70.6	17.0	1.81	2.13
68	20.0	21.13	8.270	0.685	0.430	1480	140	8.60	64.7	15.7	1.80	2.12
62	18.3	20.99	8.240	0.615	0.400	1330	127	8.54	57.5	13.9	1.77	2.10
W21 x **57**	16.7	21.06	6.555	0.650	0.405	1170	111	8.36	30.6	9.35	1.35	1.64
50	14.7	20.83	6.530	0.535	0.380	984	94.5	8.18	24.9	7.64	1.30	1.60
44	13.0	20.66	6.500	0.450	0.350	843	81.6	8.06	20.7	6.36	1.26	1.57

All shapes on these pages have parallel-faced flanges.

Extracted from Bethlehem Steel *Structural Shapes Handbook.* Reprinted with permission of Bethlehem Steel Corporation.

Table B.1 *(cont'd.)*

WIDE FLANGE SHAPES

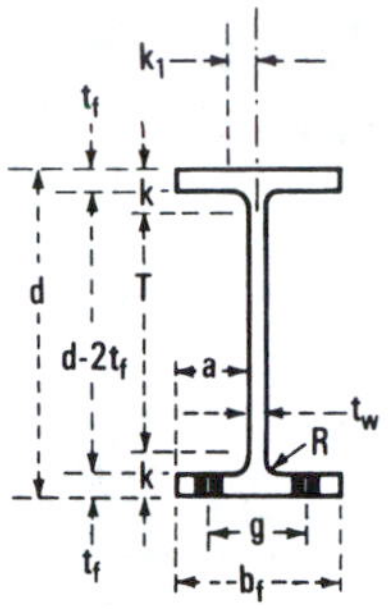

Approximate Dimensions for **Detailing**

Section Number	Weight per Foot	Depth of Section d	Flange Width b_f	Flange Thickness t_f	Web Thickness t_w	Half Web Thickness $\frac{t_w}{2}$	$d-2t_f$	a	T	k	k_1	R	Usual Flange Gage g
	lb	in.	in.	in.	in.	in.	in.	in.	in.	in.	in.	in.	in.
W27 x	**178**	27 3/4	14 1/8	1 3/16	3/4	3/8	25 7/16	6 5/8	24	1 7/8	1 1/16	0.60	5 1/2
	161	27 5/8	14	1 1/16	11/16	3/8	25 7/16	6 5/8	24	1 13/16	1	0.60	5 1/2
	146	27 3/8	14	1	5/8	5/16	25 7/16	6 5/8	24	1 11/16	1	0.60	5 1/2
W27 x	**114**	27 1/4	10 1/8	15/16	9/16	5/16	25 7/16	4 3/4	24	1 5/8	15/16	0.60	5 1/2
	102	27 1/8	10	13/16	1/2	1/4	25 7/16	4 3/4	24	1 9/16	15/16	0.60	5 1/2
	94	26 7/8	10	3/4	1/2	1/4	25 7/16	4 3/4	24	1 7/16	15/16	0.60	5 1/2
	84	26 3/4	10	5/8	7/16	1/4	25 7/16	4 3/4	24	1 3/8	15/16	0.60	5 1/2
W24 x	**162**	25	13	1 1/4	11/16	3/8	22 9/16	6 1/8	21	2	1 1/16	0.50	5 1/2
	146	24 3/4	12 7/8	1 1/16	5/8	5/16	22 9/16	6 1/8	21	1 7/8	1 1/16	0.50	5 1/2
	131	24 1/2	12 7/8	15/16	5/8	5/16	22 9/16	6 1/8	21	1 3/4	1 1/16	0.50	5 1/2
	117	24 1/4	12 3/4	7/8	9/16	5/16	22 9/16	6 1/8	21	1 5/8	1	0.50	5 1/2
	104	24	12 3/4	3/4	1/2	1/4	22 9/16	6 1/8	21	1 1/2	1	0.50	5 1/2
W24 x	**94**	24 1/4	9 1/8	7/8	1/2	1/4	22 9/16	4 1/4	21	1 5/8	1	0.50	5 1/2
	84	24 1/8	9	3/4	1/2	1/4	22 9/16	4 1/4	21	1 9/16	15/16	0.50	5 1/2
	76	23 7/8	9	11/16	7/16	1/4	22 9/16	4 1/4	21	1 7/16	15/16	0.50	5 1/2
	68	23 3/4	9	9/16	7/16	1/4	22 9/16	4 1/4	21	1 3/8	15/16	0.50	5 1/2
W24 x	**62**	23 3/4	7	9/16	7/16	1/4	22 9/16	3 1/4	21	1 3/8	15/16	0.50	3 1/2
	55	23 5/8	7	1/2	3/8	3/16	22 9/16	3 1/4	21	1 5/16	15/16	0.50	3 1/2
W21 x	**147**	22	12 1/2	1 1/8	3/4	3/8	19 3/4	5 7/8	18 1/4	1 7/8	1 1/16	0.50	5 1/2
	132	21 7/8	12 1/2	1 1/16	5/8	5/16	19 3/4	5 7/8	18 1/4	1 13/16	1	0.50	5 1/2
	122	21 5/8	12 3/8	15/16	5/8	5/16	19 3/4	5 7/8	18 1/4	1 11/16	1	0.50	5 1/2
	111	21 1/2	12 3/8	7/8	9/16	5/16	19 3/4	5 7/8	18 1/4	1 5/8	15/16	0.50	5 1/2
	101	21 3/8	12 1/4	13/16	1/2	1/4	19 3/4	5 7/8	18 1/4	1 9/16	15/16	0.50	5 1/2
W21 x	**93**	21 5/8	8 3/8	15/16	9/16	5/16	19 3/4	3 7/8	18 1/4	1 11/16	1	0.50	5 1/2
	83	21 3/8	8 3/8	13/16	1/2	1/4	19 3/4	3 7/8	18 1/4	1 9/16	15/16	0.50	5 1/2
	73	21 1/4	8 1/4	3/4	7/16	1/4	19 3/4	3 7/8	18 1/4	1 1/2	15/16	0.50	5 1/2
	68	21 1/8	8 1/4	11/16	7/16	1/4	19 3/4	3 7/8	18 1/4	1 7/16	7/8	0.50	5 1/2
	62	21	8 1/4	5/8	3/8	3/16	19 3/4	3 7/8	18 1/4	1 3/8	7/8	0.50	5 1/2
W21 x	**57**	21	6 1/2	5/8	3/8	3/16	19 3/4	3 1/8	18 1/4	1 3/8	7/8	0.50	3 1/2
	50	20 7/8	6 1/2	9/16	3/8	3/16	19 3/4	3 1/8	18 1/4	1 5/16	7/8	0.50	3 1/2
	44	20 5/8	6 1/2	7/16	3/8	3/16	19 3/4	3 1/8	18 1/4	1 3/16	7/8	0.50	3 1/2

Extracted from Bethlehem Steel *Structural Shapes Handbook*. Reprinted with permission of Bethlehem Steel Corporation.

WIDE FLANGE SHAPES

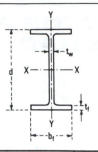

Theoretical Dimensions and Properties for **Designing**

Section Number	Weight per Foot	Area of Section	Depth of Section	Flange		Web Thick-ness	Axis X-X			Axis Y-Y			r_T
				Width	Thick-ness		I_x	S_x	r_x	I_y	S_y	r_y	
		A	d	b_f	t_f	t_w							
	lb	in.²	in.	in.	in.	in.	in.⁴	in.³	in.	in.⁴	in.³	in.	in.
W18 x 119		35.1	18.97	11.265	1.060	0.655	2190	231	7.90	253	44.9	2.69	3.02
106		31.1	18.73	11.200	0.940	0.590	1910	204	7.84	220	39.4	2.66	3.00
97		28.5	18.59	11.145	0.870	0.535	1750	188	7.82	201	36.1	2.65	2.99
86		25.3	18.39	11.090	0.770	0.480	1530	166	7.77	175	31.6	2.63	2.97
76		22.3	18.21	11.035	0.680	0.425	1330	146	7.73	152	27.6	2.61	2.95
W18 x 71		20.8	18.47	7.635	0.810	0.495	1170	127	7.50	60.3	15.8	1.70	1.98
65		19.1	18.35	7.590	0.750	0.450	1070	117	7.49	54.8	14.4	1.69	1.97
60		17.6	18.24	7.555	0.695	0.415	984	108	7.47	50.1	13.3	1.69	1.96
55		16.2	18.11	7.530	0.630	0.390	890	98.3	7.41	44.9	11.9	1.67	1.95
50		14.7	17.99	7.495	0.570	0.355	800	88.9	7.38	40.1	10.7	1.65	1.94
W18 x 46		13.5	18.06	6.060	0.605	0.360	712	78.8	7.25	22.5	7.43	1.29	1.54
40		11.8	17.90	6.015	0.525	0.315	612	68.4	7.21	19.1	6.35	1.27	1.52
35		10.3	17.70	6.000	0.425	0.300	510	57.6	7.04	15.3	5.12	1.22	1.49
W16 x 100		29.4	16.97	10.425	0.985	0.585	1490	175	7.10	186	35.7	2.52	2.81
89		26.2	16.75	10.365	0.875	0.525	1300	155	7.05	163	31.4	2.49	2.79
77		22.6	16.52	10.295	0.760	0.455	1110	134	7.00	138	26.9	2.47	2.77
67		19.7	16.33	10.235	0.665	0.395	954	117	6.96	119	23.2	2.46	2.75
W16 x 57		16.8	16.43	7.120	0.715	0.430	758	92.2	6.72	43.1	12.1	1.60	1.86
50		14.7	16.26	7.070	0.630	0.380	659	81.0	6.68	37.2	10.5	1.59	1.84
45		13.3	16.13	7.035	0.565	0.345	586	72.7	6.65	32.8	9.34	1.57	1.83
40		11.8	16.01	6.995	0.505	0.305	518	64.7	6.63	28.9	8.25	1.57	1.82
36		10.6	15.86	6.985	0.430	0.295	448	56.5	6.51	24.5	7.00	1.52	1.79
W16 x 31		9.12	15.88	5.525	0.440	0.275	375	47.2	6.41	12.4	4.49	1.17	1.39
26		7.68	15.69	5.500	0.345	0.250	301	38.4	6.26	9.59	3.49	1.12	1.36

All shapes on these pages have parallel-faced flanges.

Extracted from Bethlehem Steel *Structural Shapes Handbook*. Reprinted with permission of Bethlehem Steel Corporation.

WIDE FLANGE SHAPES

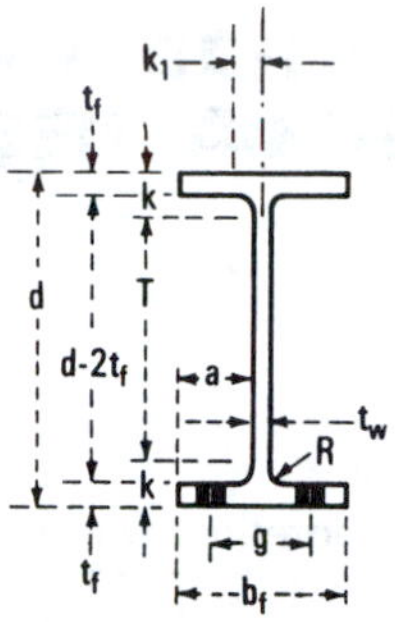

Approximate Dimensions for **Detailing**

Section Number	Weight per Foot	Depth of Section	Flange		Web Thickness	Half Web Thickness	$d-2t_f$	a	T	k	k_1	R	Usual Flange Gage
			Width	Thickness									
		d	b_f	t_f	t_w	$\frac{t_w}{2}$							g
	lb	in.	in.	in.	in.	in.	in.	in.	in.	in.	in.	in.	in.
W18 x	**119**	19	11¼	1 1/16	⅝	5/16	16⅞	5¼	15½	1¾	15/16	0.40	5½
	106	18¾	11¼	15/16	9/16	5/16	16⅞	5¼	15½	1⅝	15/16	0.40	5½
	97	18⅝	11⅛	⅞	9/16	5/16	16⅞	5¼	15½	1 9/16	⅞	0.40	5½
	86	18⅜	11⅛	¾	½	¼	16⅞	5¼	15½	1 7/16	⅞	0.40	5½
	76	18¼	11	11/16	7/16	¼	16⅞	5¼	15½	1⅜	13/16	0.40	5½
W18 x	**71**	18½	7⅝	13/16	½	¼	16⅞	3⅝	15½	1½	⅞	0.40	3½
	65	18⅜	7⅝	¾	7/16	¼	16⅞	3⅝	15½	1 7/16	⅞	0.40	3½
	60	18¼	7½	11/16	7/16	¼	16⅞	3⅝	15½	1⅜	13/16	0.40	3½
	55	18⅛	7½	⅝	⅜	3/16	16⅞	3⅝	15½	1 5/16	13/16	0.40	3½
	50	18	7½	9/16	⅜	3/16	16⅞	3⅝	15½	1¼	13/16	0.40	3½
W18 x	**46**	18	6	⅝	⅜	3/16	16⅞	2⅞	15½	1¼	13/16	0.40	3½
	40	17⅞	6	½	5/16	3/16	16⅞	2⅞	15½	1 3/16	13/16	0.40	3½
	35	17¾	6	7/16	5/16	3/16	16⅞	2⅞	15½	1⅛	¾	0.40	3½
W16 x	**100**	17	10⅜	1	9/16	5/16	15	4⅞	13⅝	1 11/16	15/16	0.40	5½
	89	16¾	10⅜	⅞	½	¼	15	4⅞	13⅝	1 9/16	⅞	0.40	5½
	77	16½	10¼	¾	7/16	¼	15	4⅞	13⅝	1 7/16	⅞	0.40	5½
	67	16⅜	10¼	11/16	⅜	3/16	15	4⅞	13⅝	1⅜	13/16	0.40	5½
W16 x	**57**	16⅜	7⅛	11/16	7/16	¼	15	3⅜	13⅝	1⅜	⅞	0.40	3½
	50	16¼	7⅛	⅝	⅜	3/16	15	3⅜	13⅝	1 5/16	13/16	0.40	3½
	45	16⅛	7	9/16	⅜	3/16	15	3⅜	13⅝	1¼	13/16	0.40	3½
	40	16	7	½	5/16	3/16	15	3⅜	13⅝	1 3/16	13/16	0.40	3½
	36	15⅞	7	7/16	5/16	3/16	15	3⅜	13⅝	1⅛	¾	0.40	3½
W16 x	**31**	15⅞	5½	7/16	¼	⅛	15	2⅝	13⅝	1⅛	¾	0.40	2¾
	26	15¾	5½	⅜	¼	⅛	15	2⅝	13⅝	1 1/16	¾	0.40	2¾

WIDE FLANGE SHAPES

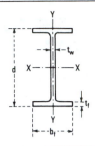

Theoretical Dimensions and Properties for **Designing**

Section Number	Weight per Foot	Area of Section	Depth of Section	Flange		Web Thick-ness	Axis X-X			Axis Y-Y			r_T
				Width	Thick-ness		I_x	S_x	r_x	I_y	S_y	r_y	
		A	d	b_f	t_f	t_w							
	lb	in.²	in.	in.	in.	in.	in.⁴	in.³	in.	in.⁴	in.³	in.	in.
W14 x	730*	215	22.42	17.890	4.910	3.070	14300	1280	8.17	4720	527	4.69	4.99
	665*	196	21.64	17.650	4.520	2.830	12400	1150	7.98	4170	472	4.62	4.92
	605*	178	20.92	17.415	4.160	2.595	10800	1040	7.80	3680	423	4.55	4.85
	550*	162	20.24	17.200	3.820	2.380	9430	931	7.63	3250	378	4.49	4.79
	500*	147	19.60	17.010	3.500	2.190	8210	838	7.48	2880	339	4.43	4.73
	455*	134	19.02	16.835	3.210	2.015	7190	756	7.33	2560	304	4.38	4.68
W14 x	426	125	18.67	16.695	3.035	1.875	6600	707	7.26	2360	283	4.34	4.64
	398	117	18.29	16.590	2.845	1.770	6000	656	7.16	2170	262	4.31	4.61
	370	109	17.92	16.475	2.660	1.655	5440	607	7.07	1990	241	4.27	4.57
	342	101	17.54	16.360	2.470	1.540	4900	559	6.98	1810	221	4.24	4.54
	311	91.4	17.12	16.230	2.260	1.410	4330	506	6.88	1610	199	4.20	4.50
	283	83.3	16.74	16.110	2.070	1.290	3840	459	6.79	1440	179	4.17	4.46
	257	75.6	16.38	15.995	1.890	1.175	3400	415	6.71	1290	161	4.13	4.43
	233	68.5	16.04	15.890	1.720	1.070	3010	375	6.63	1150	145	4.10	4.40
	211	62.0	15.72	15.800	1.560	0.980	2660	338	6.55	1030	130	4.07	4.37
	193	56.8	15.48	15.710	1.440	0.890	2400	310	6.50	931	119	4.05	4.35
	176	51.8	15.22	15.650	1.310	0.830	2140	281	6.43	838	107	4.02	4.32
	159	46.7	14.98	15.565	1.190	0.745	1900	254	6.38	748	96.2	4.00	4.30
	145	42.7	14.78	15.500	1.090	0.680	1710	232	6.33	677	87.3	3.98	4.28
W14 x	132	38.8	14.66	14.725	1.030	0.645	1530	209	6.28	548	74.5	3.76	4.05
	120	35.3	14.48	14.670	0.940	0.590	1380	190	6.24	495	67.5	3.74	4.04
	109	32.0	14.32	14.605	0.860	0.525	1240	173	6.22	447	61.2	3.73	4.02
	99	29.1	14.16	14.565	0.780	0.485	1110	157	6.17	402	55.2	3.71	4.00
	90	26.5	14.02	14.520	0.710	0.440	999	143	6.14	362	49.9	3.70	3.99
W14 x	82	24.1	14.31	10.130	0.855	0.510	882	123	6.05	148	29.3	2.48	2.74
	74	21.8	14.17	10.070	0.785	0.450	796	112	6.04	134	26.6	2.48	2.72
	68	20.0	14.04	10.035	0.720	0.415	723	103	6.01	121	24.2	2.46	2.71
	61	17.9	13.89	9.995	0.645	0.375	640	92.2	5.98	107	21.5	2.45	2.70
W14 x	53	15.6	13.92	8.060	0.660	0.370	541	77.8	5.89	57.7	14.3	1.92	2.15
	48	14.1	13.79	8.030	0.595	0.340	485	70.3	5.85	51.4	12.8	1.91	2.13
	43	12.6	13.66	7.995	0.530	0.305	428	62.7	5.82	45.2	11.3	1.89	2.12

*These shapes have a 1°-00' (1.75%) flange slope. Flange thicknesses shown are average thicknesses.
Properties shown are for a parallel flange section.

All other shapes on these pages have parallel-faced flanges.

Table B.1 *(cont'd.)*

WIDE FLANGE SHAPES

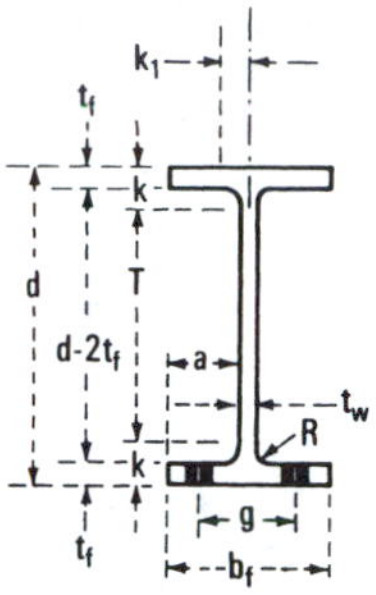

Approximate Dimensions for **Detailing**

Section Number	Weight per Foot	Depth of Section	Flange		Web Thickness	Half Web Thickness	$d-2t_f$	a	T	k	k_1	R	Usual Flange Gage
			Width	Thickness									
		d	b_f	t_f	t_w	$\frac{t_w}{2}$							g
	lb	in.	in.	in.	in.	in.	in.	in.	in.	in.	in.	in.	in.
W14 x	730	22 3/8	17 7/8	4 15/16	3 1/16	1 9/16	12 5/8	7 3/8	11 1/4	5 9/16	2 3/16	0.60	3-(7 1/2)-3
	665	21 5/8	17 5/8	4 1/2	2 13/16	1 7/16	12 5/8	7 3/8	11 1/4	5 3/16	2 1/16	0.60	3-(7 1/2)-3
	605	20 7/8	17 3/8	4 3/16	2 5/8	1 5/16	12 5/8	7 3/8	11 1/4	4 13/16	1 15/16	0.60	3-(7 1/2)-3
	550	20 1/4	17 1/4	3 13/16	2 3/8	1 3/16	12 5/8	7 3/8	11 1/4	4 1/2	1 13/16	0.60	3-(7 1/2)-3
	500	19 5/8	17	3 1/2	2 3/16	1 1/8	12 5/8	7 3/8	11 1/4	4 3/16	1 3/4	0.60	3-(7 1/2)-3
	455	19	16 7/8	3 3/16	2	1	12 5/8	7 3/8	11 1/4	3 7/8	1 5/8	0.60	3-(7 1/2)-3
W14 x	426	18 5/8	16 3/4	3 1/16	1 7/8	15/16	12 5/8	7 3/8	11 1/4	3 11/16	1 9/16	0.60	3-(5 1/2)-3
	398	18 1/4	16 5/8	2 7/8	1 3/4	7/8	12 5/8	7 3/8	11 1/4	3 1/2	1 1/2	0.60	3-(5 1/2)-3
	370	17 7/8	16 1/2	2 11/16	1 5/8	13/16	12 5/8	7 3/8	11 1/4	3 5/16	1 7/16	0.60	3-(5 1/2)-3
	342	17 1/2	16 3/8	2 1/2	1 9/16	13/16	12 5/8	7 3/8	11 1/4	3 1/8	1 3/8	0.60	3-(5 1/2)-3
	311	17 1/8	16 1/4	2 1/4	1 7/16	3/4	12 5/8	7 3/8	11 1/4	2 15/16	1 5/16	0.60	3-(5 1/2)-3
	283	16 3/4	16 1/8	2 1/16	1 5/16	11/16	12 5/8	7 3/8	11 1/4	2 3/4	1 1/4	0.60	3-(5 1/2)-3
	257	16 3/8	16	1 7/8	1 3/16	5/8	12 5/8	7 3/8	11 1/4	2 9/16	1 3/16	0.60	3-(5 1/2)-3
	233	16	15 7/8	1 3/4	1 1/16	9/16	12 5/8	7 3/8	11 1/4	2 3/8	1 3/16	0.60	3-(5 1/2)-3
	211	15 3/4	15 3/4	1 9/16	1	1/2	12 5/8	7 3/8	11 1/4	2 1/4	1 1/8	0.60	3-(5 1/2)-3
	193	15 1/2	15 3/4	1 7/16	7/8	7/16	12 5/8	7 3/8	11 1/4	2 1/8	1 1/16	0.60	3-(5 1/2)-3
	176	15 1/4	15 5/8	1 5/16	13/16	7/16	12 5/8	7 3/8	11 1/4	2	1 1/16	0.60	3-(5 1/2)-3
	159	15	15 5/8	1 3/16	3/4	3/8	12 5/8	7 3/8	11 1/4	1 7/8	1	0.60	3-(5 1/2)-3
	145	14 3/4	15 1/2	1 1/16	11/16	3/8	12 5/8	7 3/8	11 1/4	1 3/4	1	0.60	3-(5 1/2)-3
W14 x	132	14 5/8	14 3/4	1	5/8	5/16	12 5/8	7	11 1/4	1 11/16	15/16	0.60	5 1/2
	120	14 1/2	14 5/8	15/16	9/16	5/16	12 5/8	7	11 1/4	1 5/8	15/16	0.60	5 1/2
	109	14 3/8	14 5/8	7/8	1/2	1/4	12 5/8	7	11 1/4	1 9/16	7/8	0.60	5 1/2
	99	14 1/8	14 5/8	3/4	1/2	1/4	12 5/8	7	11 1/4	1 7/16	7/8	0.60	5 1/2
	90	14	14 1/2	11/16	7/16	1/4	12 5/8	7	11 1/4	1 3/8	7/8	0.60	5 1/2
W14 x	82	14 1/4	10 1/8	7/8	1/2	1/4	12 5/8	4 3/4	11	1 5/8	1	0.60	5 1/2
	74	14 1/8	10 1/8	13/16	7/16	1/4	12 5/8	4 3/4	11	1 9/16	15/16	0.60	5 1/2
	68	14	10	3/4	7/16	1/4	12 5/8	4 3/4	11	1 1/2	15/16	0.60	5 1/2
	61	13 7/8	10	5/8	3/8	3/16	12 5/8	4 3/4	11	1 7/16	15/16	0.60	5 1/2
W14 x	53	13 7/8	8	11/16	3/8	3/16	12 5/8	3 7/8	11	1 7/16	15/16	0.60	5 1/2
	48	13 3/4	8	5/8	5/16	3/16	12 5/8	3 7/8	11	1 3/8	7/8	0.60	5 1/2
	43	13 5/8	8	1/2	5/16	3/16	12 5/8	3 7/8	11	1 5/16	7/8	0.60	5 1/2

Extracted from Bethlehem Steel *Structural Shapes Handbook.* Reprinted with permission of Bethlehem Steel Corporation.

WIDE FLANGE SHAPES

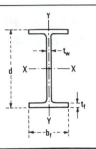

Theoretical Dimensions and Properties for **Designing**

Section Number	Weight per Foot	Area of Section	Depth of Section	Flange		Web Thickness	Axis X-X			Axis Y-Y			r_T
				Width	Thickness		I_x	S_x	r_x	I_y	S_y	r_y	
		A	d	b_f	t_f	t_w							
	lb	in.²	in.	in.	in.	in.	in.⁴	in.³	in.	in.⁴	in.³	in.	in.
W14 x	38	11.2	14.10	6.770	0.515	0.310	385	54.6	5.88	26.7	7.88	1.55	1.77
	34	10.0	13.98	6.745	0.455	0.285	340	48.6	5.83	23.3	6.91	1.53	1.76
	30	8.85	13.84	6.730	0.385	0.270	291	42.0	5.73	19.6	5.82	1.49	1.74
W14 x	26	7.69	13.91	5.025	0.420	0.255	245	35.3	5.65	8.91	3.54	1.08	1.28
	22	6.49	13.74	5.000	0.335	0.230	199	29.0	5.54	7.00	2.80	1.04	1.25
W12 x	190	55.8	14.38	12.670	1.735	1.060	1890	263	5.82	589	93.0	3.25	3.50
	170	50.0	14.03	12.570	1.560	0.960	1650	235	5.74	517	82.3	3.22	3.47
	152	44.7	13.71	12.480	1.400	0.870	1430	209	5.66	454	72.8	3.19	3.44
	136	39.9	13.41	12.400	1.250	0.790	1240	186	5.58	398	64.2	3.16	3.41
	120	35.3	13.12	12.320	1.105	0.710	1070	163	5.51	345	56.0	3.13	3.38
	106	31.2	12.89	12.220	0.990	0.610	933	145	5.47	301	49.3	3.11	3.36
	96	28.2	12.71	12.160	0.900	0.550	833	131	5.44	270	44.4	3.09	3.34
	87	25.6	12.53	12.125	0.810	0.515	740	118	5.38	241	39.7	3.07	3.32
	79	23.2	12.38	12.080	0.735	0.470	662	107	5.34	216	35.8	3.05	3.31
	72	21.1	12.25	12.040	0.670	0.430	597	97.4	5.31	195	32.4	3.04	3.29
	65	19.1	12.12	12.000	0.605	0.390	533	87.9	5.28	174	29.1	3.02	3.28
W12 x	58	17.0	12.19	10.010	0.640	0.360	475	78.0	5.28	107	21.4	2.51	2.72
	53	15.6	12.06	9.995	0.575	0.345	425	70.6	5.23	95.8	19.2	2.48	2.71
W12 x	50	14.7	12.19	8.080	0.640	0.370	394	64.7	5.18	56.3	13.9	1.96	2.17
	45	13.2	12.06	8.045	0.575	0.335	350	58.1	5.15	50.0	12.4	1.94	2.15
	40	11.8	11.94	8.005	0.515	0.295	310	51.9	5.13	44.1	11.0	1.93	2.14
W12 x	35	10.3	12.50	6.560	0.520	0.300	285	45.6	5.25	24.5	7.47	1.54	1.74
	30	8.79	12.34	6.520	0.440	0.260	238	38.6	5.21	20.3	6.24	1.52	1.73
	26	7.65	12.22	6.490	0.380	0.230	204	33.4	5.17	17.3	5.34	1.51	1.72
W12 x	22	6.48	12.31	4.030	0.425	0.260	156	25.4	4.91	4.66	2.31	0.848	1.02
	19	5.57	12.16	4.005	0.350	0.235	130	21.3	4.82	3.76	1.88	0.822	0.997
	16	4.71	11.99	3.990	0.265	0.220	103	17.1	4.67	2.82	1.41	0.773	0.963
	14	4.16	11.91	3.970	0.225	0.200	88.6	14.9	4.62	2.36	1.19	0.753	0.946

All shapes on these pages have parallel-faced flanges.

Extracted from Bethlehem Steel *Structural Shapes Handbook*. Reprinted with permission of Bethlehem Steel Corporation.

WIDE FLANGE SHAPES

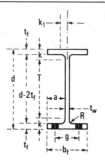

Approximate Dimensions for **Detailing**

Section Number	Weight per Foot	Dept of Section d	Flange Width b_f	Flange Thick-ness t_f	Web Thick-ness t_w	Half Web Thick-ness $\frac{t_w}{2}$	$d-2t_f$	a	T	k	k_1	R	Usual Flange Gage g
	lb	in.	in.	in.	in.	in.	in.	in.	in.	in.	in.	in.	in.
W14 x	38	14⅛	6¾	½	⁵⁄₁₆	³⁄₁₆	13¹⁄₁₆	3¼	12	1¹⁄₁₆	⅝	0.40	3½
	34	14	6¾	⁷⁄₁₆	⁵⁄₁₆	³⁄₁₆	13¹⁄₁₆	3¼	12	1	⅝	0.40	3½
	30	13⅞	6¾	⅜	¼	⅛	13¹⁄₁₆	3¼	12	¹⁵⁄₁₆	⅝	0.40	3½
W14 x	26	13⅞	5	⁷⁄₁₆	¼	⅛	13¹⁄₁₆	2⅜	12	¹⁵⁄₁₆	⁹⁄₁₆	0.40	2¾
	22	13¾	5	⁵⁄₁₆	¼	⅛	13¹⁄₁₆	2⅜	12	⅞	⁹⁄₁₆	0.40	2¾
W12 x	190	14⅜	12⅝	1¾	1¹⁄₁₆	⁹⁄₁₆	10¹⁵⁄₁₆	5¾	9½	2⁷⁄₁₆	1³⁄₁₆	0.60	5½
	170	14	12⅝	1⁹⁄₁₆	¹⁵⁄₁₆	½	10¹⁵⁄₁₆	5¾	9½	2¼	1⅛	0.60	5½
	152	13¾	12½	1⅜	⅞	⁷⁄₁₆	10¹⁵⁄₁₆	5¾	9½	2⅛	1¹⁄₁₆	0.60	5½
	136	13⅜	12⅜	1¼	¹³⁄₁₆	⁷⁄₁₆	10¹⁵⁄₁₆	5¾	9½	1¹⁵⁄₁₆	1	0.60	5½
	120	13⅛	12⅜	1⅛	¹¹⁄₁₆	⅜	10¹⁵⁄₁₆	5¾	9½	1¹³⁄₁₆	1	0.60	5½
	106	12⅞	12¼	1	⅝	⁵⁄₁₆	10¹⁵⁄₁₆	5¾	9½	1¹¹⁄₁₆	¹⁵⁄₁₆	0.60	5½
	96	12¾	12⅛	⅞	⁹⁄₁₆	⁵⁄₁₆	10¹⁵⁄₁₆	5¾	9½	1⅝	⅞	0.60	5½
	87	12½	12⅛	¹³⁄₁₆	½	¼	10¹⁵⁄₁₆	5¾	9½	1½	⅞	0.60	5½
	79	12⅜	12⅛	¾	½	¼	10¹⁵⁄₁₆	5¾	9½	1⁷⁄₁₆	⅞	0.60	5½
	72	12¼	12	¹¹⁄₁₆	⁷⁄₁₆	¼	10¹⁵⁄₁₆	5¾	9½	1⅜	⅞	0.60	5½
	65	12⅛	12	⅝	⅜	³⁄₁₆	10¹⁵⁄₁₆	5¾	9½	1⁵⁄₁₆	¹³⁄₁₆	0.60	5½
W12 x	58	12¼	10	⅝	⅜	³⁄₁₆	10¹⁵⁄₁₆	4⅞	9½	1⅜	¹³⁄₁₆	0.60	5½
	53	12	10	⁹⁄₁₆	⅜	³⁄₁₆	10¹⁵⁄₁₆	4⅞	9½	1¼	¹³⁄₁₆	0.60	5½
W12 x	50	12¼	8⅛	⅝	⅜	³⁄₁₆	10¹⁵⁄₁₆	3⅞	9½	1⅜	¹³⁄₁₆	0.60	5½
	45	12	8	⁹⁄₁₆	⁵⁄₁₆	³⁄₁₆	10¹⁵⁄₁₆	3⅞	9½	1¼	¹³⁄₁₆	0.60	5½
	40	12	8	½	⁵⁄₁₆	³⁄₁₆	10¹⁵⁄₁₆	3⅞	9½	1¼	¾	0.60	5½
W12 x	35	12½	6½	½	⁵⁄₁₆	³⁄₁₆	11⁷⁄₁₆	3⅛	10½	1	⁹⁄₁₆	0.30	3½
	30	12⅜	6½	⁷⁄₁₆	¼	⅛	11⁷⁄₁₆	3⅛	10½	¹⁵⁄₁₆	½	0.30	3½
	26	12¼	6½	⅜	¼	⅛	11⁷⁄₁₆	3⅛	10½	⅞	½	0.30	3½
W12 x	22	12¼	4	⁷⁄₁₆	¼	⅛	11⁷⁄₁₆	1⅞	10½	⅞	½	0.30	2¼
	19	12⅛	4	⅜	¼	⅛	11⁷⁄₁₆	1⅞	10½	¹³⁄₁₆	½	0.30	2¼
	16	12	4	¼	¼	⅛	11⁷⁄₁₆	1⅞	10½	¾	½	0.30	2¼
	14	11⅞	4	¼	³⁄₁₆	⅛	11⁷⁄₁₆	1⅞	10½	¹¹⁄₁₆	½	0.30	2¼

WIDE FLANGE SHAPES

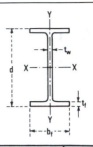

Theoretical Dimensions and Properties for **Designing**

Section Number	Weight per Foot	Area of Section	Depth of Section	Flange		Web Thick-ness	Axis X-X			Axis Y-Y			r_T
				Width	Thick-ness		I_x	S_x	r_x	I_y	S_y	r_y	
		A	d	b_f	t_f	t_w							
	lb	in.²	in.	in.	in.	in.	in.⁴	in.³	in.	in.⁴	in.³	in.	in.
W10 x 112	112	32.9	11.36	10.415	1.250	0.755	716	126	4.66	236	45.3	2.68	2.88
100	100	29.4	11.10	10.340	1.120	0.680	623	112	4.60	207	40.0	2.65	2.85
88	88	25.9	10.84	10.265	0.990	0.605	534	98.5	4.54	179	34.8	2.63	2.83
77	77	22.6	10.60	10.190	0.870	0.530	455	85.9	4.49	154	30.1	2.60	2.80
68	68	20.0	10.40	10.130	0.770	0.470	394	75.7	4.44	134	26.4	2.59	2.79
60	60	17.6	10.22	10.080	0.680	0.420	341	66.7	4.39	116	23.0	2.57	2.77
54	54	15.8	10.09	10.030	0.615	0.370	303	60.0	4.37	103	20.6	2.56	2.75
49	49	14.4	9.98	10.000	0.560	0.340	272	54.6	4.35	93.4	18.7	2.54	2.74
W10 x 45	45	13.3	10.10	8.020	0.620	0.350	248	49.1	4.33	53.4	13.3	2.01	2.18
39	39	11.5	9.92	7.985	0.530	0.315	209	42.1	4.27	45.0	11.3	1.98	2.16
33	33	9.71	9.73	7.960	0.435	0.290	170	35.0	4.19	36.6	9.20	1.94	2.14
W10 x 30	30	8.84	10.47	5.810	0.510	0.300	170	32.4	4.38	16.7	5.75	1.37	1.55
26	26	7.61	10.33	5.770	0.440	0.260	144	27.9	4.35	14.1	4.89	1.36	1.54
22	22	6.49	10.17	5.750	0.360	0.240	118	23.2	4.27	11.4	3.97	1.33	1.51
W10 x 19	19	5.62	10.24	4.020	0.395	0.250	96.3	18.8	4.14	4.29	2.14	0.874	1.03
17	17	4.99	10.11	4.010	0.330	0.240	81.9	16.2	4.05	3.56	1.78	0.845	1.01
15	15	4.41	9.99	4.000	0.270	0.230	68.9	13.8	3.95	2.89	1.45	0.810	0.987
12	12	3.54	9.87	3.960	0.210	0.190	53.8	10.9	3.90	2.18	1.10	0.785	0.965
W8 x 67	67	19.7	9.00	8.280	0.935	0.570	272	60.4	3.72	88.6	21.4	2.12	2.28
58	58	17.1	8.75	8.220	0.810	0.510	228	52.0	3.65	75.1	18.3	2.10	2.26
48	48	14.1	8.50	8.110	0.685	0.400	184	43.3	3.61	60.9	15.0	2.08	2.23
40	40	11.7	8.25	8.070	0.560	0.360	146	35.5	3.53	49.1	12.2	2.04	2.21
35	35	10.3	8.12	8.020	0.495	0.310	127	31.2	3.51	42.6	10.6	2.03	2.20
31	31	9.13	8.00	7.995	0.435	0.285	110	27.5	3.47	37.1	9.27	2.02	2.18
W8 x 28	28	8.25	8.06	6.535	0.465	0.285	98.0	24.3	3.45	21.7	6.63	1.62	1.77
24	24	7.08	7.93	6.495	0.400	0.245	82.8	20.9	3.42	18.3	5.63	1.61	1.76
W8 x 21	21	6.16	8.28	5.270	0.400	0.250	75.3	18.2	3.49	9.77	3.71	1.26	1.41
18	18	5.26	8.14	5.250	0.330	0.230	61.9	15.2	3.43	7.97	3.04	1.23	1.39
W8 x 15	15	4.44	8.11	4.015	0.315	0.245	48.0	11.8	3.29	3.41	1.70	0.876	1.03
13	13	3.84	7.99	4.000	0.255	0.230	39.6	9.91	3.21	2.73	1.37	0.843	1.01
10	10	2.96	7.89	3.940	0.205	0.170	30.8	7.81	3.22	2.09	1.06	0.841	0.994

All shapes on these pages have parallel-faced flanges.

Extracted from Bethlehem Steel *Structural Shapes Handbook*. Reprinted with permission of Bethlehem Steel Corporation.

Table B.1 *(cont'd.)*

WIDE FLANGE SHAPES

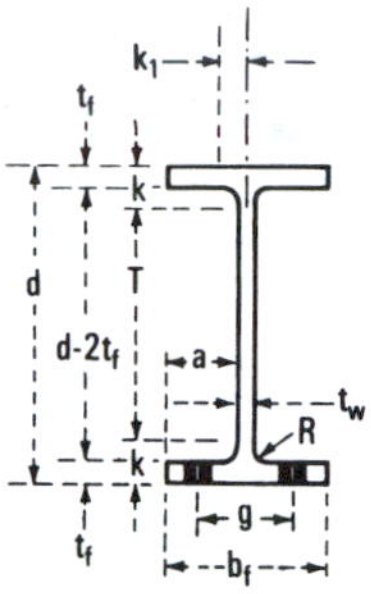

Approximate Dimensions for **Detailing**

Section Number	Weight per Foot	Depth of Section d	Flange Width b_f	Flange Thickness t_f	Web Thickness t_w	Half Web Thickness $\frac{t_w}{2}$	d-2t_f	a	T	k	k_1	R	Usual Flange Gage g
	lb	in.	in.	in.	in.	in.	in.	in.	in.	in.	in.	in.	in.
W10 x	**112**	11 3/8	10 3/8	1 1/4	3/4	3/8	8 7/8	4 7/8	7 5/8	1 7/8	15/16	0.50	5 1/2
	100	11 1/8	10 3/8	1 1/8	11/16	3/8	8 7/8	4 7/8	7 5/8	1 3/4	7/8	0.50	5 1/2
	88	10 7/8	10 1/4	1	5/8	5/16	8 7/8	4 7/8	7 5/8	1 5/8	13/16	0.50	5 1/2
	77	10 5/8	10 1/4	7/8	1/2	1/4	8 7/8	4 7/8	7 5/8	1 1/2	13/16	0.50	5 1/2
	68	10 3/8	10 1/8	3/4	1/2	1/4	8 7/8	4 7/8	7 5/8	1 3/8	3/4	0.50	5 1/2
	60	10 1/4	10 1/8	11/16	7/16	1/4	8 7/8	4 7/8	7 5/8	1 5/16	3/4	0.50	5 1/2
	54	10 1/8	10	5/8	3/8	3/16	8 7/8	4 7/8	7 5/8	1 1/4	11/16	0.50	5 1/2
	49	10	10	9/16	5/16	3/16	8 7/8	4 7/8	7 5/8	1 3/16	11/16	0.50	5 1/2
W10 x	**45**	10 1/8	8	5/8	3/8	3/16	8 7/8	3 7/8	7 5/8	1 1/4	11/16	0.50	5 1/2
	39	9 7/8	8	1/2	5/16	3/16	8 7/8	3 7/8	7 5/8	1 1/8	11/16	0.50	5 1/2
	33	9 3/4	8	7/16	5/16	3/16	8 7/8	3 7/8	7 5/8	1 1/16	11/16	0.50	5 1/2
W10 x	**30**	10 1/2	5 3/4	1/2	5/16	3/16	9 7/16	2 3/4	8 5/8	15/16	1/2	0.30	2 3/4
	26	10 3/8	5 3/4	7/16	1/4	1/8	9 7/16	2 3/4	8 5/8	7/8	1/2	0.30	2 3/4
	22	10 1/8	5 3/4	3/8	1/4	1/8	9 7/16	2 3/4	8 5/8	3/4	1/2	0.30	2 3/4
W10 x	**19**	10 1/4	4	3/8	1/4	1/8	9 7/16	1 7/8	8 5/8	13/16	1/2	0.30	2 1/4
	17	10 1/8	4	5/16	1/4	1/8	9 7/16	1 7/8	8 5/8	3/4	1/2	0.30	2 1/4
	15	10	4	1/4	1/4	1/8	9 7/16	1 7/8	8 5/8	11/16	7/16	0.30	2 1/4
	12	9 7/8	4	3/16	3/16	1/8	9 7/16	1 7/8	8 5/8	5/8	7/16	0.30	2 1/4
W8 x	**67**	9	8 1/4	15/16	9/16	5/16	7 1/8	3 7/8	6 1/8	1 7/16	11/16	0.40	5 1/2
	58	8 3/4	8 1/4	13/16	1/2	1/4	7 1/8	3 7/8	6 1/8	1 5/16	11/16	0.40	5 1/2
	48	8 1/2	8 1/8	11/16	3/8	3/16	7 1/8	3 7/8	6 1/8	1 3/16	5/8	0.40	5 1/2
	40	8 1/4	8 1/8	9/16	3/8	3/16	7 1/8	3 7/8	6 1/8	1 1/16	5/8	0.40	5 1/2
	35	8 1/8	8	1/2	5/16	3/16	7 1/8	3 7/8	6 1/8	1	9/16	0.40	5 1/2
	31	8	8	7/16	5/16	3/16	7 1/8	3 7/8	6 1/8	15/16	9/16	0.40	5 1/2
W8 x	**28**	8	6 1/2	7/16	5/16	3/16	7 1/8	3 1/8	6 1/8	15/16	9/16	0.40	3 1/2
	24	7 7/8	6 1/2	3/8	1/4	1/8	7 1/8	3 1/8	6 1/8	7/8	9/16	0.40	3 1/2
W8 x	**21**	8 1/4	5 1/4	3/8	1/4	1/8	7 1/2	2 1/2	6 5/8	13/16	1/2	0.30	2 3/4
	18	8 1/8	5 1/4	5/16	1/4	1/8	7 1/2	2 1/2	6 5/8	3/4	7/16	0.30	2 3/4
W8 x	**15**	8 1/8	4	5/16	1/4	1/8	7 1/2	1 7/8	6 5/8	3/4	1/2	0.30	2 1/4
	13	8	4	1/4	1/4	1/8	7 1/2	1 7/8	6 5/8	11/16	7/16	0.30	2 1/4
	10	7 7/8	4	3/16	3/16	1/8	7 1/2	1 7/8	6 5/8	5/8	7/16	0.30	2 1/4

Extracted from Bethlehem Steel *Structural Shapes Handbook.* Reprinted with permission of Bethlehem Steel Corporation.

WIDE FLANGE SHAPES

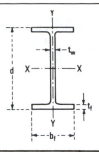

Theoretical Dimensions and Properties for **Designing**

Section Number	Weight per Foot	Area of Section	Depth of Section	Flange		Web Thick-ness	Axis X-X			Axis Y-Y			r_T
				Width	Thick-ness		I_x	S_x	r_x	I_y	S_y	r_y	
		A	d	b_f	t_f	t_w							
	lb	in.²	in.	in.	in.	in.	in.⁴	in.³	in.	in.⁴	in.³	in.	in.
W6 x	25	7.34	6.38	6.080	0.455	0.320	53.4	16.7	2.70	17.1	5.61	1.52	1.66
	20	5.87	6.20	6.020	0.365	0.260	41.4	13.4	2.66	13.3	4.41	1.50	1.64
	15	4.43	5.99	5.990	0.260	0.230	29.1	9.72	2.56	9.32	3.11	1.45	1.61
W6 x	16	4.74	6.28	4.030	0.405	0.260	32.1	10.2	2.60	4.43	2.20	0.967	1.08
	12	3.55	6.03	4.000	0.280	0.230	22.1	7.31	2.49	2.99	1.50	0.918	1.05
	9	2.68	5.90	3.940	0.215	0.170	16.4	5.56	2.47	2.20	1.11	0.905	1.03
W5 x	19	5.54	5.15	5.030	0.430	0.270	26.2	10.2	2.17	9.13	3.63	1.28	1.38
	16	4.68	5.01	5.000	0.360	0.240	21.3	8.51	2.13	7.51	3.00	1.27	1.37
†W4 x	13	3.83	4.16	4.060	0.345	0.280	11.3	5.46	1.72	3.86	1.90	1.00	1.10

MISCELLANEOUS SHAPE

Theoretical Dimensions and Properties for **Designing**

Section Number	Weight per Foot	Area of Section	Depth of Section	Flange		Web Thick-ness	Axis X-X			Axis Y-Y			r_T
				Width	Thick-ness		I_x	S_x	r_x	I_y	S_y	r_y	
		A	d	b_f	t_f	t_w							
	lb	in.²	in.	in.	in.	in.	in.⁴	in.³	in.	in.⁴	in.³	in.	in.
†M5 x	18.9	5.55	5.00	5.003	0.416	0.316	24.1	9.63	2.08	7.86	3.14	1.19	1.32

†W4 x 13 and M5 x 18.9 have flange slopes of 2.0 and 7.4 pct respectively. Flange thickness shown for these sections are average thicknesses. Properties are the same as if flanges were parallel.

All other shapes on these pages have parallel-faced flanges.

Extracted from Bethlehem Steel *Structural Shapes Handbook*. Reprinted with permission of Bethlehem Steel Corporation.

WIDE FLANGE SHAPES

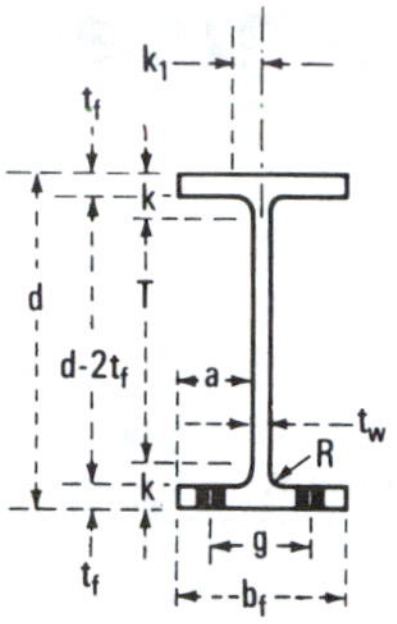

Approximate Dimensions for **Detailing**

Section Number	Weight per Foot	Depth of Section	Flange		Web Thickness	Half Web Thickness	d-2t$_f$	a	T	k	k$_1$	R	Usual Flange Gage
			Width	Thickness									
		d	b$_f$	t$_f$	t$_w$	$\frac{t_w}{2}$							g
	lb	in.	in.	in.	in.	in.	in.	in.	in.	in.	in.	in.	in.
W6 x	**25**	6 3/8	6 1/8	7/16	5/16	3/16	5 1/2	2 7/8	4 3/4	13/16	7/16	0.25	3 1/2
	20	6 1/4	6	3/8	1/4	1/8	5 1/2	2 7/8	4 3/4	3/4	7/16	0.25	3 1/2
	15	6	6	1/4	1/4	1/8	5 1/2	2 7/8	4 3/4	5/8	3/8	0.25	3 1/2
W6 x	**16**	6 1/4	4	3/8	1/4	1/8	5 1/2	1 7/8	4 3/4	3/4	7/16	0.25	2 1/4
	12	6	4	1/4	1/4	1/8	5 1/2	1 7/8	4 3/4	5/8	3/8	0.25	2 1/4
	9	5 7/8	4	3/16	3/16	1/8	5 1/2	1 7/8	4 3/4	9/16	3/8	0.25	2 1/4
W5 x	**19**	5 1/8	5	7/16	1/4	1/8	4 5/16	2 3/8	3 1/2	13/16	7/16	0.30	2 3/4
	16	5	5	3/8	1/4	1/8	4 5/16	2 3/8	3 1/2	3/4	7/16	0.30	2 3/4
W4 x	**13**	4 1/8	4	3/8	1/4	1/8	3 1/2	1 7/8	2 3/4	11/16	7/16	0.25	2 1/4

MISCELLANEOUS SHAPE

Theoretical Dimensions and Properties for **Detailing**

Section Number	Weight per Foot	Depth of Section	Flange		Web Thickness	Half Web Thickness	d-2t$_f$	a	T	k	k$_1$	R	Usual Flange Gage
			Width	Thickness									
		d	b$_f$	t$_f$	t$_w$	$\frac{t_w}{2}$							g
	lb	in.	in.	in.	in.	in.	in.	in.	in.	in.	in.	in.	in.
M5 x	**18.9**	5	5	7/16	5/16	3/16	4 3/16	2 3/8	3 1/4	7/8	1/2	0.313	2 3/4

Extracted from Bethlehem Steel *Structural Shapes Handbook.* Reprinted with permission of Bethlehem Steel Corporation.

H-PILES

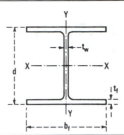

Theoretical Dimensions and Properties for **Designing**

Section Number	Weight per Foot	Area of Section	Depth of Section	Flange Width	Flange Thickness	Web Thickness	Axis X-X			Axis Y-Y			
		A	d	b_f	t_f	t_w	I_x	S_x	r_x	I_y	S_y	r_y	r_T
	lb	in.²	in.	in.	in.	in.	in.⁴	in.³	in.	in.⁴	in.³	in.	in.
HP14 x	117	34.4	14.21	14.885	0.805	0.805	1220	172	5.96	443	59.5	3.59	4.00
	102	30.0	14.01	14.785	0.705	0.705	1050	150	5.92	380	51.4	3.56	3.97
	89	26.1	13.83	14.695	0.615	0.615	904	131	5.88	326	44.3	3.53	3.94
	73	21.4	13.61	14.585	0.505	0.505	729	107	5.84	261	35.8	3.49	3.90
HP12 x	74	21.8	12.13	12.215	0.610	0.605	569	93.8	5.11	186	30.4	2.92	3.26
	63	18.4	11.94	12.125	0.515	0.515	472	79.1	5.06	153	25.3	2.88	3.23
	53	15.5	11.78	12.045	0.435	0.435	393	66.8	5.03	127	21.1	2.86	3.20
HP10 x	57	16.8	9.99	10.225	0.565	0.565	294	58.8	4.18	101	19.7	2.45	2.74
	42	12.4	9.70	10.075	0.420	0.415	210	43.4	4.13	71.7	14.2	2.41	2.69
HP8 x	36	10.6	8.02	8.155	0.445	0.445	119	29.8	3.36	40.3	9.88	1.95	2.18

H-Piles have parallel-faced flanges.

HEAVY DUTY SHEET PILING

Theoretical Dimensions and Properties

Section Number	Area of Section	Nominal Depth D	Nominal Width W	Thickness Flange t_f	Thickness Web t_w	Weight per lin. ft of bar	Weight per sq. ft of wall	Moment of Inertia	Section Modulus Single Section	Section Modulus per lin. ft of wall
	in.²	in.	in.	in.	in.	in.	lb	in.⁴	in.³	in.³
PZ22	11.86	9.0	22.0	.375	.375	40.3	22	154.7	33.1	18.1
PZ27	11.91	12.0	18.0	.375	.375	40.5	27	276.3	45.3	30.2
PZ35	19.41	14.9	22.64	.60	.50	66.0	35	681.5	91.4	48.5
PZ40	19.30	16.1	19.69	.60	.50	65.6	40	805.4	99.6	60.7
PSA23	8.99	1.3	16.0	—	.375	30.7	23	5.5	3.2	2.4
PS27.5	13.27	—	19.69	—	.40	45.1	27.5	5.3	3.3	2.0
PS31	14.96	—	19.69	—	.50	50.9	31	5.3	3.3	2.0

*Sheet piling is available in ASTM A328, A572 Grades 50 and 60, and A690.

For complete details on sheet piling, refer to the Bethlehem Steel Sheet Piling Booklet NO. 2001.

Extracted from Bethlehem Steel *Structural Shapes Handbook*. Reprinted with permission of Bethlehem Steel Corporation.

H-PILES

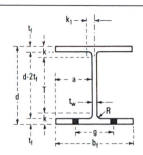

Approximate Dimensions for **Detailing**

Section Number	Weight per Foot	Depth of Section d	Flange Width b_f	Flange Thickness t_f	Web Thickness t_w	Half Web Thickness $\frac{t_w}{2}$	$d-2t_f$	a	T	k	k_1	R	Usual Flange Gage g
	lb	in.	in.	in.	in.	in.	in.	in.	in.	in.	in.	in.	in.
HP14 x	117	14 1/4	14 7/8	13/16	13/16	7/16	12 5/8	7	11 1/4	1 1/2	1 1/16	0.60	5 1/2
	102	14	14 3/4	11/16	11/16	3/8	12 5/8	7	11 1/4	1 3/8	1	0.60	5 1/2
	89	13 7/8	14 3/4	5/8	5/8	5/16	12 5/8	7	11 1/4	1 5/16	15/16	0.60	5 1/2
	73	13 5/8	14 5/8	1/2	1/2	1/4	12 5/8	7	11 1/4	1 3/16	7/8	0.60	5 1/2
HP12 x	74	12 1/8	12 1/4	5/8	5/8	5/16	10 15/16	5 3/4	9 1/2	15/16	15/16	0.60	5 1/2
	63	12	12 1/8	1/2	1/2	1/4	10 15/16	5 3/4	9 1/2	1 1/4	7/8	0.60	5 1/2
	53	11 3/4	12	7/16	7/16	1/4	10 15/16	5 3/4	9 1/2	1 1/8	7/8	0.60	5 1/2
HP10 x	57	10	10 1/4	9/16	9/16	5/16	8 7/8	4 7/8	7 5/8	1 3/16	13/16	0.50	5 1/2
	42	9 3/4	10 1/8	7/16	7/16	1/4	8 7/8	4 7/8	7 5/8	1 1/16	3/4	0.50	5 1/2
HP8 x	36	8	8 1/8	7/16	7/16	1/4	7 1/8	3 7/8	6 1/8	15/16	5/8	0.40	5 1/2

HEAVY DUTY SHEET PILING

PZ22, PZ27, PZ35 & PZ40

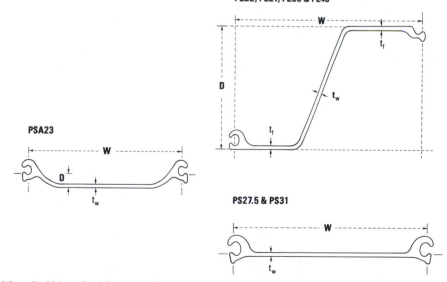

PSA23

PS27.5 & PS31

Extracted from Bethlehem Steel *Structural Shapes Handbook*. Reprinted with permission of Bethlehem Steel Corporation.

STRUCTURAL
TEES (Cut from W Shapes)

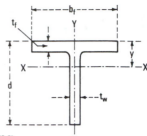

Theoretical Dimensions and Properties for **Designing**

Section Number	Weight per Foot	Area of Section	Depth of Section	Flange		Stem Thickness	Axis X-X				Axis Y-Y		
				Width	Thickness								
		A	d	b_f	t_f	t_w	I_x	S_x	r_x	y	I_y	S_y	r_y
	lb	in.²	in.	in.	in.	in.	in.⁴	in.³	in.	in.⁴	in.³	in.³	in.
WT18 x 150		44.1	18.370	16.655	1.680	0.945	1230	86.1	5.27	4.13	648	77.8	3.83
140		41.2	18.260	16.595	1.570	0.885	1140	80.0	5.25	4.07	599	72.2	3.81
130		38.2	18.130	16.550	1.440	0.840	1060	75.1	5.26	4.05	545	65.9	3.78
122.5		36.0	18.040	16.510	1.350	0.800	995	71.0	5.26	4.03	507	61.4	3.75
115		33.8	17.950	16.470	1.260	0.760	934	67.0	5.25	4.01	470	57.1	3.73
WT18 x 105		30.9	18.345	12.180	1.360	0.830	985	73.1	5.65	4.87	206	33.8	2.58
97		28.5	18.245	12.115	1.260	0.765	901	67.0	5.62	4.80	187	30.9	2.56
91		26.8	18.165	12.075	1.180	0.725	845	63.1	5.62	4.77	174	28.8	2.55
85		25.0	18.085	12.030	1.100	0.680	786	58.9	5.61	4.73	160	26.6	2.53
80		23.5	18.005	12.000	1.020	0.650	740	55.8	5.61	4.74	147	24.6	2.50
75		22.1	17.925	11.975	0.940	0.625	698	53.1	5.62	4.78	135	22.5	2.47
67.5		19.9	17.775	11.950	0.790	0.600	636	49.7	5.66	4.96	113	18.9	2.38
WT16.5 x 120.5		35.4	17.090	15.860	1.400	0.830	871	65.8	4.96	3.85	466	58.8	3.63
110.5		32.5	16.965	15.805	1.275	0.775	799	60.8	4.96	3.81	420	53.2	3.59
100.5		29.5	16.840	15.745	1.150	0.715	725	55.5	4.95	3.78	375	47.6	3.56
WT16.5 x 76		22.4	16.745	11.565	1.055	0.635	592	47.4	5.14	4.26	136	23.6	2.47
70.5		20.8	16.650	11.535	0.960	0.605	552	44.7	5.15	4.29	123	21.3	2.43
65		19.2	16.545	11.510	0.855	0.580	513	42.1	5.18	4.36	109	18.9	2.39
59		17.3	16.430	11.480	0.740	0.550	469	39.2	5.20	4.47	93.6	16.3	2.32
WT15 x 105.5		31.0	15.470	15.105	1.315	0.775	610	50.5	4.43	3.40	378	50.1	3.49
95.5		28.1	15.340	15.040	1.185	0.710	549	45.7	4.42	3.35	336	44.7	3.46
86.5		25.4	15.220	14.985	1.065	0.655	497	41.7	4.42	3.31	299	39.9	3.43
WT15 x 66		19.4	15.155	10.545	1.000	0.615	421	37.4	4.66	3.90	98.0	18.6	2.25
62		18.2	15.085	10.515	0.930	0.585	396	35.3	4.66	3.90	90.4	17.2	2.23
58		17.1	15.005	10.495	0.850	0.565	373	33.7	4.67	3.94	82.1	15.7	2.19
54		15.9	14.915	10.475	0.760	0.545	349	32.0	4.69	4.01	73.0	13.9	2.15
49.5		14.5	14.825	10.450	0.670	0.520	322	30.0	4.71	4.09	63.9	12.2	2.10
WT13.5 x 89		26.1	13.905	14.085	1.190	0.725	414	38.2	3.98	3.05	278	39.4	3.26
80.5		23.7	13.795	14.020	1.080	0.660	372	34.4	3.96	2.99	248	35.4	3.24
73		21.5	13.690	13.965	0.975	0.605	336	31.2	3.95	2.95	222	31.7	3.21
WT13.5 x 57		16.8	13.645	10.070	0.930	0.570	289	28.3	4.15	3.42	79.4	15.8	2.18
51		15.0	13.545	10.015	0.830	0.515	258	25.3	4.14	3.37	69.6	13.9	2.15
47		13.8	13.460	9.990	0.745	0.490	239	23.8	4.16	3.41	62.0	12.4	2.12
42		12.4	13.355	9.960	0.640	0.460	216	21.9	4.18	3.48	52.8	10.6	2.07

Properties shown in this table are for the full center split section.

STRUCTURAL TEES (Cut from W Shapes)

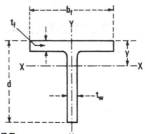

Theoretical Dimensions and Properties for **Designing**

Section Number	Weight per Foot	Area of Section	Depth of Section	Flange		Stem Thick- ness	Axis X-X				Axis Y-Y		
				Width	Thick- ness		I_x	S_x	r_x	y	I_y	S_y	r_y
		A	d	b_f	t_f	t_w							
	lb	in.²	in.	in.	in.	in.	in.⁴	in.³	in.	in.⁴	in.³	in.³	in.
WT12 x 81		23.9	12.500	12.955	1.220	0.705	293	29.9	3.50	2.70	221	34.2	3.05
73		21.5	12.370	12.900	1.090	0.650	264	27.2	3.50	2.66	195	30.3	3.01
65.5		19.3	12.240	12.855	0.960	0.605	238	24.8	3.52	2.65	170	26.5	2.97
58.5		17.2	12.130	12.800	0.850	0.550	212	22.3	3.51	2.62	149	23.2	2.94
52		15.3	12.030	12.750	0.750	0.500	189	20.0	3.51	2.59	130	20.3	2.91
WT12 x 47		13.8	12.155	9.065	0.875	0.515	186	20.3	3.67	2.99	54.5	12.0	1.98
42		12.4	12.050	9.020	0.770	0.470	166	18.3	3.67	2.97	47.2	10.5	1.95
38		11.2	11.960	8.990	0.680	0.440	151	16.9	3.68	3.00	41.3	9.18	1.92
34		10.0	11.865	8.965	0.585	0.415	137	15.6	3.70	3.06	35.2	7.85	1.87
WT12 x 31		9.11	11.870	7.040	0.590	0.430	131	15.6	3.79	3.46	17.2	4.90	1.38
27.5		8.10	11.785	7.005	0.505	0.395	117	14.1	3.80	3.50	14.5	4.15	1.34
WT10.5 x 73.5		21.6	11.030	12.510	1.150	0.720	204	23.7	3.08	2.39	188	30.0	2.95
66		19.4	10.915	12.440	1.035	0.650	181	21.1	3.06	2.33	166	26.7	2.93
61		17.9	10.840	12.390	0.960	0.600	166	19.3	3.04	2.28	152	24.6	2.92
55.5		16.3	10.755	12.340	0.875	0.550	150	17.5	3.03	2.23	137	22.2	2.90
50.5		14.9	10.680	12.290	0.800	0.500	135	15.8	3.01	2.18	124	20.2	2.89
WT10.5 x 46.5		13.7	10.810	8.420	0.930	0.580	144	17.9	3.25	2.74	46.4	11.0	1.84
41.5		12.2	10.715	8.355	0.835	0.515	127	15.7	3.22	2.66	40.7	9.75	1.83
36.5		10.7	10.620	8.295	0.740	0.455	110	13.8	3.21	2.60	35.3	8.51	1.81
34		10.0	10.565	8.270	0.685	0.430	103	12.9	3.20	2.59	32.4	7.83	1.80
31		9.13	10.495	8.240	0.615	0.400	93.8	11.9	3.21	2.58	28.7	6.97	1.77
WT10.5 x 28.5		8.37	10.530	6.555	0.650	0.405	90.4	11.8	3.29	2.85	15.3	4.67	1.35
25		7.36	10.415	6.530	0.535	0.380	80.3	10.7	3.30	2.93	12.5	3.82	1.30
22		6.49	10.330	6.500	0.450	0.350	71.1	9.68	3.31	2.98	10.3	3.18	1.26
WT9 x 59.5		17.5	9.485	11.265	1.060	0.655	119	15.9	2.60	2.03	126	22.5	2.69
53		15.6	9.365	11.200	0.940	0.590	104	14.1	2.59	1.97	110	19.7	2.66
48.5		14.3	9.295	11.145	0.870	0.535	93.8	12.7	2.56	1.91	100	18.0	2.65
43		12.7	9.195	11.090	0.770	0.480	82.4	11.2	2.55	1.86	87.6	15.8	2.63
38		11.2	9.105	11.035	0.680	0.425	71.8	9.83	2.54	1.80	76.2	13.8	2.61
WT9 x 35.5		10.4	9.235	7.635	0.810	0.495	78.2	11.2	2.74	2.26	30.1	7.89	1.70
32.5		9.55	9.175	7.590	0.750	0.450	70.7	10.1	2.72	2.20	27.4	7.22	1.69
30		8.82	9.120	7.555	0.695	0.415	64.7	9.29	2.71	2.16	25.0	6.63	1.69
27.5		8.10	9.055	7.530	0.630	0.390	59.5	8.63	2.71	2.16	22.5	5.97	1.67
25		7.33	8.995	7.495	0.570	0.355	53.5	7.79	2.70	2.12	20.0	5.35	1.65
WT9 x 23		6.77	9.030	6.060	0.605	0.360	52.1	7.77	2.77	2.33	11.3	3.72	1.29
20		5.88	8.950	6.015	0.525	0.315	44.8	6.73	2.76	2.29	9.55	3.17	1.27
17.5		5.15	8.850	6.000	0.425	0.300	40.1	6.21	2.79	2.39	7.67	2.56	1.22

Properties shown in this table are for the full center split section.

Extracted from Bethlehem Steel *Structural Shapes Handbook.* Reprinted with permission of Bethlehem Steel Corporation.

STRUCTURAL
TEES (Cut from W Shapes)

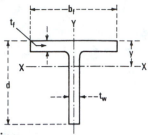

Theoretical Dimensions and Properties for **Designing**

Section Number	Weight per Foot	Area of Section	Depth of Section	Flange		Stem Thick-ness	Axis X-X				Axis Y-Y		
				Width	Thick-ness		I_x	S_x	r_x	y	I_y	S_y	r_y
		A	d	b_f	t_f	t_w							
	lb	in.²	in.	in.	in.	in.	in.⁴	in.³	in.	in.	in.⁴	in.³	in.
WT8 x	50	14.7	8.485	10.425	0.985	0.585	76.8	11.4	2.28	1.76	93.1	17.9	2.52
	44.5	13.1	8.375	10.365	0.875	0.525	67.2	10.1	2.27	1.70	81.3	15.7	2.49
	38.5	11.3	8.260	10.295	0.760	0.455	56.9	8.59	2.24	1.63	69.2	13.4	2.47
	33.5	9.84	8.165	10.235	0.665	0.395	48.6	7.36	2.22	1.56	59.5	11.6	2.46
WT8 x	28.5	8.38	8.215	7.120	0.715	0.430	48.7	7.77	2.41	1.94	21.6	6.06	1.60
	25	7.37	8.130	7.070	0.630	0.380	42.3	6.78	2.40	1.89	18.6	5.26	1.59
	22.5	6.63	8.065	7.035	0.565	0.345	37.8	6.10	2.39	1.86	16.4	4.67	1.57
	20	5.89	8.005	6.995	0.505	0.305	33.1	5.35	2.37	1.81	14.4	4.12	1.57
	18	5.28	7.930	6.985	0.430	0.295	30.6	5.05	2.41	1.88	12.2	3.50	1.52
WT8 x	15.5	4.56	7.940	5.525	0.440	0.275	27.4	4.64	2.45	2.02	6.20	2.24	1.17
	13	3.84	7.845	5.500	0.345	0.250	23.5	4.09	2.47	2.09	4.80	1.74	1.12
WT7 x	365	107	11.210	17.890	4.910	3.070	739	95.4	2.62	3.47	2360	264	4.69
	332.5	97.8	10.820	17.650	4.520	2.830	622	82.1	2.52	3.25	2080	236	4.62
	302.5	88.9	10.460	17.415	4.160	2.595	524	70.6	2.43	3.05	1840	211	4.55
	275	80.9	10.120	17.200	3.820	2.380	442	60.9	2.34	2.85	1630	189	4.49
	250	73.5	9.800	17.010	3.500	2.190	375	52.7	2.26	2.67	1440	169	4.43
	227.5	66.9	9.510	16.835	3.210	2.015	321	45.9	2.19	2.51	1280	152	4.38
WT7 x	213	62.6	9.335	16.695	3.035	1.875	287	41.4	2.14	2.40	1180	141	4.34
	199	58.5	9.145	16.590	2.845	1.770	257	37.6	2.10	2.30	1090	131	4.31
	185	54.4	8.960	16.475	2.660	1.655	229	33.9	2.05	2.19	994	121	4.27
	171	50.3	8.770	16.360	2.470	1.540	203	30.4	2.01	2.09	903	110	4.24
	155.5	45.7	8.560	16.230	2.260	1.410	176	26.7	1.96	1.97	807	99.4	4.20
	141.5	41.6	8.370	16.110	2.070	1.290	153	23.5	1.92	1.86	722	89.7	4.17
	128.5	37.8	8.190	15.995	1.890	1.175	133	20.7	1.88	1.75	645	80.7	4.13
	116.5	34.2	8.020	15.890	1.720	1.070	116	18.2	1.84	1.65	576	72.5	4.10
	105.5	31.0	7.860	15.800	1.560	0.098	102	16.2	1.81	1.57	513	65.0	4.07
	96.5	28.4	7.740	15.710	1.440	0.890	89.8	14.4	1.78	1.49	466	59.3	4.05
	88	25.9	7.610	15.650	1.310	0.830	80.5	13.0	1.76	1.43	419	53.5	4.02
	79.5	23.4	7.490	15.565	1.190	0.745	70.2	11.4	1.73	1.35	374	48.1	4.00
	72.5	21.3	7.390	15.500	1.090	0.680	62.5	10.2	1.71	1.29	338	43.7	3.98

Properties shown in this table are for the full center split section.

Extracted from Bethlehem Steel *Structural Shapes Handbook*. Reprinted with permission of Bethlehem Steel Corporation.

STRUCTURAL TEES (Cut from W Shapes)

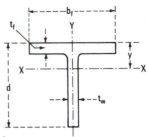

Theoretical Dimensions and Properties for **Designing**

Section Number	Weight per Foot	Area of Section	Depth of Section	Flange		Stem Thickness	Axis X-X				Axis Y-Y		
				Width	Thickness		I_x	S_x	r_x	y	I_y	S_y	r_y
		A	d	b_f	t_f	t_w							
	lb	in.²	in.	in.	in.	in.	in.⁴	in.³	in.	in.	in.⁴	in.³	in.
WT7 x	66	19.4	7.330	14.725	1.030	0.645	57.8	9.57	1.73	1.29	274	37.2	3.76
	60	17.7	7.240	14.670	0.940	0.590	51.7	8.61	1.71	1.24	247	33.7	3.74
	54.5	16.0	7.160	14.605	0.860	0.525	45.3	7.56	1.68	1.17	223	30.6	3.73
	49.5	14.6	7.080	14.565	0.780	0.485	40.9	6.88	1.67	1.14	201	27.6	3.71
	45	13.2	7.010	14.520	0.710	0.440	36.4	6.16	1.66	1.09	181	25.0	3.70
WT7 x	41	12.0	7.155	10.130	0.855	0.510	41.2	7.14	1.85	1.39	74.2	14.6	2.48
	37	10.9	7.085	10.070	0.785	0.450	36.0	6.25	1.82	1.32	66.9	13.3	2.48
	34	9.99	7.020	10.035	0.720	0.415	32.6	5.69	1.81	1.29	60.7	12.1	2.46
	30.5	8.96	6.945	9.995	0.645	0.375	28.9	5.07	1.80	1.25	53.7	10.7	2.45
WT7 x	26.5	7.81	6.960	8.060	0.660	0.370	27.6	4.94	1.88	1.38	28.8	7.16	1.92
	24	7.07	6.895	8.030	0.595	0.340	24.9	4.48	1.87	1.35	25.7	6.40	1.91
	21.5	6.31	6.830	7.995	0.530	0.305	21.9	3.98	1.86	1.31	22.6	5.65	1.89
WT7 x	19	5.58	7.050	6.770	0.515	0.310	23.3	4.22	2.04	1.54	13.3	3.94	1.55
	17	5.00	6.990	6.745	0.455	0.285	20.9	3.83	2.04	1.53	11.7	3.45	1.53
	15	4.42	6.920	6.730	0.385	0.270	19.0	3.55	2.07	1.58	9.79	2.91	1.49
WT7 x	13	3.85	6.955	5.025	0.420	0.255	17.3	3.31	2.12	1.72	4.45	1.77	1.08
	11	3.25	6.870	5.000	0.335	0.230	14.8	2.91	2.14	1.76	3.50	1.40	1.04
WT6 x	95	27.9	7.190	12.670	1.735	1.060	79.0	14.2	1.68	1.62	295	46.5	3.25
	85	25.0	7.015	12.570	1.560	0.960	67.8	12.3	1.65	1.52	259	41.2	3.22
	76	22.4	6.855	12.480	1.400	0.870	58.5	10.8	1.62	1.43	227	36.4	3.19
	68	20.0	6.705	12.400	1.250	0.790	50.6	9.46	1.59	1.35	199	32.1	3.16
	60	17.6	6.560	12.320	1.105	0.710	43.4	8.22	1.57	1.28	172	28.0	3.13
	53	15.6	6.445	12.220	0.990	0.610	36.3	6.91	1.53	1.19	151	24.7	3.11
	48	14.1	6.355	12.160	0.900	0.550	32.0	6.12	1.51	1.13	135	22.2	3.09
	43.5	12.8	6.265	12.125	0.810	0.515	28.9	5.60	1.50	1.10	120	19.9	3.07
	39.5	11.6	6.190	12.080	0.735	0.470	25.8	5.03	1.49	1.06	108	17.9	3.05
	36	10.6	6.125	12.040	0.670	0.430	23.2	4.54	1.48	1.02	97.5	16.2	3.04
	32.5	9.54	6.060	12.000	0.605	0.390	20.6	4.06	1.47	0.985	87.2	14.5	3.02

Properties shown in this table are for the full center split section.

Extracted from Bethlehem Steel *Structural Shapes Handbook.* Reprinted with permission of Bethlehem Steel Corporation.

STRUCTURAL
TEES (Cut from W Shapes)

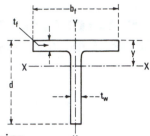

Theoretical Dimensions and Properties for **Designing**

Section Number	Weight per Foot	Area of Section	Depth of Section	Flange		Stem Thickness	Axis X-X				Axis Y-Y		
				Width	Thickness								
		A	d	b_f	t_f	t_w	I_x	S_x	r_x	y	I_y	S_y	r_y
	lb	in.²	in.	in.	in.	in.	in.⁴	in.³	in.	in.	in.⁴	in.³	in.
WT6 x	29	8.52	6.095	10.010	0.640	0.360	19.1	3.76	1.50	1.03	53.5	10.7	2.51
	26.5	7.78	6.030	9.995	0.575	0.345	17.7	3.54	1.51	1.02	47.9	9.58	2.48
WT6 x	25	7.34	6.095	8.080	0.640	0.370	18.7	3.79	1.60	1.17	28.2	6.97	1.96
	22.5	6.61	6.030	8.045	0.575	0.335	16.6	3.39	1.58	1.13	25.0	6.21	1.94
	20	5.89	5.970	8.005	0.515	0.295	14.4	2.95	1.57	1.08	22.0	5.51	1.93
WT6 x	17.5	5.17	6.250	6.560	0.520	0.300	16.0	3.23	1.76	1.30	12.2	3.73	1.54
	15	4.40	6.170	6.520	0.440	0.260	13.5	2.75	1.75	1.27	10.2	3.12	1.52
	13	3.82	6.110	6.490	0.380	0.230	11.7	2.40	1.75	1.25	8.66	2.67	1.51
WT6 x	11	3.24	6.155	4.030	0.425	0.260	11.7	2.59	1.90	1.63	2.33	1.16	0.848
	9.5	2.79	6.080	4.005	0.350	0.235	10.1	2.28	1.90	1.65	1.88	0.939	0.822
	8	2.36	5.995	3.990	0.265	0.220	8.70	2.04	1.92	1.74	1.41	0.706	0.773
	7	2.08	5.955	3.970	0.225	0.200	7.67	1.83	1.92	1.76	1.18	0.594	0.753
WT5 x	56	16.5	5.680	10.415	1.250	0.755	28.6	6.40	1.32	1.21	118	22.6	2.68
	50	14.7	5.550	10.340	1.120	0.680	24.5	5.56	1.29	1.13	103	20.0	2.65
	44	12.9	5.420	10.265	0.990	0.605	20.8	4.77	1.27	1.06	89.3	17.4	2.63
	38.5	11.3	5.300	10.190	0.870	0.530	17.4	4.05	1.24	0.990	76.8	15.1	2.60
	34	9.99	5.200	10.130	0.770	0.470	14.9	3.49	1.22	0.932	66.8	13.2	2.59
	30	8.82	5.110	10.080	0.680	0.420	12.9	3.04	1.21	0.884	58.1	11.5	2.57
	27	7.91	5.045	10.030	0.615	0.370	11.1	2.64	1.19	0.836	51.7	10.3	2.56
	24.5	7.21	4.990	10.000	0.560	0.340	10.0	2.39	1.18	0.807	46.7	9.34	2.54
WT5 x	22.5	6.63	5.050	8.020	0.620	0.350	10.2	2.47	1.24	0.907	26.7	6.65	2.01
	19.5	5.73	4.960	7.985	0.530	0.315	8.84	2.16	1.24	0.876	22.5	5.64	1.98
	16.5	4.85	4.865	7.960	0.435	0.290	7.71	1.93	1.26	0.869	18.3	4.60	1.94
WT5 x	15	4.42	5.235	5.810	0.510	0.300	9.28	2.24	1.45	1.10	8.35	2.87	1.37
	13	3.81	5.165	5.770	0.440	0.260	7.86	1.91	1.44	1.06	7.05	2.44	1.36
	11	3.24	5.085	5.750	0.360	0.240	6.88	1.72	1.46	1.07	5.71	1.99	1.33
WT5 x	9.5	2.81	5.120	4.020	0.395	0.250	6.68	1.74	1.54	1.28	2.15	1.07	0.874
	8.5	2.50	5.055	4.010	0.330	0.240	6.06	1.62	1.56	1.32	1.78	0.888	0.845
	7.5	2.21	4.995	4.000	0.270	0.230	5.45	1.50	1.57	1.37	1.45	0.723	0.810
	6	1.77	4.935	3.960	0.210	0.190	4.35	1.22	1.57	1.36	1.09	0.551	0.785

Properties shown in this table are for the full center split section.

Extracted from Bethlehem Steel *Structural Shapes Handbook*. Reprinted with permission of Bethlehem Steel Corporation.

STRUCTURAL
TEES (Cut from W Shapes)

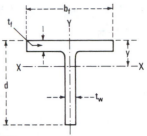

Theoretical Dimensions and Properties for **Designing**

Section Number	Weight per Foot	Area of Section	Depth of Section	Flange		Stem Thickness	Axis X-X				Axis Y-Y		
				Width	Thickness		I_x	S_x	r_x	y	I_y	S_y	r_y
		A	d	b_f	t_f	t_w							
	lb	in.²	in.	in.	in.	in.	in.⁴	in.³	in.	in.	in.⁴	in.³	in.
WT4 x	**33.5**	9.84	4.500	8.280	0.935	0.570	10.9	3.05	1.05	0.936	44.3	10.7	2.12
	29	8.55	4.375	8.220	0.810	0.510	9.12	2.61	1.03	0.874	37.5	9.13	2.10
	24	7.05	4.250	8.110	0.685	0.400	6.85	1.97	0.986	0.777	30.5	7.52	2.08
	20	5.87	4.125	8.070	0.560	0.360	5.73	1.69	0.988	0.735	24.5	6.08	2.04
	17.5	5.14	4.060	8.020	0.495	0.310	4.81	1.43	0.968	0.688	21.3	5.31	2.03
	15.5	4.56	4.000	7.995	0.435	0.285	4.28	1.28	0.968	0.668	18.5	4.64	2.02
WT4 x	**14**	4.12	4.030	6.535	0.465	0.285	4.22	1.28	1.01	0.734	10.8	3.31	1.62
	12	3.54	3.965	6.495	0.400	0.245	3.53	1.08	0.999	0.695	9.14	2.81	1.61
WT4 x	**10.5**	3.08	4.140	5.270	0.400	0.250	3.90	1.18	1.12	0.831	4.89	1.85	1.26
	9	2.63	4.070	5.250	0.330	0.230	3.41	1.05	1.14	0.834	3.98	1.52	1.23
WT4 x	**7.5**	2.22	4.055	4.015	0.315	0.245	3.28	1.07	1.22	0.998	1.70	0.849	0.876
	6.5	1.92	3.995	4.000	0.255	0.230	2.89	0.974	1.23	1.03	1.37	0.683	0.844
	5	1.48	3.945	3.940	0.205	0.170	2.15	0.717	1.20	0.953	1.05	0.532	0.841
WT3 x	**12.5**	3.67	3.190	6.080	0.455	0.320	2.29	0.886	0.789	0.610	8.53	2.81	1.52
	10	2.94	3.100	6.020	0.365	0.260	1.76	0.693	0.774	0.560	6.64	2.21	1.50
	7.5	2.21	2.995	5.990	0.260	0.230	1.41	0.577	0.797	0.558	4.66	1.56	1.45
WT3 x	**8**	2.37	3.140	4.030	0.405	0.260	1.69	0.685	0.844	0.677	2.21	1.10	0.967
	6	1.78	3.015	4.000	0.280	0.230	1.32	0.564	0.862	0.677	1.50	0.748	0.918
	4.5	1.34	2.950	3.940	0.215	0.170	0.950	0.408	0.842	0.623	1.10	0.557	0.905
WT2.5 x	**9.5**	2.77	2.575	5.030	0.430	0.270	1.01	0.485	0.605	0.487	4.56	1.82	1.28
	8	2.34	2.505	5.000	0.360	0.240	0.845	0.413	0.601	0.458	3.75	1.50	1.27
WT2 x	**6.5**	1.91	2.080	4.060	0.345	0.280	0.526	0.321	0.524	0.440	1.93	0.950	1.00

Properties shown in this table are for the full center split section

STRUCTURAL
TEE (Cut from M Shape)

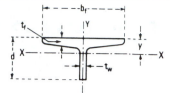

Theoretical Dimensions and Properties for **Designing**

Section Number	Weight per Foot	Area of Section	Depth of Section	Flange		Stem Thickness	Axis X-X				Axis Y-Y		
				Width	Thickness								
		A	d	b_f	t_f	t_w	I_x	S_x	r_x	y	I_y	S_y	r_y
	lb	in.²	in.	in.	in.	in.	in.⁴	in.³	in.	in.⁴	in.³	in.	in.
MT2.5 x	**9.45**	2.78	2.500	5.003	0.416*	0.316	1.05	0.527	0.615	0.511	3.93	1.57	1.19

*Average thickness

Extracted from Bethlehem Steel *Structural Shapes Handbook.* Reprinted with permission of Bethlehem Steel Corporation.

AMERICAN STANDARD SHAPES

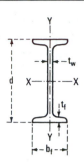

Theoretical Dimensions and Properties for **Designing**

Section Number	Weight per Foot	Area of Section A	Depth of Section d	Flange Width b_f	Flange Average Thickness t_f	Web Thickness t_w	Axis X-X I_x	Axis X-X S_x	Axis X-X r_x	Axis Y-Y I_y	Axis Y-Y S_y	Axis Y-Y r_y	r_T
	lb	in.²	in.	in.	in.	in.	in.⁴	in.³	in.	in.⁴	in.³	in.	in.
S24 x	121.0	35.6	24.50	8.050	1.090	0.800	3160	258	9.43	83.3	20.7	1.53	1.86
	106.0	31.2	24.50	7.870	1.090	0.620	2940	240	9.71	77.1	19.6	1.57	1.86
S24 x	100.0	29.3	24.00	7.245	0.870	0.745	2390	199	9.02	47.7	13.2	1.27	1.59
	90.0	26.5	24.00	7.125	0.870	0.625	2250	187	9.21	44.9	12.6	1.30	1.60
	80.0	23.5	24.00	7.000	0.870	0.500	2100	175	9.47	42.2	12.1	1.34	1.61
S20 x	96.0	28.2	20.30	7.200	0.920	0.800	1670	165	7.71	50.2	13.9	1.33	1.63
	86.0	25.3	20.30	7.060	0.920	0.660	1580	155	7.89	46.8	13.3	1.36	1.63
S20 x	75.0	22.0	20.00	6.385	0.795	0.635	1280	128	7.62	29.8	9.32	1.16	1.43
	66.0	19.4	20.00	6.255	0.795	0.505	1190	119	7.83	27.7	8.85	1.19	1.44
S18 x	70.0	20.6	18.00	6.251	0.691	0.711	926	103	6.71	24.1	7.72	1.08	1.40
	54.7	16.1	18.00	6.001	0.691	0.461	804	89.4	7.07	20.8	6.94	1.14	1.40
S15 x	50.0	14.7	15.00	5.640	0.622	0.550	486	64.8	5.75	15.7	5.57	1.03	1.30
	42.9	12.6	15.00	5.501	0.622	0.411	447	59.6	5.95	14.4	5.23	1.07	1.30
S12 x	50.0	14.7	12.00	5.477	0.659	0.687	305	50.8	4.55	15.7	5.74	1.03	1.31
	40.8	12.0	12.00	5.252	0.659	0.462	272	45.4	4.77	13.6	5.16	1.06	1.28
S12 x	35.0	10.3	12.00	5.078	0.544	0.428	229	38.2	4.72	9.87	3.89	0.980	1.20
	31.8	9.35	12.00	5.000	0.544	0.350	218	36.4	4.83	9.36	3.74	1.00	1.20

All shapes on these pages have a flange slope of 16⅔ pct.

Extracted from Bethlehem Steel *Structural Shapes Handbook.* Reprinted with permission of Bethlehem Steel Corporation.

AMERICAN STANDARD SHAPES

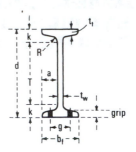

Approximate Dimensions for **Detailing**

Section Number	Weight per Foot	Depth of Section	Flange Width	Flange Average Thickness	Web Thickness	Half Web Thickness	a	T	k	R	Grip	Max Flange Fastener	Usual Flange Gage
		d	b_f	t_f	t_w	$\frac{t_w}{2}$							g
	lb	in.	in.	in.	in.	in.	in.	in.	in.	in.	in.	in.	in.
S24 x	121.0	24½	8	1 1/16	13/16	⅜	3⅝	20½	2	.60	1⅛	1	4
	106.0	24½	7⅞	1 1/16	⅝	5/16	3⅝	20½	2	.60	1⅛	1	4
S24 x	100.0	24	7¼	⅞	¾	⅜	3¼	20½	1¾	.60	⅞	1	4
	90.0	24	7⅛	⅞	⅝	5/16	3¼	20½	1¾	.60	⅞	1	4
	80.0	24	7	⅞	½	¼	3¼	20½	1¾	.60	⅞	1	4
S20 x	96.0	20¼	7¼	15/16	13/16	⅜	3¼	16¾	1¾	.60	15/16	1	4
	86.0	20¼	7	15/16	11/16	5/16	3¼	16¾	1¾	.60	15/16	1	4
S20 x	75.0	20	6⅜	13/16	⅝	5/16	2⅞	16¾	1⅝	.60	13/16	⅞	3½
	66.0	20	6¼	13/16	½	¼	2⅞	16¾	1⅝	.60	13/16	⅞	3½
S18 x	70.0	18	6¼	11/16	11/16	⅜	2¾	15	1½	.56	11/16	⅞	3½
	54.7	18	6	11/16	7/16	¼	2¾	15	1½	.56	11/16	⅞	3½
S15 x	50.0	15	5⅝	⅝	9/16	¼	2½	12¼	1⅜	.51	9/16	¾	3½
	42.9	15	5½	⅝	7/16	3/16	2½	12¼	1⅜	.51	9/16	¾	3½
S12 x	50.0	12	5½	11/16	11/16	5/16	2⅜	9⅛	1 7/16	.56	11/16	¾	3
	40.8	12	5¼	11/16	7/16	¼	2⅜	9⅛	1 7/16	.56	⅝	¾	3
S12 x	35.0	12	5⅛	9/16	7/16	3/16	2⅜	9⅝	1 3/16	.45	½	¾	3
	31.8	12	5	9/16	⅜	3/16	2⅜	9⅝	1 3/16	.45	½	¾	3

AMERICAN STANDARD SHAPES

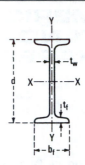

Theoretical Dimensions and Properties for **Designing**

Section Number	Weight per Foot	Area of Section	Depth of Section	Flange		Web Thickness	Axis X-X			Axis Y-Y			
				Width	Average Thickness		I_x	S_x	r_x	I_y	S_y	r_y	r_T
		A	d	b_f	t_f	t_w							
	lb	in.²	in.	in.	in.	in.	in.⁴	in.³	in.	in.⁴	in.³	in.	in.
S10 x	**35.0**	10.3	10.00	4.944	0.491	0.594	147	29.4	3.78	8.36	3.38	0.901	1.14
	25.4	7.46	10.00	4.661	0.491	0.311	124	24.7	4.07	6.79	2.91	0.954	1.12
S8 x	**23.0**	6.77	8.00	4.171	0.425	0.441	64.9	16.2	3.10	4.31	2.07	0.798	0.984
	18.4	5.41	8.00	4.001	0.425	0.271	57.6	14.4	3.26	3.73	1.86	0.831	0.969
S7 x	**15.3**	4.50	7.00	3.662	0.392	0.252	36.7	10.5	2.86	2.64	1.44	0.766	0.891
S6 x	**17.25**	5.07	6.00	3.565	0.359	0.465	26.3	8.77	2.28	2.31	1.30	0.675	0.842
	12.5	3.67	6.00	3.332	0.359	0.232	22.1	7.37	2.45	1.82	1.09	0.705	0.814

All shapes on these pages have a flange slope of 16⅔ pct.

Extracted from Bethlehem Steel *Structural Shapes Handbook*. Reprinted with permission of Bethlehem Steel Corporation.

AMERICAN STANDARD SHAPES

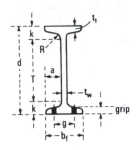

Approximate Dimensions for **Detailing**

Section Number	Weight per Foot	Depth of Section	Flange		Web Thickness	Half Web Thickness	a	T	k	R	Grip	Max Flange Fastener	Usual Flange Gage
			Width	Average Thickness									
		d	b_f	t_f	t_w	$\frac{t_w}{2}$							g
	lb	in.	in.	in.	in.	in.	in.	in.	in.	in.	in.	in.	in.
S10 x	35.0	10	5	½	⅝	5⁄16	2⅛	7¾	1⅛	.41	½	¾	2¾
	25.4	10	4⅝	½	5⁄16	⅛	2⅛	7¾	1⅛	.41	½	¾	2¾
S8 x	23.0	8	4⅛	7⁄16	7⁄16	¼	1⅞	6	1	.37	7⁄16	¾	2¼
	18.4	8	4	7⁄16	¼	⅛	1⅞	6	1	.37	7⁄16	¾	2¼
S7 x	15.3	7	3⅝	⅜	¼	⅛	1¾	5¼	⅞	.35	⅜	⅝	2¼
S6 x	17.25	6	3⅝	⅜	7⁄16	¼	1½	4⅜	13⁄16	.33	⅜	⅝	2
	12.5	6	3⅜	⅜	¼	⅛	1½	4⅜	13⁄16	.33	5⁄16	—	—

AMERICAN STANDARD CHANNELS

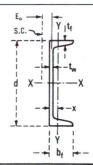

Theoretical Dimensions and Properties for **Designing**

Section Number	Weight per Foot	Area of Section	Depth of Section	Flange		Web Thickness	Axis X-X			Axis Y-Y				Shear Center Location
				Width	Average Thickness		I_x	S_x	r_x	I_y	S_y	r_y	x	E_o
		A	d	b_f	t_f	t_w								
	lb	in.²	in.	in.	in.	in.	in.⁴	in.³	in.	in.⁴	in.³	in.	in.	in.
C15 x	50.0	14.7	15.00	3.716	0.650	0.716	404	53.8	5.24	11.0	3.78	0.867	0.799	0.941
	40.0	11.8	15.00	3.520	0.650	0.520	349	46.5	5.44	9.23	3.36	0.886	0.778	1.03
	33.9	9.96	15.00	3.400	0.650	0.400	315	42.0	5.62	8.13	3.11	0.904	0.787	1.10
C12 x	30.0	8.82	12.00	3.170	0.501	0.510	162	27.0	4.29	5.14	2.06	0.763	0.674	0.873
	25.0	7.35	12.00	3.047	0.501	0.387	144	24.1	4.43	4.47	1.88	0.780	0.674	0.940
	20.7	6.09	12.00	2.942	0.501	0.282	129	21.5	4.61	3.88	1.73	0.799	0.698	1.01
C10 x	30.0	8.82	10.00	3.033	0.436	0.673	103	20.7	3.42	3.94	1.65	0.669	0.649	0.705
	25.0	7.35	10.00	2.886	0.436	0.526	91.2	18.2	3.52	3.36	1.48	0.676	0.617	0.757
	20.0	5.88	10.00	2.739	0.436	0.379	78.9	15.8	3.66	2.81	1.32	0.691	0.606	0.826
	15.3	4.49	10.00	2.600	0.436	0.240	67.4	13.5	3.87	2.28	1.16	0.713	0.634	0.916
C9 x	15.0	4.41	9.00	2.485	0.413	0.285	51.0	11.3	3.40	1.93	1.01	0.661	0.586	0.824
	13.4	3.94	9.00	2.433	0.413	0.233	47.9	10.6	3.48	1.76	0.962	0.668	0.601	0.859
C8 x	18.75	5.51	8.00	2.527	0.390	0.487	44.0	11.0	2.82	1.98	1.01	0.599	0.565	0.674
	13.75	4.04	8.00	2.343	0.390	0.303	36.1	9.03	2.99	1.53	0.853	0.615	0.553	0.756
	11.5	3.38	8.00	2.260	0.390	0.220	32.6	8.14	3.11	1.32	0.781	0.625	0.571	0.807
C7 x	12.25	3.60	7.00	2.194	0.366	0.314	24.2	6.93	2.60	1.17	0.702	0.571	0.525	0.695
	9.8	2.87	7.00	2.090	0.366	0.210	21.3	6.08	2.72	0.968	0.625	0.581	0.541	0.752

All shapes on these pages have a flange slope of 16²/₃ pct.

Extracted from Bethlehem Steel *Structural Shapes Handbook*. Reprinted with permission of Bethlehem Steel Corporation.

AMERICAN STANDARD CHANNELS

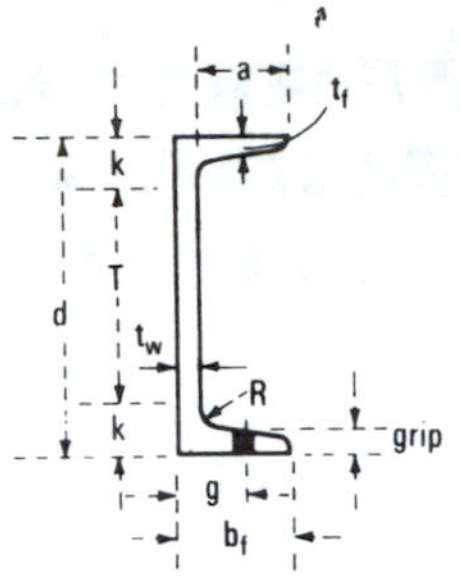

Approximate Dimensions for **Detailing**

Section Number	Weight per Foot	Depth of Section	Flange		Web Thickness	Half Web Thickness	a	T	k	R	Grip	Max Flange Fastener	Usual Flange Gage
		d	Width b_f	Average Thickness t_f	t_w	$\frac{t_w}{2}$							g
	lb	in.	in.	in.	in.	in.	in.	in.	in.	in.	in.	in.	in.
C15 x	**50.0**	15	3¾	⅝	11/16	⅜	3	12⅛	1 7/16	0.50	⅝	1	2¼
	40.0	15	3½	⅝	½	¼	3	12⅛	1 7/16	0.50	⅝	1	2
	33.9	15	3⅜	⅝	⅜	3/16	3	12⅛	1 7/16	0.50	⅝	1	2
C12 x	**30.0**	12	3⅛	½	½	¼	2⅝	9¾	1⅛	0.38	½	⅞	1¾
	25.0	12	3	½	⅜	3/16	2⅝	9¾	1⅛	0.38	½	⅞	1¾
	20.7	12	3	½	5/16	⅛	2⅝	9¾	1⅛	0.38	½	⅞	1¾
C10 x	**30.0**	10	3	7/16	11/16	5/16	2⅜	8	1	0.34	7/16	¾	1¾
	25.0	10	2⅞	7/16	½	¼	2⅜	8	1	0.34	7/16	¾	1¾
	20.0	10	2¾	7/16	⅜	3/16	2⅜	8	1	0.34	7/16	¾	1½
	15.3	10	2⅝	7/16	¼	⅛	2⅜	8	1	0.34	7/16	¾	1½
C9 x	**15.0**	9	2½	7/16	5/16	⅛	2¼	7⅛	15/16	0.33	7/16	¾	1⅜
	13.4	9	2⅜	7/16	¼	⅛	2¼	7⅛	15/16	0.33	7/16	¾	1⅜
C8 x	**18.75**	8	2½	⅜	½	¼	2	6⅛	15/16	0.32	⅜	¾	1½
	13.75	8	2⅜	⅜	5/16	⅛	2	6⅛	15/16	0.32	⅜	¾	1⅜
	11.5	8	2¼	⅜	¼	⅛	2	6⅛	15/16	0.32	⅜	¾	1⅜
C7 x	**12.25**	7	2¼	⅜	5/16	3/16	1⅞	5¼	⅞	0.31	⅜	⅝	1¼
	9.8	7	2⅛	⅜	3/16	⅛	1⅞	5¼	⅞	0.31	⅜	⅝	1¼

MISCELLANEOUS CHANNELS^w

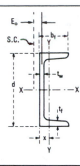

Theoretical Dimensions and Properties for **Designing**

Section Number	Weight per Foot	Area of Section A	Depth of Section d	Flange Width b_f	Flange Average Thickness t_f	Web Thickness t_w	Axis X-X I_x	S_x	r_x	Axis Y-Y I_y	S_y	r_y	x	Shear Center Location E_o
	lb	in.²	in.	in.	in.	in.	in.⁴	in.³	in.	in.⁴	in.³	in.	in.	in.
ˣMC18 x 58.0		17.1	18.00	4.200	0.625	0.700	676	75.1	6.29	17.8	5.32	1.02	0.862	1.04
51.9		15.3	18.00	4.100	0.625	0.600	627	69.7	6.41	16.4	5.07	1.04	0.858	1.10
45.8		13.5	18.00	4.000	0.625	0.500	578	64.3	6.56	15.1	4.82	1.06	0.866	1.16
42.7		12.6	18.00	3.950	0.625	0.450	554	61.6	6.64	14.4	4.69	1.07	0.877	1.19
ʸMC13 x 50.0		14.7	13.00	4.412	0.610	0.787	314	48.4	4.62	16.5	4.79	1.06	0.974	1.21
40.0		11.8	13.00	4.185	0.610	0.560	273	42.0	4.82	13.7	4.26	1.08	0.964	1.31
35.0		10.3	13.00	4.072	0.610	0.447	252	38.8	4.95	12.3	3.99	1.10	0.980	1.38
31.8		9.35	13.00	4.000	0.610	0.375	239	36.8	5.06	11.4	3.81	1.11	1.00	1.43
ᶻMC12 x 50.0		14.7	12.00	4.135	0.700	0.835	269	44.9	4.28	17.4	5.65	1.09	1.05	1.16
45.0		13.2	12.00	4.012	0.700	0.712	252	42.0	4.36	15.8	5.33	1.09	1.04	1.20
40.0		11.8	12.00	3.890	0.700	0.590	234	39.0	4.46	14.3	5.00	1.10	1.04	1.25
35.0		10.3	12.00	3.767	0.700	0.467	216	36.1	4.59	12.7	4.67	1.11	1.05	1.30
31.0		9.12	12.00	3.670	0.700	0.370	203	33.8	4.71	11.3	4.39	1.12	1.08	1.36
ˣMC10 x 41.1		12.1	10.00	4.321	0.575	0.796	158	31.5	3.61	15.8	4.88	1.14	1.09	1.26
33.6		9.87	10.00	4.100	0.575	0.575	139	27.8	3.75	13.2	4.38	1.16	1.08	1.35
28.5		8.37	10.00	3.950	0.575	0.425	127	25.3	3.89	11.4	4.02	1.17	1.12	1.43
ˣMC10 x 25.0		7.35	10.00	3.405	0.575	0.380	110	22.0	3.87	7.35	3.00	1.00	0.953	1.22
22.0		6.45	10.00	3.315	0.575	0.290	103	20.5	3.99	6.50	2.80	1.00	0.990	1.27
ˣMC9 x 25.4		7.47	9.00	3.500	0.550	0.450	88.0	19.6	3.43	7.65	3.02	1.01	0.970	1.21
23.9		7.02	9.00	3.450	0.550	0.400	85.0	18.9	3.48	7.22	2.93	1.01	0.981	1.24

ᵂAll sections on these pages are special sections. See page 3.

ˣThese channels have a flange slope of 2 degrees.
ʸThese channels have a flange slope of 8.5 degrees.
ᶻThese channels have a flange slope of 1.7 degrees.

MISCELLANEOUS CHANNELS^w

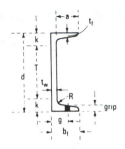

Approximate Dimensions for **Detailing**

Section Number	Weight per Foot	Depth of Section d	Flange Width b_f	Flange Average Thickness t_f	Web Thickness t_w	Half Web Thickness $\frac{t_w}{2}$	a	T	k	R	Grip	Max Flange Fastener	Usual Flange Gage g
	lb	in.	in.	in.	in.	in.	in.	in.	in.	in.	in.	in.	in.
^xMC18 x	58.0	18	4 1/4	5/8	11/16	3/8	3 1/2	15 1/4	1 3/8	0.625	5/8	1	2 1/2
	51.9	18	4 1/8	5/8	5/8	5/16	3 1/2	15 1/4	1 3/8	0.625	5/8	1	2 1/2
	45.8	18	4	5/8	1/2	1/4	3 1/2	15 1/4	1 3/8	0.625	5/8	1	2 1/2
	42.7	18	4	5/8	7/16	1/4	3 1/2	15 1/4	1 3/8	0.625	5/8	1	2 1/2
^yMC13 x	50.0	13	4 3/8	5/8	13/16	3/8	3 5/8	10 1/4	1 3/8	0.48	5/8	1	2 1/2
	40.0	13	4 1/8	5/8	9/16	1/4	3 5/8	10 1/4	1 3/8	0.48	9/16	1	2 1/2
	35.0	13	4 1/8	5/8	7/16	1/4	3 5/8	10 1/4	1 3/8	0.48	9/16	1	2 1/2
	31.8	13	4	5/8	3/8	3/16	3 5/8	10 1/4	1 3/8	0.48	9/16	1	2 1/2
^zMC12 x	50.0	12	4 1/8	11/16	13/16	7/16	3 1/4	9 3/8	15/16	0.50	11/16	1	2 1/2
	45.0	12	4	11/16	11/16	3/8	3 1/4	9 3/8	15/16	0.50	11/16	1	2 1/2
	40.0	12	3 7/8	11/16	9/16	5/16	3 1/4	9 3/8	15/16	0.50	11/16	1	2 1/2
	35.0	12	3 3/4	11/16	7/16	1/4	3 1/4	9 3/8	15/16	0.50	11/16	1	2 1/2
	31.0	12	3 5/8	11/16	3/8	3/16	3 1/4	9 3/8	15/16	0.50	11/16	1	2 1/2
^xMC10 x	41.1	10	4 3/8	9/16	13/16	3/8	3 1/2	7 1/2	1 1/4	0.575	9/16	7/8	2 1/2
	33.6	10	4 1/8	9/16	9/16	5/16	3 1/2	7 1/2	1 1/4	0.575	9/16	7/8	2 1/2
	28.5	10	4	9/16	7/16	3/16	3 1/2	7 1/2	1 1/4	0.575	9/16	7/8	2 1/2
^xMC10 x	25.0	10	3 3/8	9/16	3/8	3/16	3	7 1/2	1 1/4	0.570	9/16	7/8	2
	22.0	10	3 3/8	9/16	5/16	1/8	3	7 1/2	1 1/4	0.570	9/16	7/8	2
^xMC9 x	25.4	9	3 1/2	9/16	7/16	1/4	3	6 5/8	1 3/16	0.550	9/16	7/8	2
	23.9	9	3 1/2	9/16	3/8	3/16	3	6 5/8	1 3/16	0.550	9/16	7/8	2

MISCELLANEOUS CHANNELS^w

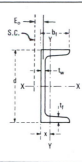

Theoretical Dimensions and Properties for **Designing**

Section Number	Weight per Foot	Area of Section	Depth of Section	Flange		Web Thick-ness	Axis X-X			Axis Y-Y				Shear Center Location
				Width	Average Thick-ness		I_x	S_x	r_x	I_y	S_y	r_y	x	E_o
		A	d	b_f	t_f	t_w								
	lb	in.²	in.	in.	in.	in.	in.⁴	in.³	in.	in.⁴	in.³	in.	in.	in.
'MC8 x	22.8	6.70	8.00	3.502	0.525	0.427	63.8	16.0	3.09	7.07	2.84	1.03	1.01	1.26
	21.4	6.28	8.00	3.450	0.525	0.375	61.6	15.4	3.13	6.64	2.74	1.03	1.02	1.28
'MC8 x	20.0	5.88	8.00	3.025	0.500	0.400	54.5	13.6	3.05	4.47	2.05	0.872	0.840	1.04
	18.7	5.50	8.00	2.978	0.500	0.353	52.5	13.1	3.09	4.20	1.97	0.874	0.849	1.07
'MC7 x	ʸ22.7	6.67	7.00	3.603	0.500	0.503	47.5	13.6	2.67	7.29	2.85	1.05	1.04	1.26
	ʸ19.1	5.61	7.00	3.452	0.500	0.352	43.2	12.3	2.77	6.11	2.57	1.04	1.08	1.33
'MC6 x	18.0	5.29	6.00	3.504	0.475	0.379	29.7	9.91	2.37	5.93	2.48	1.06	1.12	1.36
'MC6 x	15.3	4.50	6.00	3.500	0.385	0.340	25.4	8.47	2.38	4.97	2.03	1.05	1.05	1.33
'MC6 x	16.3	4.79	6.00	3.000	0.475	0.375	26.0	8.68	2.33	3.82	1.84	0.892	0.927	1.12
	15.1	4.44	6.00	2.941	0.475	0.316	25.0	8.32	2.37	3.51	1.75	0.889	0.940	1.14
'MC6 x	12.0	3.53	6.00	2.497	0.375	0.310	18.7	6.24	2.30	1.87	1.04	0.728	0.704	0.880

ᵂAll sections on these pages are special sections. See page 3.
ʸAvailable subject to inquiry.
'These channels have a flange slope of 2 degrees.

MISCELLANEOUS CHANNELS^w

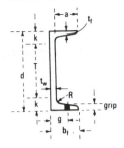

Approximate Dimensions for **Detailing**

Section Number	Weight per Foot	Depth of Section	Flange		Web Thickness	Half Web Thickness	a	T	k	R	Grip	Max Flange Fastener	Usual Flange Gage
			Width	Average Thickness									
		d	b_f	t_f	t_w	$\frac{t_w}{2}$							g
	lb	in.	in.	in.	in.	in.	in.	in.	in.	in.	in.	in.	in.
MC8 x 22.8		8	3½	½	⁷⁄₁₆	³⁄₁₆	3⅛	5⅝	1³⁄₁₆	0.525	½	⅞	2
21.4		8	3½	½	⅜	³⁄₁₆	3⅛	5⅝	1³⁄₁₆	0.525	½	⅞	2
MC8 x 20.0		8	3	½	⅜	³⁄₁₆	2⅝	5¾	1⅛	0.500	½	⅞	2
18.7		8	3	½	⅜	³⁄₁₆	2⅝	5¾	1⅛	0.500	½	⅞	2
MC7 x 22.7		7	3⅝	½	½	¼	3⅛	4¾	1⅛	0.500	½	⅞	2
19.1		7	3½	½	⅜	³⁄₁₆	3⅛	4¾	1⅛	0.500	½	⅞	2
MC6 x 18.0		6	3½	½	⅜	³⁄₁₆	3⅛	3⅞	1¹⁄₁₆	0.475	½	⅞	2
MC6 x 15.3		6	3½	⅜	⁵⁄₁₆	³⁄₁₆	3⅛	4¼	⅞	0.385	⅜	⅞	2
MC6 x 16.3		6	3	½	⅜	³⁄₁₆	2⅝	3⅞	1¹⁄₁₆	0.475	½	¾	1¾
15.1		6	3	½	⁵⁄₁₆	³⁄₁₆	2⅝	3⅞	1¹⁄₁₆	0.475	½	¾	1¾
MC6 x 12.0		6	2½	⅜	⁵⁄₁₆	⅛	2⅛	4⅜	¹³⁄₁₆	0.375	⅜	⅝	1½

STRUCTURAL TEES (Cut from S Shapes)

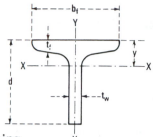

Theoretical Dimensions and Properties for **Designing**

Section Number	Weight per Foot	Area of Section	Depth of Tee	Flange		Stem Thickness	Axis X-X				Axis Y-Y		
				Width	Average Thickness		I_x	S_x	r_x	y	I_y	S_y	r_y
		A	d	b_f	t_f	t_w							
	lb	in.²	in.	in.	in.	in.	in.⁴	in.³	in.	in.	in.⁴	in.³	in.
ST12 x	60.5	17.8	12.250	8.050	1.090	0.800	259	30.1	3.82	3.63	41.7	10.4	1.53
	53	15.6	12.250	7.870	1.090	0.620	216	24.1	3.72	3.28	38.5	9.80	1.57
ST12 x	50	14.7	12.000	7.245	0.870	0.745	215	26.3	3.83	3.84	23.8	6.58	1.27
	45	13.2	12.000	7.125	0.870	0.625	190	22.6	3.79	3.60	22.5	6.31	1.30
	40	11.7	12.000	7.000	0.870	0.500	162	18.7	3.72	3.29	21.1	6.04	1.34
ST10 x	48	14.1	10.150	7.200	0.920	0.800	143	20.3	3.18	3.13	25.1	6.97	1.33
	43	12.7	10.150	7.060	0.920	0.660	125	17.2	3.14	2.91	23.4	6.63	1.36
ST10 x	37.5	11.0	10.000	6.385	0.795	0.635	109	15.8	3.15	3.07	14.9	4.66	1.16
	33	9.70	10.000	6.255	0.795	0.505	93.1	12.9	3.10	2.81	13.8	4.43	1.19
ST9 x	35	10.3	9.000	6.251	0.691	0.711	84.7	14.0	2.87	2.94	12.1	3.86	1.08
	27.35	8.04	9.000	6.001	0.691	0.461	62.4	9.61	2.79	2.50	10.4	3.47	1.14
ST7.5 x	25	7.35	7.500	5.640	0.622	0.550	40.6	7.73	2.35	2.25	7.85	2.78	1.03
	21.45	6.31	7.500	5.501	0.622	0.411	33.0	6.00	2.29	2.01	7.19	2.61	1.07
ST6 x	25	7.35	6.000	5.477	0.659	0.687	25.2	6.05	1.85	1.84	7.85	2.87	1.03
	20.4	6.00	6.000	5.252	0.659	0.462	18.9	4.28	1.78	1.58	6.78	2.58	1.06
ST6 x	17.5	5.14	6.000	5.078	0.544	0.428	17.2	3.95	1.83	1.65	4.93	1.94	0.980
	15.9	4.68	6.000	5.000	0.544	0.350	14.9	3.31	1.78	1.51	4.68	1.87	1.00
ST5 x	17.5	5.15	5.000	4.944	0.491	0.594	12.5	3.63	1.56	1.56	4.18	1.69	0.901
	12.7	3.73	5.000	4.661	0.491	0.311	7.83	2.06	1.45	1.20	3.39	1.46	0.954
ST4 x	11.5	3.38	4.000	4.171	0.425	0.441	5.03	1.77	1.22	1.15	2.15	1.03	0.798
	9.2	2.70	4.000	4.001	0.425	0.271	3.51	1.15	1.14	0.941	1.86	0.932	0.831
ST3.5 x	7.65	2.25	3.500	3.662	0.392	0.252	2.19	0.816	0.987	0.817	1.32	0.720	0.766
ST3 x	8.625	2.53	3.000	3.565	0.359	0.465	2.13	1.02	0.917	0.914	1.15	0.648	0.675
	6.25	1.83	3.000	3.332	0.359	0.232	1.27	0.552	0.833	0.691	0.911	0.547	0.705

Properties shown in this table are for the full center split section.

ANGLES
equal legs

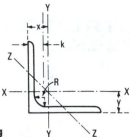

Theoretical Dimensions and Properties for **Designing**

Section Number and Size	Thickness	Weight per Foot	Area of Section	k	Axis X-X and Axis Y-Y				Axis Z-Z
					$I_{x,y}$	$S_{x,y}$	$r_{x,y}$	y or x	r_z
in.	in.	lb	in.²	in.	in.⁴	in.³	in.	in.	in.
L8 x 8 x	1⅛	56.9	16.7	1¾	98.0	17.5	2.42	2.41	1.56
R=⅝	1	51.0	15.0	1⅝	89.0	15.8	2.44	2.37	1.56
	⅞	45.0	13.2	1½	79.6	14.0	2.45	2.32	1.57
	¾	38.9	11.4	1⅜	69.7	12.2	2.47	2.28	1.58
	⅝	32.7	9.61	1¼	59.4	10.3	2.49	2.23	1.58
	⁹⁄₁₆	29.6	8.68	1³⁄₁₆	54.1	9.34	2.50	2.21	1.59
	½	26.4	7.75	1⅛	48.6	8.36	2.50	2.19	1.59
L6 x 6 x	1	37.4	11.0	1½	35.5	8.57	1.80	1.86	1.17
R=½	⅞	33.1	9.73	1⅜	31.9	7.63	1.81	1.82	1.17
	¾	28.7	8.44	1¼	28.2	6.66	1.83	1.78	1.17
	⅝	24.2	7.11	1⅛	24.2	5.66	1.84	1.73	1.18
	⁹⁄₁₆	21.9	6.43	1¹⁄₁₆	22.1	5.14	1.85	1.71	1.18
	½	19.6	5.75	1	19.9	4.61	1.86	1.68	1.18
	⁷⁄₁₆	17.2	5.06	¹⁵⁄₁₆	17.7	4.08	1.87	1.66	1.19
	⅜	14.9	4.36	⅞	15.4	3.53	1.88	1.64	1.19
ʸL5 x 5 x	⅞	27.2	7.98	1⅜	17.8	5.17	1.49	1.57	0.973
R=½	¾	23.6	6.94	1¼	15.7	4.53	1.51	1.52	0.975
ˣ	⅝	20.0	5.86	1⅛	13.6	3.86	1.52	1.48	0.978
	½	16.2	4.75	1	11.3	3.16	1.54	1.43	0.983
	⁷⁄₁₆	14.3	4.18	¹⁵⁄₁₆	10.0	2.79	1.55	1.41	0.986
	⅜	12.3	3.61	⅞	8.74	2.42	1.56	1.39	0.990
	⁵⁄₁₆	10.3	3.03	¹³⁄₁₆	7.42	2.04	1.57	1.37	0.994

Table B.1 *(cont'd.)*

ANGLES
unequal legs

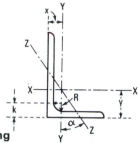

Theoretical Dimensions and Properties for **Designing**

Section Number and Size	Thick-ness	Weight per Foot	Area of Section	k	Axis X-X				Axis Y-Y				Axis Z-Z	
					I_x	S_x	r_x	y	I_y	S_y	r_y	x	r_z	Tan α
in.	in.	lb	in.²	in.	in.⁴	in.³	in.	in.	in.⁴	in.³	in.	in.	in.	
L8 x 6 x	1	44.2	13.0	1½	80.8	15.1	2.49	2.65	38.8	8.92	1.73	1.65	1.28	0.543
R=½	⅞	39.1	11.5	1⅜	72.3	13.4	2.51	2.61	34.9	7.94	1.74	1.61	1.28	0.547
	¾	33.8	9.94	1¼	63.4	11.7	2.53	2.56	30.7	6.92	1.76	1.56	1.29	0.551
	9⁄16	25.7	7.56	1 1⁄16	49.3	8.95	2.55	2.50	24.0	5.34	1.78	1.50	1.30	0.556
	½	23.0	6.75	1	44.3	8.02	2.56	2.47	21.7	4.79	1.79	1.47	1.30	0.558
	7⁄16	20.2	5.93	15⁄16	39.2	7.07	2.57	2.45	19.3	4.23	1.80	1.45	1.31	0.640
L8 x 4 x		37.4	11.0	1½	69.6	14.1	2.52	3.05	11.6	3.94	1.03	1.05	0.846	0.247
R=½	⅞	33.1	9.73	1⅜	62.5	12.5	2.53	3.00	10.5	3.51	1.04	0.999	0.848	0.253
	¾	28.7	8.44	1¼	54.9	10.9	2.55	2.95	9.36	3.07	1.05	0.953	0.852	0.258
	⅝	24.2	7.11	1⅛	46.9	9.21	2.57	2.91	8.10	2.62	1.07	0.906	0.857	0.262
	9⁄16	21.9	6.43	1 1⁄16	42.8	8.35	2.58	2.88	7.43	2.38	1.07	0.882	0.861	0.265
	½	19.6	5.75	1	38.5	7.49	2.59	2.86	6.74	2.15	1.08	0.859	0.865	0.267
	7⁄16	17.2	5.06	15⁄16	34.1	6.60	2.60	2.83	6.02	1.90	1.09	0.835	0.869	0.269
L7 x 4 x	¾	26.2	7.69	1¼	37.8	8.42	2.22	2.51	9.05	3.03	1.09	1.01	0.860	0.324
R=½	⅝	22.1	6.48	1⅛	32.4	7.14	2.24	2.46	7.84	2.58	1.10	0.963	0.865	0.329
	½	17.9	5.25	1	26.7	5.81	2.25	2.42	6.53	2.12	1.11	0.917	0.872	0.335
	7⁄16	15.8	4.62	15⁄16	23.7	5.13	2.26	2.39	5.83	1.88	1.12	0.893	0.876	0.337
	⅜	13.6	3.98	⅞	20.6	4.44	2.27	2.37	5.10	1.63	1.13	0.870	0.880	0.340
L6 x 4 x	¾	23.6	6.94	1¼	24.5	6.25	1.88	2.08	8.68	2.97	1.12	1.08	0.860	0.428
R=½	⅝	20.0	5.86	1⅛	21.1	5.31	1.90	2.03	7.52	2.54	1.13	1.03	0.864	0.435
	9⁄16	18.1	5.31	1 1⁄16	19.3	4.83	1.90	2.01	6.91	2.31	1.14	1.01	0.866	0.438
	½	16.2	4.75	1	17.4	4.33	1.91	1.99	6.27	2.08	1.15	0.987	0.870	0.440
	7⁄16	14.3	4.18	15⁄16	15.5	3.83	1.92	1.96	5.60	1.85	1.16	0.964	0.873	0.443
	⅜	12.3	3.61	⅞	13.5	3.32	1.93	1.94	4.90	1.60	1.17	0.941	0.877	0.446
	5⁄16	10.3	3.03	13⁄16	11.4	2.79	1.94	1.92	4.18	1.35	1.17	0.918	0.882	0.448
ʸ**L6 x 3½ x**	⅜	11.7	3.42	⅞	12.9	3.24	1.94	2.04	3.34	1.23	0.988	0.787	0.767	0.350
R=½	5⁄16	9.8	2.87	13⁄16	10.9	2.73	1.95	2.01	2.85	1.04	0.996	0.763	0.772	0.352

ʸAvailability—Subject to Inquiry.

Table B.2 TOTAL ALLOWABLE UNIFORMLY DISTRIBUTED LOADS AND RELATED DEFLECTION FOR SELECTED LATERALLY SUPPORTED SIMPLE BEAM SECTION PROFILES AND SPANS

Total Allowable Uniformly Distributed Loads (kip) And Maximum Deflection (in)
For Simple Beams That Are Laterally Supported
W8 and W10 STEEL BEAMS (F_y = 36 000 lb/in^2)

span ft	8	9	10	11	12	13	14	15	16	17	18	19	20	21	22	23	24	25	26	27	28	29	30	31	32	33	34	35
W8X 10	16	14	12	11	10	10	9	8	8	7	7	7	6	6	6	5	5	5	5	5	4	4	4	4	4	4	4	4
	0.20	0.25	0.31	0.38	0.45	0.53	0.62	0.71	0.81	0.91	1.02	1.14	1.26	1.39	1.52	1.67	1.81	1.97	2.13	2.29	2.47	2.65	2.83	3.03	3.22	3.43	3.64	3.86
15	24	21	19	17	16	15	13	13	12	11	10	10	9	9	9	8	8	8	7	7	7	7	6	6	6	6	6	5
	0.20	0.25	0.31	0.37	0.44	0.52	0.60	0.69	0.78	0.88	0.99	1.10	1.22	1.35	1.48	1.61	1.76	1.91	2.06	2.22	2.39	2.57	2.75	2.93	3.12	3.32	3.53	3.74
21	36	32	29	26	24	22	21	19	18	17	16	15	15	14	13	13	12	12	11	11	10	10	10	9	9	9	9	8
	0.19	0.24	0.30	0.36	0.43	0.51	0.59	0.68	0.77	0.87	0.97	1.08	1.20	1.32	1.45	1.59	1.73	1.88	2.03	2.19	2.35	2.52	2.70	2.88	3.07	3.27	3.47	3.68
28	49	43	39	35	32	30	28	26	24	23	22	20	19	19	18	17	16	16	15	14	14	13	13	13	12	12	11	11
	0.20	0.25	0.31	0.37	0.44	0.52	0.60	0.69	0.79	0.89	1.00	1.11	1.23	1.36	1.49	1.63	1.77	1.92	2.08	2.24	2.41	2.59	2.77	2.96	3.15	3.35	3.56	3.77
40	71	63	57	52	47	44	41	38	36	33	32	30	28	27	26	25	24	23	22	21	20	20	19	18	18	17	17	16
	0.19	0.24	0.30	0.37	0.43	0.51	0.59	0.68	0.77	0.87	0.98	1.09	1.21	1.33	1.46	1.60	1.74	1.89	2.04	2.20	2.37	2.54	2.72	2.90	3.09	3.29	3.49	3.70
67	121	107	97	88	81	74	69	64	60	57	54	51	48	46	44	42	40	39	37	36	35	33	32	31	30	29	28	28
	0.18	0.22	0.28	0.33	0.40	0.47	0.54	0.62	0.71	0.80	0.89	1.00	1.10	1.22	1.33	1.46	1.59	1.72	1.86	2.01	2.16	2.32	2.48	2.65	2.82	3.00	3.19	3.38
W10X 12	22	19	17	16	15	13	12	12	11	10	10	9	9	8	8	8	7	7	7	6	6	6	6	6	5	5	5	5
	0.16	0.21	0.26	0.31	0.37	0.43	0.50	0.57	0.65	0.74	0.83	0.92	1.02	1.13	1.24	1.35	1.47	1.60	1.73	1.86	2.00	2.15	2.30	2.45	2.61	2.78	2.95	3.13
19	38	33	30	27	25	23	21	20	19	18	17	16	15	14	14	13	13	12	11	11	10	10	10	9	9	9	9	9
	0.16	0.20	0.24	0.29	0.35	0.41	0.47	0.55	0.62	0.70	0.79	0.87	0.97	1.07	1.17	1.28	1.40	1.51	1.64	1.77	1.90	2.04	2.18	2.33	2.48	2.64	2.80	2.97
30	65	58	52	47	43	40	37	35	32	30	29	27	26	25	24	23	22	21	20	19	19	18	17	17	16	16	15	15
	0.15	0.19	0.24	0.29	0.34	0.40	0.46	0.53	0.61	0.68	0.77	0.85	0.95	1.04	1.15	1.25	1.36	1.48	1.60	1.72	1.85	1.99	2.13	2.27	2.42	2.58	2.74	2.90
45	98	87	79	71	65	60	56	52	49	46	44	41	39	37	36	34	33	31	30	29	28	27	26	25	25	24	23	22
	0.16	0.20	0.25	0.30	0.35	0.42	0.48	0.55	0.63	0.71	0.80	0.89	0.98	1.08	1.19	1.30	1.42	1.54	1.66	1.79	1.93	2.07	2.21	2.36	2.52	2.68	2.84	3.01
54	120	107	96	87	80	74	69	64	60	56	53	51	48	46	44	42	40	38	37	36	34	33	32	31	30	29	28	27
	0.16	0.20	0.25	0.30	0.35	0.42	0.48	0.55	0.63	0.71	0.80	0.89	0.98	1.08	1.19	1.30	1.42	1.54	1.66	1.79	1.93	2.07	2.21	2.36	2.52	2.68	2.84	3.01
77	172	153	137	125	115	106	98	92	86	81	76	72	69	65	62	60	57	55	53	51	49	47	46	44	43	42	40	39
	0.15	0.19	0.23	0.28	0.34	0.40	0.46	0.53	0.60	0.68	0.76	0.85	0.94	1.03	1.13	1.24	1.35	1.46	1.58	1.71	1.84	1.97	2.11	2.25	2.40	2.55	2.71	2.87
112	252	224	202	183	168	155	144	134	126	119	112	106	101	96	92	88	84	81	78	75	72	70	67	65	63	61	59	58
	0.14	0.18	0.22	0.26	0.31	0.37	0.43	0.49	0.56	0.63	0.71	0.79	0.87	0.96	1.06	1.16	1.26	1.37	1.48	1.59	1.71	1.84	1.97	2.10	2.24	2.38	2.53	2.68

Computed based upon flexure using data from the Bethlehem Steel *Structural Shapes Handbook*. Data used with permission of Bethlehem Steel Corporation.

Table B.2 (cont'd.)

Total Allowable Uniformly Distributed Loads (kip) And Maximum Deflection (in) For Simple Beams That Are Laterally Supported
W12 and W14 STEEL BEAMS (F_y = 36 000 lb/in²)

span ft	12	13	14	15	16	17	18	19	20	21	22	23	24	25	26	27	28	29	30	32	34	36	38	40	42	44	46	48
W12X 22	34	31	29	27	25	24	23	21	20	19	18	18	17	16	16	15	15	14	14	13	12	11	11	10	10	9	9	8
	0.29	*0.34*	*0.40*	*0.45*	*0.52*	*0.58*	*0.65*	*0.73*	*0.81*	*0.89*	*0.98*	*1.07*	*1.16*	*1.26*	*1.37*	*1.47*	*1.58*	*1.70*	*1.82*	*2.07*	*2.34*	*2.62*	*2.92*	*3.23*	*3.57*	*3.91*	*4.28*	*4.66*
35	61	56	52	49	46	43	41	38	36	35	33	32	30	29	28	27	26	25	24	23	21	20	19	18	17	17	16	15
	0.29	*0.34*	*0.39*	*0.45*	*0.51*	*0.57*	*0.64*	*0.72*	*0.79*	*0.88*	*0.96*	*1.05*	*1.14*	*1.24*	*1.34*	*1.45*	*1.56*	*1.67*	*1.79*	*2.03*	*2.30*	*2.57*	*2.87*	*3.18*	*3.50*	*3.85*	*4.20*	*4.58*
50	86	80	74	69	65	61	58	54	52	49	47	45	43	41	40	38	37	36	35	32	30	29	27	26	25	24	23	22
	0.29	*0.34*	*0.40*	*0.46*	*0.52*	*0.59*	*0.66*	*0.74*	*0.82*	*0.90*	*0.99*	*1.08*	*1.17*	*1.27*	*1.38*	*1.49*	*1.60*	*1.71*	*1.83*	*2.09*	*2.36*	*2.64*	*2.94*	*3.26*	*3.60*	*3.95*	*4.31*	*4.70*
58	104	96	89	83	78	73	69	66	62	59	57	54	52	50	48	46	45	43	42	39	37	35	33	31	30	28	27	26
	0.29	*0.34*	*0.40*	*0.46*	*0.52*	*0.59*	*0.66*	*0.74*	*0.82*	*0.90*	*0.99*	*1.08*	*1.17*	*1.27*	*1.38*	*1.49*	*1.60*	*1.71*	*1.83*	*2.09*	*2.36*	*2.64*	*2.94*	*3.26*	*3.60*	*3.95*	*4.31*	*4.70*
72	130	120	111	104	97	92	87	82	78	74	71	68	65	62	60	58	56	54	52	49	46	43	41	39	37	35	34	32
	0.29	*0.34*	*0.40*	*0.46*	*0.52*	*0.59*	*0.66*	*0.73*	*0.81*	*0.89*	*0.98*	*1.07*	*1.17*	*1.27*	*1.37*	*1.48*	*1.59*	*1.70*	*1.82*	*2.07*	*2.34*	*2.62*	*2.92*	*3.24*	*3.57*	*3.92*	*4.29*	*4.67*
96	175	161	150	140	131	123	116	110	105	100	95	91	87	84	81	78	75	72	70	66	62	58	55	52	50	48	46	44
	0.28	*0.33*	*0.38*	*0.44*	*0.50*	*0.56*	*0.63*	*0.70*	*0.78*	*0.86*	*0.94*	*1.03*	*1.12*	*1.22*	*1.32*	*1.42*	*1.53*	*1.64*	*1.76*	*2.00*	*2.26*	*2.53*	*2.82*	*3.12*	*3.44*	*3.78*	*4.13*	*4.50*
136	248	229	213	198	186	175	165	157	149	142	135	129	124	119	114	110	106	103	99	93	88	83	78	74	71	68	65	62
	0.27	*0.31*	*0.36*	*0.42*	*0.48*	*0.54*	*0.60*	*0.67*	*0.74*	*0.82*	*0.90*	*0.99*	*1.07*	*1.16*	*1.26*	*1.36*	*1.46*	*1.57*	*1.68*	*1.91*	*2.15*	*2.41*	*2.69*	*2.98*	*3.28*	*3.60*	*3.94*	*4.29*
190	351	324	301	281	263	248	234	221	210	200	191	183	175	168	162	156	150	145	140	132	124	117	111	105	100	96	91	88
	0.25	*0.29*	*0.34*	*0.39*	*0.44*	*0.50*	*0.56*	*0.62*	*0.69*	*0.76*	*0.84*	*0.91*	*0.99*	*1.08*	*1.17*	*1.26*	*1.35*	*1.45*	*1.55*	*1.77*	*2.00*	*2.24*	*2.49*	*2.76*	*3.05*	*3.34*	*3.66*	*3.98*
W14X 26	47	43	40	38	35	33	31	30	28	27	26	25	24	23	22	21	20	19	19	18	17	16	15	14	13	13	12	12
	0.26	*0.30*	*0.35*	*0.40*	*0.46*	*0.52*	*0.58*	*0.65*	*0.72*	*0.79*	*0.87*	*0.95*	*1.03*	*1.12*	*1.21*	*1.30*	*1.40*	*1.50*	*1.61*	*1.83*	*2.07*	*2.32*	*2.58*	*2.86*	*3.16*	*3.46*	*3.78*	*4.12*
38	73	67	62	58	55	51	49	46	44	42	40	38	36	34	32	31	30	29	27	26	24	23	22	21	20	19	19	18
	0.25	*0.30*	*0.35*	*0.40*	*0.45*	*0.51*	*0.57*	*0.64*	*0.70*	*0.78*	*0.85*	*0.93*	*1.01*	*1.10*	*1.19*	*1.28*	*1.38*	*1.48*	*1.58*	*1.80*	*2.03*	*2.28*	*2.54*	*2.82*	*3.11*	*3.41*	*3.73*	*4.06*
53	104	96	89	83	78	73	69	66	62	59	57	54	52	50	48	46	44	43	41	39	37	35	33	31	30	28	27	26
	0.26	*0.30*	*0.35*	*0.40*	*0.46*	*0.52*	*0.58*	*0.64*	*0.71*	*0.79*	*0.86*	*0.94*	*1.03*	*1.12*	*1.21*	*1.30*	*1.40*	*1.50*	*1.61*	*1.83*	*2.06*	*2.31*	*2.58*	*2.86*	*3.15*	*3.46*	*3.78*	*4.11*
82	164	151	141	131	123	116	109	104	98	94	89	86	82	79	76	73	70	68	66	62	58	55	52	49	47	45	43	41
	0.25	*0.29*	*0.34*	*0.39*	*0.44*	*0.50*	*0.56*	*0.62*	*0.69*	*0.76*	*0.84*	*0.92*	*1.00*	*1.08*	*1.17*	*1.26*	*1.36*	*1.46*	*1.56*	*1.77*	*2.00*	*2.24*	*2.50*	*2.77*	*3.05*	*3.35*	*3.66*	*3.99*
99	209	193	179	167	157	148	140	132	126	120	114	109	105	100	97	93	90	87	84	79	74	70	66	63	60	57	55	52
	0.25	*0.30*	*0.34*	*0.40*	*0.45*	*0.51*	*0.57*	*0.63*	*0.70*	*0.77*	*0.85*	*0.93*	*1.01*	*1.10*	*1.19*	*1.28*	*1.38*	*1.48*	*1.58*	*1.80*	*2.03*	*2.28*	*2.54*	*2.81*	*3.10*	*3.40*	*3.72*	*4.05*
132	279	257	239	223	209	197	186	176	167	159	152	145	139	134	129	124	119	115	111	105	98	93	88	84	80	76	73	70
	0.24	*0.29*	*0.33*	*0.38*	*0.43*	*0.49*	*0.55*	*0.61*	*0.68*	*0.75*	*0.82*	*0.90*	*0.98*	*1.06*	*1.15*	*1.24*	*1.33*	*1.43*	*1.53*	*1.74*	*1.96*	*2.20*	*2.45*	*2.71*	*2.99*	*3.28*	*3.59*	*3.91*
193	413	382	354	331	310	292	276	261	248	236	225	216	207	198	191	184	177	171	165	155	146	138	131	124	118	113	108	103
	0.23	*0.27*	*0.31*	*0.36*	*0.41*	*0.46*	*0.52*	*0.58*	*0.64*	*0.71*	*0.78*	*0.85*	*0.92*	*1.00*	*1.08*	*1.17*	*1.26*	*1.35*	*1.44*	*1.64*	*1.85*	*2.08*	*2.32*	*2.57*	*2.83*	*3.10*	*3.39*	*3.69*
257	553	511	474	443	415	391	369	349	332	316	302	289	277	266	255	246	237	229	221	208	195	184	175	166	158	151	144	138
	0.22	*0.26*	*0.30*	*0.34*	*0.39*	*0.44*	*0.49*	*0.55*	*0.61*	*0.67*	*0.73*	*0.80*	*0.87*	*0.95*	*1.02*	*1.10*	*1.19*	*1.27*	*1.36*	*1.55*	*1.75*	*1.96*	*2.19*	*2.42*	*2.67*	*2.93*	*3.21*	*3.49*
342	745	688	639	596	559	526	497	471	447	426	407	389	373	358	344	331	319	308	298	280	263	248	235	224	213	203	194	186
	0.20	*0.24*	*0.28*	*0.32*	*0.36*	*0.41*	*0.46*	*0.51*	*0.57*	*0.62*	*0.69*	*0.75*	*0.82*	*0.89*	*0.96*	*1.03*	*1.11*	*1.19*	*1.27*	*1.45*	*1.64*	*1.84*	*2.04*	*2.27*	*2.50*	*2.74*	*3.00*	*3.26*

Computed based upon flexure using data from the Bethlehem Steel *Structural Shapes Handbook*. Data used with permission of Bethlehem Steel Corporation.

Table B.2 (cont'd.)

Total Allowable Uniformly Distributed Loads (kip) And Maximum Deflection (in)
For Simple Beams That Are Laterally Supported
W16 and W18 STEEL BEAMS (F_y = 36 000 lb/in²)

span ft	12	13	14	15	16	17	18	19	20	21	22	23	24	25	26	27	28	29	30	32	34	36	38	40	42	44	46	48
W16X31	63	58	54	50	47	44	42	40	38	36	34	33	31	30	29	28	27	26	25	24	22	21	20	19	18	17	16	16
	0.22	0.26	0.31	0.35	0.40	0.45	0.51	0.56	0.62	0.69	0.76	0.83	0.90	0.98	1.06	1.14	1.22	1.31	1.41	1.60	1.81	2.02	2.26	2.50	2.76	3.02	3.31	3.60
45	97	89	83	78	73	68	65	61	58	55	53	51	48	47	45	43	42	40	39	36	34	32	31	29	28	26	25	24
	0.22	0.26	0.30	0.35	0.39	0.45	0.50	0.56	0.62	0.68	0.75	0.81	0.89	0.96	1.04	1.12	1.21	1.30	1.39	1.58	1.78	2.00	2.22	2.46	2.72	2.98	3.26	3.55
57	123	113	105	98	92	87	82	78	74	70	67	64	61	59	57	55	53	51	49	46	43	41	39	37	35	34	32	31
	0.22	0.26	0.30	0.34	0.39	0.44	0.49	0.55	0.60	0.67	0.73	0.80	0.87	0.94	1.02	1.10	1.18	1.27	1.36	1.55	1.75	1.96	2.18	2.42	2.66	2.92	3.20	3.48
77	179	165	153	143	134	126	119	113	107	102	97	93	89	86	82	79	77	74	71	67	63	60	56	54	51	49	47	45
	0.22	0.25	0.29	0.34	0.38	0.43	0.49	0.54	0.60	0.66	0.73	0.79	0.86	0.94	1.01	1.09	1.17	1.26	1.35	1.53	1.73	1.94	2.16	2.40	2.64	2.90	3.17	3.45
89	207	191	177	165	155	146	138	131	124	118	113	108	103	99	95	92	89	86	83	78	73	69	65	62	59	56	54	52
	0.21	0.25	0.29	0.33	0.38	0.43	0.48	0.53	0.59	0.65	0.72	0.78	0.85	0.93	1.00	1.08	1.16	1.24	1.33	1.52	1.71	1.92	2.14	2.37	2.61	2.87	3.13	3.41
100	233	215	200	187	175	165	156	147	140	133	127	122	117	112	108	104	100	97	93	88	82	78	74	70	67	64	61	58
	0.21	0.25	0.29	0.33	0.37	0.42	0.47	0.53	0.58	0.64	0.71	0.77	0.84	0.91	0.99	1.06	1.14	1.23	1.31	1.49	1.69	1.89	2.11	2.33	2.57	2.82	3.09	3.36
W18X35	77	71	66	61	58	54	51	49	46	44	42	40	38	37	35	34	33	32	31	29	27	26	24	23	22	21	20	19
	0.20	0.24	0.27	0.32	0.36	0.41	0.45	0.51	0.56	0.62	0.68	0.74	0.81	0.88	0.95	1.02	1.10	1.18	1.26	1.44	1.62	1.82	2.02	2.24	2.47	2.71	2.97	3.23
46	105	97	90	84	79	74	70	66	63	60	57	55	53	50	48	47	45	43	42	39	37	35	33	32	30	29	27	26
	0.20	0.23	0.27	0.31	0.35	0.40	0.45	0.50	0.55	0.61	0.66	0.73	0.79	0.86	0.93	1.00	1.08	1.16	1.24	1.41	1.59	1.78	1.98	2.20	2.42	2.66	2.91	3.17
60	144	133	123	115	108	102	96	91	86	82	79	75	72	69	66	64	62	60	58	54	51	48	45	43	41	39	38	36
	0.20	0.23	0.27	0.31	0.35	0.39	0.44	0.49	0.54	0.60	0.66	0.72	0.78	0.85	0.92	0.99	1.07	1.15	1.23	1.40	1.58	1.77	1.97	2.18	2.40	2.64	2.88	3.14
71	169	156	145	135	127	120	113	107	102	97	92	88	85	81	78	75	73	70	68	64	60	56	53	51	48	46	44	42
	0.19	0.23	0.26	0.30	0.34	0.39	0.44	0.49	0.54	0.59	0.65	0.71	0.78	0.84	0.91	0.98	1.06	1.13	1.21	1.38	1.56	1.75	1.95	2.16	2.38	2.61	2.85	3.10
97	251	231	215	201	188	177	167	158	150	143	137	131	125	120	116	111	107	104	100	94	88	84	79	75	72	68	65	63
	0.19	0.23	0.26	0.30	0.34	0.39	0.43	0.48	0.53	0.59	0.65	0.71	0.77	0.83	0.90	0.97	1.05	1.12	1.20	1.37	1.54	1.73	1.93	2.13	2.35	2.58	2.82	3.07
119	308	284	264	246	231	217	205	195	185	176	168	161	154	148	142	137	132	127	123	116	109	103	97	92	88	84	80	77
	0.19	0.22	0.26	0.29	0.34	0.38	0.42	0.47	0.52	0.58	0.63	0.69	0.75	0.82	0.89	0.95	1.03	1.10	1.18	1.34	1.51	1.70	1.89	2.10	2.31	2.53	2.77	3.02

Computed based upon flexure using data from the Bethlehem Steel *Structural Shapes Handbook.* Data used with permission of Bethlehem Steel Corporation.

Table B.2 (cont'd.)

Total Allowable Uniformly Distributed Loads (kip) And Maximum Deflection (in)
For Simple Beams That Are Laterally Supported
W21 and W24 STEEL BEAMS (F$_Y$ = 36 000 lb/in²)

span ft	14	16	18	20	22	24	26	28	30	32	34	36	38	40	42	44	46	48	50	52	54	56	58	60	62	64	66	68
W21X 44	93	82	73	65	59	54	50	47	44	41	38	36	34	33	31	30	28	27	26	25	24	23	23	22	21	20	20	19
	0.24	*0.31*	*0.39*	*0.48*	*0.58*	*0.69*	*0.81*	*0.94*	*1.08*	*1.23*	*1.39*	*1.56*	*1.74*	*1.92*	*2.12*	*2.33*	*2.54*	*2.77*	*3.00*	*3.25*	*3.50*	*3.77*	*4.04*	*4.33*	*4.62*	*4.92*	*5.23*	*5.56*
57	127	111	99	89	81	74	68	63	59	56	52	49	47	44	42	40	39	37	36	34	33	32	31	30	29	28	27	26
	0.23	*0.30*	*0.38*	*0.47*	*0.57*	*0.68*	*0.80*	*0.92*	*1.06*	*1.21*	*1.36*	*1.53*	*1.70*	*1.88*	*2.08*	*2.28*	*2.49*	*2.71*	*2.94*	*3.18*	*3.43*	*3.69*	*3.96*	*4.24*	*4.53*	*4.82*	*5.13*	*5.45*
73	173	151	134	121	110	101	93	86	81	76	71	67	64	60	58	55	53	50	48	46	45	43	42	40	39	38	37	36
	0.23	*0.30*	*0.38*	*0.47*	*0.57*	*0.67*	*0.79*	*0.92*	*1.05*	*1.20*	*1.35*	*1.52*	*1.69*	*1.87*	*2.07*	*2.27*	*2.48*	*2.70*	*2.93*	*3.17*	*3.42*	*3.67*	*3.94*	*4.22*	*4.50*	*4.80*	*5.10*	*5.42*
93	219	192	171	154	140	128	118	110	102	96	90	85	81	77	73	70	67	64	61	59	57	55	53	51	50	48	47	45
	0.23	*0.29*	*0.37*	*0.46*	*0.56*	*0.66*	*0.78*	*0.90*	*1.04*	*1.18*	*1.33*	*1.49*	*1.66*	*1.84*	*2.03*	*2.23*	*2.44*	*2.65*	*2.88*	*3.11*	*3.36*	*3.61*	*3.87*	*4.15*	*4.43*	*4.72*	*5.02*	*5.32*
122	312	273	243	218	199	182	168	156	146	137	128	121	115	109	104	99	95	91	87	84	81	78	75	73	70	68	66	64
	0.22	*0.29*	*0.37*	*0.46*	*0.55*	*0.66*	*0.77*	*0.90*	*1.03*	*1.17*	*1.32*	*1.48*	*1.65*	*1.83*	*2.02*	*2.22*	*2.42*	*2.64*	*2.86*	*3.10*	*3.34*	*3.59*	*3.85*	*4.12*	*4.40*	*4.69*	*4.99*	*5.29*
147	376	329	292	263	239	219	202	188	175	165	155	146	139	132	125	120	114	110	105	101	97	94	91	88	85	82	80	77
	0.22	*0.29*	*0.36*	*0.45*	*0.54*	*0.65*	*0.76*	*0.88*	*1.01*	*1.15*	*1.30*	*1.46*	*1.62*	*1.80*	*1.98*	*2.18*	*2.38*	*2.59*	*2.81*	*3.04*	*3.28*	*3.53*	*3.78*	*4.05*	*4.32*	*4.61*	*4.90*	*5.20*
W24X 62	150	131	116	105	95	87	81	75	70	66	62	58	55	52	50	48	46	44	42	40	39	37	36	35	34	33	32	31
	0.21	*0.27*	*0.34*	*0.42*	*0.51*	*0.60*	*0.71*	*0.82*	*0.94*	*1.07*	*1.21*	*1.36*	*1.51*	*1.68*	*1.85*	*2.03*	*2.22*	*2.42*	*2.62*	*2.84*	*3.06*	*3.29*	*3.53*	*3.66*	*3.91*	*4.30*	*4.57*	*4.85*
76	201	176	156	141	128	117	108	101	94	88	83	78	74	70	67	64	61	59	56	54	52	50	49	47	45	44	43	41
	0.20	*0.27*	*0.34*	*0.42*	*0.50*	*0.60*	*0.70*	*0.82*	*0.94*	*1.07*	*1.20*	*1.35*	*1.50*	*1.66*	*1.84*	*2.01*	*2.20*	*2.40*	*2.60*	*2.81*	*3.03*	*3.26*	*3.50*	*3.75*	*4.00*	*4.26*	*4.53*	*4.81*
94	254	222	197	178	161	148	137	127	118	111	104	99	93	89	85	81	77	74	71	68	66	63	61	59	57	56	54	52
	0.20	*0.26*	*0.33*	*0.41*	*0.49*	*0.59*	*0.69*	*0.80*	*0.92*	*1.05*	*1.18*	*1.32*	*1.47*	*1.63*	*1.80*	*1.98*	*2.16*	*2.35*	*2.55*	*2.76*	*2.98*	*3.20*	*3.43*	*3.67*	*3.92*	*4.18*	*4.45*	*4.72*
104	295	258	229	206	188	172	159	147	138	129	121	115	109	103	98	94	90	86	83	79	76	74	71	69	67	65	63	61
	0.20	*0.26*	*0.33*	*0.41*	*0.50*	*0.60*	*0.70*	*0.81*	*0.93*	*1.06*	*1.19*	*1.34*	*1.49*	*1.65*	*1.82*	*2.00*	*2.19*	*2.38*	*2.58*	*2.79*	*3.01*	*3.24*	*3.48*	*3.72*	*3.97*	*4.23*	*4.50*	*4.78*
131	376	329	292	263	239	219	202	188	175	165	155	146	139	132	125	120	114	110	105	101	97	94	91	88	85	82	80	77
	0.20	*0.26*	*0.33*	*0.41*	*0.49*	*0.59*	*0.69*	*0.80*	*0.91*	*1.04*	*1.17*	*1.32*	*1.47*	*1.63*	*1.79*	*1.97*	*2.15*	*2.34*	*2.54*	*2.75*	*2.96*	*3.19*	*3.42*	*3.66*	*3.91*	*4.16*	*4.43*	*4.70*
162	473	414	368	331	301	276	255	237	221	207	195	184	174	166	158	151	144	138	132	127	123	118	114	110	107	104	100	97
	0.19	*0.25*	*0.32*	*0.40*	*0.48*	*0.57*	*0.67*	*0.78*	*0.89*	*1.02*	*1.15*	*1.29*	*1.44*	*1.59*	*1.75*	*1.92*	*2.10*	*2.29*	*2.49*	*2.69*	*2.90*	*3.12*	*3.34*	*3.58*	*3.82*	*4.07*	*4.33*	*4.60*

Computed based upon flexure using data from the Bethlehem Steel *Structural Shapes Handbook*. Data used with permission of Bethlehem Steel Corporation.

Table B.2 (cont'd.)

Total Allowable Uniformly Distributed Loads (kip) And Maximum Deflection (in)
For Simple Beams That Are Laterally Supported
W27 and W30 STEEL BEAMS (F_y = 36 000 lb/in²)

span ft	18	20	22	24	26	28	30	32	34	36	38	40	42	44	46	48	50	52	54	56	58	60	62	64	66	68	70	72
W27X 84	189	170	155	142	131	122	114	107	100	95	90	85	81	77	74	71	68	66	63	61	59	57	55	53	52	50	49	47
	0.30	0.37	0.45	0.53	0.63	0.73	0.83	0.95	1.07	1.20	1.34	1.48	1.64	1.80	1.96	2.14	2.32	2.51	2.71	2.91	3.12	3.34	3.57	3.80	4.04	4.29	4.55	4.81
102	237	214	194	178	164	153	142	134	126	119	112	107	102	97	93	89	85	82	79	76	74	71	69	67	65	63	61	59
	0.30	0.37	0.44	0.53	0.62	0.72	0.82	0.94	1.06	1.19	1.32	1.46	1.62	1.77	1.94	2.11	2.29	2.48	2.67	2.87	3.08	3.30	3.52	3.75	3.99	4.23	4.49	4.75
114	266	239	217	199	184	171	159	150	141	133	126	120	114	109	104	100	96	92	89	85	82	80	77	75	72	70	68	66
	0.29	0.36	0.44	0.52	0.61	0.71	0.82	0.93	1.05	1.18	1.31	1.45	1.60	1.76	1.92	2.09	2.27	2.45	2.65	2.85	3.05	3.27	3.49	3.72	3.95	4.20	4.45	4.70
146	365	329	299	274	253	235	219	206	193	183	173	164	157	149	143	137	132	126	122	117	113	110	106	103	100	97	94	91
	0.29	0.36	0.44	0.52	0.61	0.71	0.82	0.93	1.05	1.17	1.31	1.45	1.60	1.75	1.92	2.09	2.27	2.45	2.64	2.84	3.05	3.26	3.48	3.71	3.95	4.19	4.44	4.70
161	404	364	331	303	280	260	243	228	214	202	192	182	173	165	158	152	146	140	135	130	126	121	117	114	110	107	104	101
	0.29	0.36	0.44	0.52	0.61	0.71	0.81	0.92	1.04	1.17	1.30	1.44	1.59	1.74	1.90	2.07	2.25	2.43	2.62	2.82	3.03	3.24	3.46	3.68	3.92	4.16	4.41	4.66
178	446	402	365	335	309	287	268	251	236	223	211	201	191	183	175	167	161	154	149	143	138	134	130	126	122	118	115	112
	0.29	0.36	0.43	0.51	0.60	0.70	0.80	0.91	1.03	1.16	1.29	1.43	1.57	1.73	1.89	2.05	2.23	2.41	2.60	2.80	3.00	3.21	3.43	3.65	3.88	4.12	4.37	4.62
W30X 99	239	215	196	179	166	154	143	135	127	120	113	108	102	98	94	90	86	83	80	77	74	72	69	67	65	63	61	60
	0.27	0.33	0.41	0.48	0.57	0.66	0.75	0.86	0.97	1.08	1.21	1.34	1.48	1.62	1.77	1.93	2.09	2.26	2.44	2.62	2.82	3.01	3.22	3.43	3.65	3.87	4.10	4.34
116	292	263	239	219	202	188	175	165	155	146	139	132	125	120	114	110	105	101	97	94	91	88	85	82	80	77	75	73
	0.27	0.33	0.40	0.48	0.56	0.65	0.75	0.85	0.96	1.07	1.20	1.33	1.46	1.60	1.75	1.91	2.07	2.24	2.42	2.60	2.79	2.98	3.18	3.39	3.61	3.83	4.06	4.29
132	338	304	276	253	234	217	203	190	179	169	160	152	145	138	132	127	122	117	113	109	105	101	98	95	92	89	87	84
	0.26	0.33	0.40	0.47	0.55	0.64	0.74	0.84	0.95	1.06	1.18	1.31	1.44	1.58	1.73	1.88	2.04	2.21	2.38	2.56	2.75	2.94	3.14	3.35	3.56	3.78	4.01	4.24
173	479	431	392	359	332	308	287	270	254	240	227	216	205	196	187	180	172	166	160	154	149	144	139	135	131	127	123	120
	0.26	0.33	0.39	0.47	0.55	0.64	0.73	0.84	0.94	1.06	1.18	1.31	1.44	1.58	1.73	1.88	2.04	2.21	2.38	2.56	2.74	2.94	3.14	3.34	3.55	3.77	4.00	4.23
191	532	478	435	399	368	342	319	299	281	266	252	239	228	217	208	199	191	184	177	171	165	159	154	150	145	141	137	133
	0.26	0.32	0.39	0.47	0.55	0.63	0.73	0.83	0.94	1.05	1.17	1.30	1.43	1.57	1.71	1.87	2.02	2.19	2.36	2.54	2.72	2.91	3.11	3.32	3.53	3.74	3.97	4.20
211	589	530	482	442	408	379	354	332	312	295	279	265	253	241	231	221	212	204	196	189	183	177	171	166	161	156	152	147
	0.26	0.32	0.39	0.46	0.54	0.63	0.72	0.82	0.92	1.04	1.15	1.28	1.41	1.55	1.69	1.84	2.00	2.16	2.33	2.51	2.69	2.88	3.07	3.27	3.48	3.69	3.92	4.14

Computed based upon flexure using data from the Bethlehem Steel *Structural Shapes Handbook*. Data used with permission of Bethlehem Steel Corporation.

Table B.2 (cont'd.)

Total Allowable Uniformly Distributed Loads (kip) And Maximum Deflection (in)
For Simple Beams That Are Laterally Supported
W33 and W36 STEEL BEAMS (F_y = 36 000 lb/in²)

span ft	18	20	22	24	26	28	30	32	34	36	38	40	42	44	46	48	50	52	54	56	58	60	62	64	66	68	70	72
W33X 118	319	287	261	239	221	205	191	180	169	160	151	144	137	131	125	120	115	110	106	103	99	96	93	90	87	84	82	80
	0.24	*0.30*	*0.37*	*0.44*	*0.51*	*0.59*	*0.68*	*0.77*	*0.87*	*0.98*	*1.09*	*1.21*	*1.33*	*1.46*	*1.60*	*1.74*	*1.89*	*2.04*	*2.20*	*2.37*	*2.54*	*2.72*	*2.90*	*3.09*	*3.29*	*3.49*	*3.70*	*3.92*
141	398	358	326	299	276	256	239	224	211	199	189	179	171	163	156	149	143	138	133	128	124	119	116	112	109	105	102	100
	0.24	*0.30*	*0.36*	*0.43*	*0.50*	*0.59*	*0.67*	*0.76*	*0.86*	*0.97*	*1.08*	*1.19*	*1.32*	*1.45*	*1.58*	*1.72*	*1.87*	*2.02*	*2.18*	*2.34*	*2.51*	*2.69*	*2.87*	*3.06*	*3.25*	*3.45*	*3.66*	*3.87*
152	433	390	354	325	300	278	260	244	229	216	205	195	186	177	169	162	156	150	144	139	134	130	126	122	118	115	111	108
	0.24	*0.30*	*0.36*	*0.43*	*0.50*	*0.58*	*0.67*	*0.76*	*0.86*	*0.96*	*1.07*	*1.19*	*1.31*	*1.43*	*1.57*	*1.71*	*1.85*	*2.00*	*2.16*	*2.32*	*2.49*	*2.67*	*2.85*	*3.03*	*3.23*	*3.43*	*3.63*	*3.84*
201	608	547	497	456	421	391	365	342	322	304	288	274	261	249	238	228	219	210	203	195	189	182	177	171	166	161	156	152
	0.24	*0.30*	*0.36*	*0.43*	*0.50*	*0.58*	*0.66*	*0.76*	*0.85*	*0.96*	*1.07*	*1.18*	*1.30*	*1.43*	*1.56*	*1.70*	*1.85*	*2.00*	*2.15*	*2.32*	*2.48*	*2.66*	*2.84*	*3.02*	*3.22*	*3.41*	*3.62*	*3.83*
221	673	606	551	505	466	433	404	379	356	336	319	303	288	275	263	252	242	233	224	216	209	202	195	189	184	178	173	168
	0.24	*0.29*	*0.36*	*0.42*	*0.50*	*0.58*	*0.66*	*0.75*	*0.85*	*0.95*	*1.06*	*1.17*	*1.30*	*1.42*	*1.55*	*1.69*	*1.84*	*1.99*	*2.14*	*2.30*	*2.47*	*2.64*	*2.82*	*3.01*	*3.20*	*3.39*	*3.60*	*3.81*
241	737	663	603	553	510	474	442	415	390	368	349	332	316	301	288	276	265	255	246	237	229	221	214	207	201	195	189	184
	0.23	*0.29*	*0.35*	*0.42*	*0.49*	*0.57*	*0.65*	*0.74*	*0.84*	*0.94*	*1.05*	*1.16*	*1.28*	*1.40*	*1.53*	*1.67*	*1.81*	*1.96*	*2.11*	*2.27*	*2.44*	*2.61*	*2.79*	*2.97*	*3.16*	*3.35*	*3.55*	*3.76*
W36X 135	390	351	319	293	270	251	234	220	207	195	185	176	167	160	153	146	140	135	130	125	121	117	113	110	106	103	100	98
	0.23	*0.28*	*0.34*	*0.40*	*0.47*	*0.55*	*0.63*	*0.72*	*0.81*	*0.91*	*1.01*	*1.12*	*1.23*	*1.35*	*1.48*	*1.61*	*1.75*	*1.89*	*2.04*	*2.19*	*2.35*	*2.52*	*2.69*	*2.86*	*3.04*	*3.23*	*3.42*	*3.62*
160	482	434	394	361	334	310	289	271	255	241	228	217	206	197	189	181	173	167	161	155	150	145	140	136	131	128	124	120
	0.22	*0.28*	*0.33*	*0.40*	*0.47*	*0.54*	*0.62*	*0.71*	*0.80*	*0.89*	*1.00*	*1.10*	*1.22*	*1.34*	*1.46*	*1.59*	*1.73*	*1.87*	*2.01*	*2.16*	*2.32*	*2.48*	*2.65*	*2.83*	*3.01*	*3.19*	*3.38*	*3.58*
182	554	498	453	415	383	356	332	312	293	277	262	249	237	227	217	208	199	192	185	178	172	166	161	156	151	147	142	138
	0.22	*0.27*	*0.33*	*0.39*	*0.46*	*0.54*	*0.62*	*0.70*	*0.79*	*0.89*	*0.99*	*1.10*	*1.21*	*1.33*	*1.45*	*1.58*	*1.71*	*1.85*	*2.00*	*2.15*	*2.30*	*2.46*	*2.63*	*2.80*	*2.98*	*3.16*	*3.35*	*3.55*
210	639	575	523	479	442	411	383	360	338	320	303	288	274	261	250	240	230	221	213	205	198	192	186	180	174	169	164	160
	0.22	*0.27*	*0.33*	*0.39*	*0.46*	*0.53*	*0.61*	*0.69*	*0.78*	*0.88*	*0.98*	*1.08*	*1.19*	*1.31*	*1.43*	*1.56*	*1.69*	*1.83*	*1.97*	*2.12*	*2.27*	*2.43*	*2.60*	*2.77*	*2.95*	*3.13*	*3.31*	*3.51*
230	744	670	609	558	515	478	446	419	394	372	352	335	319	304	291	279	268	258	248	239	231	223	216	209	203	197	191	186
	0.22	*0.28*	*0.34*	*0.40*	*0.47*	*0.54*	*0.62*	*0.71*	*0.80*	*0.90*	*1.00*	*1.11*	*1.22*	*1.34*	*1.47*	*1.60*	*1.73*	*1.87*	*2.02*	*2.17*	*2.33*	*2.49*	*2.66*	*2.84*	*3.02*	*3.20*	*3.39*	*3.59*
260	847	762	693	635	586	545	508	477	448	424	401	381	363	347	331	318	305	293	282	272	263	254	246	238	231	224	218	212
	0.22	*0.27*	*0.33*	*0.39*	*0.46*	*0.54*	*0.62*	*0.70*	*0.79*	*0.89*	*0.99*	*1.09*	*1.21*	*1.32*	*1.45*	*1.58*	*1.71*	*1.85*	*1.99*	*2.14*	*2.30*	*2.46*	*2.63*	*2.80*	*2.98*	*3.16*	*3.35*	*3.55*
300	987	888	807	740	683	634	592	555	522	493	467	444	423	404	386	370	355	342	329	317	306	296	286	278	269	261	254	247
	0.22	*0.27*	*0.33*	*0.39*	*0.46*	*0.53*	*0.61*	*0.70*	*0.78*	*0.88*	*0.98*	*1.09*	*1.20*	*1.31*	*1.44*	*1.56*	*1.70*	*1.84*	*1.98*	*2.13*	*2.28*	*2.44*	*2.61*	*2.78*	*2.96*	*3.14*	*3.33*	*3.52*

Computed based upon flexure using data from the Bethlehem Steel *Structural Shapes Handbook*. Data used with permission of Bethlehem Steel Corporation.

Table B.2 (cont'd.)

Total Allowable Uniformly Distributed Loads (kip) And Maximum Deflection (in)
For Simple Beams That Are Laterally Supported
W8 and W10 Steel Beams (F_Y = 50 000 lb/in²)

span ft	8	9	10	11	12	13	14	15	16	17	18	19	20	21	22	23	24	25	26	27	28	29	30	31	32	33	34	35
W8X 10	22	19	17	16	14	13	12	12	11	10	10	9	9	8	8	8	7	7	7	6	6	6	6	6	5	5	5	5
	0.28	0.35	0.44	0.53	0.63	0.74	0.86	0.98	1.12	1.26	1.42	1.58	1.75	1.93	2.12	2.31	2.52	2.73	2.96	3.19	3.43	3.68	3.93	4.20	4.48	4.76	5.05	5.36
15	33	29	26	24	22	20	19	17	16	15	15	14	13	12	12	11	11	10	10	10	9	9	9	8	8	8	8	7
	0.27	0.34	0.42	0.51	0.61	0.72	0.83	0.95	1.08	1.22	1.37	1.53	1.70	1.87	2.05	2.24	2.44	2.65	2.86	3.09	3.32	3.56	3.81	4.07	4.34	4.62	4.90	5.19
21	51	45	40	37	34	31	29	27	25	24	22	21	20	19	18	18	17	16	16	15	14	14	13	13	13	12	12	12
	0.27	0.34	0.42	0.50	0.60	0.70	0.82	0.94	1.07	1.20	1.35	1.50	1.67	1.84	2.02	2.20	2.40	2.60	2.82	3.04	3.27	3.50	3.75	4.00	4.27	4.54	4.82	5.10
28	67	60	54	49	45	42	39	36	34	32	30	28	27	26	25	23	22	22	21	20	19	19	18	17	17	16	16	15
	0.27	0.35	0.43	0.52	0.62	0.72	0.84	0.96	1.09	1.24	1.39	1.54	1.71	1.89	2.07	2.26	2.46	2.67	2.89	3.12	3.35	3.60	3.85	4.11	4.38	4.66	4.94	5.24
40	99	88	79	72	66	61	56	53	49	46	44	42	39	38	36	34	33	32	30	29	28	27	26	25	24	23	23	23
	0.27	0.34	0.42	0.51	0.60	0.71	0.82	0.94	1.07	1.21	1.36	1.51	1.68	1.85	2.03	2.22	2.41	2.62	2.83	3.06	3.29	3.53	3.77	4.03	4.29	4.56	4.85	5.13
67	168	149	134	122	112	103	96	89	84	79	75	71	67	64	61	58	56	54	52	50	48	46	45	43	42	41	39	38
	0.25	0.31	0.38	0.46	0.55	0.65	0.75	0.86	0.98	1.11	1.24	1.38	1.53	1.69	1.85	2.03	2.21	2.39	2.59	2.79	3.00	3.22	3.45	3.68	3.92	4.17	4.43	4.69
W10X 12	30	27	24	22	20	19	17	16	15	14	13	13	12	12	11	11	10	10	9	9	9	8	8	8	8	7	7	7
	0.23	0.29	0.35	0.43	0.51	0.60	0.69	0.80	0.91	1.02	1.15	1.28	1.42	1.56	1.72	1.88	2.04	2.22	2.40	2.58	2.78	2.98	3.19	3.41	3.63	3.86	4.10	4.34
19	52	46	42	38	35	32	30	28	26	25	23	22	21	20	19	18	17	17	16	15	15	14	14	13	13	13	12	12
	0.22	0.27	0.34	0.41	0.48	0.57	0.66	0.76	0.86	0.97	1.09	1.21	1.35	1.48	1.63	1.78	1.94	2.10	2.28	2.45	2.64	2.83	3.03	3.23	3.45	3.67	3.89	4.12
30	90	80	72	65	60	55	51	48	45	42	40	38	36	34	33	31	30	29	28	27	26	25	24	23	22	22	21	21
	0.21	0.27	0.33	0.40	0.47	0.56	0.64	0.74	0.84	0.95	1.06	1.19	1.31	1.45	1.59	1.74	1.89	2.05	2.22	2.40	2.58	2.76	2.96	3.16	3.36	3.58	3.80	4.02
45	136	121	109	99	91	84	78	73	68	64	61	57	55	52	50	47	45	44	42	40	39	38	36	35	34	33	32	31
	0.22	0.28	0.34	0.41	0.49	0.58	0.67	0.77	0.87	0.99	1.11	1.23	1.37	1.51	1.65	1.81	1.97	2.13	2.31	2.49	2.68	2.87	3.07	3.28	3.50	3.72	3.95	4.18
54	167	148	133	121	111	103	95	89	83	78	74	70	67	63	61	58	56	53	51	49	48	46	44	43	42	40	39	38
	0.22	0.28	0.34	0.41	0.49	0.58	0.67	0.77	0.87	0.99	1.11	1.23	1.37	1.51	1.65	1.81	1.97	2.13	2.31	2.49	2.68	2.87	3.07	3.28	3.50	3.72	3.95	4.18
77	239	212	191	174	159	147	136	127	119	112	106	100	95	91	87	83	80	76	73	71	68	66	64	62	60	58	56	55
	0.21	0.26	0.33	0.39	0.47	0.55	0.64	0.73	0.83	0.94	1.05	1.17	1.30	1.44	1.58	1.72	1.87	2.03	2.20	2.37	2.55	2.74	2.93	3.13	3.33	3.54	3.76	3.99
112	350	311	280	255	233	215	200	187	175	165	156	147	140	133	127	122	117	112	108	104	100	97	93	90	87	85	82	80
	0.19	0.25	0.30	0.37	0.44	0.51	0.59	0.68	0.78	0.88	0.98	1.10	1.21	1.34	1.47	1.60	1.75	1.90	2.05	2.21	2.38	2.55	2.73	2.92	3.11	3.30	3.51	3.72

Computed based upon flexure using data from the Bethlehem Steel *Structural Shapes Handbook*. Data used with permission of Bethlehem Steel Corporation.

Table B.2 (cont'd.)

Total Allowable Uniformly Distributed Loads (kip) And Maximum Deflection (in) For Simple Beams That Are Laterally Supported W12 and W14 Steel Beams (F_y = 50 000 lb/in²)

span ft	12	13	14	15	16	17	18	19	20	21	22	23	24	25	26	27	28	29	30	32	34	36	38	40	42	44	46	48
W12X 22	47	43	40	38	35	33	31	30	28	27	26	25	24	23	22	21	20	19	19	18	17	16	15	14	13	13	12	12
	0.40	0.47	0.55	0.63	0.72	0.81	0.91	1.01	1.12	1.24	1.36	1.48	1.62	1.75	1.90	2.05	2.20	2.36	2.53	2.87	3.24	3.64	4.05	4.49	4.95	5.43	5.94	6.47
35	84	78	72	68	63	60	56	53	51	48	46	44	42	41	39	38	36	35	34	32	30	28	27	25	24	23	22	21
	0.40	0.47	0.54	0.62	0.71	0.80	0.89	1.00	1.10	1.22	1.34	1.46	1.59	1.72	1.86	2.01	2.16	2.32	2.48	2.82	3.19	3.57	3.98	4.41	4.87	5.34	5.84	6.36
50	120	111	103	96	90	85	80	76	72	68	65	63	60	58	55	53	51	50	48	45	42	40	38	36	34	33	31	30
	0.41	0.48	0.55	0.64	0.72	0.82	0.92	1.02	1.13	1.25	1.37	1.50	1.63	1.77	1.91	2.06	2.22	2.38	2.55	2.90	3.27	3.67	4.09	4.53	4.99	5.48	5.99	6.52
58	144	133	124	116	108	102	96	91	87	83	79	75	72	69	67	64	62	60	58	54	51	48	46	43	41	39	38	36
	0.41	0.48	0.55	0.64	0.72	0.82	0.92	1.02	1.13	1.25	1.37	1.50	1.63	1.77	1.91	2.06	2.22	2.38	2.55	2.90	3.27	3.67	4.09	4.53	4.99	5.48	5.99	6.52
72	180	166	155	144	135	127	120	114	108	103	98	94	90	87	83	80	77	75	72	68	64	60	57	54	52	49	47	45
	0.41	0.48	0.55	0.63	0.72	0.81	0.91	1.02	1.13	1.24	1.36	1.49	1.62	1.76	1.90	2.05	2.21	2.37	2.53	2.88	3.25	3.65	4.06	4.50	4.96	5.45	5.95	6.48
96	243	224	208	194	182	171	162	153	146	139	132	127	121	116	112	108	104	100	97	91	86	81	77	73	69	66	63	61
	0.39	0.46	0.53	0.61	0.69	0.78	0.88	0.98	1.08	1.20	1.31	1.43	1.56	1.69	1.83	1.98	2.13	2.28	2.44	2.78	3.13	3.51	3.91	4.34	4.78	5.25	5.74	6.25
136	344	318	295	276	258	243	230	218	207	197	188	180	172	165	159	153	148	143	138	129	122	115	109	103	98	94	90	86
	0.37	0.44	0.51	0.58	0.66	0.75	0.84	0.93	1.03	1.14	1.25	1.37	1.49	1.62	1.75	1.89	2.03	2.17	2.33	2.65	2.99	3.35	3.73	4.14	4.56	5.01	5.47	5.96
190	487	450	417	390	365	344	325	308	292	278	266	254	243	234	225	216	209	202	195	183	172	162	154	146	139	133	127	122
	0.35	0.41	0.47	0.54	0.61	0.69	0.78	0.87	0.96	1.06	1.16	1.27	1.38	1.50	1.62	1.75	1.88	2.02	2.16	2.46	2.77	3.11	3.46	3.84	4.23	4.64	5.08	5.53
W14X 26	65	60	56	52	49	46	44	41	39	37	36	34	33	31	30	29	28	27	26	25	23	22	21	20	19	18	17	16
	0.36	0.42	0.49	0.56	0.64	0.72	0.80	0.90	0.99	1.10	1.20	1.31	1.43	1.55	1.68	1.81	1.95	2.09	2.24	2.54	2.87	3.22	3.59	3.97	4.38	4.81	5.26	5.72
38	101	93	87	81	76	71	67	64	61	58	55	53	51	49	47	45	43	42	40	38	36	34	32	30	29	28	26	25
	0.35	0.41	0.48	0.55	0.63	0.71	0.79	0.88	0.98	1.08	1.18	1.29	1.41	1.53	1.65	1.78	1.92	2.06	2.20	2.50	2.83	3.17	3.53	3.91	4.31	4.73	5.17	5.63
53	144	133	123	115	108	102	96	91	86	82	79	75	72	69	66	64	62	60	58	54	51	48	45	43	41	39	38	36
	0.36	0.42	0.49	0.56	0.63	0.72	0.80	0.89	0.99	1.09	1.20	1.31	1.43	1.55	1.68	1.81	1.94	2.09	2.23	2.54	2.87	3.21	3.58	3.97	4.37	4.80	5.25	5.71
82	228	210	195	182	171	161	152	144	137	130	124	119	114	109	105	101	98	94	91	85	80	76	72	68	65	62	59	57
	0.35	0.41	0.47	0.54	0.62	0.69	0.78	0.87	0.96	1.06	1.16	1.27	1.38	1.50	1.63	1.75	1.88	2.02	2.16	2.46	2.78	3.12	3.47	3.85	4.24	4.65	5.09	5.54
99	291	268	249	233	218	205	194	184	174	166	159	152	145	140	134	129	125	120	116	109	103	97	92	87	83	79	76	73
	0.35	0.41	0.48	0.55	0.62	0.70	0.79	0.88	0.98	1.08	1.18	1.29	1.40	1.52	1.65	1.78	1.91	2.05	2.19	2.50	2.82	3.16	3.52	3.90	4.30	4.72	5.16	5.62
132	387	357	332	310	290	273	258	244	232	221	211	202	193	186	179	172	166	160	155	145	137	129	122	116	111	106	101	97
	0.34	0.40	0.46	0.53	0.60	0.68	0.76	0.85	0.94	1.04	1.14	1.25	1.36	1.47	1.59	1.72	1.85	1.98	2.12	2.41	2.72	3.05	3.40	3.77	4.15	4.56	4.98	5.43
193	574	530	492	459	431	405	383	363	344	328	313	299	287	276	265	255	246	238	230	215	203	191	181	172	164	157	150	144
	0.32	0.38	0.44	0.50	0.57	0.64	0.72	0.80	0.89	0.98	1.08	1.18	1.28	1.39	1.51	1.62	1.75	1.87	2.00	2.28	2.57	2.89	3.22	3.56	3.93	4.31	4.71	5.13
257	768	709	659	615	576	542	512	485	461	439	419	401	384	369	355	342	329	318	307	288	271	256	243	231	220	210	200	192
	0.30	0.36	0.41	0.47	0.54	0.61	0.68	0.76	0.84	0.93	1.02	1.11	1.21	1.32	1.42	1.53	1.65	1.77	1.89	2.15	2.43	2.73	3.04	3.37	3.71	4.07	4.45	4.85
342	1035	955	887	828	776	731	690	654	621	591	565	540	518	497	478	460	444	428	414	388	365	345	327	311	296	282	270	259
	0.28	0.33	0.39	0.44	0.50	0.57	0.64	0.71	0.79	0.87	0.95	1.04	1.13	1.23	1.33	1.43	1.54	1.65	1.77	2.01	2.27	2.55	2.84	3.15	3.47	3.81	4.16	4.53

Computed based upon flexure using data from the Bethlehem Steel *Structural Shapes Handbook*. Data used with permission of Bethlehem Steel Corporation.

Table B.2 (cont'd.)

Total Allowable Uniformly Distributed Loads (kip) And Maximum Deflection (in)
For Simple Beams That Are Laterally Supported
W16 and W18 Steel Beams (F_y = 50 000 lb/in²)

span ft	12	13	14	15	16	17	18	19	20	21	22	23	24	25	26	27	28	29	30	32	34	36	38	40	42	44	46	48
W16X 31	87	81	75	70	66	62	58	55	52	50	48	46	44	42	40	39	37	36	35	33	31	29	28	26	25	24	23	22
	0.31	0.37	0.43	0.49	0.56	0.63	0.70	0.78	0.87	0.96	1.05	1.15	1.25	1.36	1.47	1.58	1.70	1.82	1.95	2.22	2.51	2.81	3.13	3.47	3.83	4.20	4.59	5.00
45	135	124	115	108	101	95	90	85	81	77	73	70	67	65	62	60	58	56	54	50	48	45	43	40	38	37	35	34
	0.31	0.36	0.42	0.48	0.55	0.62	0.69	0.77	0.86	0.94	1.04	1.13	1.23	1.34	1.45	1.56	1.68	1.80	1.92	2.19	2.47	2.77	3.09	3.42	3.77	4.14	4.53	4.93
57	171	158	146	137	128	121	114	108	102	98	93	89	85	82	79	76	73	71	68	64	60	57	54	51	49	47	45	43
	0.30	0.35	0.41	0.47	0.54	0.61	0.68	0.76	0.84	0.92	1.01	1.11	1.21	1.31	1.42	1.53	1.64	1.76	1.89	2.15	2.42	2.72	3.03	3.36	3.70	4.06	4.44	4.83
77	248	229	213	198	186	175	165	157	149	142	135	129	124	119	115	110	106	103	99	93	88	83	78	74	71	68	65	62
	0.30	0.35	0.41	0.47	0.53	0.60	0.67	0.75	0.83	0.92	1.01	1.10	1.20	1.30	1.41	1.52	1.63	1.75	1.87	2.13	2.41	2.70	3.01	3.33	3.67	4.03	4.40	4.80
89	287	265	246	230	215	203	191	181	172	164	157	150	144	138	132	128	123	119	115	108	101	96	91	86	82	78	75	72
	0.30	0.35	0.40	0.46	0.53	0.59	0.67	0.74	0.82	0.91	0.99	1.09	1.18	1.28	1.39	1.50	1.61	1.73	1.85	2.10	2.38	2.66	2.97	3.29	3.63	3.98	4.35	4.74
100	324	299	278	259	243	229	216	205	194	185	177	169	162	156	150	144	139	134	130	122	114	108	102	97	93	88	85	81
	0.29	0.34	0.40	0.46	0.52	0.59	0.66	0.73	0.81	0.89	0.98	1.07	1.17	1.27	1.37	1.48	1.59	1.70	1.82	2.07	2.34	2.62	2.92	3.24	3.57	3.92	4.28	4.67
W18X 35	107	98	91	85	80	75	71	67	64	61	58	56	53	51	49	47	46	44	43	40	38	36	34	32	30	29	28	27
	0.28	0.33	0.38	0.44	0.50	0.56	0.63	0.70	0.78	0.86	0.94	1.03	1.12	1.22	1.32	1.42	1.53	1.64	1.75	1.99	2.25	2.52	2.81	3.12	3.43	3.77	4.12	4.49
46	146	135	125	117	109	103	97	92	88	83	80	76	73	70	67	65	63	60	58	55	51	49	46	44	42	40	38	36
	0.27	0.32	0.37	0.43	0.49	0.55	0.62	0.69	0.76	0.84	0.92	1.01	1.10	1.19	1.29	1.39	1.50	1.60	1.72	1.95	2.21	2.47	2.76	3.05	3.37	3.69	4.04	4.40
60	200	185	171	160	150	141	133	126	120	114	109	104	100	96	92	89	86	83	80	75	71	67	63	60	57	55	52	50
	0.27	0.32	0.37	0.43	0.48	0.55	0.61	0.68	0.76	0.83	0.92	1.00	1.09	1.18	1.28	1.38	1.48	1.59	1.70	1.94	2.19	2.45	2.73	3.03	3.34	3.66	4.00	4.36
71	235	217	202	188	176	166	157	149	141	134	128	123	118	113	109	105	101	97	94	88	83	78	74	71	67	64	61	59
	0.27	0.32	0.37	0.42	0.48	0.54	0.61	0.68	0.75	0.83	0.91	0.99	1.08	1.17	1.27	1.36	1.47	1.57	1.68	1.92	2.16	2.43	2.70	2.99	3.30	3.62	3.96	4.31
97	348	321	298	278	261	246	232	220	209	199	190	182	174	167	161	155	149	144	139	131	123	116	110	104	99	95	91	87
	0.27	0.31	0.36	0.42	0.47	0.54	0.60	0.67	0.74	0.82	0.90	0.98	1.07	1.16	1.25	1.35	1.45	1.56	1.67	1.90	2.14	2.40	2.67	2.96	3.27	3.59	3.92	4.27
119	428	395	367	342	321	302	285	270	257	244	233	223	214	205	197	190	183	177	171	160	151	143	135	128	122	117	112	107
	0.26	0.31	0.36	0.41	0.47	0.53	0.59	0.66	0.73	0.80	0.88	0.96	1.05	1.14	1.23	1.33	1.43	1.53	1.64	1.86	2.10	2.36	2.63	2.91	3.21	3.52	3.85	4.19

Computed based upon flexure using data from the Bethlehem Steel *Structural Shapes Handbook*. Data used with permission of Bethlehem Steel Corporation.

Table B.2 (cont'd.)

Total Allowable Uniformly Distributed Loads (kip) And Maximum Deflection (in) For Simple Beams That Are Laterally Supported
W21 and W24 Steel Beams (F_Y = 50 000 lb/in²)

span ft	14	16	18	20	22	24	26	28	30	32	34	36	38	40	42	44	46	48	50	52	54	56	58	60	62	64	66	68
W21X 44	130	113	101	91	82	76	70	65	60	57	53	50	48	45	43	41	39	38	36	35	34	32	31	30	29	28	27	27
	0.33	0.43	0.54	0.67	0.81	0.96	1.13	1.31	1.50	1.71	1.93	2.16	2.41	2.67	2.94	3.23	3.53	3.84	4.17	4.51	4.87	5.23	5.61	6.01	6.41	6.84	7.27	7.72
57	176	154	137	123	112	103	95	88	82	77	73	69	65	62	59	56	54	51	49	47	46	44	43	41	40	39	37	36
	0.32	0.42	0.53	0.65	0.79	0.94	1.11	1.28	1.47	1.67	1.89	2.12	2.36	2.62	2.89	3.17	3.46	3.77	4.09	4.42	4.77	5.13	5.50	5.89	6.29	6.70	7.12	7.56
73	240	210	186	168	153	140	129	120	112	105	99	93	88	84	80	76	73	70	67	65	62	60	58	56	54	52	51	49
	0.32	0.42	0.53	0.65	0.79	0.94	1.10	1.28	1.46	1.67	1.88	2.11	2.35	2.60	2.87	3.15	3.44	3.75	4.07	4.40	4.74	5.10	5.47	5.86	6.25	6.66	7.09	7.52
93	305	267	237	213	194	178	164	152	142	133	125	119	112	107	102	97	93	89	85	82	79	76	74	71	69	67	65	63
	0.31	0.41	0.52	0.64	0.77	0.92	1.08	1.25	1.44	1.64	1.85	2.07	2.31	2.56	2.82	3.10	3.38	3.68	4.00	4.32	4.66	5.01	5.38	5.76	6.15	6.55	6.97	7.39
122	433	379	337	303	276	253	233	217	202	190	178	169	160	152	144	138	132	126	121	117	112	108	105	101	98	95	92	89
	0.31	0.41	0.52	0.64	0.77	0.92	1.07	1.25	1.43	1.63	1.84	2.06	2.30	2.54	2.80	3.08	3.36	3.66	3.98	4.30	4.64	4.99	5.35	5.72	6.11	6.51	6.93	7.35
147	522	457	406	366	332	305	281	261	244	228	215	203	192	183	174	166	159	152	146	141	135	131	126	122	118	114	111	108
	0.31	0.40	0.51	0.62	0.76	0.90	1.06	1.22	1.41	1.60	1.81	2.02	2.26	2.50	2.76	3.02	3.31	3.60	3.91	4.22	4.56	4.90	5.26	5.62	6.01	6.40	6.81	7.22
W24X 62	208	182	162	146	132	121	112	104	97	91	86	81	77	73	69	66	63	61	58	56	54	52	50	49	47	45	44	43
	0.29	0.37	0.47	0.58	0.71	0.84	0.98	1.14	1.31	1.49	1.68	1.89	2.10	2.33	2.57	2.82	3.08	3.36	3.64	3.94	4.25	4.57	4.90	5.25	5.60	5.97	6.35	6.74
76	279	244	217	196	178	163	150	140	130	122	115	109	103	98	93	89	85	81	78	75	72	70	67	65	63	61	59	58
	0.28	0.37	0.47	0.58	0.70	0.83	0.98	1.13	1.30	1.48	1.67	1.87	2.09	2.31	2.55	2.80	3.06	3.33	3.61	3.91	4.21	4.53	4.86	5.20	5.55	5.92	6.29	6.68
94	352	308	274	247	224	206	190	176	164	154	145	137	130	123	117	112	107	103	99	95	91	88	85	82	80	77	75	73
	0.28	0.36	0.46	0.57	0.69	0.82	0.96	1.11	1.28	1.45	1.64	1.84	2.05	2.27	2.50	2.74	3.00	3.27	3.54	3.83	4.13	4.45	4.77	5.10	5.45	5.81	6.17	6.55
104	409	358	318	287	261	239	220	205	191	179	169	159	151	143	136	130	125	119	115	110	106	102	99	96	92	90	87	84
	0.28	0.37	0.46	0.57	0.69	0.83	0.97	1.12	1.29	1.47	1.66	1.86	2.07	2.30	2.53	2.78	3.04	3.31	3.59	3.88	4.18	4.50	4.83	5.17	5.52	5.88	6.25	6.63
131	522	457	406	366	332	305	281	261	244	228	215	203	192	183	174	166	159	152	146	141	135	131	126	122	118	114	111	108
	0.28	0.36	0.46	0.56	0.68	0.81	0.95	1.11	1.27	1.44	1.63	1.83	2.04	2.26	2.49	2.73	2.99	3.25	3.53	3.82	4.11	4.42	4.75	5.08	5.42	5.78	6.15	6.52
162	657	575	511	460	418	383	354	329	307	287	271	256	242	230	219	209	200	192	184	177	170	164	159	153	148	144	139	135
	0.27	0.35	0.45	0.55	0.67	0.80	0.93	1.08	1.24	1.41	1.60	1.79	1.99	2.21	2.44	2.67	2.92	3.18	3.45	3.73	4.03	4.33	4.64	4.97	5.31	5.65	6.01	6.38

Computed based upon flexure using data from the Bethlehem Steel *Structural Shapes Handbook*. Data used with permission of Bethlehem Steel Corporation.

Table B.2 (cont'd.)

Total Allowable Uniformly Distributed Loads (kip) And Maximum Deflection (in)
For Simple Beams That Are Laterally Supported
W27 and W30 Steel Beams (F_y = 50 000 lb/in²)

span ft	18	20	22	24	26	28	30	32	34	36	38	40	42	44	46	48	50	52	54	56	58	60	62	64	66	68	70	72
W27X 84	263	237	215	197	182	169	158	148	139	131	125	118	113	108	103	99	95	91	88	85	82	79	76	74	72	70	68	66
	0.42	0.52	0.62	0.74	0.87	1.01	1.16	1.32	1.49	1.67	1.86	2.06	2.27	2.49	2.73	2.97	3.22	3.48	3.76	4.04	4.33	4.64	4.95	5.28	5.61	5.96	6.31	6.68
102	330	297	270	247	228	212	198	185	174	165	156	148	141	135	129	124	119	114	110	106	102	99	96	93	90	87	85	82
	0.41	0.51	0.62	0.73	0.86	1.00	1.14	1.30	1.47	1.65	1.84	2.03	2.24	2.46	2.69	2.93	3.18	3.44	3.71	3.99	4.28	4.58	4.89	5.21	5.54	5.88	6.23	6.59
114	369	332	302	277	256	237	221	208	195	185	175	166	158	151	144	138	133	128	123	119	115	111	107	104	101	98	95	92
	0.41	0.50	0.61	0.73	0.85	0.99	1.13	1.29	1.46	1.63	1.82	2.02	2.22	2.44	2.67	2.90	3.15	3.41	3.68	3.95	4.24	4.54	4.84	5.16	5.49	5.83	6.18	6.53
146	507	457	415	381	351	326	304	285	269	254	240	228	217	208	199	190	183	176	169	163	157	152	147	143	138	134	130	127
	0.41	0.50	0.61	0.72	0.85	0.99	1.13	1.29	1.45	1.63	1.82	2.01	2.22	2.44	2.66	2.90	3.15	3.40	3.67	3.95	4.23	4.53	4.84	5.15	5.48	5.82	6.17	6.52
161	562	506	460	421	389	361	337	316	297	281	266	253	241	230	220	211	202	194	187	181	174	169	163	158	153	149	144	140
	0.40	0.50	0.60	0.72	0.84	0.98	1.12	1.28	1.44	1.62	1.80	2.00	2.20	2.42	2.64	2.88	3.12	3.38	3.64	3.92	4.20	4.50	4.80	5.12	5.44	5.78	6.12	6.48
178	620	558	507	465	429	398	372	349	328	310	294	279	266	254	242	232	223	215	207	199	192	186	180	174	169	164	159	155
	0.40	0.50	0.60	0.71	0.84	0.97	1.11	1.27	1.43	1.60	1.79	1.98	2.18	2.40	2.62	2.85	3.10	3.35	3.61	3.88	4.16	4.46	4.76	5.07	5.39	5.72	6.07	6.42
W30X 99	332	299	272	249	230	213	199	187	176	166	157	149	142	136	130	125	120	115	111	107	103	100	96	93	91	88	85	83
	0.38	0.46	0.56	0.67	0.79	0.91	1.05	1.19	1.34	1.51	1.68	1.86	2.05	2.25	2.46	2.68	2.91	3.14	3.39	3.64	3.91	4.18	4.47	4.76	5.06	5.37	5.70	6.03
116	406	366	332	305	281	261	244	228	215	203	192	183	174	166	159	152	146	141	135	131	126	122	118	114	111	108	104	102
	0.37	0.46	0.56	0.66	0.78	0.90	1.04	1.18	1.33	1.49	1.66	1.84	2.03	2.23	2.43	2.65	2.88	3.11	3.35	3.61	3.87	4.14	4.42	4.71	5.01	5.32	5.64	5.96
132	469	422	384	352	325	302	281	264	248	235	222	211	201	192	184	176	169	162	156	151	146	141	136	132	128	124	121	117
	0.37	0.45	0.55	0.65	0.77	0.89	1.02	1.16	1.31	1.47	1.64	1.82	2.00	2.20	2.40	2.62	2.84	3.07	3.31	3.56	3.82	4.09	4.36	4.65	4.95	5.25	5.56	5.89
173	665	599	544	499	461	428	399	374	352	333	315	299	285	272	260	250	240	230	222	214	206	200	193	187	181	176	171	166
	0.37	0.45	0.55	0.65	0.77	0.89	1.02	1.16	1.31	1.47	1.64	1.81	2.00	2.19	2.40	2.61	2.83	3.06	3.30	3.55	3.81	4.08	4.36	4.64	4.94	5.24	5.55	5.87
191	738	664	604	554	511	475	443	415	391	369	350	332	316	302	289	277	266	256	246	237	229	221	214	208	201	195	190	185
	0.36	0.45	0.54	0.65	0.76	0.88	1.01	1.15	1.30	1.46	1.62	1.80	1.98	2.18	2.38	2.59	2.81	3.04	3.28	3.53	3.78	4.05	4.32	4.60	4.90	5.20	5.51	5.83
211	818	737	670	614	567	526	491	460	433	409	388	368	351	335	320	307	295	283	273	263	254	246	238	230	223	217	210	205
	0.36	0.44	0.54	0.64	0.75	0.87	1.00	1.14	1.28	1.44	1.60	1.78	1.96	2.15	2.35	2.56	2.77	3.00	3.24	3.48	3.73	3.99	4.27	4.55	4.83	5.13	5.44	5.75

Computed based upon flexure using data from the Bethlehem Steel *Structural Shapes Handbook*. Data used with permission of Bethlehem Steel Corporation.

Table B.2 (cont'd.)

Total Allowable Uniformly Distributed Loads (kip) And Maximum Deflection (in)
For Simple Beams That Are Laterally Supported
W33 and W36 Steel Beams (F_y = 50 000 lb/in²)

span ft	18	20	22	24	26	28	30	32	34	36	38	40	42	44	46	48	50	52	54	56	58	60	62	64	66	68	70	72
W33X118	443	399	363	332	307	285	266	249	235	222	210	199	190	181	173	166	160	153	148	142	138	133	129	125	121	117	114	111
	0.34	*0.42*	*0.51*	*0.60*	*0.71*	*0.82*	*0.94*	*1.07*	*1.21*	*1.36*	*1.51*	*1.68*	*1.85*	*2.03*	*2.22*	*2.42*	*2.62*	*2.84*	*3.06*	*3.29*	*3.53*	*3.78*	*4.03*	*4.30*	*4.57*	*4.85*	*5.14*	*5.44*
141	553	498	452	415	383	356	332	311	293	277	262	249	237	226	216	207	199	191	184	178	172	166	161	156	151	146	142	138
	0.34	*0.41*	*0.50*	*0.60*	*0.70*	*0.81*	*0.93*	*1.06*	*1.20*	*1.34*	*1.50*	*1.66*	*1.83*	*2.01*	*2.19*	*2.39*	*2.59*	*2.80*	*3.02*	*3.25*	*3.49*	*3.73*	*3.99*	*4.25*	*4.52*	*4.79*	*5.08*	*5.37*
152	601	541	492	451	416	386	361	338	318	301	285	271	258	246	235	225	216	208	200	193	187	180	175	169	164	159	155	150
	0.33	*0.41*	*0.50*	*0.59*	*0.70*	*0.81*	*0.93*	*1.05*	*1.19*	*1.33*	*1.49*	*1.65*	*1.81*	*1.99*	*2.18*	*2.37*	*2.57*	*2.78*	*3.00*	*3.23*	*3.46*	*3.70*	*3.96*	*4.21*	*4.48*	*4.76*	*5.04*	*5.33*
201	844	760	691	633	585	543	507	475	447	422	400	380	362	345	330	317	304	292	281	271	262	253	245	237	230	224	217	211
	0.33	*0.41*	*0.50*	*0.59*	*0.69*	*0.80*	*0.92*	*1.05*	*1.19*	*1.33*	*1.48*	*1.64*	*1.81*	*1.99*	*2.17*	*2.36*	*2.56*	*2.77*	*2.99*	*3.22*	*3.45*	*3.69*	*3.94*	*4.20*	*4.47*	*4.74*	*5.02*	*5.32*
221	934	841	765	701	647	601	561	526	495	467	443	421	400	382	366	350	336	323	311	300	290	280	271	263	255	247	240	234
	0.33	*0.41*	*0.50*	*0.59*	*0.69*	*0.80*	*0.92*	*1.04*	*1.18*	*1.32*	*1.47*	*1.63*	*1.80*	*1.97*	*2.16*	*2.35*	*2.55*	*2.76*	*2.97*	*3.20*	*3.43*	*3.67*	*3.92*	*4.18*	*4.44*	*4.71*	*5.00*	*5.29*
241	1023	921	837	768	708	658	614	576	542	512	485	461	439	419	400	384	368	354	341	329	318	307	297	288	279	271	263	256
	0.33	*0.40*	*0.49*	*0.58*	*0.68*	*0.79*	*0.91*	*1.03*	*1.16*	*1.30*	*1.45*	*1.61*	*1.78*	*1.95*	*2.13*	*2.32*	*2.52*	*2.72*	*2.93*	*3.16*	*3.39*	*3.62*	*3.87*	*4.12*	*4.38*	*4.65*	*4.93*	*5.22*
W36X135	542	488	443	406	375	348	325	305	287	271	257	244	232	222	212	203	195	188	181	174	168	163	157	152	148	143	139	135
	0.31	*0.39*	*0.47*	*0.56*	*0.66*	*0.76*	*0.87*	*0.99*	*1.12*	*1.26*	*1.40*	*1.55*	*1.71*	*1.88*	*2.05*	*2.24*	*2.43*	*2.62*	*2.83*	*3.04*	*3.26*	*3.49*	*3.73*	*3.97*	*4.23*	*4.49*	*4.75*	*5.03*
160	669	602	547	502	463	430	401	376	354	335	317	301	287	274	262	251	241	232	223	215	208	201	194	188	182	177	172	167
	0.31	*0.38*	*0.46*	*0.55*	*0.65*	*0.75*	*0.86*	*0.98*	*1.11*	*1.24*	*1.38*	*1.53*	*1.69*	*1.86*	*2.03*	*2.21*	*2.40*	*2.59*	*2.79*	*3.01*	*3.22*	*3.45*	*3.68*	*3.93*	*4.17*	*4.43*	*4.70*	*4.97*
182	769	692	629	577	532	494	461	433	407	385	364	346	330	315	301	288	277	266	256	247	239	231	223	216	210	204	198	192
	0.31	*0.38*	*0.46*	*0.55*	*0.64*	*0.75*	*0.86*	*0.97*	*1.10*	*1.23*	*1.37*	*1.52*	*1.68*	*1.84*	*2.01*	*2.19*	*2.38*	*2.57*	*2.77*	*2.98*	*3.20*	*3.42*	*3.65*	*3.89*	*4.14*	*4.39*	*4.66*	*4.93*
210	888	799	726	666	614	571	533	499	470	444	420	399	380	363	347	333	320	307	296	285	275	266	258	250	242	235	228	222
	0.30	*0.38*	*0.45*	*0.54*	*0.63*	*0.74*	*0.85*	*0.96*	*1.09*	*1.22*	*1.36*	*1.50*	*1.66*	*1.82*	*1.99*	*2.16*	*2.35*	*2.54*	*2.74*	*2.94*	*3.16*	*3.38*	*3.61*	*3.85*	*4.09*	*4.34*	*4.60*	*4.87*
230	1033	930	845	775	715	664	620	581	547	517	489	465	443	423	404	387	372	358	344	332	321	310	300	291	282	274	266	258
	0.31	*0.38*	*0.47*	*0.55*	*0.65*	*0.75*	*0.87*	*0.99*	*1.11*	*1.25*	*1.39*	*1.54*	*1.70*	*1.86*	*2.04*	*2.22*	*2.40*	*2.60*	*2.81*	*3.02*	*3.24*	*3.46*	*3.70*	*3.94*	*4.19*	*4.45*	*4.71*	*4.99*
260	1176	1059	963	882	814	756	706	662	623	588	557	529	504	481	460	441	424	407	392	378	365	353	342	331	321	311	303	294
	0.31	*0.38*	*0.46*	*0.55*	*0.64*	*0.74*	*0.85*	*0.97*	*1.10*	*1.23*	*1.37*	*1.52*	*1.68*	*1.84*	*2.01*	*2.19*	*2.37*	*2.57*	*2.77*	*2.98*	*3.19*	*3.42*	*3.65*	*3.89*	*4.14*	*4.39*	*4.65*	*4.92*
300	1370	1233	1121	1028	949	881	822	771	725	685	649	617	587	561	536	514	493	474	457	440	425	411	398	385	374	363	352	343
	0.31	*0.38*	*0.46*	*0.54*	*0.64*	*0.74*	*0.85*	*0.97*	*1.09*	*1.22*	*1.36*	*1.51*	*1.66*	*1.82*	*1.99*	*2.17*	*2.36*	*2.55*	*2.75*	*2.96*	*3.17*	*3.39*	*3.62*	*3.86*	*4.11*	*4.36*	*4.62*	*4.89*

Computed based upon flexure using data from the Bethlehem Steel *Structural Shapes Handbook*. Data used with permission of Bethlehem Steel Corporation.

Table B.3 ALLOWABLE AXIAL LOADS FOR SELECTED SECTION PROFILES AND EFFECTIVE COLUMN LENGTHS (KL)

Total Allowable Load (kip) For Columns That Are Axially Loaded
W8, W10 and W12 Steel Columns (F_y = 36 000 lb/in^2)

length (KL), ft	8	9	10	11	12	13	14	15	16	17	18	19	20	22	24	26	28	30	32	34	36	38	40	42	44	46	48	50
W8X 10	33	27	22	18	15	13	11	–	–	–	–	–	–	–	–	–	–	–	–	–	–	–	–	–	–	–	–	–
15	52	44	35	29	25	21	18	–	–	–	–	–	–	–	–	–	–	–	–	–	–	–	–	–	–	–	–	–
21	97	91	84	76	68	60	52	45	40	35	31	28	25	–	–	–	–	–	–	–	–	–	–	–	–	–	–	–
28	144	138	132	125	118	111	103	95	87	78	69	62	56	46	39	33	–	–	–	–	–	–	–	–	–	–	–	–
40	218	212	205	199	192	184	177	169	160	152	143	134	124	104	88	75	64	56	49	44	–	–	–	–	–	–	–	–
67	370	360	350	339	328	316	304	292	279	265	251	236	221	190	159	136	117	102	90	79	–	–	–	–	–	–	–	–
W10X 12	35	28	23	19	16	13	–	–	–	–	–	–	–	–	–	–	–	–	–	–	–	–	–	–	–	–	–	–
19	66	55	45	37	31	26	23	–	–	–	–	–	–	–	–	–	–	–	–	–	–	–	–	–	–	–	–	–
30	145	137	128	119	109	99	88	76	67	60	53	48	43	36	–	–	–	–	–	–	–	–	–	–	–	–	–	–
45	247	240	232	224	216	208	199	190	180	170	160	149	138	115	97	82	71	62	54	–	–	–	–	–	–	–	–	–
54	306	300	294	288	281	274	267	259	251	243	235	226	217	199	179	158	137	119	105	93	83	74	67	61	–	–	–	–
77	439	431	422	413	404	394	384	373	362	351	339	327	315	289	261	232	202	176	155	137	122	110	99	90	–	–	–	–
112	642	631	619	606	593	579	565	550	535	519	503	486	469	433	395	355	313	272	239	212	189	170	153	139	127	–	–	–
W12X 22	73	60	48	40	34	29	25	–	–	–	–	–	–	–	–	–	–	–	–	–	–	–	–	–	–	–	–	–
35	177	169	161	151	142	132	121	110	99	88	78	70	63	52	44	–	–	–	–	–	–	–	–	–	–	–	–	–
40	217	210	203	196	188	180	172	163	154	144	135	124	114	94	79	67	58	51	45	–	–	–	–	–	–	–	–	–
50	271	263	254	246	236	226	216	206	195	183	171	159	146	121	102	87	75	65	57	–	–	–	–	–	–	–	–	–
58	329	322	315	308	301	293	285	276	268	259	249	240	230	209	187	164	142	123	108	96	86	77	69	63	–	–	–	–
72	418	412	406	399	392	385	377	369	361	353	344	336	326	308	288	267	245	222	197	175	156	140	126	115	104	96	88	81
96	560	552	544	535	526	516	506	496	486	475	464	452	440	416	390	362	334	304	272	242	215	193	175	158	144	132	121	112
136	795	784	772	760	747	734	721	707	693	678	662	647	630	597	561	524	485	444	402	357	319	286	258	234	213	195	179	165
190	1115	1100	1084	1068	1051	1034	1016	997	978	958	937	916	894	849	802	752	700	646	588	529	472	423	382	346	316	289	265	244

Computed using data from the Bethlehem Steel *Structural Shapes Handbook*. Data used with permission of Bethlehem Steel Corporation.

Table B.3 (cont'd.)

Total Allowable Load (kip) For Columns That Are Axially Loaded
W14 Steel Columns (F_y = 36 000 lb/in²)

length (KL), ft	8	9	10	11	12	13	14	15	16	17	18	19	20	22	24	26	28	30	32	34	36	38	40	42	44	46	48	50
W14X 26	133	127	120	114	107	99	-	-	-	-	-	-	-	-	-	-	-	-	-	-	-	-	-	-	-	-	-	-
30	150	143	135	127	119	110	100	91	83	75	66	59	53	48	40	-	-	-	-	-	-	-	-	-	-	-	-	-
38	193	184	175	165	155	144	133	121	109	97	86	77	70	58	48	-	-	-	-	-	-	-	-	-	-	-	-	-
48	258	250	242	233	224	214	204	193	182	171	159	146	133	110	93	79	68	59	-	-	-	-	-	-	-	-	-	-
53	286	277	268	258	248	237	226	215	202	190	177	163	149	123	104	88	76	66	58	-	-	-	-	-	-	-	-	-
61	345	338	330	322	314	306	297	288	278	268	258	248	237	214	190	165	142	124	109	96	86	77	70	-	-	-	-	-
74	421	412	403	394	384	374	363	352	341	329	317	305	292	265	236	206	177	154	136	120	107	96	87	-	-	-	-	-
82	465	456	446	435	425	413	402	389	377	364	351	337	323	293	261	227	196	171	150	133	119	107	96	-	-	-	-	-
99	589	582	575	568	561	554	546	538	529	521	512	503	494	475	454	433	411	388	365	340	314	287	260	235	215	196	180	166
109	647	640	633	626	618	609	601	592	583	574	564	554	544	523	501	478	454	429	403	376	348	319	289	262	238	218	200	185
120	714	707	699	690	682	673	663	654	644	633	623	612	601	578	554	528	502	475	446	416	385	353	320	290	264	242	222	205
132	775	767	759	749	740	730	720	710	699	688	677	665	653	628	602	575	547	517	487	455	421	387	351	318	290	265	244	225
145	869	860	851	842	832	822	812	801	790	779	767	755	743	718	691	663	634	604	573	540	507	472	435	398	362	331	304	281
159	950	941	931	921	911	900	889	877	865	853	840	827	814	786	758	727	696	663	629	594	558	520	480	439	400	362	336	310
176	1054	1044	1034	1022	1011	999	987	974	961	947	933	919	904	874	842	809	775	739	701	662	622	580	537	492	448	410	377	347
193	1157	1146	1134	1122	1110	1097	1083	1069	1055	1040	1025	1010	994	961	927	891	853	814	774	732	688	643	596	547	499	457	419	386
211	1263	1251	1239	1226	1212	1198	1183	1168	1153	1137	1121	1104	1087	1051	1014	975	934	892	848	803	755	707	656	603	550	503	462	426
233	1396	1383	1370	1355	1340	1325	1309	1293	1276	1258	1241	1222	1204	1165	1124	1081	1037	991	943	893	842	788	733	675	617	564	518	478
257	1542	1528	1513	1497	1481	1464	1447	1429	1410	1391	1372	1352	1331	1289	1244	1198	1149	1099	1047	993	936	878	818	755	691	632	580	535
283	1700	1685	1668	1651	1634	1615	1597	1577	1557	1536	1515	1494	1471	1425	1377	1326	1274	1219	1162	1104	1043	980	914	846	776	710	652	601
311	1867	1850	1832	1813	1794	1774	1754	1733	1711	1689	1666	1642	1618	1568	1515	1460	1403	1344	1283	1219	1153	1084	1013	940	864	790	726	669
342	2064	2045	2026	2006	1985	1963	1941	1918	1894	1870	1845	1819	1793	1738	1681	1621	1559	1495	1428	1358	1286	1212	1135	1055	972	890	817	753
370	2229	2209	2188	2167	2144	2121	2097	2073	2047	2021	1995	1967	1939	1881	1820	1756	1690	1621	1549	1475	1398	1319	1237	1151	1063	974	895	824
398	2394	2373	2351	2328	2304	2280	2255	2229	2202	2174	2146	2117	2087	2025	1961	1893	1823	1750	1675	1596	1515	1431	1344	1254	1161	1065	978	902
426	2559	2536	2513	2489	2464	2438	2411	2384	2356	2326	2296	2266	2234	2169	2100	2029	1955	1878	1798	1715	1630	1541	1449	1354	1255	1154	1060	977
500	3013	2988	2961	2933	2905	2875	2845	2813	2781	2748	2714	2678	2642	2568	2490	2409	2324	2236	2145	2051	1954	1853	1748	1640	1529	1413	1298	1197
550	3323	3296	3267	3237	3206	3175	3142	3108	3073	3037	3000	2962	2923	2842	2758	2670	2579	2484	2386	2284	2179	2070	1958	1841	1721	1597	1470	1355
605	3655	3625	3594	3562	3529	3494	3459	3422	3384	3345	3306	3265	3223	3136	3045	2951	2853	2751	2645	2535	2422	2305	2184	2059	1930	1797	1659	1529
665	4028	3996	3963	3928	3892	3855	3817	3777	3737	3695	3652	3608	3563	3469	3372	3270	3164	3055	2941	2823	2702	2576	2446	2312	2173	2030	1882	1735
730	4423	4388	4352	4315	4277	4237	4196	4153	4110	4065	4019	3971	3923	3823	3718	3609	3496	3378	3256	3130	3000	2865	2726	2582	2434	2281	2123	1962

Computed using data from the Bethlehem Steel *Structural Shapes Handbook*. Data used with permission of Bethlehem Steel Corporation.

Table B.3 *(cont'd.)*

Total Allowable Load (kip) For Columns That Are Axially Loaded
W8, W10 and W12 Steel Columns (F_y = 50 000 lb/in²)

length (KL), ft	8	9	10	11	12	13	14	15	16	17	18	19	20	22	24	26	28	30	32	34	36	38	40	42	44	46	48	50
W8X 10	34	27	22	18	15	13	11	-	-	-	-	-	-	-	-	-	-	-	-	-	-	-	-	-	-	-	-	-
15	55	44	35	29	25	21	18	-	-	-	-	-	-	-	-	-	-	-	-	-	-	-	-	-	-	-	-	-
21	122	110	97	84	70	60	52	45	40	35	31	28	25	-	-	-	-	-	-	-	-	-	-	-	-	-	-	-
28	188	178	166	154	142	128	114	100	88	78	69	62	56	46	39	33	-	-	-	-	-	-	-	-	-	-	-	-
40	290	279	268	256	244	231	217	203	188	172	156	139	121	104	88	75	64	56	49	44	-	-	-	-	-	-	-	-
67	494	477	459	440	420	399	378	355	331	307	281	254	226	167	159	136	117	102	90	79	-	-	-	-	-	-	-	-
W10X 12	35	28	23	19	16	13	-	-	-	-	-	-	-	-	-	-	-	-	-	-	-	-	-	-	-	-	-	-
19	70	55	45	37	31	26	23	-	-	-	-	-	-	-	-	-	-	-	-	-	-	-	-	-	-	-	-	-
30	185	170	154	137	119	102	88	76	67	60	53	48	43	36	-	-	-	-	-	-	-	-	-	-	-	-	-	-
45	328	316	303	289	274	259	243	227	209	191	172	154	139	115	97	82	71	62	54	-	-	-	-	-	-	-	-	-
54	414	403	392	381	369	356	343	330	316	301	286	271	255	221	186	159	137	119	105	93	83	74	67	61	-	-	-	-
77	593	579	564	548	531	513	495	476	457	437	416	394	371	324	275	234	202	176	155	137	122	110	99	90	-	-	-	-
112	869	848	827	805	782	757	732	706	679	651	622	592	560	495	425	362	313	272	239	212	189	170	153	139	127	-	-	-
W12X 22	74	60	48	40	34	29	25	-	-	-	-	-	-	-	-	-	-	-	-	-	-	-	-	-	-	-	-	-
35	230	215	200	184	167	148	129	113	99	88	78	70	63	52	44	-	-	-	-	-	-	-	-	-	-	-	-	-
40	288	276	264	251	237	222	207	191	175	158	141	126	114	94	79	67	58	51	45	-	-	-	-	-	-	-	-	-
50	360	346	331	315	298	281	262	243	223	202	181	162	146	121	102	87	75	65	57	-	-	-	-	-	-	-	-	-
58	443	432	420	407	394	380	365	351	335	319	302	285	267	229	193	164	142	123	108	96	-	-	-	-	-	-	-	-
72	569	558	547	535	522	509	496	482	468	453	438	422	406	372	336	297	258	225	197	175	156	140	126	115	104	96	88	81
96	762	748	733	718	701	685	667	649	630	611	591	570	549	505	458	408	356	310	273	242	215	193	175	158	144	132	121	112
136	1082	1062	1042	1021	999	976	952	927	901	875	848	819	790	730	666	598	527	459	403	357	319	286	258	234	213	195	179	165
190	1518	1492	1465	1437	1407	1376	1344	1311	1276	1241	1204	1167	1128	1047	962	872	776	679	597	529	472	423	382	346	316	289	265	244

Computed using data from the Bethlehem Steel *Structural Shapes Handbook*. Data used with permission of Bethlehem Steel Corporation.

Table B.3 *(cont'd.)*

Total Allowable Load (kip) For Columns That Are Axially Loaded
W14 Steel Columns (F_y = 50 000 lb/in²)

length (KL), ft	8	9	10	11	12	13	14	15	16	17	18	19	20	22	24	26	28	30	32	34	36	38	40	42	44	46	48	50
W14X 26	172	161	150	138	125	112	98	85	75	66	59	53	48	40	33	-	-	-	-	-	-	-	-	-	-	-	-	-
30	194	181	167	153	137	120	104	91	80	71	63	56	51	42	35	-	-	-	-	-	-	-	-	-	-	-	-	-
38	251	235	219	201	183	163	142	124	109	97	86	77	70	58	48	-	-	-	-	-	-	-	-	-	-	-	-	-
48	343	329	313	298	281	263	245	226	206	185	165	148	133	110	93	79	68	59	-	-	-	-	-	-	-	-	-	-
53	380	364	348	330	312	293	273	251	229	206	184	165	149	123	104	88	76	66	58	-	-	-	-	-	-	-	-	-
61	464	452	439	425	410	395	379	363	346	328	310	291	272	230	193	165	142	124	109	96	86	77	70	-	-	-	-	-
74	567	552	536	520	502	484	465	446	426	405	383	360	337	287	241	206	177	154	136	120	107	96	87	-	-	-	-	-
82	627	610	593	575	555	535	515	493	471	447	423	398	372	318	267	227	196	171	150	133	119	106	96	-	-	-	-	-
99	805	793	782	769	757	743	730	715	701	685	670	654	637	603	567	529	489	448	404	359	320	288	260	235	215	196	180	166
109	885	873	860	847	833	818	803	788	772	755	738	721	703	665	626	585	541	496	449	399	356	320	289	262	238	218	200	185
120	977	963	949	935	919	903	887	870	852	834	815	796	776	735	692	647	599	549	497	443	395	355	320	290	264	242	222	205
132	1060	1046	1031	1015	998	981	964	945	926	907	887	866	845	800	754	705	654	600	544	486	433	389	351	318	290	265	244	225
145	1189	1174	1159	1142	1125	1108	1090	1071	1051	1031	1011	990	968	923	875	825	773	719	662	603	541	486	438	398	362	331	304	281
159	1301	1285	1268	1250	1232	1213	1193	1173	1152	1130	1108	1085	1061	1012	960	906	850	791	729	665	598	537	484	439	400	366	336	310
176	1444	1426	1407	1388	1368	1347	1325	1302	1279	1255	1231	1205	1179	1125	1069	1009	947	882	815	744	670	601	543	492	448	410	377	347
193	1585	1565	1545	1524	1502	1479	1455	1431	1406	1380	1353	1326	1298	1239	1178	1113	1046	976	902	826	745	669	604	548	499	457	419	386
211	1730	1709	1687	1665	1641	1616	1590	1564	1537	1509	1480	1450	1419	1356	1289	1220	1147	1071	991	908	822	738	666	604	550	503	462	426
233	1913	1890	1866	1841	1815	1788	1760	1731	1701	1671	1639	1607	1573	1504	1431	1355	1275	1192	1106	1015	921	827	746	677	617	564	518	478
257	2113	2088	2062	2034	2006	1976	1946	1914	1882	1848	1814	1778	1742	1666	1587	1504	1417	1326	1232	1133	1030	926	836	758	691	632	580	535
283	2330	2303	2274	2245	2214	2182	2148	2114	2079	2042	2005	1966	1927	1845	1759	1668	1574	1476	1373	1266	1155	1040	939	852	776	710	652	601
311	2558	2529	2498	2465	2432	2397	2361	2324	2285	2246	2205	2163	2120	2031	1938	1840	1738	1631	1520	1404	1283	1158	1045	948	864	790	726	669
342	2829	2797	2763	2728	2692	2654	2614	2574	2532	2489	2445	2399	2352	2255	2153	2047	1935	1819	1698	1572	1441	1304	1177	1067	973	890	817	753
370	3055	3021	2985	2947	2908	2868	2826	2782	2738	2692	2644	2596	2546	2442	2333	2220	2101	1977	1848	1713	1573	1427	1288	1168	1065	974	895	824
398	3282	3246	3208	3168	3126	3083	3039	2993	2946	2897	2847	2795	2743	2633	2518	2397	2272	2141	2004	1862	1714	1560	1409	1278	1164	1065	978	902
426	3509	3470	3430	3388	3344	3298	3251	3203	3153	3101	3048	2994	2938	2822	2700	2573	2440	2302	2158	2008	1852	1689	1526	1384	1261	1154	1060	977
500	4134	4090	4043	3995	3945	3893	3840	3784	3727	3668	3608	3546	3482	3350	3211	3066	2915	2758	2594	2423	2246	2061	1870	1696	1545	1414	1298	1197
550	4561	4513	4463	4411	4357	4301	4243	4183	4121	4058	3992	3925	3856	3713	3564	3407	3244	3074	2897	2714	2522	2324	2117	1920	1749	1601	1470	1355
605	5017	4965	4911	4855	4797	4736	4674	4609	4543	4474	4404	4331	4257	4103	3942	3774	3598	3415	3225	3028	2822	2609	2387	2166	1974	1806	1659	1529
665	5531	5475	5417	5356	5293	5228	5161	5092	5020	4946	4870	4793	4713	4547	4374	4193	4004	3807	3603	3391	3170	2941	2703	2459	2241	2050	1883	1735
730	6074	6014	5951	5887	5819	5749	5677	5602	5525	5446	5365	5282	5196	5018	4832	4638	4436	4225	4006	3779	3543	3298	3044	2780	2533	2318	2129	1962

Computed using data from the Bethlehem Steel *Structural Shapes Handbook*. Data used with permission of Bethlehem Steel Corporation.

Joist designation	8K1	10K1	12K1	12K3	12K5	14K1	14K3	14K4	14K6	16K2	16K3	16K4	16K5	16K6	16K7	16K9
Depth (in)	8	10	12	12	12	14	14	14	14	16	16	16	16	16	16	16
Approx. weight (lb/ft of joist)	5.1	5.0	5.0	5.7	7.1	5.2	6.0	6.7	7.7	5.5	6.3	7.0	7.5	8.1	8.6	10.0
Span (ft) ↓ 10	550	550														
	480	550														
12	444	550	550	550	550											
	288	455	550	550	550											
14	324	412	500	550	550	550	550	550	550							
	179	289	425	463	463	550	550	550	550							
16	246	313	380	476	550	448	550	550	550	550	550	550	550	550	550	550
	119	192	282	351	396	390	467	467	467	550	550	550	550	550	550	550
18		246	299	374	507	352	441	530	550	456	508	550	550	550	550	550
		134	197	245	317	272	339	397	408	409	456	490	490	490	490	490
20		199	241	302	409	284	356	428	525	368	410	493	550	550	550	550
		97	142	177	230	197	246	287	347	297	330	386	426	426	426	426
22			199	249	337	234	293	353	432	303	337	406	456	498	550	550
			106	132	172	147	184	215	259	222	247	289	323	351	385	385
24			166	208	282	196	245	295	362	254	283	340	384	418	465	550
			81	101	132	113	141	165	199	170	189	221	248	269	298	346
26						166	209	251	308	216	240	289	326	355	395	474
						88	110	129	156	133	148	173	194	211	233	276
28						143	180	216	265	186	207	249	281	306	340	408
						70	88	103	124	106	118	138	155	168	186	220
30										161	180	216	244	266	296	355
										86	96	112	126	137	151	178
32										142	158	190	214	233	259	311
										71	79	92	103	112	124	147

Table B.4 *(cont'd.)*

Joist designation	18K3	18K4	18K5	18K6	18K7	18K9	18K10	20K3	20K4	20K5	20K6	20K7	20K9	20K10	22K4	22K5	22K6	22K7	22K9	22K10	22K11
Depth (in)	18	18	18	18	18	18	18	20	20	20	20	20	20	20	22	22	22	22	22	22	22
Approx. weight (lb/ft of joist)	6.6	7.2	7.7	8.5	9.0	10.2	11.7	6.7	7.6	8.2	8.9	9.3	10.8	12.2	8.0	8.8	9.2	9.7	11.3	12.6	13.8
Span (ft) ↓																					
18	550	550	550	550	550	550	550														
	550	550	550	550	550	550	550														
20	463	550	550	550	550	550	550	517	550	550	550	550	550	550							
	423	490	490	490	490	490	490	517	550	550	550	550	550	550							
22	382	460	518	550	550	550	550	426	514	550	550	550	550	550	550	550	550	550	550	550	550
	316	370	414	438	438	438	438	393	431	490	490	490	490	490	548	548	548	548	548	548	548
24	320	385	434	473	526	550	550	357	430	485	528	550	550	550	475	536	550	550	550	550	550
	242	284	318	345	382	396	396	302	353	396	430	448	448	448	431	483	495	495	495	495	495
26	272	328	369	402	448	538	550	304	366	412	449	500	550	550	404	455	496	550	550	550	550
	190	222	249	271	299	354	361	236	277	310	337	373	405	405	338	379	411	454	454	454	454
28	234	282	318	346	385	463	548	261	315	355	386	430	517	550	348	392	427	475	550	550	550
	151	177	199	216	239	282	331	189	221	248	269	298	353	375	270	302	328	364	413	413	413
30	203	245	276	301	335	402	477	227	274	308	336	374	450	533	302	341	371	413	497	550	550
	123	144	161	175	194	229	269	153	179	201	218	242	286	336	219	245	266	295	349	385	385
32	178	215	242	264	294	353	418	199	240	271	295	328	395	468	265	299	326	363	436	517	549
	101	118	132	144	159	188	221	126	147	165	179	199	235	276	180	201	219	242	287	337	355
34	158	190	214	233	260	312	370	176	212	239	261	290	349	414	235	265	288	321	386	458	516
	84	98	110	120	132	156	184	105	122	137	149	165	195	229	149	167	182	202	239	280	314
36	141	169	191	208	232	278	330	157	189	213	232	259	311	369	209	236	257	286	344	408	467
	70	82	92	101	111	132	154	88	103	115	125	139	164	193	126	141	153	169	201	236	269
38								141	170	191	208	232	279	331	187	211	230	256	308	366	419
								74	87	98	106	118	139	164	107	119	130	144	170	200	228
40								127	153	172	188	209	251	298	169	190	207	231	278	330	377
								64	75	84	91	101	119	140	91	102	111	123	146	171	195
42															153	173	188	209	252	299	342
															79	83	96	106	126	148	168
44															139	157	171	191	229	272	311
															68	76	83	92	109	128	146

Table B.4 (cont'd.)

Joist designation	24K4	24K5	24K6	24K7	24K8	24K9	24K10	24K12	26K5	26K6	26K7	26K8	26K9	26K10	26K12	28K6	28K7	28K8	28K9	28K10	28K12
Depth (in)	24	24	24	24	24	24	24	24	26	26	26	26	26	26	26	28	28	28	28	28	28
Approx. weight (lb/ft of joist)	8.4	9.3	9.7	10.1	11.5	12.0	13.1	16.0	9.8	10.6	10.9	12.1	12.2	13.8	16.6	11.4	11.8	12.7	13.0	14.3	17.1
Span (ft)↓	550	550	550	550	550	550	550	550													
24	516	544	544	544	544	544	544	544													
26	442	499	543	550	550	550	550	550	542	550	550	550	550	550	550						
	405	453	493	499	499	499	499	499	535	541	541	541	541	541	541						
28	381	429	467	521	550	550	550	550	466	508	550	550	550	550	550	548	550	550	550	550	550
	323	362	393	436	456	456	456	456	427	464	501	501	501	501	501	541	543	543	543	543	543
30	331	373	406	453	500	544	550	550	405	441	492	544	550	550	550	477	531	550	550	550	550
	262	293	319	353	387	419	422	422	346	377	417	457	459	459	459	439	486	500	500	500	500
32	290	327	357	397	439	478	549	549	356	387	432	477	519	549	549	418	466	515	549	549	549
	215	241	262	290	318	344	393	393	285	309	343	375	407	431	431	361	400	438	463	463	463
34	257	290	315	351	388	423	502	516	315	343	382	422	459	516	516	370	412	456	496	516	516
	179	201	218	242	264	286	337	344	237	257	285	312	338	378	378	300	333	364	395	410	410
36	229	258	281	313	346	377	447	487	280	305	340	376	409	486	487	330	367	406	442	487	487
	150	169	183	203	222	241	283	306	199	216	240	263	284	334	334	252	280	306	332	366	366
38	205	231	252	281	310	338	401	461	251	274	305	337	367	436	461	296	329	384	396	461	461
	128	143	156	172	189	204	240	275	169	184	204	223	241	284	299	214	237	260	282	325	325
40	185	208	227	253	280	304	361	438	227	247	275	304	331	393	438	266	297	328	357	424	438
	109	122	133	148	161	175	206	247	145	157	174	191	207	243	269	183	203	222	241	284	291
42	168	189	206	229	253	276	327	417	205	224	249	275	300	356	417	241	269	297	324	384	417
	94	106	115	127	139	151	177	224	125	136	150	164	178	210	244	158	175	192	208	245	264
44	153	172	187	209	231	251	298	387	187	204	227	251	273	324	398	220	245	271	295	350	398
	82	92	100	110	121	131	154	199	108	118	131	143	155	182	222	137	152	167	181	212	240
46	138	157	171	191	211	230	272	354	171	186	207	229	250	296	380	201	224	248	270	320	380
	71	80	87	97	106	114	135	174	95	103	114	125	135	159	203	120	133	146	158	186	219
48	128	144	157	175	194	211	250	325	157	171	190	210	229	272	353	184	208	227	247	294	365
	63	70	77	85	93	101	118	153	83	90	100	110	119	140	180	105	117	128	139	163	201
50									144	157	175	194	211	250	325	170	189	209	228	270	350
									73	80	89	97	105	124	159	93	103	113	123	144	185
52									133	145	162	179	195	231	301	157	175	193	210	250	325
									65	71	79	86	93	110	142	83	92	100	109	128	165
54																145	162	179	195	232	301
																74	82	89	97	114	147
56																135	151	166	181	215	280
																66	73	80	87	102	132

26	← Joist span, in ft. Measured from the center line of the joist girder or steel girder upon which the joist end is bearing or 4 in (102 mm) in from the wall surface upon which the joist is bearing.
28	381 ← Maximum total load, in lb/ft of joist span. Multiply by 14.59 to convert to N/m. *323* ← Maximum live load for deflection not to exceed 1/360 of span, in lb/ft of joist span. Multiply by 14.59 to convert to N/m.

Source: Adapted and condensed from Steel Joist Institute Tables.

Table B.5 REQUIRED NUMBER OF ROWS OF BRIDGING FOR K-SERIES OPEN-WEB STEEL JOISTS BASED UPON SPAN LENGTH (FEET) AND WEB SECTION NUMBER[a] (*Source:* Steel Joist Institute, Inc., K-Series Bridging Requirements)

Web section number[b]	Number of rows				
	1	2	3	4[c]	5[c]
1	Up through 16′	Over 16′ through 24′	Over 24′ through 28′	—	—
2	Up through 17′	Over 17′ through 25′	Over 25′ through 32′	—	—
3	Up through 18′	Over 18′ through 28′	Over 28′ through 38′	Over 38′ through 40′	—
4	Up through 19′	Over 19′ through 28′	Over 28′ through 38′	Over 38′ through 48′	—
5	Up through 19′	Over 19′ through 29′	Over 29′ through 39′	Over 39′ through 50′	Over 50′ through 52′
6	Up through 19′	Over 19′ through 29′	Over 29′ through 39′	Over 39′ through 51′	Over 51′ through 56′
7	Up through 20′	Over 20′ through 33′	Over 33′ through 45′	Over 45′ through 58′	Over 58′ through 60′
8	Up through 20′	Over 20′ through 33′	Over 33′ through 45′	Over 45′ through 58′	Over 58′ through 60′
9	Up through 20′	Over 20′ through 33′	Over 33′ through 46′	Over 46′ through 59′	Over 59′ through 60′
10	Up through 20′	Over 20′ through 37′	Over 37′ through 51′	Over 51′ through 60′	—
11	Up through 20′	Over 20′ through 38′	Over 38′ through 53′	Over 53′ through 60′	—
12	Up through 20′	Over 20′ through 39′	Over 39′ through 53′	Over 53′ through 60′	—

[a]Additional bridging may be required for wind uplift.

[b]Last digit of joist designation shown in joist load tables. For example, for a 12K5 joist, the web section number is 5.

[c]Where 4 or 5 rows of bridging are required, a row nearest the mid-span of the joist shall be diagonal bridging with bolted connections at chords and intersection.

Table B.6 RADII OF GYRATION (r) OF SELECTED SHAPES FOR USE AS BRIDGING FOR STEEL JOISTS

Shape	Weight lb/ft	Radius, r (in) Axis x–x	Axis y–y	Axis z–z
Angles				
L3 $\times$ 3 $\times \frac{1}{2}$	9.40	0.898	0.898	0.584
L2$\frac{1}{2}\times$ 2$\frac{1}{2}\times \frac{1}{2}$	7.70	0.739	0.739	0.487
L2 $\times$ 2 $\times \frac{1}{4}$	3.19	0.609	0.609	0.391
L1$\frac{3}{4}\times$ 1$\frac{3}{4}\times \frac{1}{4}$	2.77	0.529	0.529	0.341
L1$\frac{1}{2}\times$ 1$\frac{1}{2}\times \frac{1}{4}$	2.34	0.449	0.449	0.292
L1$\frac{1}{4}\times$ 1$\frac{1}{4}\times \frac{1}{4}$	1.98	0.369	0.369	0.243
L1 $\times$ 1 $\times \frac{1}{8}$	0.800	0.304	0.304	0.196
Pipes, standard weight				
$\frac{1}{2}$	0.85	0.261	0.261	0.261
$\frac{3}{4}$	1.13	0.334	0.334	0.334
1	1.68	0.421	0.421	0.421
1$\frac{1}{4}$	2.27	0.540	0.540	0.540
1$\frac{1}{2}$	2.72	0.623	0.623	0.623
2	3.65	0.787	0.787	0.787
Square bars				
$\frac{1}{2}\times \frac{1}{2}$	0.821	0.144	0.144	—
$\frac{3}{4}\times \frac{3}{4}$	1.914	0.217	0.217	—
1 $\times$ 1	3.403	0.289	0.289	—
Round bars				
$\frac{1}{2}$	0.668	0.125	0.125	0.125
$\frac{3}{4}$	1.503	0.188	0.188	0.188
1	2.673	0.250	0.250	0.250
Channels				
C6 $\times$ 13	13.00	2.13	0.642	—
C5 $\times$ 9	9.00	1.83	0.450	—
C4 $\times$ 7.25	7.25	1.47	0.343	—
C3 $\times$ 6	6.00	1.08	0.268	—

Table B.7 DEFORMED STEEL REINFORCING BAR SIZES

Bar size no.		Nominal diameter			Cross-sectional area			
					Number of bars			
U.S.	Metric (SI)	in	mm	Units	1	2	3	4
3	10	0.375	9.5	in^2	0.11	0.22	0.33	0.44
				mm^2	71	142	213	284
4	13	0.500	12.7	in^2	0.20	0.40	0.60	0.80
				mm^2	129	258	387	516
5	16	0.625	15.9	in^2	0.31	0.62	0.93	1.24
				mm^2	199	398	597	796
6	19	0.750	19.1	in^2	0.44	0.88	1.032	1.75
				mm^2	284	568	852	1136
7	22	0.875	22.2	in^2	0.60	1.20	1.80	2.41
				mm^2	387	774	1161	1548
8	25	1.000	25.4	in^2	0.79	1.58	2.37	3.16
				mm^2	510	1020	1530	2040
9	29	1.128	28.7	in^2	1.00	2.00	3.00	4.00
				mm^2	645	1290	1935	2580
10	32	1.270	32.3	in^2	1.27	2.54	3.81	5.08
				mm^2	819	1638	2457	3276
11	36	1.410	35.8	in^2	1.56	3.12	4.68	6.24
				mm^2	1006	2012	3018	4024
14	43	1.639	43.0	in^2	2.25	4.50	6.75	9.00
				mm^2	1452	2904	4356	5808
18	57	2.257	57.3	in^2	4.00	8.00	12.00	16.00
				mm^2	2581	5162	7743	10 324

Appendix C
PROPERTIES OF WOOD MEMBERS AND SPAN TABLES FOR JOISTS AND RAFTERS

Table C.1 SECTION PROPERTIES OF SOLID SAWN LUMBER ABOUT THE x AXIS

Nominal size		Surfaced (actual) size		Section area, A (in^2)	Section modulus, S (in^3)	Moment of inertia, I (in^4)
Breadth, b (in)	Depth, d (in)	Breadth, b (in)	Depth, d (in)			
2	2	1.5	1.5	2.25	0.563	0.422
	3	1.5	2.5	3.75	1.56	1.95
	4	1.5	3.5	5.25	3.06	5.36
	6	1.5	5.5	8.25	7.56	20.80
	8	1.5	7.25	10.88	13.14	47.63
	10	1.5	9.25	13.88	21.39	98.93
	12	1.5	11.25	16.88	31.64	177.98
	14	1.5	13.25	19.88	43.89	290.78
3	3	2.5	2.5	6.25	2.60	3.26
	4	2.5	3.5	8.75	5.10	8.93
	6	2.5	5.5	13.75	12.60	34.66
	8	2.5	7.25	18.12	21.90	79.39
	10	2.5	9.25	23.12	35.65	164.89
	12	2.5	11.25	28.12	52.73	296.63
	14	2.5	13.25	33.12	73.15	484.63
	16	2.5	15.25	38.12	96.90	738.87
4	4	3.5	3.5	12.25	7.15	12.51
	6	3.5	5.5	19.25	17.65	48.53
	8	3.5	7.25	25.38	30.66	111.15
	10	3.5	9.25	32.38	49.91	230.84
	12	3.5	11.25	39.38	73.83	415.28
	14	3.5	13.25	46.38	102.41	678.48
	16	3.5	15.25	53.38	135.66	1034.42
6	6	5.5	5.5	30.25	27.73	76.26
	8	5.5	7.5	41.25	51.56	193.36
	10	5.5	9.5	52.25	82.73	392.96
	12	5.5	11.5	63.25	121.23	697.07
	14	5.5	13.5	74.25	167.06	1 127.67
	16	5.5	15.5	85.25	220.23	1 706.78
	18	5.5	17.5	96.25	280.73	2 456.38
	20	5.5	19.5	107.25	348.56	3 398.48

Table C.1 *(cont'd.)*

Nominal size		Surfaced (actual) size		Section area, A (in^2)	Section modulus, S (in^3)	Moment of inertia, I (in^4)
Breadth, b (in)	Depth, d (in)	Breadth, b (in)	Depth, d (in)			
8	8	7.5	7.5	56.25	70.31	263.67
	10	7.5	9.5	71.25	112.81	535.86
	12	7.5	11.5	86.25	165.31	950.55
	14	7.5	13.5	101.25	227.81	1537.73
	16	7.5	15.5	116.25	300.31	2327.42
	18	7.5	17.5	131.25	382.81	3349.61
	20	7.5	19.5	146.25	475.31	4634.30
	22	7.5	21.5	161.25	577.81	6211.48
	24	7.5	23.5	176.25	690.31	8111.17
10	10	9.5	9.5	90.25	142.90	678.76
	12	9.5	11.5	109.25	209.40	1204.03
	14	9.5	13.5	128.25	288.56	1947.80
	16	9.5	15.5	147.25	380.40	2948.07
	18	9.5	17.5	166.25	484.90	4242.84
	20	9.5	19.5	185.25	602.06	5870.11
	22	9.5	21.5	204.25	731.90	7867.88
12	12	11.5	11.5	132.25	253.48	1457.51
	14	11.5	13.5	155.25	349.31	2357.86
	16	11.5	15.5	178.25	460.48	3568.71
	18	11.5	17.5	201.25	586.98	5136.07
	20	11.5	19.5	224.25	728.81	7 105.92
	22	11.5	21.5	247.25	885.98	9 524.28
	24	11.5	23.5	270.25	1 058.48	12 437.13

Table C.2 SECTION PROPERTIES OF SELECTED GLUED-LAMINATED LUMBER ABOUT THE X AXIS

Surfaced (actual) size		Number of laminations $1\frac{1}{2}$ in thick	Section area, A (in^2)	Section modulus, S (in^3)	Moment of inertia, I (in^4)
Breadth, b (in)	Depth, d (in)				
$3\frac{1}{8}$	4.50	3	14.1	10.5	23.7
	6.00	4	18.8	18.8	56.3
	7.50	5	23.4	29.3	109.9
	9.00	6	28.1	42.2	189.8
	10.50	7	32.8	57.4	301.5
	12.00	8	37.5	75.0	450.0
	13.50	9	42.2	94.9	640.7
	15.00	10	46.9	117.2	878.9
	16.50	11	51.6	141.8	1 169.8
	18.00	12	56.3	168.8	1 518.8
	19.50	13	60.9	198.0	1 931.0
	21.00	14	65.6	229.7	2 411.7
	22.50	15	70.3	263.7	2 966.3
	24.00	16	75.0	300.0	3 600.0
$5\frac{1}{8}$	6.00	4	30.8	30.8	92.3
	7.50	5	38.4	48.0	180.2
	9.00	6	46.1	69.2	311.3
	10.50	7	53.8	94.2	494.4
	12.00	8	61.5	123.0	738.0
	13.50	9	69.2	155.7	1 050.8
	15.00	10	76.9	192.2	1 441.4
	16.50	11	84.6	232.5	1 918.5
	18.00	12	92.3	276.8	2 490.8
	19.50	13	99.9	324.8	3 166.8
	21.00	14	107.6	376.7	3 955.2
	22.50	15	115.3	432.4	4 864.7
	24.00	16	123.0	492.0	5 904.0
	25.50	17	130.7	555.4	7 081.6
	27.00	18	138.4	622.7	8 406.3
	28.50	19	146.1	693.8	9 886.6
	30.00	20	153.8	768.8	11 531.3
$6\frac{3}{4}$	7.50	5	50.6	63.3	237.3
	9.00	6	60.8	91.1	410.1
	10.50	7	70.9	124.0	651.2
	12.00	8	51.0	162.0	972.0
	13.50	9	91.1	205.0	1 384.0
	15.00	10	101.3	253.1	1 898.4
	16.50	11	111.4	306.2	2 526.8
	18.00	12	121.5	364.5	3 280.5
	19.50	13	131.6	427.8	4 170.9
	21.00	14	141.8	496.1	5 209.3
	22.50	15	151.9	569.5	6 407.2
	24.00	16	162.0	648.0	7 776.0
	25.50	17	172.1	731.5	9 327.0
	27.00	18	182.3	820.1	11 071.7
	28.50	19	192.4	913.8	13 021.4

Table C.2 *(cont'd.)*

Surfaced (actual) size		Number of laminations $1\frac{1}{2}$ in thick	Section area, A (in^2)	Section modulus, S (in^3)	Moment of inertia, I (in^4)
Breadth, b (in)	Depth, d (in)				
$6\frac{3}{4}$	30.00	20	202.5	1 012.5	15 178.5
	31.50	21	212.6	1 116.3	17 581.4
	33.00	22	222.8	1 255.1	20 214.6
	34.50	23	232.9	1 339.0	23 098.3
	36.00	24	243.0	1 458.0	26 244.0
	37.50	25	253.1	1 582.0	29 663.1
	39.00	26	263.3	1 711.1	33 366.9
	40.50	27	273.4	1 845.3	37 367.0
	42.00	28	283.5	1 984.5	41 674.5
$8\frac{3}{4}$	10.50	7	91.9	160.8	844.1
	12.00	8	105.0	210.0	1 260.0
	13.50	9	118.1	265.8	1 794.0
	15.00	10	131.3	328.1	2 460.9
	16.50	11	144.4	397.0	3 275.5
	18.00	12	157.5	472.5	4 252.5
	19.50	13	170.6	554.5	5 406.7
	21.00	14	183.8	643.1	6 752.8
	22.50	15	196.9	738.3	8 305.7
	24.00	16	210.0	840.0	10 080.0
	25.50	17	223.1	948.3	12 090.6
	27.00	18	236.3	1 063.1	14 352.2
	28.50	19	249.4	1 184.5	16 879.6
	30.00	20	262.5	1 312.5	19 687.5
	31.50	21	275.6	1 447.0	22 790.7
	33.00	22	288.8	1 588.1	26 204.1
	34.50	23	301.9	1 735.8	29 942.2
	36.00	24	315.0	1 890.0	34 020.0
	37.50	25	328.1	2 050.8	38 452.2
	39.00	26	341.3	2 218.1	43 253.4
	40.50	27	354.4	2 392.0	48 438.6
	42.00	28	367.5	2 572.5	54 022.5
	43.50	29	380.6	2 759.5	60 019.8
	45.00	30	393.8	2 953.1	66 445.3
	46.50	31	406.9	3 153.3	73 313.8
	48.00	32	420.0	3 360.0	80 640.0
$10\frac{3}{4}$	12.00	8	129.0	258.0	1 548.0
	13.50	9	145.1	326.5	2 204.1
	15.00	10	161.3	403.1	3 023.4
	16.50	11	177.4	487.8	4 024.2
	18.00	12	193.5	580.5	5 224.5
	19.50	13	209.6	681.3	6 642.5
	21.00	14	225.8	790.1	8 296.3
	22.50	15	241.9	907.0	10 204.1
	24.00	16	258.0	1 032.0	12 384.0
	25.50	17	274.1	1 165.0	14 854.1
	27.00	18	290.3	1 306.1	17 632.7
	28.50	19	306.4	1 455.3	20 737.8

Table C.2 *(cont'd.)*

$10\frac{3}{4}$	30.00	20	322.5	1 612.5	24 187.5
	31.50	21	338.6	1 777.8	28 000.1
	33.00	22	354.8	1 951.1	32 193.6
	34.50	23	370.9	2 132.5	36 786.2
	36.00	24	387.0	2 322.0	41 796.0
	37.50	25	403.1	2 519.5	47 241.2
	39.00	26	419.3	2 725.1	53 139.9
	40.50	27	435.4	2 938.8	59 510.3
	42.00	28	451.5	3 160.5	66 370.5
	43.50	29	467.6	3 390.3	73 738.6
	45.00	30	483.8	3 628.1	81 632.8
	46.50	31	499.9	3 874.0	90 071.2
	48.00	32	516.0	4 128.0	99 072.0
	49.50	33	532.1	4 390.0	108 653.3
	51.00	34	548.3	4 660.1	118 833.2
	52.50	35	564.4	4 938.3	129 629.9
	54.00	36	580.5	5 224.5	141 061.5
	55.50	37	596.6	5 518.8	153 146.2
	57.00	38	612.8	5 821.1	165 902.1
	58.50	39	628.9	6 131.5	179 347.3
	60.00	40	645.0	6 450.0	193 500.0

Table C.3 SECTION PROPERTIES OF SELECTED LAMINATED-VENEER LUMBER (LVL) MEMBERS ABOUT THE x AXIS

Actual size		Section area, A (in^2)	Section modulus, S (in^3)	Moment of inertia, I (in^4)
Breadth, b (in)	Depth, d (in)			
$1\frac{3}{4}$	$5\frac{1}{2}$	9.625	8.82	24.26
	$7\frac{1}{4}$	16.688	15.33	55.57
	$9\frac{1}{4}$	16.188	24.96	115.42
	$9\frac{1}{2}$	16.625	26.32	125.03
	$11\frac{1}{4}$	19.688	36.91	207.64
	$11\frac{7}{8}$	20.781	41.13	244.21
	14	24.500	57.17	400.17
	16[a]	28.000	74.67	597.33
	18[a]	31.500	94.50	850.50

[a]Only for use in multiple beam members.

Table C.4 SECTION PROPERTIES OF SELECTED PARALLEL STRAND LUMBER (PSL) MEMBERS ABOUT THE x AXIS

Actual size		Section area, A (in^2)	Section modulus, S (in^3)	Moment of inertia, I (in^4)
Breadth, b (in)	Depth, d (in)			
$3\frac{1}{2}$	14	49.00	114.33	800.33
	16	56.00	149.33	1 194.67
	18	63.00	189.00	1 701.00
	20	70.00	233.33	2 333.33
	22	77.00	282.33	3 105.67
	24	84.00	336.00	4 032.00
	26	91.00	394.33	5 126.33
	28	98.00	457.33	6 402.67
	30	105.00	525.00	7 875.00
	32	112.00	597.33	9 557.33
	34	119.00	674.33	11 463.67
	36	126.00	756.00	13 608.00
	38	133.00	842.33	16 004.33
	40	140.00	933.33	18 666.67
	42	147.00	1 029.00	21 609.00
$4\frac{3}{8}$	18	78.57	236.25	2 126.25
	20	87.50	291.67	2 916.67
	22	96.25	352.92	3 882.08
	24	105.00	420.00	5 040.00
	26	113.75	492.92	6 407.92
	28	122.50	571.67	8 003.33
	30	131.25	656.25	9 843.75
	32	140.00	746.67	11 946.67
	34	148.75	842.92	14 329.58
	36	157.50	945.00	17 010.00
	38	166.25	1 052.92	20 005.42
	40	175.00	1 166.67	23 333.33
	42	183.75	1 286.25	27 011.25
	44	192.50	1 411.67	31 056.67
	46	201.25	1 542.92	35 487.08
$5\frac{1}{4}$	22	115.50	423.50	4 658.50
	24	126.00	504.00	6 048.00
	26	136.50	591.50	7 689.50
	28	147.00	686.00	9 604.00
	30	157.50	787.50	11 812.50
	32	168.00	896.00	14 336.00
	34	178.50	1 011.50	17 195.50
	36	189.00	1 134.00	20 412.00
	38	199.50	1 263.50	24 006.50
	40	210.00	1 400.00	28 000.00
	42	220.50	1 543.50	32 413.50
	44	231.00	1 694.00	37 268.00
	46	241.50	1 851.50	42 584.50
	48	252.00	2 016.00	48 384.00
	50	262.50	2 187.50	54 687.50

Table C.4 *(cont'd.)*

Actual size		Section area, A (in^2)	Section modulus, S (in^3)	Moment of inertia, I (in^4)
Breadth, b (in)	Depth, d (in)			
7	26	182.00	788.67	10 252.67
	28	196.00	914.67	12 805.33
	30	210.00	1 050.00	15 750.00
	32	224.00	1 194.67	19 114.67
	34	238.00	1 348.67	22 927.33
	36	252.00	1 512.00	27 216.00
	38	266.00	1 684.67	32 008.67
	40	280.00	1 866.67	37 333.33
	42	294.00	2 058.00	43 218.00
	44	308.00	2 258.67	49 690.67
	46	322.00	2 468.67	56 779.33
	48	336.00	2 688.00	64 512.00
	50	350.00	2 916.67	72 916.67
	52	364.00	3 154.67	82 021.33
	54	378.00	3 402.00	91 854.00

Table C.5 SECTION PROPERTIES OF SELECTED LAMINATED STRAND LUMBER (LSL) MEMBERS ABOUT THE x AXIS

Actual size		Section area, A (in^2)	Section modulus, S (in^3)	Moment of inertia, I (in^4)
Breadth, b (in)	Depth, d (in)			
$3\frac{1}{2}$	$4\frac{3}{8}$	15.313	11.17	24.42
	$5\frac{1}{2}$	19.250	17.65	48.53
	$7\frac{1}{4}$	25.375	30.66	111.15
	$8\frac{5}{8}$	30.188	43.39	187.14
	$9\frac{1}{2}$	33.250	52.65	250.07
	$11\frac{1}{4}$	39.375	73.83	415.28
	$11\frac{7}{8}$	41.563	82.26	488.41
	14	49.000	114.33	800.33
	16	56.000	149.33	1 194.67
	18	63.000	189.00	1 701.00

Table C.6 BASE DESIGN VALUES AND ADJUSTMENT FACTORS FOR VISUALLY GRADED DIMENSION LUMBER (2 TO 4 IN THICK) FOR SELECTED GRADES AND SPECIES OF WOOD

Species and commercial grade	Base design values (lb/in^2) (For lumber with a moisture content not exceeding 19%)					
	F_b	F_t	F_v	$F_{c\perp}$	F_c	E
Douglas Fir-Larch (North), No. 2	825	500	95	625	1350	1 600 000
Hemlock-Fir, No. 2	850	500	75	405	1250	1 300 000
Redwood, No. 2, open grain	725	425	80	425	700	1 000 000
Spruce-Pine-Fir (South), No. 2	750	325	70	335	975	1 100 000

Repetitive member factor, C_r	
(Applied to the extreme fiber stress in bending (F_b) for 2 inch to 4 inch thick members such as joists, rafters, truss chords, studs, planks, decking or similar members that are three or more in number and in contact or spaced no more than 24 inches on center and joined by floor, roof or other load distributing elements adequate to support the design load.)	$C_r = 1.15$ (for F_b only)

		F_b			
		Thickness (in)			
	Width (in)	2 and 3	4	F_t	F_c
Size factor, C_F	2, 3, and 4	1.5	1.5	1.5	1.15
(No. 2 Grade only)	5	1.4	1.4	1.4	1.10
	6	1.3	1.3	1.3	1.10
	8	1.2	1.3	1.2	1.05
	10	1.1	1.2	1.1	1.00
	12	1.0	1.1	1.0	1.00
	14 and wider	0.9	1.0	0.9	0.90

Wet service factor, C_M	F_b	F_t	F_v	$F_{c\perp}$	F_c	E
(For use with a moisture content above 19% for an extended time period)	0.85	1.00	0.97	0.67	0.80	0.90

		Thickness (in)	
	Width (in)	2 and 3	4
Flat-use factor, C_{fu}	2 and 3	1.00	—
(Applied to the extreme fiber stress in	4	1.10	1.00
bending (F_b) for dimension lumber when used with the	5	1.10	1.05
load applied to the flat face)	6	1.15	1.05
	8	1.15	1.05
	10 and wider	1.20	1.10

Source: Condensed and adapted from the Uniform Building Code, Table 23-I-A-1 (1997).

Table C.7 BASE DESIGN VALUES AND ADJUSTMENT FACTORS FOR VISUALLY GRADED TIMBERS (5 IN BY 5 IN AND LARGER) FOR SELECTED GRADES AND SPECIES OF WOOD

Species and commercial grade	Base design values (lb/in^2) (For lumber with a moisture content not exceeding 19%)					
	F_b	F_t	F_v	$F_c\perp$	F_c	E
Douglas Fir-Larch (North), No. 2	875	425	85	625	600	1 300 000
Hemlock-Fir, No. 2, beams and stringers	675	325	70	405	475	1 100 000
Hemlock-Fir, No. 2, posts and timbers	525	350	70	405	375	1 300 000
Redwood, No. 2	975	650	95	650	900	1 100 000
Spruce-Pine-Fir (South) , No. 2, beams and stringers	600	300	65	425	425	1 000 000
Spruce-Pine-Fir (South), No. 2, posts and timbers	500	325	65	425	500	1 000 000

	Nominal depth (in)	Actual depth, d (in)	Size factor, C_F
	Not to exceed 12	Not to exceed 12	1.000
	14	13½	0.987
	16	15½	0.972
	18	17½	0.959
	20	19½	0.947
	22	21½	0.937
	24	23½	0.928
	26	25½	0.920
Size factor, C_F	28	27½	0.912
[Applied to the extreme fiber stress in bending (F_b) for	30	29½	0.905
timbers, beams, stringers, or posts greater than 12 in. in	32	31½	0.898
depth (d); values in this table are based on	34	33½	0.892
$C_F = (12/d)^{1/9}$]	36	35½	0.887
	38	37½	0.881
	40	39½	0.876
	42	41½	0.871
	44	43½	0.867
	46	45½	0.862
	48	47½	0.858

Wet service factor, C_M (For use with a moisture content above 19% for an extended time period)	F_b	F_t	F_v	$F_c\perp$	F_c	E
	1.00	1.00	1.00	0.67	0.91	1.00

Source: Condensed and adapted from the Uniform Building Code, Table 23-I-A-4 (1997).

Table C.8 BASE DESIGN VALUES AND ADJUSTMENT FACTORS FOR SELECTED
ENGINEERED WOOD PRODUCTS

Wood product	Base design values (lb/in^2) (For dry applications)					
	F_b	F_t	F_v	$F_c\perp$	F_c	E
Glued-laminated timber (gluelams)	2000	—	165	400		2 000 000
Laminated veneer lumber (LVL)						
1.8 E WS Microllam LVL	2600	—	285	750	2460	1 800 000
2.0 E WS Microllam LVL	2900	—	290	650	2900	2 000 000
Parallel strand lumber (PSL)						
Parallam PSL commercial beam	2900	—	290	750	2900	2 000 000
Laminated strand lumber (LSL)[a]						
3½" - 1.3E TimberStrand LSL	1700	—	285	650	1400	1 300 000
3½" - 1.5E TimberStrand LSL	2250	—	285	650	1950	1 500 000

	Nominal depth (in)	Actual depth, d (in)	Size factor, C_F
	Not to exceed 12	Not to exceed 12	1.000
	14	13½	0.987
	16	15½	0.972
	18	17½	0.959
	20	19½	0.947
	22	21½	0.937
	24	23½	0.928
	26	25½	0.920
Size factor, C_F	28	27½	0.912
[Applied to the extreme fiber	30	29½	0.905
stress in bending (F_b) for	32	31½	0.898
members greater than 12 in. in	34	33½	0.892
depth (d); values in this table	36	35½	0.887
are based on $C_F = (12/d)^{1/9}$]	38	37½	0.881
	40	39½	0.876
	42	41½	0.871
	44	43½	0.867
	46	45½	0.862
	48	47½	0.858

Source: Extracted from manufacturers' technical literature.

Table C.9 VALUES FOR VISUALLY GRADED JOISTS AND RAFTERS FOR SELECTED GRADES AND SPECIES OF WOOD

<table>
<tr><td rowspan="4"></td><td rowspan="4"></td><td colspan="4" align="center">Design values (lb/in²)
(For lumber with a moisture content not exceeding 19%)</td></tr>
<tr><td colspan="3" align="center">F_b [a]</td><td></td></tr>
<tr><td align="center">Species and
commercial grade</td><td align="center">Size
(in)</td><td align="center">Normal
duration</td><td align="center">Snow
loading</td><td align="center">7-Day
loading</td><td align="center">E</td></tr>
<tr><td>Douglas Fir-Larch, No. 2</td><td>2 × 4</td><td>1510</td><td>1735</td><td>1885</td><td></td></tr>
<tr><td></td><td>2 × 6</td><td>1310</td><td>1505</td><td>1635</td><td></td></tr>
<tr><td></td><td>2 × 8</td><td>1210</td><td>1390</td><td>1510</td><td>1 600 000</td></tr>
<tr><td></td><td>2 × 10</td><td>1105</td><td>1275</td><td>1385</td><td></td></tr>
<tr><td></td><td>2 × 12</td><td>1005</td><td>1155</td><td>1260</td><td></td></tr>
<tr><td>Hemlock-Fir, No. 2</td><td>2 × 4</td><td>1725</td><td>1985</td><td>2155</td><td></td></tr>
<tr><td></td><td>2 × 6</td><td>1495</td><td>1720</td><td>1870</td><td></td></tr>
<tr><td></td><td>2 × 8</td><td>1380</td><td>1585</td><td>1725</td><td>1 600 000</td></tr>
<tr><td></td><td>2 × 10</td><td>1265</td><td>1455</td><td>1580</td><td></td></tr>
<tr><td></td><td>2 × 12</td><td>1150</td><td>1325</td><td>1440</td><td></td></tr>
<tr><td>Redwood, No. 2, open grain</td><td>2 × 4</td><td>1250</td><td>1440</td><td>1565</td><td></td></tr>
<tr><td></td><td>2 × 6</td><td>1085</td><td>1245</td><td>1355</td><td></td></tr>
<tr><td></td><td>2 × 8</td><td>1000</td><td>1150</td><td>1250</td><td>1 000 000</td></tr>
<tr><td></td><td>2 × 10</td><td>915</td><td>1055</td><td>1145</td><td></td></tr>
<tr><td></td><td>2 × 12</td><td>835</td><td>960</td><td>1040</td><td></td></tr>
<tr><td>Mixed Southern Pine, No 2.</td><td>2 × 4</td><td>1500</td><td>1720</td><td>1870</td><td></td></tr>
<tr><td></td><td>2 × 6</td><td>1320</td><td>1520</td><td>1650</td><td></td></tr>
<tr><td></td><td>2 × 8</td><td>1210</td><td>1390</td><td>1510</td><td>1 400 000</td></tr>
<tr><td></td><td>2 × 10</td><td>1060</td><td>1220</td><td>1330</td><td></td></tr>
<tr><td></td><td>2 × 12</td><td>1010</td><td>1160</td><td>1260</td><td></td></tr>
</table>

Source: Condensed and adapted from the Uniform Building Code, Table 23-I-X-1 (1997).

[a]Tabulated design values for F_b and F_v for *normal-duration loading* apply to normal occupancy (live load) conditions over a 10-year period. These F_b values are for use with repetitive members spaced not more than 24 in O.C. For wider spacing, these F_b values should be reduced by 13%. A load-duration factor (C_D) is used when additional intermittent loading is anticipated, such as under snow and construction loading conditions. *Snow loading* ($C_D = 1.15$) is for a two-month snow loading condition. The *7-day loading condition* ($C_D = 1.25$) represents loading that may occur during construction. Design values for snow loading and 7-day loading are provided in this table for use with joists and rafters only.

Table C.10 ALLOWABLE SPANS FOR FLOOR JOISTS WITH $l/360$ DEFLECTION LIMITS FOR 40-LB/FT² LIVE LOAD PLUS 10-LB/FT² DEAD LOAD.[a]

Joist size (in)	Spacing (in)	Required modulus of elasticity, E (1 000 000 lb/in²)													
		0.8	0.9	1.0	1.1	1.2	1.3	1.4	1.5	1.6	1.7	1.8	1.9	2.0	2.1
2 × 6	12.0	8-6	8-10	9-2	9-6	9-9	10-0	10-3	10-6	10-9	10-11	11-2	11-4	11-7	11-9
	16.0	7-9	8-0	8-4	8-7	8-10	9-1	9-4	9-6	9-9	9-11	10-2	10-4	10-6	10-8
	19.2	7-3	7-7	7-10	8-1	8-4	8-7	8-9	9-0	9-2	9-4	9-6	9-8	9-10	10-0
	24.0	6-9	7-0	7-3	7-6	7-9	7-11	8-2	8-4	8-6	8-8	8-10	9-0	9-2	9-4
2 × 8	12.0	11-3	11-8	12-1	12-6	12-10	13-2	13-6	13-10	14-2	14-5	14-8	15-0	15-3	15-6
	16.0	10-2	10-7	11-0	11-4	11-8	12-0	12-3	12-7	12-10	13-1	13-4	13-7	13-10	14-1
	19.2	9-7	10-0	10-4	10-8	11-0	11-3	11-7	11-10	12-1	12-4	12-7	12-10	13-0	13-3
	24.0	8-11	9-3	9-7	9-11	10-2	10-6	10-9	11-0	11-3	11-5	11-8	11-11	12-1	12-3
2 × 10	12.0	14-4	14-11	15-5	15-11	16-5	16-10	17-3	17-8	18-0	18-5	18-9	19-1	19-5	19-9
	16.0	13-0	13-6	14-0	14-6	14-11	15-3	15-8	16-0	16-5	16-9	17-0	17-4	17-8	17-11
	19.2	12-3	12-9	13-2	13-7	14-0	14-5	14-9	15-1	15-5	15-9	16-0	16-4	16-7	16-11
	24.0	11-4	11-10	12-3	12-8	13-0	13-4	13-8	14-0	14-4	14-7	14-11	15-2	15-5	15-8
2 × 12	12.0	17-5	18-1	18-9	19-4	19-11	20-6	21-0	61-6	21-11	22-5	22-10	23-3	23-7	24-0
	16.0	15-10	16-5	17-0	17-7	18-1	18-7	19-1	19-6	19-11	20-4	20-9	21-1	21-6	21-10
	19.2	14-11	15-6	16-0	16-7	17-0	17-6	17-11	18-4	18-9	19-2	19-6	19-10	20-2	20-6
	24.0	13-10	14-4	14-11	15-4	15-10	16-3	16-8	17-0	17-5	17-9	19-1	18-5	18-9	19-1
Required F_b (lb/in²)	12.0	718	777	833	888	941	993	1043	1092	1140	1187	1233	1278	1323	1367
	16.0	790	855	917	977	1036	1093	1148	1202	1255	1306	1357	1407	1456	1504
	19.2	840	909	975	1039	1101	1161	1220	1277	1333	1388	1442	1495	1547	1598
	24.0	905	979	1050	1119	1186	1251	1314	1376	1436	1496	1554	1611	1667	1722

Source: Condensed and adapted from the Uniform Building Code, Table 23-I-V-J-1 (1997).

[a]Spans are expressed in feet-inches. Design values for F_b and E should come from Table C.9 (Uniform Building Code Table 23-I-X-1) only.

Table C.11 ALLOWABLE SPANS FOR CEILING JOISTS WITH $l/240$ DEFLECTION LIMITS FOR 10-LB/FT² LIVE LOAD PLUS 10-LB/FT² DEAD LOAD[a]

Joist size (in)	Spacing (in)	Required modulus of elasticity, E (1 000 000 lb/in²)													
		0.8	0.9	1.0	1.1	1.2	1.3	1.4	1.5	1.6	1.7	1.8	1.9	2.0	2.1
2 × 4	12.0	9-10	10-3	10-7	10-11	11-3	11-7	11-10	12-2	12-5	12-8	12-11	13-2	13-4	13-7
	16.0	8-11	9-4	9-8	9-11	10-3	10-6	10-9	11-0	11-3	11-6	11-9	11-11	12-2	12-4
	19.2	8-5	8-9	9-1	9-4	9-8	9-11	10-2	10-4	10-7	10-10	11-0	11-13	11-5	11-7
	24.0	7-10	8-1	8-5	8-8	8-11	9-2	9-5	9-8	9-10	10-0	10-3	10-5	10-7	10-9
2 × 6	12.0	15-6	16-1	16-8	17-2	17-8	18-2	18-8	19-1	19-6	19-11	20-3	20-8	21-0	21-4
	16.0	14-1	14-7	15-2	15-7	16-1	16-6	16-11	17-4	17-8	18-1	18-5	18-9	19-1	19-5
	19.2	13-3	13-9	14-3	14-8	15-2	15-7	15-11	16-4	16-8	17-0	17-4	17-8	17-11	18-3
	24.0	12-3	12-9	13-3	13-8	14-1	14-5	14-9	15-2	15-6	15-9	16-1	16-4	16-8	16-11
2 × 8	12.0	20-5	21-2	21-11	22-8	23-4	24-0	24-7	25-2	25-8					
	16.0	18-6	19-3	19-11	20-7	21-2	21-9	22-4	22-10	23-4	23-10	24-3	24-8	25-2	25-7
	19.2	17-5	18-1	18-9	19-5	19-11	20-6	21-0	21-6	21-11	22-5	22-10	23-3	23-8	24-0
	24.0	16-2	16-10	17-5	18-0	18-6	19-1	19-6	19-11	20-5	20-10	21-2	21-7	21-11	22-4
2 × 10	12.0	26-0													
	16.0	23-8	24-7	25-5											
	19.2	22-3	23-1	23-11	24-9	25-5									
	24.0	20-8	21-6	22-3	22-11	23-8	24-3	24-10	25-5	26-0					
Required F_b (lb/in²)		711	769	825	880	932	983	1033	1082	1129	1176	1221	1266	1310	1354
		783	847	909	968	1026	1082	1137	1191	1243	1294	1344	1394	1422	1490
		832	900	965	1029	1090	1150	1208	1265	1321	1375	1429	1481	1583	1583
		896	969	1040	1108	1174	1239	1302	1363	1423	1481	1595	1595	1651	1706

Source: Condensed and adapted from the Uniform Building Code, Table 23-I-V-J-3 (1997).

[a]Spans are expressed in feet-inches. Design values for F_b and E should come from Table C.9 (Uniform Building Code Table 23-I-X-1) only.

Table C.12 ALLOWABLE SPANS FOR RAFTERS WITH $l/240$ DEFLECTION LIMITS FOR 30-LB/FT² LIVE LOAD PLUS 15-LB/FT² DEAD LOAD[a]

Rafter size (in)	Spacing (in)	Required bending design value, F_b (lb/in²)													
		700	800	900	1000	1100	1200	1300	1400	1500	1600	1700	1800	1900	2000
2 × 6	12.0	9-6	10-0	10-0	10-7	11-1	11-7	12-1	12-6	13-0	13-5	13-10	14-2	14-7	15-0
	16.0	8-2	8-8	8-8	9-2	9-7	10-0	10-5	10-10	11-3	11-7	11-11	12-4	12-8	13-0
	19.2	7-6	7-11	7-11	8-4	8-9	9-2	9-6	9-11	10-3	10-7	10-11	11-3	11-6	11-10
	24.0	6-8	7-1	7-1	7-8	7-10	8-2	8-6	8-10	9-2	9-6	9-9	10-0	10-4	10-7
2 × 8	12.0	12-6	13-3	13-3	13-11	14-8	15-3	15-11	16-6	17-1	17-8	18-2	18-9	19-3	19-9
	16.0	10-10	11-6	11-8	12-1	12-8	13-3	13-9	14-4	14-10	15-3	15-9	16-3	16-8	17-1
	19.2	9-10	10-6	10-6	11-0	11-7	12-1	12-7	13-1	13-6	13-11	14-5	14-10	15-2	15-7
	24.0	8-10	9-4	9-4	9-10	10-4	10-10	11-3	11-8	12-1	12-6	12-10	13-3	13-7	13-11
2 × 10	12.0	15-11	16-11	16-11	17-10	18-8	19-6	20-4	21-1	21-10	22-6	23-3	23-11	24-8	25-2
	16.0	13-9	14-8	14-8	15-5	16-2	16-11	17-7	18-3	18-11	19-6	20-1	20-8	21-3	21-10
	19.2	12-7	13-4	13-4	14-1	14-9	15-5	16-1	16-8	17-3	17-10	18-4	18-11	19-5	19-11
	24.0	11-3	11-11	11-11	12-7	13-2	13-9	14-4	14-11	15-5	15-11	16-5	16-11	17-4	17-10
2 × 12	12.0	19-4	20-6	20-6	21-8	22-8	23-9	24-8	25-7						
	16.0	16-9	17-9	17-9	18-9	19-8	20-6	21-5	22-2	23-0	23-9	24-5	25-2	25-10	
	19.2	15-4	16-3	16-3	17-1	17-11	18-9	19-6	20-3	21-0	21-8	22-4	23-0	23-7	24-2
	24.0	13-8	14-6	14-6	15-4	16-1	16-9	17-5	18-1	18-9	19-4	20-0	20-6	21-1	21-8
Required E (1 000 000 lb/in²)		0.44	0.55	0.66	0.77	0.89	1.01	1.14	1.28	1.41	1.56	1.71	1.86	2.02	2.18
		0.38	0.48	0.57	0.67	0.77	0.88	0.99	1.10	1.22	1.35	1.48	1.61	1.75	1.89
		0.35	0.44	0.52	0.61	0.70	0.80	0.90	1.01	1.12	1.23	1.35	1.47	1.59	1.72
		0.31	0.39	0.46	0.54	0.63	0.72	0.81	0.90	1.00	1.10	1.21	1.31	1.43	1.54

Source: Condensed and adapted from the Uniform Building Code Table 23-I-V-R-4 (1997).

[a] Spans are expressed in feet-inches. Design values for F_b and E should come from Table C.9 (Uniform Building Code Table 23-I-X-1) only.

Appendix D
MATH REVIEW

In this appendix fundamental mathematics principles related to the engineering concepts discussed in the text are reviewed.

D.I   FRACTION/DECIMAL EQUIVALENTS AND CONVERSIONS

A measurement can be expressed in fractions or decimals. Frequently, it is necessary to convert from decimals to fractions or fractions to decimals. Table D.1 provides the decimal equivalents of fractions. Converting from fractions to decimals involves simply dividing the fraction. Converting from decimals to fractions is easily accomplished by relying on the table. It is also possible to compute the fraction equivalent, as demonstrated in Example D.1.

■ EXAMPLE D.1

(a) Convert $\frac{31}{64}$ inch to its decimal equivalent.
(b) Convert 0.2813 inch to its fraction equivalent by using data in Table D.1.
(c) Convert 0.2813 inch to its fraction equivalent by computing the fraction to $\frac{1}{64}$-in accuracy.

Solution   (a) By dividing 31 by 64, the decimal equivalent is found:

$$\frac{31}{64} \text{ in} = 0.484375 \text{ in}$$

(b) Upon review of Table D.1, 0.28125 or $\frac{9}{32}$ is very nearly 0.2813. Therefore, 0.2813 is very nearly $\frac{9}{32}$ in.

(c) By multiplying 0.2813 by 64ths of an inch, it is found that 0.2813 is nearly $\frac{18}{64}$:

$$0.2813 \cdot 64 \text{ per 64th} = \frac{18.003}{64} = \frac{9}{32}$$

However, $\frac{18}{64}$ reduces to $\frac{9}{32}$. Therefore, 0.2813 in is $\frac{9}{32}$ in at $\frac{1}{64}$-in accuracy.

Table D.1 FRACTION AND DECIMAL EQUIVALENTS

Fraction	Decimal	Fraction	Decimal
$\frac{1}{64}$	0.015 625	$\frac{33}{64}$	0.515 625
$\frac{1}{32}$	0.031 25	$\frac{17}{32}$	0.531 25
$\frac{3}{64}$	0.046 875	$\frac{35}{64}$	0.546 875
$\frac{1}{16}$	0.062 5	$\frac{9}{16}$	0.562 5
$\frac{5}{64}$	0.078 125	$\frac{37}{64}$	0.578 125
$\frac{3}{32}$	0.093 75	$\frac{19}{32}$	0.593 75
$\frac{7}{64}$	0.109 375	$\frac{39}{64}$	0.609 375
$\frac{1}{8}$	0.125	$\frac{5}{8}$	0.625
$\frac{9}{64}$	0.140 625	$\frac{41}{64}$	0.640 625
$\frac{5}{32}$	0.156 25	$\frac{21}{32}$	0.656 25
$\frac{11}{64}$	0.171 875	$\frac{43}{64}$	0.671 875
$\frac{3}{16}$	0.187 5	$\frac{11}{16}$	0.687 5
$\frac{13}{64}$	0.203 125	$\frac{45}{64}$	0.703 125
$\frac{7}{32}$	0.218 75	$\frac{23}{32}$	0.718 75
$\frac{15}{64}$	0.234 375	$\frac{47}{64}$	0.734 375
$\frac{1}{4}$	0.25	$\frac{3}{4}$	0.75
$\frac{17}{64}$	0.265 625	$\frac{49}{64}$	0.765 625
$\frac{9}{32}$	0.281 25	$\frac{25}{32}$	0.781 25
$\frac{19}{64}$	0.296 875	$\frac{51}{64}$	0.796 875
$\frac{5}{16}$	0.312 5	$\frac{13}{16}$	0.812 5
$\frac{21}{64}$	0.328 125	$\frac{53}{64}$	0.828 125
$\frac{11}{32}$	0.343 75	$\frac{27}{32}$	0.843 75
$\frac{23}{64}$	0.359 375	$\frac{55}{64}$	0.859 375
$\frac{3}{8}$	0.375	$\frac{7}{8}$	0.875
$\frac{25}{64}$	0.390 625	$\frac{57}{64}$	0.890 625
$\frac{13}{32}$	0.406 25	$\frac{29}{32}$	0.906 25
$\frac{27}{64}$	0.421 875	$\frac{59}{64}$	0.921 875
$\frac{7}{16}$	0.437 5	$\frac{15}{16}$	0.937 5
$\frac{29}{64}$	0.453 125	$\frac{61}{64}$	0.953 125
$\frac{15}{32}$	0.468 75	$\frac{31}{32}$	0.968 75
$\frac{31}{64}$	0.484 375	$\frac{63}{64}$	0.984 375
$\frac{1}{2}$	0.5	1	1.0

D.2 SCALAR AND VECTOR QUANTITIES

A *scalar quantity* can be expressed in terms of magnitude only, such as miles, dollars, years, square feet, and so on. Scalar quantities can be added, subtracted, multiplied, and divided following the rules of basic arithmetic. A *vector quantity* is described by magnitude and direction (line of action and sense). A force is an example of a vector quantity. Vector quantities cannot be added or subtracted arithmetically.

For example, two forces of equal magnitude that pull against each other result in a combined force of zero; the opposing forces cancel one another. Vector quantities must be added graphically or by using vector mathematics. A force is an example of a vector quantity.

A *vector* is a graphic presentation of a vector quantity. Direction is shown as a line on a drawing with an arrowhead indicating the sense placed at the end of the line. Magnitude is indicated with a line drawn to a selected scale (i.e., 1 in = 100 lb, 10 mm = 100 N, etc.). The sum of two or more vectors is called the resultant. A *resultant* is a single vector that can replace two or more other vectors acting in a force system such that it produces the same effect. Two or more vector quantities may be added together and replaced by a single resultant.

■ EXAMPLE D.2

One morning, a loaded coal train traveled easterly from the coal mine to the power plant, a one-way distance of 210 miles. That afternoon, the unloaded train returned to the coal mine by traveling westerly along the same route.
(a) Determine the miles traveled by the train (a scalar quantity) during that day.
(b) Determine the resultant of travel by the train (the vector resultant) during that day.

Solution

(a) 210-mile trip to town + 210-mile return trip = 420 miles
(b) 210 miles of easterly travel + 210 miles of westerly travel = 0 miles

---

The coal train in Example D.2 traveled 420 miles on its round trip (a scalar quantity). However, the ultimate result of its travel was that it ended up at its starting point; that is, it returned to its original location. The scalar quantity of miles traveled was 420 miles, but the resultant of the trip was zero miles.

■ EXAMPLE D.3

An automobile leaves point A and travels 100 km in an easterly direction. The automobile turns and heads north for an additional 175 km to point B.
(a) Determine the miles traveled by the automobile (a scalar quantity).
(b) Determine the resultant of travel by the automobile (the vector resultant).

Solution

(a) 100 km + 175 km = 275 km

(b) The Pythagorean theorem, $a^2 + b^2 = c^2$, is used to find the hypotenuse of a right triangle. The hypotenuse represents the resultant of travel.

$$(100 \text{ km})^2 + (175 \text{ km})^2 = (40\ 625 \text{ km})^2 = 201.6 \text{ km}$$

The automobile in Example D.3 traveled 275 km. The resultant of travel is measured by the distance between points A and B. The scalar quantity of distance traveled was 275 km, but the resultant quantity of the trip was 201.6 km, the distance between points A and B.

D.3 USE OF SYMBOLS

A *symbol* is a character that represents a variable quantity in a mathematical equation or engineering formula, much like the familiar variables x, y, and z used in the study of algebraic equations such as $y = 4x + 13z$. Symbols represent a variable quantity with specific units. For example, the symbol L may represent length in feet or meters, W may represent weight in newtons or pounds, and so on. The user of an engineering formula must recognize the meaning and limitations of each symbol before using the formula in analysis and relying on analysis results. Much of the study of engineering principles is exploring the meaning and understanding the limitations of the variables and therefore analysis results.

Unfortunately, use of a specific symbol varies from field to field and even within specific disciplines. For example, the Greek lowercase letter sigma (σ) is universally accepted as the symbol for stress in science and engineering formulas. However, in structural engineering analysis, the letter F or f is used as the symbol for stress. The symbol F can also identify a force. Different symbols related to type of forces are also used instead of F, such as P for a concentrated load, w for a uniformly distributed load, and W for a total uniformly distributed load. It is important to grow familiar with standardized symbols used in the engineering field.

Because of its wide acceptance and universal use in general engineering, the Greek lowercase letter sigma (σ) is used to symbolize a variable quantity for stress in the first 17 chapters of this book. An exception in the use of σ is made in latter chapters where symbols related to a material standard in a specific area of the structural engineering profession are used: for example, f_s and f_c for steel-reinforced concrete, F_b for steel and wood, and so on. The symbol F is used to identify a variable quantity for a concentrated force. Different symbols related to type of loads are also used instead of F, such as P for a concentrated load, w for a uniformly distributed load, and W for a total uniformly distributed load. It is important to grow familiar with standardized symbols used in the engineering field.

D.4 ALGEBRAIC EQUATIONS

An *equation* is a statement that two quantities are equal, such as $a = b$, $ax = by$, $a/x = b/y$, and so on. An *axiom* is a statement that is assumed to be true without proof. The following axioms can be used as the basis for solving equations:

Axiom 1: The reflexive property of equality states that any quantity is equal to itself: $a = a$, $b = b$, $c = c$, and so on.

Axiom 2: The symmetric property of equality states that the sides of an equation can be reversed: If $a = b$, then $b = a$.

Axiom 3: The transitive property of equality states that if $a = b$ and $b = c$, then $a = c$.

Axiom 4: The same quantity may be added or subtracted from both sides of an equation: If $a = b$, then $x + a = x + b$ and $x - a = x - b$.

Axiom 5: Both sides of an equation may be multiplied or divided by the same quantity: If $a = b$, then $ax = bx$ and $a/x = b/x$ as long as x does not equal zero.

Axiom 6: A quantity may be substituted for its equal in an equation: 5280 feet/hour may be substituted for 1 mile/hour, 7.48 gallons for 1 cubic foot, and 60 minutes for 1 hour.

As demonstrated in Example D.4, solving for an unknown quantity, whether it is x, q, σ, or R_1, is accomplished by using the axioms noted above.

■ EXAMPLE D.4

Solve for n in the following equations:
(a) $n + 3 = 7$
(b) $n - 3 = 7$
(c) $7 - m = n - 3$ where $n = m$

Solution (a) By Axiom 4:

$$n + 3 - 3 = 7 - 3$$
$$n = 4$$

(b) By Axiom 4:

$$n - 3 + 3 = 7 + 3$$
$$n = 10$$

(c) By Axiom 2:

$$n - 3 = 7 - m$$

By Axiom 3:

$$n - 3 = 7 - n$$
$$n - 3 + n = 7 - n + n$$
$$2n - 3 = 7$$

By Axiom 4:

$$2n - 3 + 3 = 7 + 3$$
$$2n = 10$$

By Axiom 5:

$$\frac{2n}{2} = \frac{10}{2}$$
$$n = 5$$

■ EXAMPLE D.5

With the equation $S = bd^2/6$, rearrange the terms to find b in terms of S and d.

Solution By Axiom 5:

$$6S = 6\left(\frac{bd^2}{6}\right)$$
$$6S = bd^2$$
$$\frac{6S}{d^2} = \frac{bd^2}{d^2}$$
$$\frac{6S}{d^2} = b$$

By Axiom 2:

$$b = \frac{6S}{d^2}$$

D.5 QUADRATIC FORMULA

A *quadratic equation* is a second-degree equation involving an unknown quantity expressed with a second power (i.e., x^2, $3x^2$, $\frac{1}{2}d^2$, l^2, etc.) but no higher power. The standard form of a quadratic equation is $ax^2 + bx + c = 0$, where a, b, and c are known quantities ($a \neq 0$) and x is an unknown quantity. Variations of the quadratic equation include $x^2 + 16x + 140 = 0$, $2M^2 + 10M = 0$, and $ax^2 + c = 0$.

Although several methods of solving for the unknown variable in quadratic equations exist, a common method includes use of the *quadratic formula:*

$$\text{quadratic formula, } x = \frac{-b \pm \sqrt{b^2 - 4ac}}{2a}$$

where a, b, and c are known quantities ($a \neq 0$) and x is an unknown quantity as expressed in the form of $ax^2 + bx + c = 0$. Because of the $\pm$ sign in the formula, there are two solutions to the unknown quantity:

$$x_1 = \frac{-b + \sqrt{b^2 - 4ac}}{2a}$$
$$x_2 = \frac{-b - \sqrt{b^2 - 4ac}}{2a}$$

■ EXAMPLE D.6

Use the quadratic formula to solve for s in the equation $2s^2 - 7s = -6$.

Solution It is first necessary to place the equation in the form of $ax^2 + bx + c = 0$:

$$2s^2 - 7s = -6$$
$$2s^2 - 7s + 6 = 0$$

With the terms rearranged in the form of a quadratic equation $2s^2 - 7s + 6 = 0$, it is found that the known quantities are $a = 2$, $b = -7$, and $c = 6$.

$$x_1 = \frac{-b + \sqrt{b^2 - 4ac}}{2a}$$

$$= \frac{-(-7) + \sqrt{(-7)^2 - 4(2)(6)}}{2(2)}$$

$$= \frac{7 + \sqrt{49 - 48}}{4} = 2$$

$$x_2 = \frac{-b - \sqrt{b^2 - 4ac}}{2a}$$

$$= \frac{-(-7) - \sqrt{(-7)^2 - 4(2)(6)}}{2(2)}$$

$$= \frac{7 - \sqrt{49 - 48}}{4} = 1.5$$

In the solution of engineering problems there is generally only one real solution to a problem involving the quadratic formula.

■ EXAMPLE D.7

A drilled pier is a slender columnlike foundation that transfers the building load to lower soils or bedrock. The load imposed must be supported by the skin friction of the perimeter of the pier in the bedrock and the end pressure of the pier on the bedrock; end pressure + skin friction = load imposed. The 50-kip load on the pier and the diameter of the pier (D) in feet are related by the equation

$$\text{end pressure} + \text{skin friction} = \text{load imposed}$$

$$\frac{20\pi D^2}{4} + 2\pi D = 50$$

A large-diameter drilled pier can support a greater load than can a pier with a small diameter, but a large-diameter pier is more costly. Use the quadratic formula to solve for the minimum diameter (D) of the drilled pier.

Solution It is first necessary to place the equation in the form of $ax^2 + bx + c = 0$:

$$\frac{20\pi D^2}{4} + 2\pi D = 50$$

$$5\pi d^2 + 2\pi - 50 = 0$$

With the terms rearranged in the form of a quadratic equation $5\pi d^2 + 2\pi d - 50 = 0$, it is found that the known quantities are: $a = 5\pi$, $b = 2\pi$, and $c = -50$.

$$d_1 = \frac{-b + \sqrt{b^2 - 4ac}}{2a}$$

$$= \frac{-(2\pi) + \sqrt{(2\pi)^2 - 4(5\pi)(-50)}}{2(5\pi)} = 1.595$$

$$d_2 = \frac{-b - \sqrt{b^2 - 4ac}}{2a}$$

$$= \frac{-(2\pi) - \sqrt{(2\pi)^2 - 4(5\pi)(-50)}}{2(5\pi)} = -1.995$$

The drilled pier must have a minimum diameter of 1.595 ft.

It is helpful to check the solution by entering the result into the original equation:

$$\frac{20\pi D^2}{4} + 2\pi d = 50 \quad \text{where} \quad d = 1.595$$

$$\frac{20\pi(1.595)^2}{4} + 2\pi(1.595) = 50$$

$$39.96 + 10.02 = 50$$

$$49.98 = 50$$

Note: A slight rounding error exists.

In Example D.7 there are two solutions for the unknown quantity (D) in the equation $20\pi D^2/4 + 2\pi d = 50$: $d_1 = 1.595$ and $d_2 = -1.995$. However, because the drilled pier must have a diameter expressed as a positive number, the minimum diameter is 1.595 ft (19.1 in).

D.6 ANGLES

An *angle* is formed when two lines meet at a common endpoint. An angle is expressed in units of *degrees* (°) or *radians* (rad). One radian is equal to $180/\pi$ degrees or roughly 57.3° (1 rad = 57.3°). As shown in Figure D.1, an angle is measured in a counterclockwise direction about the left endpoint of a line resting

Figure D.1 An angle is expressed in units of degrees (°) or radians (rad). An angle is measured in a counterclockwise direction about the left endpoint of a line resting horizontally.

horizontally. An *acute angle* is less than 90°; an *obtuse angle* is greater than 90° but less than 180°; and a *reflex angle* is greater than 180°.

D.7 TRIANGLES

A *triangle* is composed of three sides and three angles. The sum of the internal angles of a triangle is always 180°. A *right angle* is an angle that measures exactly 90°. It is the type of angle that makes up the corner of a square. A *right triangle* is a triangle that has one right angle. The side opposite the right triangle is called the *hypotenuse*. The other sides of a triangle are called *legs*. As shown in Figure D.2, the *adjacent leg* is adjacent to the angle under consideration; it touches the angle under consideration. The *opposite leg* is opposite to the angle under consideration; it does not touch the angle under consideration.

Pythagorean Theorem

The *Pythagorean theorem* states that in a right triangle, the sum of the squares of two legs is equal to the square of the hypotenuse. This theorem is expressed mathematically as

$$\text{Pythagorean theorem, } a^2 + b^2 = c^2$$

where a and b are the length of the legs and c is the length of the hypotenuse of a right triangle, as shown in Figure D.3. If a, b, or c are unknown, they may be found by rearranging the Pythagorean theorem algebraically:

$$c = \sqrt{a^2 + b^2}$$
$$a = \sqrt{c^2 - b^2}$$
$$b = \sqrt{c^2 - a^2}$$

■ EXAMPLE D.8

Determine the length of the missing side of each triangle.
(a) A right triangle has legs measuring 3 ft and 4 ft, as shown in Figure D.4a. Find the length of the hypotenuse.

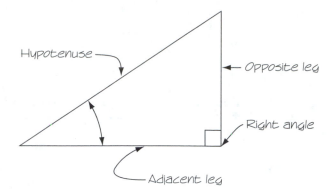

Figure D.2 Location of angles and legs of a right triangle.

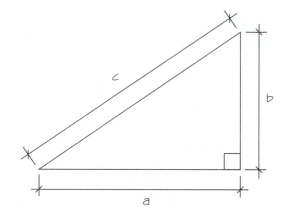

Figure D.3 With the Pythagorean theorem, *a* and *b* are the length of the legs and *c* is the length of the hypotenuse of a right triangle.

(b) A right triangle has a leg measuring 5 m and a hypotenuse measuring 13 m, as shown in Figure D.4b. Find the length of the unknown leg.

(c) A right triangle has a leg measuring 8 m and a hypotenuse measuring 17 m, as shown in Figure D.4c. Find the length of the unknown leg.

Solution (a) With legs *a* and *b* known, the hypotenuse (*c*) is found by

$$c = \sqrt{a^2 + b^2} = \sqrt{3^2 + 4^2} = \sqrt{9 + 16} = \sqrt{25} = 5 \text{ ft}$$

(b) With leg *a* and the hypotenuse (*c*) known, leg *b* is found by

$$b = \sqrt{c^2 - a^2} = \sqrt{13^2 - 5^2} = \sqrt{169 - 25} = \sqrt{144} = 12 \text{ m}$$

(c) With leg *a* and the hypotenuse (*c*) known, leg *b* is found by

$$b = \sqrt{c^2 - a^2} = \sqrt{17^2 - 8^2} = \sqrt{289 - 64} = \sqrt{225} = 15 \text{ m}$$

As shown in Example D.8, right-triangle relationships may be found in common *a–b–c* (leg–leg–hypotenuse) increments, such as 3–4–5, 5–12–13, and 8–15–17. So if the legs are 3 and 4 units long, the hypotenuse must be equal to 5 units because of the 3–4–5 relationship. Similarly, a right triangle having 12- and 16-unit legs has a hypotenuse equal to 20, because $3 \cdot 4 = 12$ units, $4 \cdot 4 = 16$ units, and thus $5 \cdot 4 = 20$ units.

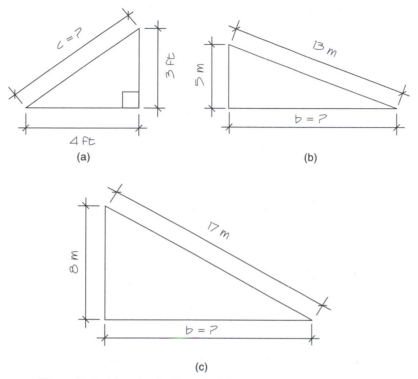

Figure D.4 Triangles for Example D.8.

D.8   RIGHT ANGLE TRIGONOMETRY

The subject of *trigonometry* is the study of the relationship between the sides of a triangle and the angles related to the sides. In fact, the term *trigonometry* comes from the Greek words *trigonon,* meaning triangle, and *metria,* meaning measurement. *Right-angle trigonometry* is the study of the relationships between the sides of a right triangle and the angles related to the sides. Right-angle "trig" is important in solving many basic engineering problems.

Trigonometric functions relate an angle to a ratio of the lengths of the sides of a right triangle, as shown in Figure D.2. Three trigonometric functions are important: *sine,* pronounced "sign" and abbreviated *sin; cosine,* pronounced "co-sign" and abbreviated *cos;* and *tangent,* pronounced "tan-jent" and abbreviated *tan.* These trigonometric functions expressed as ratios of length of appropriate sides are:

$$\sin \theta = \frac{o}{h} = \text{ratio of the length of the opposite leg to the length of the hypotenuse}$$

D.8   Right Angle Trigonometry   **639**

Table D.2 TRIGONOMETRIC FUNCTIONS AND THE RATIO THAT RELATES TO COMMON ANGLES

Angle (deg)	Sine: opposite leg / hypotenuse	Cosine: adjacent leg / hypotenuse	Tangent: opposite leg / adjacent leg
0	0.000	1.000	0.000
15	0.259	0.966	0.268
26.565	0.447	0.894	0.500
30	0.500	0.866	0.577
45	0.707	0.707	1.000
60	0.866	0.500	1.732
63.435	0.894	0.447	2.000
75	0.966	0.259	3.732
90	1.000	0.000	∞
180	0.000	−1.000	0.000
270	−1.000	0.000	∞
360	0.000	1.000	0.000

$$\cos \theta = \frac{a}{h} = \text{ratio of the length of the adjacent leg to the length of the hypotenuse}$$

$$\tan \theta = \frac{o}{a} = \text{ratio of the length of the opposite leg to the length of the adjacent leg}$$

Trigonometric functions and the trigonometric ratios that relate to common angles are provided in Table D.2. Expanded trig tables are found in trigonometry texts and engineering resources. Additionally, the trigonometric ratio of any angle can also be found by entering the angle in a calculator and depressing the relate key. For example, sin 50° is 0.766 and is found by entering the number 50 in a calculator and depressing the sin key; cos 50° is 0.643 and is found by entering the number 50 and depressing the cos key; tan 50° is 1.192 and is found by entering the number 50 and depressing the tan key; and so on.

A trigonometric function is used to find the ratio of the lengths of the sides of a right triangle that relates to an angle within the triangle. When an angle is known, the trigonometric function (sin, cos, or tan) is used to find the ratio of the legs. For example, the trig function $\tan \theta = o/a$ is used to find the ratio of the lengths of the legs of a right triangle with an angle of 30°:

$$\tan \theta = \frac{o}{a}$$

$$\tan 30° = \frac{\text{ratio of length of the opposite leg}}{\text{length of the adjacent leg}}$$

$$= 0.500$$

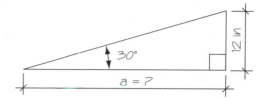

Figure D.5   Triangle for Example D.9.

The Mathematical Symbol Theta (θ)

Theta (θ) is a Greek letter typically used to represent an angle, just as x is used to express a variable in an equation.

The ratio of 0.500 means that the ratio of the length of the opposite leg to the length of the adjacent leg of the right triangle under consideration is 0.500; that is, the opposite leg is one-half (0.500) the length of the leg adjacent to the 30° angle.

■ EXAMPLE D.9

A right triangle has an interior angle of 30° opposite a 12-in leg, as shown in Figure D.5. Determine the length of the adjacent leg of the triangle.

Solution   The opposite angle and length of the opposite leg are known. The length of the adjacent leg is desired but unknown. The tangent of an angle (tan θ) is the relationship of the length of the opposite leg (o) to the length of the adjacent leg (a); tan θ = o/a. The tangent of 30° is found in Table D.2 or by entering 30 into a calculator and depressing the tan key:

$$\tan 30° = 0.577$$

The length of the adjacent leg is found by rearranging terms in the expression tan θ = o/a:

$$\tan θ = \frac{o}{a}$$

$$a = \frac{a}{\tan θ}$$

$$= 12 \text{ in } (0.577) = 6.92 \text{ in}$$

■ EXAMPLE D.10

A right triangle has an interior angle of 26.565° contacting the adjacent leg as shown in Figure D.6. The adjacent leg of the triangle is 2.00 m long. Determine the length of the opposite leg of the triangle.

D.8   Right Angle Trigonometry

641

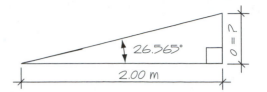

Figure D.6 Triangle for Example D.10.

Solution The adjacent angle and length of the adjacent leg are known. The length of the opposite leg is desired but unknown. The tangent of an angle (tan θ) is the relationship of length of the opposite leg (o) to the length of the adjacent leg (a); tan θ = o/a. The tangent of 26.565° is found in Table D.2 or by entering 26.565 into a calculator and depressing the tan key:

$$\tan 26.565° = 0.500$$

The length of the opposite leg is found by rearranging terms in the expression tan θ = o/a:

$$\tan θ = \frac{o}{a}$$
$$o = a(\tan θ)$$
$$= 2.00 \text{ m}(0.500) = 1.00 \text{ m}$$

In Example D.10 the length of the opposite leg of the right triangle is 1.00 m. It is 0.500 (one-half) the length of the adjacent leg.

An *inverse function* does exactly the opposite of the original function. *Inverse trigonometric functions* relate a ratio of the lengths of the sides of a right triangle to an angle. Three inverse trigonometric functions are important: *arcsine*, abbreviated sin^{-1}; *arccosine*, abbreviated cos^{-1}; and *arctangent*, abbreviated tan^{-1}. An inverse trigonometric function is used to find the angle that relates to the ratio of the lengths of the sides of a right triangle. When the ratio of the legs is known, an inverse trigonometric function (sin^{-1}, cos^{-1}, and tan^{-1}) is used to find the angle. It is like toggling back and forth through a trig table such as that in Table D.2.

■ EXAMPLE D.11

A right triangle has an adjacent leg that is 3 ft long and an opposite leg that is 4 ft long, as shown in Figure D.7. Determine the angle contacting the adjacent leg of the triangle.

Solution The lengths of the adjacent and opposite legs are known. The tangent of an angle (tan θ) is the relationship of length of the opposite leg (o) to the length of the adjacent leg (a); tan θ = o/a.

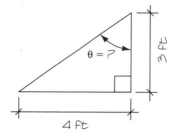

Figure D.7 Triangle for Example D.11.

$$\tan \theta = \frac{o}{a}$$

$$\theta = \tan^{-1} \frac{o}{a}$$

$$= \tan^{-1} \frac{4 \text{ ft}}{3 \text{ ft}}$$

$$= \tan^{-1} (1.333)$$

$$= 53.13°$$

Basic Trigonometric Identities

Trigonometric identities are equations that relate trigonometric functions. Many trigonometric identities are available for use in solving for angles and lengths of the side of triangles that are not right triangles. Two important identities are known as the sine law and the cosine law.

The *sine law* states that *the ratio between any angle of a triangle and the side opposite that angle is a constant.* This law is expressed mathematically as

$$\text{sine law,} \quad \frac{a}{\sin \theta_a} = \frac{b}{\sin \theta_b} = \frac{c}{\sin \theta_c}$$

where angle θ_a is opposite side a, angle θ_b is opposite side b, and angle θ_c is opposite side c.

■ EXAMPLE D.12

As shown in Figure D.8, the length of one leg of a triangle is 2.50 m. The angle at one end of the leg is 30° and at the other end is 40°. Determine the length of the unknown legs of the triangle.

Solution Assume that angle θ_a is 30° and angle θ_b is 40°, then leg c is 2.50 m. The unknown angle (θ_c) is found by subtracting the known angles from 180°:

$$\theta_c = 180° - 30° - 40° = 110°$$

D.8   Right Angle Trigonometry

643

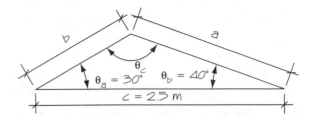

Figure D.8 Triangle for Example D.12.

Using the sine law, leg a may be found:

$$\frac{a}{\sin \theta_a} = \frac{c}{\sin \theta_c}$$

$$a = \frac{c}{\sin \theta_c} \sin \theta_a$$

$$= (2.50 \text{ m/sin } 110°)\sin 30°$$

$$= 1.33 \text{ m}$$

Using the sine law, leg b may be found:

$$\frac{b}{\sin \theta_b} = \frac{c}{\sin \theta_c}$$

$$b = \frac{c}{\sin \theta_c \sin \theta_b}$$

$$= (2.50 \text{ m/sin } 110°)\sin 40°$$

$$= 1.71 \text{ m}$$

The *cosine law* states that *the square of any side of a triangle is equal to the sum of the squares of the other two sides decreased by twice the product of these two sides multiplied by the cosine of the angle included between the two sides.* This law is expressed mathematically as

$$\text{cosine law, } b^2 = a^2 + c^2 - 2ac \cos \theta_b$$

In other forms, the cosine law is expressed as

$$c^2 = a^2 + b^2 - 2ab \cos \theta_c$$
$$a^2 = b^2 + c^2 - 2bc \cos \theta_a$$

where angle θ_a is opposite side a, angle θ_b is opposite side b, and angle θ_c is opposite side c.

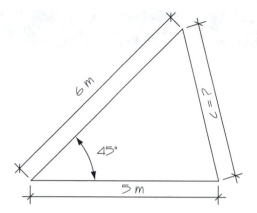

Figure D.9 Triangle for Example D.13.

■ EXAMPLE D.13

As shown in Figure D.9, the length of leg a of a triangle is 5 m and leg b is 6 m. Angle θ_c is 45°. Determine the length of the leg opposite angle θ_c (leg c).

Solution   Using the cosine law, leg c may be found:

$$
\begin{aligned}
c^2 &= a^2 + b^2 - 2ab \cos \theta_c \\
&= (5 \text{ m})^2 + (6 \text{ m})^2 - 2(5 \text{ m})(6 \text{ m})\cos 45° \\
&= 25 + 36 - 60(0.707) \\
&= 15.57 \\
c &= 4.31 \text{ m}
\end{aligned}
$$

ANSWERS

(Odd numbered questions only)

1.9   a. 4.27 mm
       b. 393.7 in
       c. 337.2 lb
       d. 10 500 lb
       e. 186 000 lb-in
       f. 165.5 MN/m^2
       g. 2.36 rad
       h. 35.3 ft^2
       i. 667.2 kN
       j. 125.0 lb/in
       k. 4047 m^2
       l. 0.002 78 m^4

1.11   a. 3.25
       b. 55 000
       c. 17.0
       d. 1000
       e. 0.000 999
       f. 0.534

2.1   5100 N 101° $\measuredangle$
2.3   5.48 kN 98.6° $\measuredangle$
2.5 F_x = 1.80 kip←
      F_y = 2.40 kip ↑
2.7 F_x = 165.3 N→
      F_y = 47.4 N↑
2.9 F_x = 846 N←
      F_y = 308 N↓
2.11   1309 N 106.9° $\measuredangle$
2.13   3197.7 N 276.3° $\measuredangle$
2.15   a. 2.83 kN 88.2° $\measuredangle$
       b. 2.83 kN 268.2° $\measuredangle$

      c. F_x = 0.09 kN←
        F_y = 2.83 kN↓

3.1   a. 360 lb-ft
       b. 4320 lb-in
       c. 0.36 kip-ft
       d. 4.32 kip-in
3.3 R_{By} = 18.75 kN (tension cable)
      R_{Ay} = 6.25 kN (reaction A)
3.5   a. 15.0 kN-m
       b. R_A = 4.24 kN 135° $\measuredangle$
       c. R_B = 3.0 kN→
3.7   b. 83.3 kip 143.1° $\measuredangle$
       c. 83.3 kip
3.9   b. M_A = 680.6 kip-ft
        R_{Ay} = 82.5 kip
       c. M_A = 683.4 kip-ft
        R_{Ay} = 83.0 kip
3.11   b. R_{By} = 12.5 MN
        R_{Ay} =12.5 MN
3.13   b. 3.55 kip 44.1° $\measuredangle$
       c. 2.54 kip

4.1 R_1 = R_2 = 2080 lb↑
4.3 R_1 = R_2 = 3080 lb↑
4.5 R_1 = R_2 = 20 000 lb↑
4.7 R_1 = 4800 lb↑
      R_2 = 7200 lb↑
4.9 R_1 = 3700 lb↑
      R_2 = 10 300 lb↑

4.11　$R_1 = 4683.33$ lb↑

　　　$R_2 = 7816.67$ lb↑

5.1　$F_{AC} = 20.83$ kN compression

　　$F_{AB} = 16.67$ kN tension

　　$F_{BC} = 20.83$ kN compression

5.3　$F_{AE} = 20.83$ kN compression

　　$F_{AD} = 16.67$ kN tension

　　$F_{DE} = 0$

　　$F_{EC} = 20.83$ kN compression

　　$F_{CD} = 25$ kN tension

5.5　$F_{AD} = 39$ kip compression

　　$F_{AF} = 36$ kip tension

　　$F_{CD} = 31.35$ kip compression

　　$F_{DF} = 9.99$ kip compression

　　$F_{CE} = 31.35$ kip compression

　　$F_{CF} = 14.12$ kip compression

　　$F_{EF} = 9.99$ kip compression

　　$F_{BF} = 36$ kip tension

　　$F_{BE} = 39$ kip compression

5.7　$F_{AC} = 15$ kip tension

　　$F_{AE} = 12$ kip compression

　　$F_{CE} = 6$ kip compression

　　$F_{EF} = 12$ kip compression

　　$F_{CF} = 5$ kip compression

　　$F_{CH} = 16$ kip tension

　　$F_{DH} = 16$ kip tension

　　$F_{FH} = 0$

　　$F_{DF} = 5$ kip compression

　　$F_{GF} = 12$ kip compression

　　$F_{DG} = 6$ kip compression

　　$F_{BG} = 12$ kip tension

　　$F_{BD} = 15$ kip tension

5.9　$F_{AG} = 5.66$ kip tension

　　$F_{AC} = 4.0$ kip compression

　　$F_{CG} = 5.66$ kip compression

　　$F_{GH} = 8.0$ kip tension

　　$F_{CH} = 2.83$ kip tension

　　$F_{CD} = 10.0$ kip compression

　　$F_{DH} = 2.83$ kip compression

　　$F_{HI} = 12.0$ kip tension

　　$F_{DI} = 0$

　　$F_{DE} = 12.0$ kip compression

　　$F_{EI} = 0$

　　$F_{IJ} = 12.0$ kip tension

　　$F_{EJ} = 2.83$ kip compression

　　$F_{EF} = 2.0$ kip compression

　　$F_{FJ} = 2.83$ kip compression

　　$F_{JK} = 8.0$ kip tension

　　$F_{FK} = 5.66$ kip compression

　　$F_{BF} = 4.0$ kip compression

　　$F_{BK} = 5.66$ kip tension

5.11　30.4 lb

5.13　621.2 lb

5.15　a. 5583 lb

　　　b. 1 979 345 lb

5.17　35.25 ft

6.1　15.9 MPa

6.3　4997 lb/in^2

6.5　a. 7130 lb/in^2

　　b. 4074 lb/in

6.7　180 mm by 180 mm plate

6.9　a. 384 lb/in^2

　　b. 328 lb/in^2

　　c. No

　　d. No

　　e. 12 300 lb

6.11　a. 40.0 mm

　　　b. 50.4 mm

　　　c. 90.4 mm

6.13　14 mm

7.1　a. 0.000 310 m/m

　　b. 0.031%

7.3　0.875 mm

7.5　9.00 mm

7.7　a. 2.29 mm

　　b. 2.29 mm

7.9 a. 0.000 476 m/m
     b. 99.95 mm
     c. 0.9997 in
7.11 a. 2.500 665 in
     b. No

8.1 a. 800 in-lb
     b. 9658 lb/in^2
     c. 2.38°
8.3 a. 800 in-lb
     b. 1207 lb/in^2
     c. 0.1490°
8.5 18.34 MPa
8.7 0.56°

9.1 a. 15.3 in
     b. 11.7 in
9.3 a. 0.90 in
     b. 6.03 in
9.5 a. 1.66 in
     b. 7 expansion joints
     c. 32'-0"
9.7 a. 29'-9" (expansion)
     b. 21'-11" (contraction)
9.9 a. 7770 lb/in^2
     b. 17 920 lb/in^2
     c. 28 070 lb/in^2
9.11 a. 350 lb/in^2
     b. 365 lb/in^2

10.1 c. 40 000 lb/in^2
     d. 80 000 lb/in^2
10.3 c. 4350 lb/in^2
10.5 Steel in tension is 10%
     stronger than aluminum.
     Concrete in compression
     has 58% the strength of wood.

10.7 Ranked in order of stiffness:
     steel: 29 000 000 lb/in^2
     copper: 16 000 000 lb/in^2
     aluminum: 10 000 000 lb/in^2
     concrete: 2 000 000 lb/in^2
     wood: 1 600 000 lb/in^2
10.9 a. 41.7 kN
     b. 62.6 kN

11.1 V_{MAX} = 2080 lb
      V=0 @ 8ft from R_1
11.3 V_{MAX} = 3080 lb
      V=0 @ 8ft from R_1
11.5 V_{MAX} = -/+20 000 lb
      V=0 @ 9.33ft from R_1
11.7 V_{MAX} = -5200 lb
      V=0 @ 9.6ft from R_1
11.9 V_{MAX} = -5300 lb
      V=0 @ 3ft from R_1 and R_2
11.11 V_{MAX} = 5000 lb
      V=0 @ R_1, 5.36ft to right of
      R_1, and R_2

12.1 M_{MAX} = 8320 ft-lb
12.3 M_{MAX} = 16 320 ft-lb
12.5 M_{MAX} = 101 333 ft-lb
12.7 M_{MAX} = 23 040 ft-lb
12.9 M_{MAX} = 16 000 ft-lb
      Inflection point: 6.354 ft
12.11 M_{MAX} = 16 000 ft-lb
      Inflection points: 12.95 ft, 5.79 ft

13.1 x = 1.5 in
      y = 4.0 in
      z = 3.0 in
13.3 x = 0 in
      y = 2.19 in
      z = 0 in

13.5 $x = 3.04$ m
$y = 4.71$ m
$z = 0.1$ m

13.7 $x = 75.0$ mm
$y = 25.6$ mm

13.9 $x = 0$
$y = 4.34$ in

14.1 a. 169.78 in^4
b. 170.67 in^4

14.3 $4\,643\,437.5$ mm^4

14.5 $7\,966\,608$ mm^4

14.7 a. 12.51 in^4
b. 48.53 in^4
c. 111.15 in^4
d. 230.84 in^4
e. 415.28 in^4

14.9 a. 7.15 in^3
b. 17.65 in^3
c. 30.66 in^3
d. 49.91 in^3
e. 73.83 in^3

14.11 a. 1.01 in
b. 1.59 in
c. 2.10 in
d. 2.67 in
e. 3.25 in

14.13 a. 5.78 mm
b. 11.56 mm
c. 17.34 mm
d. 23.12 mm
e. 28.90 mm

14.15 a. $I = 196\%$ greater
$S = 197\%$ greater
$r = 72\%$ greater
b. x-x
c. x-x
d. x-x

15.3 a. 1560 kN
b. 2400 kN
c. 3600 kN

15.5 72.5 lb/ft^2

15.7 a. @ 12 in: 16.0 ft^2
@ 16 in: 21.33 ft^2
@ 19.2 in: 25.6 ft^2
b. @ 12 in: 1040 lb
@ 16 in: 1386 lb
@ 19.2 in: 1664 lb
c. @ 12 in: $R_A = R_B = 520$ lb
@ 16 in: $R_A = R_B = 693$ lb
@ 19.2 in: $R_A = R_B = 832$ lb

15.9 b. Reactions for B1:
$R_B = 160$ kN
$R_A = 160$ kN
Reactions for G1:
$R_B = 160$ kN
$R_A = 160$ kN

16.1 a. 16.00 in^3
b. 24.00 in^3
c. 320.00 in^3

16.3 a. M@8 ft $= 8320$ ft-lb
b. S@8 ft $= 66.56$ in^3
c. d @8 ft $= 10.68$ in

16.5 a. 1.0 in^3
b. 0.44 in^3
c. 1.0 in^3

16.7 15.38 MPa

16.9 Acceptable

16.11 $\Delta_{max} = 0.747$ in, not acceptable

16.13 2.22 MPa

17.1 a. $11\,550$ lb
b. 2949 kip
c. 1397 kip

17.3 a. 576 kip
     b. 1082 kip

17.5 Perpendicular to the widest
     (4 in) surface

18.1 a. 203.2 mm
     b. 7.2 mm
     c. 203.1 mm
     d. 11.0 mm

18.3 a. 355.9 mm
     b. 17.9 mm
     c. 375.5 mm
     d. 17.9 mm

18.5 a. 229 in^4
     b. 9.87 in^4 - 4.3% of I_x
     c. 38.2 in^3
     d. 3.89 in^3 - 10.2% of S_y
     e. 4.72 in
     f. 0.980 in - 20.8% of r_x

18.7 1.84 in

18.9 F_y = 50 000 lb/in^2: 345 MPa
     F_u = 70 000 lb/in^2: 483 MPa

19.1 a. W16X31
     b. W10X45

19.3 W24X76

19.5 W36X210

19.7 W18X35

19.9 W24X55

19.11 6X7X$\frac{5}{16}$ bearing plate

20.1 a. 22K4
     b. 20K5
     c. 3 rows of horizontal
       bridging

20.3 a. 14K1
     b. 16K2
     c. 20K3

     d. @12 in (1 ft) spacing:
       31 joists
       @16 in (1.33 ft) spacing:
       23 joists
       @24 in (2 ft) spacing:
       15 joists
     e. 15 20K3 joists @24 in (2 ft)
       spacing

20.5 a. 5 rows of bridging (4 hori-
       zontal and 1 diagonal)
     b. 48 in
     c. horizontal: r > 0.160 in
       diagonal: r > 0.139 in
     d. L1X1X$\frac{1}{8}$

21.1 242 kip

21.3 284 kip

21.5 a. allowable load increases
     b. allowable load decreases
     c. allowable load increases

21.7 a. 583 kip and 582.6 kip
     b. allowable load for K=1.2 is
       93.6% of load for K=1.0

21.9 W14X159

21.11 PL12X16X1

21.13 a. W14X90
      b. PL15X15X1

22.1 For visually-graded dimension
     lumber (2″ to 4″ thick):
     Hemlock-Fir, Douglas
     Fir-Larch, Southern Pine,
     Redwood
     For visually-graded timbers
     (5″ by 5″ and larger) - beams &
     stringers:
     Redwood, Douglas Fir-Larch,
     Hemlock-Fir, Southern
     Pine

22.3 For visually-graded dimension lumber (2″ to 4″ thick): Douglas Fir-Larch, Redwood, Hemlock-Fir, Southern Pine. For visually-graded timbers (5″ by 5″ and larger) - beams & stringers: Redwood, Douglas Fir-Larch, Hemlock-Fir, Southern Pine.

22.5 For visually-graded dimension lumber (2″ to 4″ thick): Douglas Fir-Larch, Hemlock-Fir, Redwood, Southern Pine. For visually-graded timbers (5″ by 5″ and larger) - beams & stringers: Douglas Fir-Larch, Hemlock-Fir, Redwood, Southern Pine.

22.7 851 lb/in^2

22.9 997 lb/in^2

22.11 1242 lb/in^2

23.1 4X16

23.3 two-$1\frac{3}{4}$ X $9\frac{1}{4}$

23.5 $3\frac{1}{2}$ X $9\frac{1}{2}$

23.7 9.63 in

23.9 a. 1.10 in
  b. 0.36 in
  c. beam ends should be held in position.

24.1 a. 2X6: 10′-9″
  2X8: 14′-2″
  2X10: 18′-0″
  2X12: 21′-11″
  b. 2X6: 9′-9″
  2X8: 12′-10″
  2X10: 16′-5″
  2X12: 19′-1″

c. 2X6: 9′-2″
  2X8: 12′-1″
  2X10: 14′-9″
  2X12: 17′-0″

24.3 a. 2X6: 10′-3″
  2X8: 13′-6″
  2X10: 17′-3″
  2X12: 20′-6″
  b. 2X6: 9′-4″
  2X8: 12′-3″
  2X10: 14′-11″
  2X12: 17′-7″
  c. 2X6: 8′-9″
  2X8: 11′-3″
  2X10: 13′-7″
  2X12: 16′-0″

24.5 2X10

24.7 2X8

25.1 a. 4998 lb
  b. 4692 lb

25.3 9776 lb

25.5 2573 lb

25.7 26 476 lb

25.9 8 ft: 6995 lb
  10 ft: 4692 lb
  12 ft: 3332 lb
  14 ft: 2475 lb
  16 ft: 1911 lb
  18 ft: 1419 lb
  20 ft: 1237 lb
  22 ft: 1017 lb
  24 ft: 858 lb

25.11 8 ft: 3687 lb
  10 ft: 2891 lb
  12 ft: 1691 lb
  14 ft: 1507 lb
  16 ft: 1188 lb
  18 ft: 943 lb
  20 ft: 772 lb

22 ft: 637 lb
24 ft: 539 lb

26.1 4965 lb/in^2
26.3 a. 3834 lb/in^2
      b. No
26.5 No. 3: 9.5 mm vs. No. 10
      No.4: 12.7 mm vs. No. 13
      No. 5: 15.9 mm vs. No. 16
      No. 6: 19.1 mm vs. No. 19
      No. 7: 22.2 mm vs. No. 22
      No. 8: 25.4 mm vs. No. 25
      No. 9: 28.7 mm vs. No. 29
      No. 10: 32.3 mm vs. No. 32
      No. 11: 35.8 mm vs. No. 36
      No. 14: 41.6 mm vs. No. 43
      No. 18: 57.3 mm vs. No. 57

27.1 33.47 in
27.3 a. 5'-6$\frac{1}{2}$"
      b. 8.5 in O.C. spacing
27.5 ℓ_d = 59.2 in; sufficient
      development length is provided

28.1 Unacceptable.
28.3 a. 290 kip
      b. 12 in
      c. l$_d$ = 16.5 in

28.5 a. 378 kip
      b. 482 kip
      c. Eight No. 10 bars

29.1 16 ft
29.3 17'-4"
29.5 11$\frac{1}{8}$ in
29.7 a. 5$\frac{5}{8}$ in
      b. 24.0 in^2/ft
      c. 67.5 in^2/ft

30.1 285 lb/in^2
30.3 2032 kPa
30.5 a. 266 lb/in^2
      b. 7706 lb/ft of wall
      c. Yes
30.7 a. P = 15 250 lb/ft of wall;
      wall can safely carry load
      b. F_b - F_c = 7.2 lb/in^2; allow-
      able flexure stress (f_b) is not
      exceeded

INDEX

Web shear, 361
Web, in steel joists, 369
Welded wire fabric (WWF), 452
Welded wire mesh (WWM), 452
Wet service factor, 408
Wind loads, 284–285
Wind stagnation pressure, 284
Woolworth Building, 4

Working stress design, 181–183, 454–455, 523
World Trade Center, 5, 7
Wythe, 508

Yield point, 176
Yield strength, 181
Young's modulus, 137

It's Not Just the Latest Code,
It's the Best Code

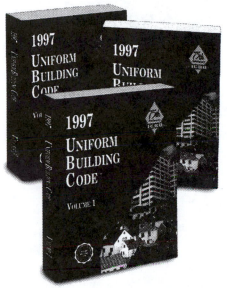

The 1997 *Uniform Building Code*™ (UBC) is the one code book that no building official or design professional should be without. In addition to being the predecessor of the 2000 *International Building Code*™, the UBC reflects the latest technological advances in building design. New or revised provisions in the 1997 UBC cover topics such as means of egress, concrete, seismic isolation, accessibility, structural forces, roofing and much more.

Volume One contains administrative, fire- and life-safety, and field-inspection provisions, including all nonstructural provisions and those structural provisions necessary for field inspections.

Volume Two contains provisions for structural engineering design, including those design standards formerly published as UBC standards.

Volume Three contains the remaining material, testing and installation standards.

The 1997 UBC is also available on CD-ROM for Windows. Fast, accurate and comprehensive, this convenient format will help you to:

◆ Search for any section that contains a certain provision.
◆ Cut and paste code provisions into correspondence, reports, etc.
◆ Obtain printouts of sections of the code.
◆ Obtain a list of pages on which a provision appears.

		Item No.	ICBO Members	Nonmembers
UBC Volumes 1, 2 and 3	soft cover	099S97	$144.55	$180.70
	loose leaf	099L97	166.25	207.80
UBC on CD-ROM (contains Volumes 1, 2, and 3)*		001C97	149.00	186.25
UBC soft cover and UBC on CD-ROM		099CSC	222.25	278.25
UBC loose leaf and UBC on CD-ROM		099CLL	244.25	305.30

*Does not include NFPA standards.
Please call ICBO for individual book prices or quantity discounts.
Prices subject to change.

To order, please call
ICBO toll-free at

(800) 284-4406

**International
Conference of
Building Officials**
www.icbo.org

Publications • Seminars • Certification • Software • Evaluation Service • Videos • Membership